ELECTRICAL ENGINEERING DEVELOPMENTS

ELECTRODEPOSITION

PROPERTIES, PROCESSES AND APPLICATIONS

ELECTRICAL ENGINEERING DEVELOPMENTS

Additional books in this series can be found on Nova's website
under the Series tab.

Additional E-books in this series can be found on Nova's website
under the E-book tab.

ELECTRICAL ENGINEERING DEVELOPMENTS

ELECTRODEPOSITION

PROPERTIES, PROCESSES AND APPLICATIONS

UDIT SURYA MOHANTY
EDITOR

Nova Science Publishers, Inc.
New York

For permission to use material from this book please contact us:
Telephone 631-231-7269; Fax 631-231-8175
Web Site: http://www.novapublishers.com

Additional color graphics may be available in the e-book version of this book.

Library of Congress Cataloging-in-Publication Data

Electrodeposition : properties, processes, and applications / editor, Udit Surya Mohanty.
p. cm.
Includes bibliographical references and index.
ISBN 978-1-61470-826-1 (hardcover)
1. Alloy plating. 2. Electroplating. I. Mohanty, Udit Surya.
TS693.E44 2011
671.7'32--dc23
2011026365

Published by Nova Science Publishers, Inc. † New York

CONTENTS

PREFACE

The present book has been contributed by prominent experts in the field of electrochemical science and technology. It identifies gaps in current knowledge and barriers to applications, and recommends research areas that need to be addressed to enable the rapid development of technologies. Some of the key areas covered in this book are electrochemical synthesis of non-stoichiometric hydroxyapatite-nanodiamond composite coating; electrodeposited biomimetic hydroxyapatite functionalized by lactoferrin for its potential use as bone-implantable biomaterials; development of new types of layered manganese oxides intercalated with cationic polymers applicable for the manufacturing of pseudocapacitor electrodes, electrochromic materials, catalysts, and biosensors and preparation of CuInSe2 thin films and its potential application in photovoltaic devices.

Chapter 1 - The stable crystal structure of nickel is face centred cubic (FCC) but by special techniques it can be produced as a metastable hexagonal close packed (HCP) structure. The conventional process to produce nickel is by reduction of nickel ore at high temperature. This nickel has FCC structure and its morphology cannot be controlled. The electrodeposition of nickel from both aqueous and ionic liquid electrolytes results in controlled morphology and crystal structure. The morphology and structure of nickel depend on the electrolyte and operating parameters of the electrolysis. Although HCP nickel can be produced by hydrothermal, thermal decomposition and electrochemical techniques, the last one has advantages over the other two. The magnetic and thermodynamic property of nickel is dependent on the crystal structure. Both forms of crystal structure show a difference in ferromagnetic property. Lastly, nickel is used in superalloys, batteries and can be used as hydrogen storage material.

Chapter 2 - Results of investigations in the field of electrochemical composite coatings (ECC) are presented. Deposition peculiarities, functional properties and structure of the main types of ECC are also discussed.

Chapter 3 - A review on pulse electrodeposition of Au-Sn alloys from non-toxic and environmentally green electrolytes at ambient temperature is presented based on two main approaches. The first approach is focused on the electrodeposition of Au-Sn alloy films from a single solution. This portion starts with electrolyte preparation and is followed by solution stability and characterization using turbidity measurements, SEM, TEM, XRD, UV/Vis spectroscopy and Cryo-XPS. The microstructures of individual as-deposited films (Au_5Sn and AuSn phases), composite layers formed by alternatively combining Au_5Sn and AuSn phases and reflowed solders are then shown. Factors that affect the compositions of electrodeposited films on patterned wafers are discussed as well. Finally, a process to recover Au from the waste solutions is presented; the Au salt is then used to prepare fresh Au-Sn electrolytes. The second approach is focused on pulse electrodeposition of Au-Sn alloy films using a sequential procedure with pure Au films and pure Sn films from two separate solutions. Within this portion, the development of electrolytes for pure Au deposition and pure Sn deposition are described. The microstructures of sequentially electrodeposited Sn-rich, Au-Sn eutectic solders and reflowed solders are also presented.

Chapter 4 - Pb free solders has become increasingly important for industrial applications, since an international effort has been made to develop and characterize lead free solders for replacing traditional Pb-Sn solders as a consequence of governmental regulations and marketing pressure concerning health and environmental hazards of lead. The corrosion behaviour of a set of Pb free solder alloys were investigated in 3.5 % NaCl solution by employing potentiodynamic polarisation techniques. The various Pb free solders used in the present study were Sn-XAg-0.5 cu (X= 1-4 wt %); Sn-8.5Zn-XAg-0.1 Al-0.05 Ga (X=0.1-2 wt %), Sn-8.5Zn-XAl-0.5 Ga (X=0.02-0.5 wt %) and Sn-8.5Zn-XAg-0.1 Al-0.5 Ga where X varies from 0.1 to 2 wt %. Polarisation curves revealed that Sn-XAg-0.5 Cu alloys with higher Ag content (> 2wt %) exhibited a stronger tendency towards passivation and the passivation behaviour was ascribed to the presence of both SnO and SnO_2 on the cathode surface. An increase in the Ag content from 1 to 4 wt % decreased the corrosion current density (Icorr) and shifted the corrosion potential (E_{corr}) towards more negative values. These changes were clearly reflected in the corrosion rate and linear polarization resistance of the solder alloy. Nevertheless for Sn-8.5 Zn-XAg-0.1Al-0.05Ga alloy, an increase in the Ag content from 0.1 to 2 wt % resulted in a progressive increase in the corrosion current density and insignificant improvement in the passivation behaviour of the solder alloy. On the other hand the four element Sn-8.5Zn-XAl-0.5 Ga alloy showed decrease in both the corrosion current density and corrosion rate with increase in Al content from 0.02 to 0.05 wt %. The best results were observed for Sn-8.5Zn-XAg-0.1Al-0.5Ga solder alloy. The corrosion current density underwent a significant decrease and the corrosion potential shifted towards more noble values with increase in Ag content from 0.1 to 1.5 wt %. SE micrographs established that the oxides and hydroxides of zinc were responsible for the formation of passive film and the presence of Ag atoms in the oxide layer contributed towards the passivation behaviour of solders to a certain extent. SEM and EDX analysis also confirmed than $SnCl_2$ was the major corrosion product formed after the electrochemical experiments in all the investigated alloys.

Chapter 5 - Coatings with less defects and high corrosion resistance can be obtained by pulse electrodeposition, which can be divided into two categories according to the applied current waveform. (1) Pulse current electrodeposition (PC) employed the pulses in one direction; (2) Pulse reverse current electrodeposition (PRC) employed the anodic and cathodic pulses. The results on the pulse electrodeposition studies carried in our lab in recent years has been summarized in this chapter. In one of the investigations, a nickel coating with high density nano-scale twins (NT) was synthesized by using pulsed electrodeposition technique. The passive films formed on nickel coatings showed higher pitting corrosion resistance and a bi-layer semi conducting structure distribution, compared with that of industrial electrodeposited nickel. The passive films behaved like p-type semiconductors at lower potentials, but they behaved like n-type semiconductors at higher potentials. The coating exhibits high potential as a substitute material in the processing industry, power plant. In another study the inhibition of hydrogen blister on Cu–Sn alloy coatings was demonstrated. The Cu–Sn alloy coatings were synthesized on 27SiMn steel by direct current (DC), pulse current (PC) and pulse reverse current (PRC) electrodeposition techniques from pyrophosphate-based electrolyte. The hydrogen permeation amount of different electrodeposition techniques decreased in the order: DC > PC > PRC. Frequency and duty cycle have significant effect on the sub-surface concentration of atomic hydrogen and hydrogen permeation amount. The appropriate PRC technique can prevent or greatly decrease the occurrence of hydrogen blister during the industrial electrodeposition process.

Chapter 6 - Polycrystalline Cu_2O nanowire arrays have been grown via porous alumina membranes using three-electrode electrochemical deposition. The effect of electrolyte, pH value, deposition potential, annealing temperature, and annealing atmosphere on the growth of Cu_2O nanowire arrays has been investigated. On the other hand, ordered ZnO/AZO/PAM nanowire arrays have been prepared by seed layer assisted electrochemical deposition. The comprehensive of electrochemical process adopted (e.g. electrolytes, deposition parameters); various characterization techniques and the outcome of the results have been summarized in details. It is observed that the ZnO nanowire arrays are assembled into the nanochannel of porous alumina template with diameter of 120~140 nm. The crystalline structure of single ZnO nanowire is dependent on AZO seed layer. The nucleation and growth process of ZnO/AZO/PAM nanowires are interpreted by seed layer assisted growth mechanism.

Chapter 7 - This chapter describes our findings over the last few years, concerning the use of electrodeposition as a facile technique for the fabrication of several thin films of noble metals, nanostructures and nanorods onto various substrates. This includes three tasks. Firstly: the electrodeposition of metallic Ni or Cu thin films on reticulated vitreous carbon (RVC), and Pd or black Ni onto copper screens for use as electro-catalytically active cathodes for the hydrogen evolution reaction (HER). Moreover, the effects of some hydrodynamic and solution parameters concerning the performance of the electrodeposition process of heavy metal ions (e.g., lead ions) from flowing wastewater are briefly discussed. Secondly: the fabrication (via electrodeposition) of metallic (e.g., Au) and metal oxide (e.g., manganese oxide) nanostructures for application in fuel cell catalysis including the oxygen reduction and evolution reactions in addition to formic acid oxidation. Details of the morphological and electrochemical characterizations of the thus-electrodeposited nanostructures are outlined. This includes scanning electron microscope imaging (SEM), electron back scatter diffraction (EBSD) and X-ray diffraction (XRD) patterns. Thirdly: the fabrication of one-dimensional nanostructures, i.e., nanowires, for vital electronic, chemical, biological, and medical applications. Several methodologies are outlined for this sake; however, the electrodeposition in templates seems more interesting in terms of cost and potential for high volume production. The recent work combining the electrodeposition and the template growth approach for nanofabrication is introduced. Special emphasis is dedicated to assembling metal-semiconductor nanowire contacts in which the contact interface is formed along the cross-section of the wire.

Chapter 8 - The refractory carbides are characterized by their high melting points, high values of hardness, high tensile strength and chemical stability. Due to these remarkable properties they have a wide range of applications such as cutting and grinding tools, bearings, textile-machinery parts and oxidation-resistant gas burners. Metal carbide coatings are mainly used to improve the mechanical (wear resistance, strength and hardness) and chemical (corrosion and oxidation resistance) properties of metallic parts. The principal coating processes for refractory carbides are chemical-vapor deposition, physical-vapor deposition and thermal spray, but such coatings can also be obtained by electroplating. In the present work molybdenum carbide layers were electrodeposited from the LiF-NaF-KF eutectic melt containing K_2CO_3 and Na_2MoO_4 salts. The coatings were obtained by both direct and pulse electroplating processes. The influence of bath composition, temperature and electrical parameters of electrodeposition on the morphology of the deposits was studied using scanning electron microscopy. The composition of the coatings was determined by X-ray diffractometry. The nucleation mechanism of the molybdenum carbides was analyzed by chronoamperometry.

Chapter 9 - This chapter outlines the electrodeposition of non-ferrous metals such as copper, zinc, nickel, cobalt, chromium and lead from their aqueous solutions. Effects of organic and inorganic impurities generally present in the aqueous solutions during electrodeposition are discussed. Energy reduction techniques during electrodeposition and augmentation of mass transfer in electrochemical cells are also included.

Chapter 10 - In medical devices such as orthopaedic metallic bone implants or coronary stents, mechanical strength can only be achieved by metals, yet these lack the required biocompatibility and bio-integration. Surface treatments to improve the metal prosthesis interface with biological tissues have been extensively studied. The electrochemically-assisted surface deposition on the metallic prosthesis of a biomimetic coating appears as the most interesting resolution. This method allows both to overcome the difficulty of depositing protein components by plasma spray or physical vapour deposition, and to control the coating process easily. In this chapter we review the obtained results and the future perspective about the electrodeposited biomimetic hydroxyapatite. Coatings of hydroxyapatite nanocrystals mimicking bone nanocrystals in composition, structure, morphology, nano size and bioactivity have been obtained on a titanium surface, by an electrochemically assisted deposition in order to improve the surface bioactivity. With a view to reduce the thrombogenic potential of artificial blood-contact devices and natural tissues, a new anti-thrombogenic coating, consisting of an hydroxyapatite nanocrystals-heparin conjugate, has been electrodeposited on metallic coronary stents. It is also possible to obtain calcium-phosphate/collagen coating on a titanium surface with an electrochemical cell containing a slightly acidic collagen molecule suspension in a Ca^{2+} and PO_4^{3-} ions aqueous solution. In such a process, the collagen/calcium phosphate composite formation involves the self-assembly of collagen molecules into reconstituted fibrils during the contemporary crystallization of calcium phosphate mineral on the electrode surface. Following the biomimetic approach, it is possible to realize a nano-structured hydroxyapatite/collagen biomimetic coating. This allows the formation of a coating of reconstituted collagen microfibers inside calcified by hydroxyapatite nanocrystals, closely resembling bone calcified collagen fibres. Natural wood templates have been selected as a starting material to obtain open-pore geometries with wide surface area and microstructures allowing cell in-growth and reorganization and providing the necessary space for vascularisation. In fact, the alternation of fibre bundles and channel-like porous areas makes wood an elective material to be used as template in starting the development of new bone substitute biomaterial. Particularly, Biomorphic Silicon Carbide (BioSiC) wood-derived structures have been optimized to be employed as bio inert bone scaffolds representing a new generation of light, tough and strong material for biomedical applications.The good biocompatibility and biological response of BioSiC coated by electrochemically-assisted surface deposition of nano-structured biomimetic hydroxyapatite allow to consider this porous material both as a bone substitute in orthopaedic, odontoiatric, dental and maxillo-facial implantation. Here we report a detailed description of the electrochemical assisted deposition of a biomimetic apatite coating on prosthetic surface in order to show its potentiality in improving tissue integration and in realising bioactive molecules with kinetic determination studies.

Chapter 11 - Hydroxyapatite (HA) is the main component of human bones, a highly bioactive and biocompatible material; however, it has poor mechanical properties. Carbon-based coatings are found to significantly improve the mechanical properties of apatite, increase its adhesion, prevent metal ion release from metal implants and inhibit the formation of fibrous tissue and blood clotting upon implantation. In this chapter, homogeneous nanodiamond-reinforced hydroxyapatite (HA-ND) composite coating with improved mechanical strength and ductility was developed to enhance the biological properties of metal surfaces (stainless steel and titanium). The CO^3- and HPO^4-containing HA was deposited by electrodeposition from simulated body fluid with dispersed nanodiamond particles. Study of the initial interaction of osteoblast-like MG-63 cells revealed that cells attached well on all plain samples (HA-ND, pure HA and stainless steel). However, pre-coating with fibronectin (FN) even at low adsorption concentrations (1mg/ml) strongly improved cell adhesion and preferentially spreading on the HA-ND samples as indicated by the flattened cell morphology and pronounced vinculin positive focal adhesions. This effect correlates with the observed higher affinity for FN. Moreover, osteoblasts tended to rearrange both adsorbed and secreted FN in a fibril-like pattern, suggesting improved FN matrix organization on HA-ND samples.

Chapter 12 - Copper indium chalcogenides are materials under permanent investigation due to their potential use for photovoltaic applications, given that they show a high absorption coefficient, coupled to a band gap energy value that matches that of the solar radiation. Many techniques have been reported for preparation of $CuInSe_2$ (CISe) as a thin film. Among them, electrodeposition presents many advantages: it is cost-effective, mainly because it does not require high vacuum, it may be used to copy intricate geometries while being applied even on top of flexible substrates and it can be scale-up into industrial production. $CuInSe_2$ (CISe) thin films have been prepared by electrodeposition from a single bath on top of conductive glass as well as on conductive glass coated with thin layers of TiO_2 and In_2S_3. Potentiostatic and pulsed electrodeposition produced films with different properties. The electrodeposition conditions were decided after carrying out cyclic voltammetry in acidic electrolytes. CISe precursors films need to be subjected to different thermal treatments to enhance the crystallinity of the deposit. The presence of sulfur vapor during annealing leads to substantial changes in the composition of the chalcogenide. The crystallinity, morphology and stoichiometry of the annealed films are characterized by XRD, microRaman spectroscopy and SEM, coupled with Energy dispersive scanning (EDS). Etching the films in KCN solution is a key step, enabling a final adjustment in the stoichiometry. Photoelectrochemical measurements are employed to obtain semiconductor properties of the CISe films and to elucidate the influence of the post-deposition treatments.

Chapter 13 - This chapter discusses the electrochemical deposition of tin on copper in an additives-free acid methanesulfonate electrolyte, in a controlled flow condition using a rotating disc electrode. The electrochemistry of Sn^{2+}/Sn are characterised by electrochemical measurements: cyclic voltammograms (at a static electrolyte), linear sweep voltammograms (at a controlled rotating flow condition) and electrodeposition at a constant electrode potential. Effects of fluid flow and concentration of stannous ion (Sn^{2+}) on the electrodeposition of tin are investigated. Surface microstructures of tin deposits are imaged using a scanning electron microscope. Randles-Sevčik, Levich and Koutécky-Levich equations are used to estimate the diffusion coefficient of Sn^{2+}. It was found that the electrodeposition of tin on copper occurred on a very fast kinetic in a diffusion controlled environment. The charge transfer, mixed control and mass transport controlled region were dependent on the electrolysis process parameters. The electrodeposition of tin on copper followed the expressions based on the Scharifker-Hills equations for a three dimensional instantaneous nucleation mechanisms; the growth of tin deposits followed a diffusion controlled process. Tin coatings showed a matte surface appearance and adherent to the copper substrate. A wide range of surface microstructures of tin deposits, starting from compact to dendritic deposits and porous morphology were readily electrodeposited and dependent on the applied electrode potential (or current ~ 5 to 50 mA cm^{-2}). It was found that the electrode potential window for the limiting mass transport controlled region became narrower and the hydrogen evolution on tin occurred at a more electro-positive potential at a higher concentration of stannous ion and an increasing rotating flow condition.

Chapter 14 - Manganese oxide materials, especially manganese dioxide (MnO_2), are very attractive because of their distinctive structures and physicochemical properties as well as environmental compatibility and cost effectiveness. Among various MnO_2 polymorphs, birnessite has attracted particular attention for its unique adsorptive, catalytic, ion-exchange, and electrochemical properties. Birnessite (δ-MnO_2) is characterized by a two-dimensional layered structure that consists of edge shared MnO_6 octahedra with cations and water molecules occupying the interlayer space. In our previous studies we have presented an electrochemical route to construct birnessite-type layered manganese oxides intercalated with alkaline metals or alkylammonium ions in a thin film form. The process involves anodic oxidation of aqueous manganese (II) ions in the presence of guest cations. The deposition mechanism consists of anodic formation of MnO_2 and simultaneous assembly of electrolyte cations to balance the negative charges on the deposited MnO_2 layers. This method is quite versatile because the inorganic host (MnO_2) can adjust itself to accommodate guest molecules during electrodeposition. This has enabled us to design various types of new MnO_2-based layered structures. The guest molecules influence the stacking structure of MnO_2 layers, resulting in unique catalytic, ion-exchange, and electrochemical properties. This chapter reviews the electrochemical formation, properties, and applications of a variety of layered manganese oxides, mainly derived from our own studies during the last eight years.

In: Electrodeposition: Properties, Processes and Applications ISBN: 978-1-61470-826-1
Editor: Udit Surya Mohanty

Chapter 1

A REVIEW ON THE ELECTRODEPOSITION OF NICKEL: SYNTHESIS, MAGNETIC, THERMODYNAMIC PROPERTIES AND ITS POTENTIAL APPLICATIONS

Abhishek Lahiri*
World Premier International Research Center, Advanced Institute for Materials Research (WPI-AIMR),Tohoku University, Sendai, Japan

ABSTRACT

The stable crystal structure of nickel is face centred cubic (FCC) but by special techniques it can be produced as a metastable hexagonal close packed (HCP) structure. The conventional process to produce nickel is by reduction of nickel ore at high temperature. This nickel has FCC structure and its morphology cannot be controlled. The electrodeposition of nickel from both aqueous and ionic liquid electrolytes results in controlled morphology and crystal structure. The morphology and structure of nickel depend on the electrolyte and operating parameters of the electrolysis. Although HCP nickel can be produced by hydrothermal, thermal decomposition and electrochemical techniques, the last one has advantages over the other two. The magnetic and thermodynamic property of nickel is dependent on the crystal structure. Both forms of crystal structure show a difference in ferromagnetic property. Lastly, nickel is used in superalloys, batteries and can be used as hydrogen storage material.

INTRODUCTION

Nickel is a transition metal element and is silvery white in colour. Nickel occurs in nature both as sulfide as well as oxide minerals. The principle sulphide mineral is pentlandite [$(NiFe)_9S_8$] and oxide minerals are known as nickeliferous laterite. The extraction of nickel from nickel ore concentrate is mostly done by pyro-metallurgical process whereby an impure metal is produced. The metal is then purified by Carbonyl process or Mond process.

---

*E-mail address: lahiri.abhishek@gmail.com.

In the Mond process, the impure nickel is then treated with carbon monoxide to produce nickel carbonyl ($NiCO_4$) which on heating at temperatures of 250°C forms nickel with 99.999% purity. The Mond's process was also used to deposit nickel onto different substrates. However, due to the toxicity of CO, the process was abandoned for industrial use.

Nickel is widely used as a coating material which is done both by electroless and electrodeposition techniques. In the electroless nickel plating, a reducing agent such as sodium hypophosphite ($NaPO_2H_2.H_2O$) is used to reduce the nickel ions in the aqueous solutions to produce nickel films. The advantage of electroless plating is the uniformity of the deposit and can be utilised in plating irregular shaped objects like holes, valves, thread etc. In some circumstances, controlling the morphology and structure becomes important wherein the electrodeposition process becomes an easier technique compared to electroless process. Nanostructured coatings have high corrosion resistance properties. Furthermore it has been found that the coating had enhanced hardness, improved ductility and electrical resistivity and had superior magnetic properties [1].

1.1. Crystal Structures of Nickel

Nickel exists in two crystallographic forms, namely, the face centred cubic (FCC) and hexagonal closed pack (HCP). The FCC nickel is the stable structure. However, there are still arguments regarding the existence of HCP nickel phase which is generally considered metastable. Yang [2] initially showed the possibility of depositing a mixed phase of FCC and HCP nickel. It was found that on using the nickel sulphate-boric acid bath, only FCC phase formed, whereas on using chloride-boric acid bath, both FCC and HCP phases formed. Cullity listed HCP nickel in his textbook on X-ray diffraction [3]. The formation of HCP nickel was believed to be due to the presence of hydrides, nitrides and carbides. Hemenger and Weik [4] produced pure HCP nickel phase by depositing a thin nickel film (200 Å) by evaporation technique in 10^{-5} Torr atmosphere. They calculated the lattice parameters (a= 2.622 Å, c= 4.320 Å, c/a= 1.648) using electron diffraction and compared the data with NiO and Ni_3N. A difference in lattice parameter led them to conclude that the phase formed by evaporation technique was HCP nickel. Wright and Goddhard [5] later showed that HCP nickel could be electrodeposited by having a precise electrolyte bath composition. The lattice parameter was found to be a=2.5±0.01 Å, c=3.98±0.06 Å. A list of lattice parameters (a, c, c/a) of HCP nickel obtained by various synthesis techniques are given in table 1.1 which shows significant difference in lattice parameters. This difference may be related to the presence of impurities or may be allotropes of HCP nickel itself. However, the question of pure HCP nickel is still being questioned and argued. Confusion between HCP nickel, Ni_3C, NiO and Ni_3N continue to remain as the lattice parameters of all these materials are almost similar [6].

Crystallographic transformation can also be obtained by solid-state techniques such as high-energy particle irradiation, ion-beam mixing, annealing of diffusion couples, hydrogen charging and high-energy ball milling techniques [18, 19]. In all these techniques, the atoms get displaced from equilibrium lattice sites resulting in lattice strain and forming nanocrystalline grains having special type of grain boundary. Experiments have shown that nanoparticles of a number of metals such as Nb, Mo, Co, W, Ta, and some of less common metals such as Y, Gd, Tb, Dy, Ho, and Er have structures different from their stable state

[20]. By ion irradiation, it was demonstrated that FCC nickel could be transformed to HCP nickel [21]. The metastable phase was formed on ion irradiating $Ni_{75}Nb_{25}$ and $Ni_{70}Nb_{30}$ alloy with a dose of $5*10^{15}$ Xe^{+} cm^{-1} and $3*10^{14}$ Xe^{+} cm^{-1} respectively. The lattice parameters (a=2.53 Å and c=4.04 Å) were found to be similar to that observed by Wright and Goddhard.

Table 1.1. Literature data for HCP nickel structures

Reference	Synthesis technique	a (nm)	c (nm)	c/a
[4]	Evaporation of Ni	0.2622	0.4320	1.648
[5]	Electrodeposition of Ni	0.2500	0.3975	1.590
[7]	$Ni(acac)_2$ +K-B alloy	0.2653	0.4348	1.639
[8]	Ni(II) salt +NaH + t-BuOH	0.2440	0.3960	1.623
[9]	$Ni(OAc)_2$ +methylhydrazine + 2-propanol	0.2649	0.4333	1.636
[10]	Evaporation of Ni	0.2440	0.4220	1.730
[11]	Nickel chloride reduced by KBH_4 in autoclave	0.2653	0.4348	1.639
[12]	Reduction of $Ni(NO_3)_2$ in PEG	0.265	0.433	1.63
[13]	$Ni(acac)_2$ + hydrazine in tetrahydrafuran	0.2665	0.4300	1.613
[14]	Thermal decomposition of Ni(II) glycinate	0.2493	0.4084	1.638
[15]	Ni salt, Mg salt in citrate	0.2652	0.4334	1.634
[16]	Thermal decomposition of Ni(II) acetate without surfactants	0.2647	0.4333	1.637
[16]	Thermal decomposition of Ni(II) acetate with TOPO as surfactant	0.2650	0.4336	1.636
[16]	Thermal decomposition of Ni(II) acetate with TOP as surfactant	0.2655	0.4354	1.639
[17]	Molecular beam epitaxy growth of HCP nickel	0.249	0.398	1.60

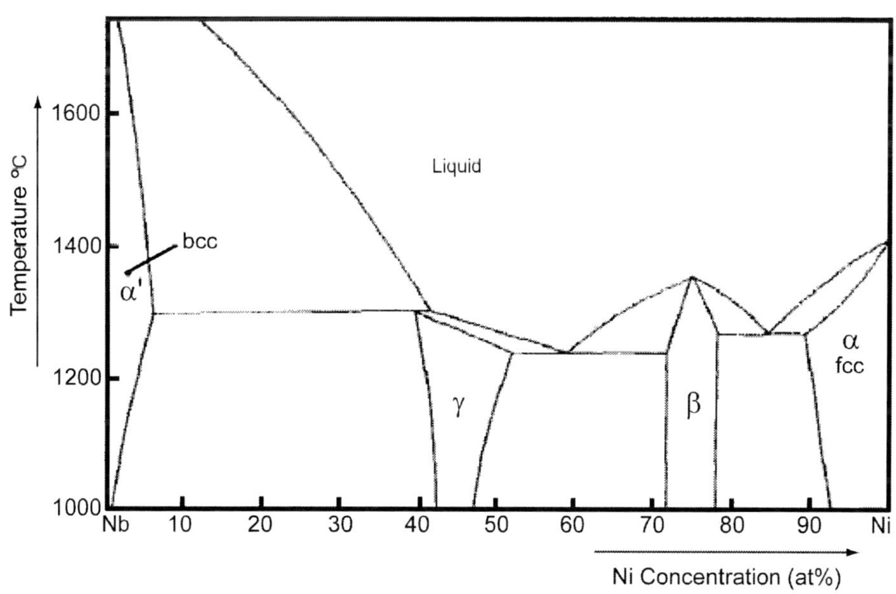

Figure 1.1.(a). Equilibrium phase diagram of Ni-Nb. (Figure obtained from ref [21] with permission from AIP).

Zhang et al. [21] explained the formation of HCP nickel thermodynamically. From the phase diagram in figure 1.1 a, it is evident that $Ni_{75}Nb_{25}$ is the β phase whereas $Ni_{70}Nb_{30}$ lie between the γ and β phases, closer to β. They showed that by thermal heating and ion irradiation techniques, the metastable HCP nickel phase was obtained. They argued that ion irradiation relaxes the film to lower energy states and as it is a non equilibrium procedure with a cooling rate of 10^{13}-10^{14} K sec^{-1}, it resulted in the formation of metastable states rather than thermodynamically favorable phases. Zhang et al. [21] showed that on low irradiation dose for $Ni_{70}Nb_{30}$ resulted in formation of HCP nickel whereas high radiation dose was required for $Ni_{75}Nb_{25}$ which tallies well with the free energy composition diagram.

1.2. SYNTHESIS

1.2.1. FCC Nickel

Pure nickel can be easily produced by reduction route wherein nickel oxide is reduced in hydrogen [22], methane [23] or carbon [24]. The overall reduction reaction for nickel oxide to nickel is given in equation 1.

$$NiO\ (s) + H_2(g) = Ni(s) + H_2O(g) \tag{1}$$

However, it is generally accepted that in a gas-solid reduction reaction, the following four steps are involved [22];

1. Mass transport of reactant and product gases between gas bulk and reacting oxide particles
2. Chemical reactions between reactant and oxygen atoms of the oxide
3. Mass transport of reactants and products in the form of gases, atoms, cations, or anions throughout the particle, including diffusion in condensed phases and porous media;
4. Nucleation and growth of metal product.

In the high temperature reduction process, only the formation of FCC nickel takes place. Besides, the morphology cannot be controlled. Electrodeposition technique gives a much better control over the morphology, thickness of deposit and uniformity of the deposition. Therefore, this process is widely used in developing functional materials. The reaction mechanism of the electrodeposition process from various electrolytes was generalised' by Saraby-Reintjes and Fleischmann [25]. The mechanism involves two consecutive one-electron charge transfer along with the participation of anion and the formation of adsorbed complex. Reactions 2-4 summarises the process.

$$Ni^{2+} + X^- = NiX^+ \tag{2}$$

$$NiX^+ + e^- = NiX_{ads} \tag{3}$$

$$NiX_{ads} = Ni + X^- \quad (4)$$

The anion X^- can be OH^-, Cl^- or SO_4^{2-}. In case of Watts bath ($NiSO_4$+NaCl+H_3BO_3), the anion X^- is thought to be Cl^- ion and the rate determining step is the first electron transfer according to reaction 3 [25]. There were other studies which proposed that nickel monohydroxide ($NiOH^+$) ion is an important species in charge transfer step in aqueous solutions. Cui and Lee [26] studied the nickel deposition from chloride solution in presence and absence of oxygen. They found that in presence of oxygen, nickel initially deposited as $Ni(OH)_2$ after which nickel deposition initiated. Further studies performed by Orinakova et al. [27] and Supicova et al. [28] gave much more insight into the electrodeposition mechanism using Watts solution and chloride solution. The results indicated that the electrodeposition process takes place in three steps in the Watts electrolyte solution: (i) a chemical reaction preceding an electrochemical reaction, (ii) the occurrence of surface reactions with the adsorption of intermediates onto the electrode and (iii) a reaction of the electroactive substance transported to the electrode by diffusion. Their reaction mechanism supported the Bokris [29] generalized model as given by equations 5-8.

$$Ni^{2+} + H_2O = Ni(OH)^+_{aq} + H^+ \quad (5)$$

$$Ni(OH)^+_{aq} \leftrightarrow Ni(OH)^+_{ads} \quad (6)$$

$$Ni(OH)^+_{ads} + e^- = NiOH \quad (7)$$

$$NiOH + H^+ + e^- = Ni + H_2O \quad (8)$$

The reaction mechanism of the Ni^{2+} reduction from acid sulphate, chloride and Watts electrolytes was also studied extensively by Epelboin and Wiart [30-34]. From impedance study of nickel electrodeposition, it was found that the electrode kinetics depends on the type of anion [32]. In chloride electrolytes, the predominant steps are the slow electrode activation with cathodic polarization. Further impedance analysis showed that electrolyte composition influenced the electrodeposition of nickel. From the impedance measurement, Chassaing et al. [35] proposed a reaction mechanism for electrolyte of pH 2- 4 as shown in equations 9-11.

$$2(H^+ + e^-) = H_2 \quad (9)$$

$$Ni^{2+} + e^- = Ni^+_{ads} \quad (10)$$

$$Ni^+_{ads} + e^- = Ni \quad (11)$$

Nickel electrodeposition on carbon was studied at two pH values of 3 and 5 using electrochemical impedance spectroscopy (EIS) by Proud and Muller [36]. They found the adsorption of a species in both chloride and sulphate baths depended on the pH of the bath at potentials far from the deposition potential for nickel. An overall reaction mechanism based on EIS studies proposed by Wiart [34] is given below. He proposed that reaction 12 was the initial step for the electrodeposition of nickel at pH 5 followed by reaction 13.

$$Ni(OH)^+_{aq} \leftrightarrow Ni(OH)^+_{ads} \tag{12}$$

$$Ni(OH)^+_{ads} + e^- = Ni(OH)_{ads} \tag{13}$$

However, on changing the pH to 3, initially a different reaction takes place as given by equation 14.

$$Ni(OH)^+ + e^- = Ni(OH)_{ads} \tag{14}$$

The adsorbed nickel hydroxide then decomposes according to equation 15.

$$Ni(OH)_{ads} + e^- \rightarrow Ni + (OH)^- \tag{15}$$

During nickel deposition, there is simultaneous generation of proton [34] which then results in the formation of hydrogen containing nickel as given by equations 16-18.

$$H^+ + e^- \rightarrow H_{ads} \tag{16}$$

$$2H_{ads} \rightarrow H_2 \tag{17}$$

$$Ni(OH)_{ads} + H_{ads} + e^- \rightarrow Ni + H_{incl} + (OH)^- \tag{18}$$

Many studies on nickel electrodeposition were performed by various researchers both by using cyclic voltammetry (CV) and Electrochemical Impedance Spectroscopy (EIS) techniques. A very good review on nickel electrodeposition from various aqueous electrolytes have been published by Orinakova et al. [37] in 2006 which covers mostly the electrodeposition process of nickel and it alloys.

As there is hydrogen evolution in aqueous electrolyte as shown by equation 16-18, the electrodeposition process gets complicated. Recent work on electrodeposition of nickel using ionic liquids has generated lot of interest in the scientific community as it possesses a higher electrochemical window. Besides, room temperature ionic liquid (RTILs) have good physical and chemical properties which include high chemical and thermal stability, high ionic conductivity, low vapor pressure and good solubility of many organic and inorganic compounds. Lot of research has been performed for electrodepositing metals using RTILs [38, 39].

Guo and Sun [40] first showed the possibility of deposition of nickel from an ionic liquid. It was shown that nickel chloride easily dissolves into 1-ethyl-3-methylimidazolium chloride (EMIC) ionic liquid forming $NiCl_4^-$ complex (figure 1.2 a inset) [41, 42]. However, no deposition of nickel occurred. On addition of zinc chloride in the ionic liquid, the potential was shifted positively which resulted in the deposition of nickel. Figures 1.2a and 1.2b compares the CVs of nickel chloride in EMIC in presence and absence of zinc chloride.

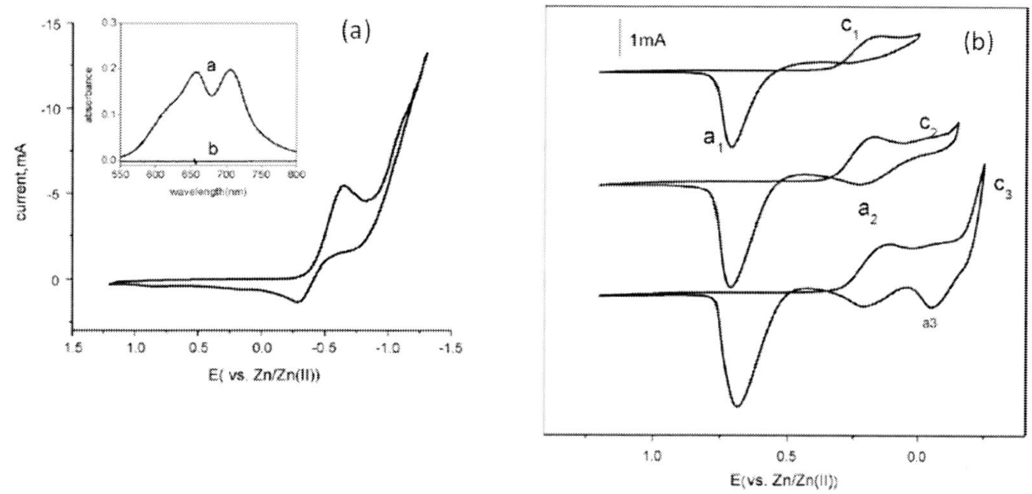

Figure 1.2. (a). CV of Ni (II) by dissolution of $NiCl_2$ in EMIC ionic liquid (EMIC: $NiCl_2$= 98.3mol%: 1.7mol%) at tungsten electrode at 150°C. Inset shows the absorbance spectrum of the solution in EMIC + $EMIBF_4$ (inset *(a)*) and EMIC + $EMIBF_4$+$ZnCl_2$. (b) CV of NiCl2 in $ZnCl_2$-EMIC (36-64 mol%) at tungsten electrode at 150°C (Figures obtained from ref [40] with permission from Elsevier).

$NiCl_2$-EMIC in fig 1.2a shows clear reduction peak at -0.63 V and a corresponding oxidation peak at -0.27 V. However, experiments performed at -0.75 V did not result in nickel deposition. In comparison, on addition of $ZnCl_2$, considerable changes were observed. Figure 1.2b shows reduction peak c_1, with a peak potential at 0.18V and a coupled anodic wave, a_1, with a peak potential at 0.70V. In addition, a "nucleation loop" was also observed which represents the nucleation/ growth controlled electrodeposition process. Guo and Sun [40] further performed constant potential electrodeposition at c_1 and using SEM, XRD and EDS found that nickel deposition takes place at this potential. Thus, waves c_1 and a_1 was attributed to the reduction of Ni (II) to Ni metal and the re-oxidation of the deposited Ni. They also proposed that $ZnCl_2$ changes the coordination of Ni (II) which then assists in the deposition process. They also could produce Ni-Zn alloy using the same ionic liquid by performing constant potential deposition between 0.1V and -0.1V. Abbott et al. [43] used a eutectic based Choline Chloride (ChCl) ionic liquid to electrodeposit nickel from nickel chloride dihydrate electrolyte. The ionic liquid was a eutectic mixture of ChCl: 2 ethylene glycol (EG) and ChCl:2 urea. The CV of $NiCl_2.6H_2O$ in the two eutectic based ionic liquids is compared in figure 1.3. A clear difference in the reversibility is observed which was due to the different ligand attached to the nickel ions. On constant potential deposition from ChCl: 2EG, Abbott et al. [43] found that the deposition of nickel consisted of nodular shaped morphology and was less rough compared to the deposition form ChCl: 2 urea ionic liquid. They also studied the effect of addition of brighteners from which uniform nickel deposition could be obtained.

Later, Deng et al. [44] showed the possibility of depositing nickel from a water and air stable ionic liquid without the addition of additives. They used 1-ethyl-3-methylimidazolium dicyanamide (EMI-DCA) a room-temperature ionic liquid and suggested that Ni (II) reacted with DCA^- anions formed $[Ni(DCA)_4]^{2-}$ complex which then reduces via a single step electron transfer process to deposit nickel. From chronoamperometric studies they concluded that the deposition process proceeds via a three-dimensional progressive nucleation with diffusion-controlled growth on both glassy and copper substrates.

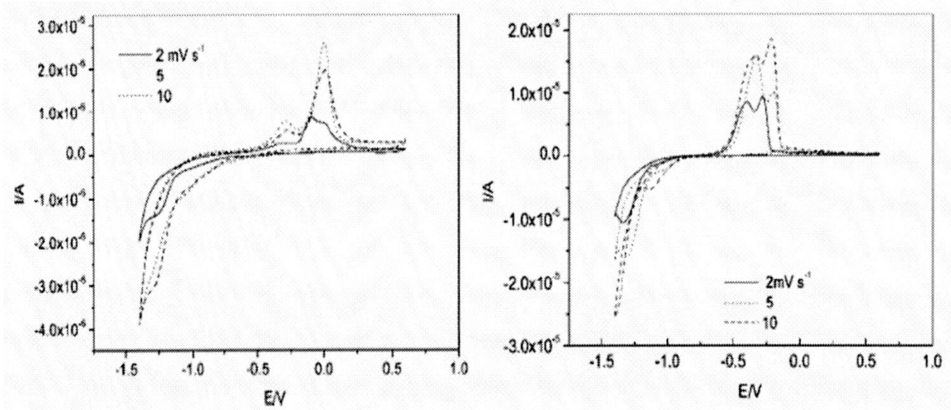

a b

Figure 1.3. (a). CV of 0.2M $NiCl_2.6H_2O$ in ChCl: 2EG on Pt electrode with varying sweep rate at 20°C (b) in ChCl: 2 urea. (Figures obtained from ref [43] with permission from Maney publishing).

The deposited nickel formed at constant potential experiments resulted in compact dense nodular morphology. It was also shown that by varying the deposit potential the roughness of the deposit could be controlled. For example, the roughness on depositing nickel at -1.4 V was found to be 16 nm whereas on deposting at -1.6 V was 90 nm.

1.2.2. HCP Nickel

Compared to the studies on FCC nickel, the studies on HCP nickel has been limited and there are no review papers exclusively dealing with HCP nickel. Therefore, here the review on HCP nickel will be presented in considerable detail. However, most of the synthesis of HCP nickel was by thermal decomposition process and morphology of nickel obtained by these processes is not same. So some details of these processes will be explained.

The presence of HCP nickel was initially observed by Yang [2] during electrodeposition of nickel from sulphate boric bath. However, on depositing from a chloride boric bath a mixed phase of FCC and HCP structures are obtained. There has been a lot of controversy about the presence of HCP nickel. It was stated by Hemminger and Weik [4] in their 1965 paper that conclusive proof for HCP nickel was not reported earlier. Hemminger and Weik later showed that on depositing 200 Å nickel film by evaporation at 10^{-5} torr, HCP nickel could be produced. They clarified the structure using selection area diffraction (SAED) technique. The HCP nickel was further confirmed by Wright and Goddard [5] in 1965 wherein they electrodeposited it from nickel salt. However, they reported that very precise electrolyte bath concentration and experimental conditions has to be maintained to obtain the HCP phase.

In 1988, Carturan et al. [7] showed that by decomposing nickel acetylacetonate with K/B alloy at 200° C under Ar for 2 hours, a black powder of HCP nickel was obtained. The X-ray diffraction pattern of the powder showed the presence of HCP nickel phase. The authors also performed small angle X-ray scattering (SAXS), BET and thermogravimetric analysis to evaluate its various properties. From BET, they showed that the surface are of the HCP nickel produced was 45 m^2 gm^{-1}. From the SAXS analysis, they observed that the particle size was around 19 nm. LECO analysis technique showed the presence of 0.72% carbon which was related to the presence of some solvent during the synthesis process. Thermal analysis of the HCP nickel obtained by Carturan et al. is given in Fig 1.4. The endothermic peak between 220 and 280 °C was related to the evolution of volatile matter and corresponded with the 4% weight loss in the thermogravimetric analysis. They further confirmed this by doing an isothermal run of the black powder at 310 °C, wherein HCP nickel phase was still found and thus relegated the possibility for the formation of Ni_3C. The exothermic peak at 350°C was related to the phase transformation from HCP nickel to FCC nickel as there was no weight loss. They further heat treated nickel and using line broadening analysis technique concluded the phase transformation phenomena.

In the last one decade, lot of research has been focused on evaluating the properties of HCP nickel. New synthesis techniques have been developed from different routes. Most of the formation of HCP nickel was attributed to the presence of impurities. Datta et al. [45] found that on milling $Ni_{95}Si_5$ for 30 hours, there is a structural change from FCC nickel to HCP. Figure 1.5 (a) compares the XRD patterns of $Ni_{95}Si_5$ during milling. It is evident from figure 1.5 (a) that after 2 hours of milling the FCC phase is predominant. On milling for 10 and 20 hours, the crystallinity of the FCC phase decreases. After 30 hours of milling, the HCP phase starts to form and after 50 hours low crystalline HCP phase is evident. However, when pure nickel was milled for 50 hours, no change in crystal structure was observed. They also found that the crystal size on milling $Ni_{95}Si_5$ was around 10 nm after 50 hours whereas for nickel was 21 nm. From the above observations and equation of state calculation, they predicted that a pressure of 8.7 GPa was generated which led to the transformation of FCC to HCP state. To further confirm the phase change, Datta et al. [45] performed DSC analysis on the 50 hours milled sample which is shown in figure 1.5 b. An exothermic reaction at 663K is evident which they related to the phase transformation from HCP to FCC phase. Furthermore, they performed isothermal heat treatment to confirm the observed transformation in DSC measurements.

Syukri et al. [9] showed a route to produce HCP nickel by decomposing a nickel acetate tetrahydrate complex at 300°C in nitrogen atmosphere. They observed that drying the sample at higher temperatures led to the formation of FCC nickel.

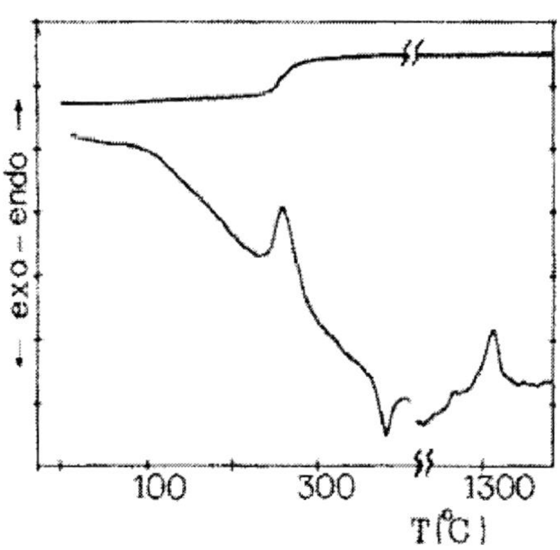

Figure 1.4. Thermogravimetric analysis (upper curve) and differential thermal analysis of HCP nickel powder (Figure obtained from ref [7] with permission from Elsevier).

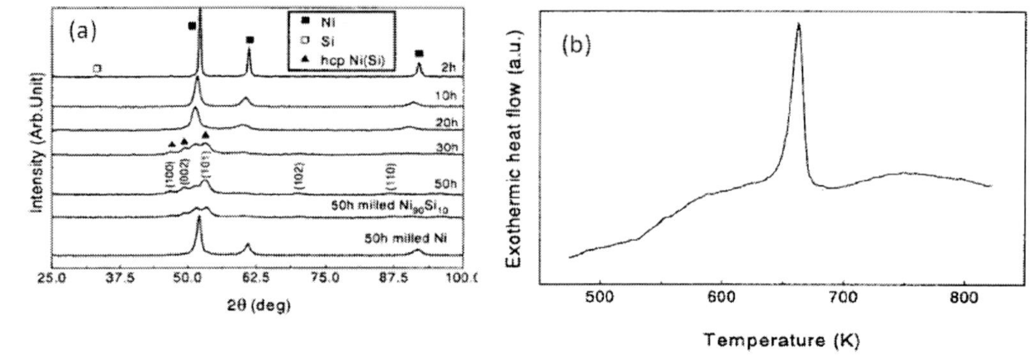

Figure 1.5. (a). Comparison of XRD of $Ni_{95}Si_5$ at different milling times (b) differential scanning calorimetry (DSC) of the 50 hours milled sample (figure obtained from [45] with permission from Cambridge University press).

By analysing the formation of nickel from other salts such as nickel oxalate, they concluded that the HCP nickel was formed due to the presence of carbon atoms in the nickel complex. Furthermore, Syukri et al. [9] concluded that HCP nickel phase transforms to FCC phase above 400 °C. Chinnasamy et al. [46] used the same Nickel acetate tetrahydrate $[Ni(OAc)_2{\cdot}4H_2O]$ as used by Syukri et al, and by ultrasonication and refluxing obtained nickel nanoparticles. They observed that by varying the reaction temperature and Ni/polyol concentration, HCP and FCC crystal structure could be controlled. Figure 1.6 a and 1.6 b compares the microstructure and the diffraction pattern of the samples obtained at different reaction conditions.

From the microstructures it is evident that at different synthesis temperatures, the particle size is different. At 563 K, the particle size is smallest and the corresponding XRD shows that the phase formed is HCP nickel. In comparison, the particle size at 473 K is about 500 nm and spherical in shape and the corresponding XRD shows the presence of both FCC and HCP phases. On changing the concentration of $[Ni(OAc)_2{\cdot}4H_2O]$, pure FCC nickel was achieved at 473 K. Thus, from the microstructure and XRD, Chinnasamy et al. [46] concluded that the formation of smaller particles, led to the production of HCP nickel.

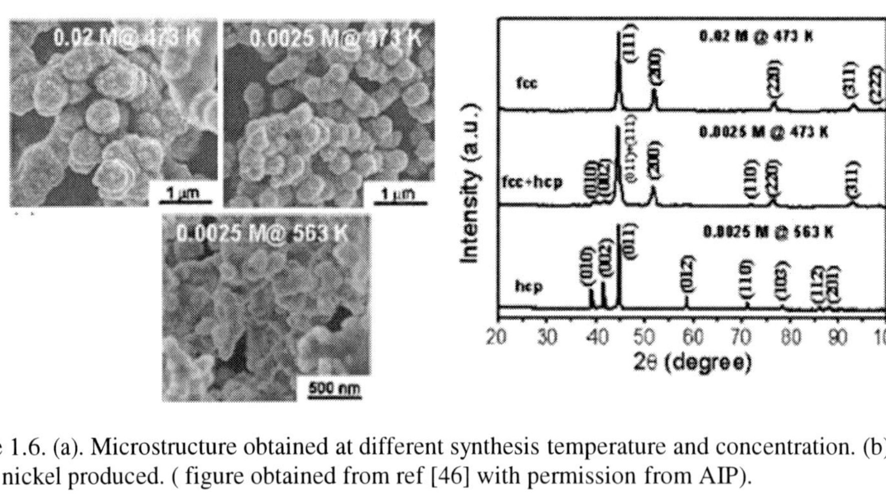

Figure 1.6. (a). Microstructure obtained at different synthesis temperature and concentration. (b) XRD of the nickel produced. (figure obtained from ref [46] with permission from AIP).

They further confirmed their analogy by reducing the particle size by the addition of Pt as seeds during the reflux process and found that even at higher concentrations of nickel acetate tetrahydrate, HCP nickel could be achieved at 563 K. Mi et al. [11] in 2005 showed that the same black colour powder obtained by Carturan et al. [7] can be produced by taking nickel chloride, KBH_4 and ethylenediamine in an autoclave and maintaining a temperature of 400°C. They characterised the sample using XRD and TEM from which the HCP nickel phase was confirmed. However, no thermal analysis or chemical analysis techniques were used so as to evaluate the impurities present during the synthesis technique. HCP nickel nanoparticles could also be obtained from a organic complex of nickel nitrate, $Ni(NO_3)_2$, by precipitating in ethanol [12]. On characterising the nanoparticles by XRD, it was found that a mixed phase of FCC and HCP nickel was found on using commercially available PEG-100 grade solution, whereas other grades gave HCP phase. This suggests that impurities in the chemicals also determine the phase formation.

A one-pot synthesis route by decomposing a Ni oleate complex for the controlled growth of HCP nickel phase was shown by Han et al. [47]. They showed that by changing the dodecylamine concentration during the synthesis and heating rate, FCC Ni phase could be obtained, thus showing a technique to control the formation of the two phases. Chen et al. [48] decomposed a nickel acetylacetonate complex between temperature of 215 and 285 °C to form HCP nickel. They found that the growth of HCP nickel completely depended on the decomposition temperature. On thermal decomposition of nickel acetylacetonate with oleylamine at 215 and 240 °C, only FCC nickel phase was present. However, on decomposing at 260 °C, HCP nickel formed which indicates the different reaction mechanism takes place during the decomposition process. Similarly, on thermal decomposition of nickel acetylacetonate in presence of tri-n-octylamine, HCP nickel was formed at 285°C whereas FCC was formed at 265 °C. The formation of HCP nickel at higher temperature was attributed to the favorable thermodynamic conditions. Also, Chen et al. [48] showed that the formation of such structure depended on kinetic parameters at slow heating rates (3 °C min^{-1}), FCC was predominant phase whereas on increasing the heating rate the structure changed to HCP.

Richard-Plouet et al. [49] demonstrated a route to show the presence of HCP nickel by decomposition of a complex organically modified nickel phyllosilicates structure, $Ni_3Si_{2.13}$-$(C_3H_6NH_3)_{1.98}(O_{5.28},OH_{1.05},F_{0.69})(CH_3COO)_{2.25}.1.76H_2O$, labeled as S20 and $Ni_3Si(C_3H_6NH_3)(O_{1.9},OH_{4.45},F_{0.65})$-$(CH_3COO)_{1.1.}$ $0.38H_2O$, labeled as S50 and found that the Si-O network helps in stabilizing the HCP nickel form between temperatures of 423 and 623 K. By reacting nickel nitrate, $Ni(NO_3)_2$ with citrate $(C_6H_8O_6)$ at different stages of 50, 80 and 120 °C and finally, heat treating in an argon gas at 300 °C, Gong et al. [15] showed the formation of HCP nickel. They also observed that on increasing the temperature of heat treatment leads to the formation of the more stable FCC phase. Thus, all the above analysis leads us to the conclusion that the stability of HCP phase is low and is preferentially formed at low temperatures.

A one-pot synthesis route similar to Chen et al. [48] was also performed by Luo et al. [16] to form HCP nanoparticles. It was observed that by addition of surfactant, the size and shape of the nanoparticles could be controlled. HCP nickel was formed by decomposing nickel acetate tetrahydrate complex at 240 °C in presence and absence of 1mmol of alkylphosphine surfactants (TOPO or TOP). The TEM images of the nanoparticles obtained in presence and absence of surfactants are given in fig 1.7.

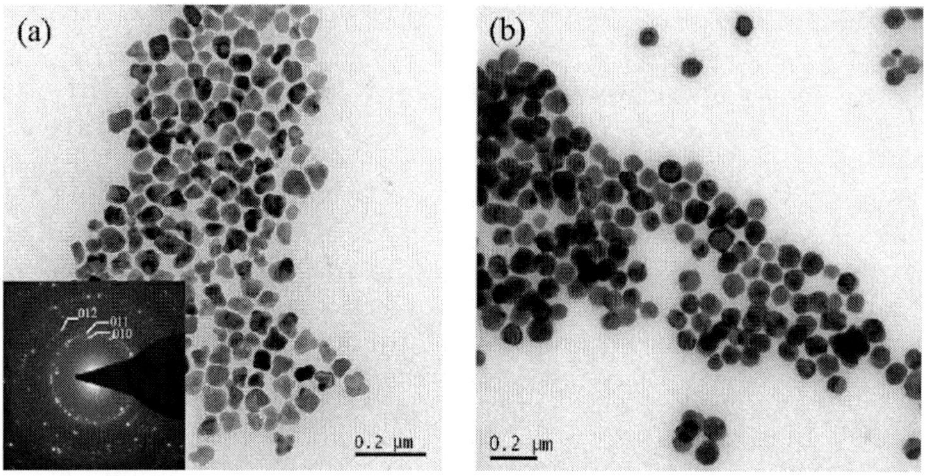

Figure 1.7. (a). TEM of HCP nickel prepared without surfactant (b) in presence of 1mmol TOPO (figure obtained from ref [16] with permission from Elsevier).

Figure 1.7 a shows the formation of irregular HCP structure with an average particle size of 61.8±9.0 nm. The HCP phase is confirmed by the SAED. On addition of 1mmol of surfactant the change in morphology (figure 1.7 b) is evident which shows almost uniform spherical structure having average particle size of 81.5±8.1 nm. It was also shown that using a different surfactant the average particle size decreased to 6.5 ±0.6 nm. From this study concluded that the surfactant inhibits the crystal growth without affecting the crystal structure. Also, from their experimental results they found that concentration of amine relates to the formation of HCP nickel and proposed a reaction mechanism. They proposed that on using oleylamine, the Ni nuclei are surrounded by the amine molecules which forms a steric barrier and slows the growth rate of the crystal and leads to the generation of thermodynamically stable cubic phase. However, on using octadecene as solvent, lower concentration of amine molecules are present, which allows Ni atoms to quickly reach the surface of the nuclei, facilitating the formation of metastable hcp phase.

Bolokang and Pasha [50] showed that the metastable HCP phase can be produced just by quenching nickel from 1100 °C. They found that quenching nickel from 950 °C led to formation of NiO phase from which they proposed that HCP is more stable in oxygen atmosphere. As most of the routes, involve decomposition of nickel complex in organic solvents, there is always the possibility of forming Ni_3N or Ni_3C whose lattice parameters are similar with HCP nickel. Therefore Tian et al. [10] grew the nickel onto MgO (001) in an UHV chamber to avoid the possibility of any impurity and studied the growth using insitu TEM technique. On performing selected area diffraction (SAED) studies, they found that initially the growth of nickel contained HCP phase and when the deposition thickness reached 2.5 nm, the phase changed to the more stable FCC structure. From their analysis, they related the formed HCP phase to the presence of (001) MgO plane which they believe to have created strain in the lattice.

Ohtake et al. [17] used molecular beam epitaxy technique to deposit thin films of HCP nickel onto Au (100) substrate. The deposition was made in a vacuum chamber by evaporating pure nickel at a vacuum pressure of 3×10^{-8} Pa. HCP nickel was formed when the evaporation temperature was below 300 °C. Also, it was found that on growing thicker films, the nickel atoms rearranged to form FCC nickel which is consistent with the observation of Tian et al. [10]. Fewer studies by electrodeposition technique for the synthesis of HCP nickel has been performed recently. Using a fast scan voltammetry Tehrani and Ghani [51] obtained HCP nickel electrochemically from a solution of 5.0 mM $NiCl_2.6H_2O$ and 1.0 M NH_4Cl. A three electrode setup was used wherein the working electrode was graphite with reference and counter as Ag/AgCl and Pt, respectively. On performing a fast scan of 6500 mV sec^{-1} HCP nickel was formed. They found that by changing the deposition potential and using fast scan rate, nickel nanoparticles with particle size can be formed as shown in table 1.2. It is observed from table 1.2, that the average particle size is small at low initial electrode potential. On increasing the electrode potential to -1.6 V, the particle size increases. The reported that due to hydrogen evolution and prolonged scan time, the particle size of electrodeposited nickel could grow. However, from the XRD results obtained, it looked like a mixture of FCC and HCP nickel phase.

Table 1.2. The different experimental parameters to obtain different nickel particle size (table obtained from ref [51] with permission from Elsevier)

E_1initial (mV)	E_2 final (mV)	Scan rate (mV sec^{-1})	Number of scans	Scan time (ms)	Deposition time (sec)	Average size (nm)	Distribution standard deviation
-1.3	-0.5	6500	3415	123	420	62	
-1.4	-0.5	6500	870	138	120	13.6	±2.7
-1.45	-0.5	6500	822	146	120	11.9	±2.6
-1.5	-0.5	6500	588	153	90	11.8	±2.5
-1.5	-0.5	6500	784	153	120	9.7	±2.3
-1.5	-0.5	6500	1176	153	180	14.4	±3.3
-1.5	-0.5	6500	1960	153	300	17.3	±3.8
-1.5	-0.5	6500	2745	153	420	90	
-1.55	-0.5	6500	745	161	120	15	±2.8
-1.6	-0.5	6500	710	169	120	15	±4.8
-1.7	-0.5	6500	2270	185	420	345	
-1.7	-0.5	6500	648	185	120	126	
-1.8	-0.5	6500	1826	230	420	122	

They argued that by applying a suitable high overpotential over a very short period of time, the reaction became kinetically favourable for the growth of HCP nickel. Also they found that by changing the deposition potential, the size of the nanocrystals changed from between 20 and 100 nm. Recently, it was shown that electrodepositing nickel from ionic liquid (1-ethyl-3-methylimidazolium chloride) could also produce HCP nickel. Lahiri et al. [52] observed that on electrodepositing nickel from a ionic liquid melt of 1-ethyl-3-methylimidazolium chloride at 5.4 mole% of $NiCl_2.6H_2O$ at 160°C, HCP nickel forms. On performing the same experiment at 150°C, led to the formation of FCC phase. From this they concluded that a secondary reaction must have taken place at 160°C to obtain the HCP phase. Using Fourier transform infrared spectroscopy FTIR, Lahiri and Das [53] later showed the presence of water was essential in the formation of HCP phase, as the water was getting electrolysed and generating hydrogen at the cathode which then changed the crystal structure. However they postulated that the concentration of water in the electrolyte is critical for the formation of the HCP phase.

1.3. Magnetic and Thermodynamic Properties

Nickel nanoparticles posses excellent magnetic properties and is being exploited in the application of magnetic sensors, memory devices, catalysts and biomolecular applications. The stable FCC nickel phase is ferromagnetic. However, it has been reported that impurities effect the magnetism of FCC structure. It has been shown that chemisorbed hydrogen changes the magnetic moment in nickel clusters [54]. Fournier and Salahub [55] demonstrated that three factors affect the change in magnetism by hydrogen atoms. (i) The extra electron brought in with the hydrogen goes into a down-spin Ni d level which reduces the moment. (ii) The reduced moment is accompanied by a reduced exchange splitting and consequently some up-spin of d electrons, not directly involved in the bonding to hydrogen, are transferred to lower lying down-spin d orbitals. (iii) For atoms close to the adsorbate, d character in the local density of states is pushed above the Fermi level through antibonding interactions with the hydrogen, further reducing the moments of these atoms. CO was also shown to reduce the magnetic moments of nickel [56]. A comparison of various adsorbates on the magnetic moment of nickel has been shown by Knickelbein [57].

Changing the nickel film thickness from 1-2 monolayer to 100 monolayer also changes the magnetic properties which has been described in detail by Baberschke [58]. Hirako et al. [59] studied the magnetism in nickel thin films by modifying the films using a redox reaction (Ni (0)/Ni (II)) in an alkaline solution and found that the magnetic properties changed. They related this change to the presence of some nickel compounds during the redox reaction on the nickel thin film.

In comparison with FCC nickel, the saturation magnetization of HCP nickel is very low. Table 1.3 presents the values of saturation magnetization and remanence ratio's of various FCC and HCP nickel. In general, the change in magnetic properties between 5 and 300K in table 1.3 can be broadly linked to the distribution of Ni particles in terms of size and localisation.

From the table, it is can be seen that nanoparticles in general have lower M_s values compared to bulk nickel. This phenomenon was related to three main arguments; (i) In small particles (nanoparticles), the presence of large number of surface atoms decrease the magnetic moment leading to a low saturation magnetization. (ii) Due to poor cystallinity of nickel nanoparticles. (iii) Due to the presence of impurities or non crystalline layers on the nickel particle.

Furthermore, it is observed that the saturation magnetization M_s for HCP nickel is around 7 times lower than the bulk FCC phase. Also, the Ms of HCP nickel and nickel carbide (Ni_3C) is almost similar, which suggests that a small amount of impurity present in the HCP changes its magnetic properties. On decreasing the temperature to 5K, a decrease in saturation magnetization is observed with decrease in the particle size. It was also reported in ref [14] that the high surface to volume ratio (nanoparticles) makes the saturation of magnetisation difficult even at low temperatures of 10K.

Table 1.3. Literature data on the magnetic properties of nickel taken from ref [14] with permission from Elsevier

T(K)	System	Size (nm)	M_s(emu/gm_{Ni})	Remanence ratio	H_c (O_e)
300K	Ni (fcc)	75000	51.3	0.13	104
	Ni (fcc)	3000	55	0.05	10
	Ni (fcc) in C nanotubes covered with C	100	nd	0.25	384
	Ni (fcc) coated with C	30–40	49.6	0.32	199
	Ni (fcc) in C nanotubes	19	nd	0.31	525
	Ni (fcc) covered with O	10–15	43.7	0.38	265
	Ni (fcc) coated with C	10–15	22.5	0.06	7
	Ni (fcc)	12	32	0.16	40
	Ni_3C (hcp)	10	0.8	0.12	70
	Ni (hcp)	nd	7.4	0.12	94
10K	Ni (fcc) coated with C	10–15	25.2	0.29	316
	Ni (fcc) covered with O	10–15	46.8	0.44	500
	Ni_3C (hcp)	10	1.4	0.40	500
5K	Ni (fcc)	3000	57.6		
	Ni (fcc)	59	53.3	0.37	280
	Ni (fcc)	46	47.8	0.34	340
	Ni (fcc)	22	28.7	0.37	352
	Ni (fcc)	4	27.7	0.05	200

Figure 1.8 compares the variable field magnetization data of FCC and HCP nickel nanocrystals. It was observed that with a small external magnetic field, both the cubic phase and hexagonal phases Ni nanocrystals rapidly reach magnetic saturation. The saturation magnetization (*M*s), remanent magnetization (*M*r), coercivity (*H*c) was for the FCC nickel nanoparticles was found to be 28.5 emu gm^{-1}, 6.20 emu gm^{-1}, and 80.83 Oe. In comparison, for hexagonal nickel phase the values are 0.819 emu gm^{-1}, 0.050 emu gm^{-1} and 40.69 Oe respectively. The coercivity of both the cubic and hexagonal phase Ni nanocrystals was found to be larger than that of bulk Ni (0.7 Oe). From all the above data Han et al. [47] concluded that the magnetic properties of the cubic phase Ni NCs are stronger than the HCP phase.

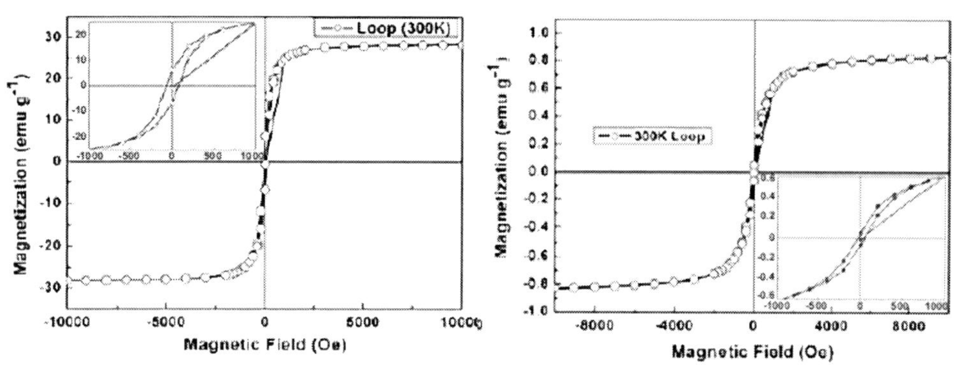

Figure 1.8. (a). Variable field magnetization data of pure cubic nickel nanocrystals (b) HCP nickel nanocrystals (Figure obtained from ref [47] with permission from Wiley).

Jeon et al. [13] studied the magnetic properties of the HCP nickel and a mixture of FCC and HCP nickel. They synthesized the HCP nickel by a method similar to that described by Carturan et al. [7]. Figure 1.9 compares the hysteresis loops of three different samples. From the M-H curve, it is evident that the magnetic property changes from sample 1 to sample 3 as there is a difference in the hysteresis loop. From the XRD phase analysis, Jeon et al. [13] found that sample 3 contains higher concentration of FCC nickel and therefore the M-H curve shows better ferromagnetic property.

The coercivity (H_c) values for samples 1-3 were 302, 353 and 546 Oe, respectively. This value corresponds well with the XRD phase analysis. From the above analysis they concluded that magnetization of HCP nickel nanoparticles is much smaller than FCC nickel nanoparticles and the HCP nickel does not show perfect ferromagnetic behavior. However comparing the magnetic property results of Jeon et al. and Han et al, it is evident that there is a difference in the magnetization. Many more studies on ferromagnetic property of HCP nickel have been made by various research groups [11, 12, 15, 46, 48, 49, 60, 61] by changing the synthesis technique of HCP nickel and different coercivity and magnetization values have been obtained. It appears that the different synthetic route have produced different types of impurities such as hydrogen, carbon or other inorganic materials in HCP nickel and these have affected the magnetic properties.

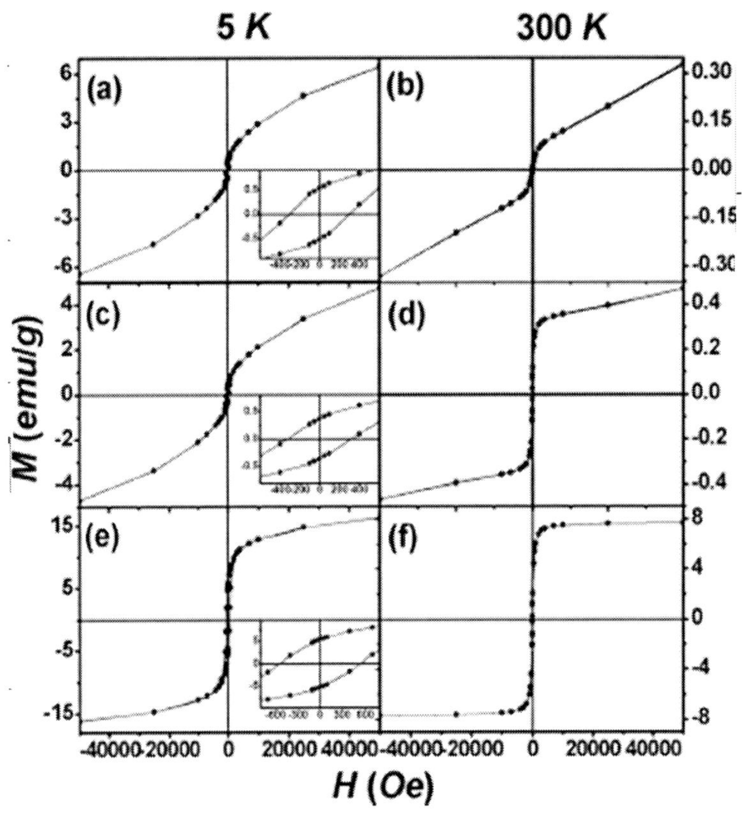

Figure 1.9. Comparison of the hysteresis loops of different samples, (a.b) sample 1, (c,d) sample 2 and (e,f) sample 3 at 5 and 300K (Figure obtained from ref [13] with permission from ACS).

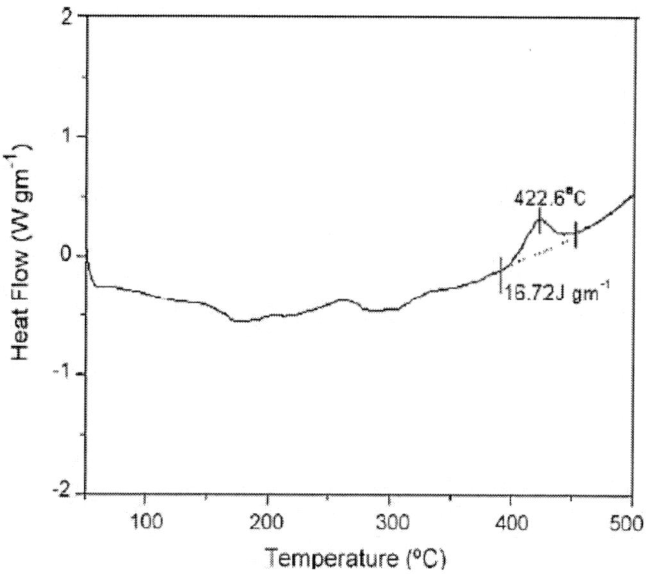

Figure 1.10. DSC of HCP nickel obtained by electrodeposition in ionic liquid (Figure obtained from [60] with permission from Elsevier).

Thermodynamic property of FCC nickel is well known. No transformation takes place in the cubic phase. The specific heat (C_p) of FCC nickel is 0.444 J $gm^{-1}K^{-1}$ at 298K. In comparison, very less information about the thermodynamic properties of HCP nickel has been reported. Carturan et al. [7] found a transformation from HCP to FCC phase at 623K whereas Datta et al. [45] observed the transformation at slightly higher temperature of 663K. The difference could be due to the presence of silicon in laters' sample. Lahiri and Tadisina [60] performed differential scanning calorimetry (DSC) studies on the HCP nickel obtained by electrodeposition from ionic liquids. The DSC plot is shown in figure 1.10.

The transformation observed by them was at 422.6°C (695.2 K) which is much higher than that obtained by other investigators. The difference in transformation temperature can be attributed to the level and type of impurity in HCP nickel. The specific heat value for HCP nickel obtained from DSC was high and was reported to be 1.425 J $gm^{-1}K^{-1}$ at 298K. They argued that the high specific heat value was due to the presence of hydrogen in the HCP nickel lattice. The C_p constants a, b and c for the C_p expression $a + bT + cT^2$ was found to be 1.226, 7.6×10^{-4} and 9.94×10^{-7}, respectively, where T was the temperature.

1.4. Applications

Nickel and nickel alloys are used in a wide variety of applications, such as, corrosion and heat resistance materials, aircraft gas turbines, steam turbine power plants, medical applications, nuclear power systems, and the chemical and petrochemical industries. The alloys of nickel and iron group metals often have a high hardness, high internal tension, and valuable magnetic properties. The hardness and strength of the electrolytic deposited coatings is better than alloys prepared by conventional metallurgical procedures. They have a protective function and are resistant against wear and corrosion. They can be used for decorative purposes as well due to high metallic lustre and colour [37].

Ni–Co films are widely used in protective and decorative plating applications and in aircraft turbine engines. They are also used as permanent magnetic memories as they have a high commutation speed. Cu-Ni alloys due to its low corrosion are widely used in piping, heat exchangers and condensers in sea water [62]. It is also used as thermocouples and resistors. Good magnetic property with minimum film resistance is essential for magnetic materials such as magnetoresistive sensors and magnetic recording devices. These can usually be best fabricated by electrodeposition technique. Ni/Cu multilayers with iron and cobalt have exhibited giant magnetoresistance [63].

Ni-Ti alloys are also called shape memory alloys as they can undergo deformation at one temperature and recover its original shape on heating. Due to its corrosion resistant properties and biocompatibility, it is vastly being exploited in colorectal surgery [64] and material for orthodontics [65]. Ni-Al alloys or Raney nickel has very good catalytic properties and is widely used in hydrogenation reactions. Other nickel alloys such as Ni-Zr, Ni-P, Ni-W-P/SiO_2 also have been used as catalyst to study the hydrogenation reactions [66, 67], in particular the catalytic hydrogenation reaction of benzene to cyclohexane. A very good review of Raney nickel and its uses in various hydrogenation reactions has been reported by Fouilloux [68].Electroless deposition of nickel has also been used in fabricating dynamic random access memory (DRAM) [69]. The use of nickel and its compounds in spintronic materials have also interested the scientific community [70, 71].

Nickel and its compounds have attracted lot of attention in the field of batteries and supercapacitors. In batteries, particular attention is given to nickel metal hydride batteries and nickel hydrogen battery.

The working principle for the nickel metal hydride battery is given by the electrochemical reactions below. The overall reaction can be written as [72];

$$MH + NiOOH = M + Ni(OH)_2 \tag{19}$$

In the above equation, M represents an intermetallic alloy. The reaction kinetics of these batteries is fast and therefore the charging/recharging is improved. The nickel metal hydride batteries was commercialised by Ovanic batteries. A very good review of nickel hydride batteries and its commercial use is presented by Dhar et al. [72]. Nickel cadmium batteries are very similar to the nickel metal hydride batteries.

In nickel-hydrogen battery, at the nickel electrode, nickel reduction/oxidation and water reduction/ oxidation takes place as given by equations 20 and 21 [73].

$$NiOOH + H_2O + e^- \leftrightarrow Ni(OH)_2 + OH^- \tag{20}$$

$$\frac{1}{2}O_2 + H_2O + 2e^- \leftrightarrow 2OH^- \tag{21}$$

At the platinum electrode, hydrogen reaction and oxygen reduction reactions, take place.

$$\frac{1}{2}H_2 + OH^- \leftrightarrow H_2O + e^- \tag{22}$$

$$\frac{1}{2}O_2 + H_2O + 2e^- \rightarrow 2OH^- \quad (23)$$

Although the energy density of nickel hydrogen battery is low compared to other batteries like lithium ion battery, but they have a long life and can handle more than 20000 charge cycles [74]. Nickel oxide is vastly studied in supercapacitor applications. Many authors have developed different morphologies of nickel oxide such as mesoporous type, nanoporous electrodeposits, hexagonal nanoporous, nanoplatelets, films and nanowhiskers, all of which have shown high capacitance [75, 76]. Besides, nickel oxides composites such as activated carbon/nickel oxide, nickel based mixed rare earth oxide, $Ni(OH)_2$/CNT, and cobalt nickel oxide/CNT have also shown remarkable specific capacitance in the range of greater than 2000 F gm^{-1} [77-79].

One of the most important application of nickel could be its used as a hydrogen storage material. The nickel hydrogen phase diagram shows that at high pressures of greater than 100Mpa there is considerable solubility of hydrogen [80, 81]. At extremely high pressure of 1600 to 1900Mpa, nickel forms a low hydrogen content α-Ni solid solution and a hydrogen rich nonstoichiometric β-Ni hydride phase [19, 80]. It was recently shown by Lahiri et al. [52], that HCP nickel contained 1.2wt% of hydrogen, thus making it a potential hydrogen storage material.

Conclusion

It can be concluded from the review that the nickel is an important metal and has vast amount of applications in daily life, from corrosion resistant coatings to memory devices and batteries. Work on nickel has been going on for past 60 years and new functional properties are still to emerge, especially in the field of energy devices. Since last 20 years, lot of research has been focussed on developing energy storage material for supercapacitor application and batteries.

The synthesis route for FCC nickel has been well known and electrodeposition technique is preferred to control the film thickness and deposition rate. Comparatively, less research has been performed on HCP nickel and is mostly synthesised by the decomposition of nickel complex. Electrodeposition technique is an alternate for HCP nickel synthesis by which the particle size and the thickness of deposit could be controlled, but it needs further exploitation. The use of HCP nickel also has been limited and a potential use as hydrogen storage material is shown. Certainly, the work has just begun in exploiting the functionality of HCP nickel and future application of this material in science and engineering needs to be explored.

References

[1] B. Injeti Gurrappa and Leo, *Sci.Technol.Adv.Mater.*, 9, 4, 043001 (2008).

[2] L. Yang, *J. Electrochem.Soc.*.97,9, 241 (1950).

[3] B. D. Cullity, *Elements of X-ray diffraction.* 1956: Addison-Wesley Publishing Company, Inc.

[4] P. Hemenger and H. Weik, *Acta Crystallogra.*, 19, 4, 690 (1965).

[5] J. G. Wright and J. Goddard, *Phil.Mag.*, 11,111, 485,(1965)
[6] L. He, *J.Magn.Magn. Mater.*, 322, 14, 1991, (2010).
[7] G. Carturan, G. Cocco, S. Enzo, R. Ganzerla, and M. Lenarda, *Mater. Lett.*,7, 1-2, 47 (1988)
[8] S. Illy, O. Tillement, F. Machizaud, J. M. Dubois, F. Massicot, Y. Fort, and J. Ghanbaja, *Phil.Mag.*,.79, 5,1021. (1999)
[9] Syukri, T. Ban, Y. Ohya, and Y. Takahashi, *Mat. Chem. Phys.*, 78, 3. 645-649. (2003)
[10] W. Tian, H. P. Sun, X. Q. Pan, J. H. Yu, M. Yeadon, C. B. Boothroyd, Y. P. Feng, R. A. Lukaszew, and R. Clarke, *Appl. Phys. Lett.*, 86, 13, 131915 (2005).
[11] Y. Mi, D. Yuan, Y. Liu, J. Zhang, and Y. Xiao, *Mat. Chem. Phys.*,.89, 359. (2005)
[12] V. Tzitzios and et al., *Nanotechnol.*,.17, 15, 3750 (2006).
[13] Y. T. Jeon, J. Y. Moon, G. H. Lee, J. Park, and Y. Chang, *J. Phys. Chem. B*,.110, 3, 1187 (2005)
[14] V. Rodríguez-González, E. Marceau, P. Beaunier, M. Che, and C. Train, *J. Solid State Chem.*, 180, 1, 22. (2007).
[15] J. Gong, L. L. Wang, Y. Liu, J. H. Yang, and Z. G. Zong, *J. Alloys. Compd.*,. 457, 1-2, 6 (2008)
[16] X. Luo, Y. Chen, G.-H. Yue, D.-L. Peng, and X. Luo, *J. Alloys. Compd*,. 476, 1-2,864 (2009)
[17] M. Ohtake, T. Tanaka, F. Kirino, and M. Futamoto, *J. Phys.: Conf. Ser.*,. 200, 7, 072072. (2010)
[18] C. C. Koch, O. B. Cavin, C. G. McKamey, and J. O. Scarbrough, *Appl. Physl. Lett*,. 43, 11, 1017, (1983)
[19] W. L. Johnson, M. Li, and C. E. Krill Iii, *J. Non-Crystal. Sol.*,,156-158, 481. (1993)
[20] V. G. Gryaznov and L. I. Trusov, *Prog. Mater. Sci.* 37, 4, 289, (1993)
[21] Z. J. Zhang, H. Y. Bai, Q. L. Qiu, T. Yang, K. Tao, and B. X. Liu, *J. Appl. Phys.*73, 4, 1702. (1993)
[22] T. Hidayat, M. A. Rhamdhani, E. Jak, and P. C. Hayes, *Metal. Mater. Trans.*B,.40, 1,1. (2009)
[23] R. Alizadeh, E. Jamshidi, and H. Ale Ebrahim, *Chem. Eng. Technol.* 30, 8, 1123 (2007)
[24] S. K. Sharma, F. J. Vastola, and P. L. Walker, *Carbon*,. 35, 4, 535 (1997)
[25] A. Saraby-Reintjes and M. Fleischmann, *Electrochim. Acta,* 29, 4, 557. (1984).
[26] C. Q. Cui and J. Y. Lee, *Electrochim. Acta*,. 40, 11, 1653 (1995)
[27] R. Orináková, L. Trnková, M. Gálová, and M. Supicová, *Electrochim. Acta*,. 49, 21, 3587 (2004)
[28] M. Šupicová, *R. Rozik, L. Trnková, R. Oriňáková, and M. Gálová, J. Solid. State Electrochem* . 10, 2, 61 (2006)
[29] J. O. M. Bockris, D. Drazic, and A. R. Despic, *Electrochim. Acta.* 4, 325 (1961)
[30] I. Epelboin and R. Wiart, *J. Electrochem. Soc.* 118, 10,1577 (1971)
[31] I. Epelboin, M. Joussellin, and R. Wiart, *J. Electroanal. Chem.* 101, 2, 281 (1979)
[32] I. Epelboin, M. Joussellin, and R. Wiart, *J. Electroanal. Chem.* 119, 1, 61, (1981)
[33] M. Froment and R. Wiart, *Electrochim. Acta.* 8, 481. (1963)
[34] R. Wiart, *Electrochim. Acta*,. 35, 10, 1587 (1990).
[35] E. Chassaing, M. Joussellin, and R. Wiart, *J. Electroanal. Chem.* 157, 1, 75, (1983)
[36] W. G. Proud and C. Müller, *Electrochim. Acta* 38, 405, (1993)

[37] R. Oriňáková, A. Turoňová, D. Kladeková, M. Gálová, and R. Smith, *J. Appl. Electrochem.* 36, 9, 957 (2006).
[38] T. Torimoto, T. Tsuda, K.-i. Okazaki, and S. Kuwabata, *Adv. Mater.* 22, 11, 1196, (2010)
[39] M. Armand, F. Endres, D. R. MacFarlane, H. Ohno, and B. Scrosati, *Nat. Mater.* 8, 8, 621, (2009)
[40] S. P. Gou and I. W. Sun, *Electrochim. Acta* 53, 5, 2538, (2008)
[41] R. J. Gale, B. Gilbert, and R. A. Osteryoung, *Inorg. Chem.*18, 10, 2723, (1979)
[42] T. M. Laher and C. L. Hussey, *Inorg. Chem.* 21, 11, 4079 (1982)
[43] A. P. Abbott, K. El Ttaib, K. S. Ryder, and E. L. Smith, *Trans. Inst. Met. Fin.* 86, 4, 234. (2008)
[44] M. J. Deng, I. W. Sun, P. Y. Chen, J. K. Chang, and W. T. Tsai, Electrochim. Acta,. 53, 19, 5812, (2008)
[45] M. K. Datta, S. K. Pabi, and B. S. Murty, *J. Mater. Res.*15, 7, 1429, (2000).
[46] C. N. Chinnasamy, B. Jeyadevan, K. Shinoda, K. Tohji, A. Narayanasamy, K. Sato, and S. Hisano, *J. Appl. Phys.* 97, 10, 10J309. (2005).
[47] M. Han, Q. Liu, J. He, Y. Song, Z. Xu, and J. M. Zhu, *Adv. Mater.* 19, 8, 1096, (2007)
[48] Y. Chen, D. L. Peng, D. Lin, and X. Luo, *Nanotechnol.* 18, 505703, (2007).
[49] M. Richard-Plouet, M. Guillot, S. Vilminot, C. d. Leuvrey, C. EstournÃ¨s, and M. Kurmoo, *Chem. Mater.* 19, 4, 865. (2007)
[50] A. S. Bolokang and M. J. Phasha, *Mater. Lett.* 65, 1, 59, (2011)
[51] R. Mohamed Ali Tehrani and S. Ab Ghani, *J. Coll. Inter.Sci.*339, 1, 125, (2009)
[52] A. Lahiri, R. Das, and R. G. Reddy, *J. Pow. Sour.*195, 6, 1688, (2010)
[53] A. Lahiri and R. Das, *J. Appl. Electrochem.*40, 11, 1991, (2010)
[54] J. K. Blum and W. Göpel, *Thin Solid Films* 42, 1, 7, (1977)
[55] R. Fournier and D. R. Salahub, *Int. J. Quant. Chem.* 29, 5, 1077, (1986)
[56] B. K. Mark, *J. Chem. Phys.* 115, 5, 1983, (2001)
[57] B. K. Mark, *J. Chem. Phys.*116, 22, 9703, (2002)
[58] K. Baberschke, *Appl. Phys. A*, 62, 5, 417, (1996)
[59] N. Hiraoka, Y. Oba, T. Watanab, H. Maki, Y. Einaga, and T. Sato, e-*J. Surf. Sci. Nanotechnol.* 7, 787, -790. (2009)
[60] A. Lahiri and Z. Tadisina, *Mat. Chem. Phys.* 124, 1, 41, (2010)
[61] J. Yang, B. Feng, Y. Liu, Y. Zhang, L. Yang, Y. Wang, M. Wei, J. Lang, and D. Wang, *J. Alloys. Compd.* 467, L21-L25.(2009)
[62] I. Miloševand M. Metikoš-huković, *J. Appl. Electrochem.* 29, 3, 393, (1999)
[63] Q. Huang, D. P. Young, and E. J. Podlaha, *J. Appl. Phys.* 94, 3, 1864, (2003)
[64] D. Stewart, S. Hunt, R. Pierce, M. Dongli, M. Frisella, K. Cook, B. Starcher, and *J. Flesh. Surg. Innov.,*. 14, 4, 252. (2007)
[65] F. Miura, M. Mogi, and Y. Ohura, . *Europ. J. Orthodont.* 10, 1, 187, (1988).
[66] T. Takahashi, S. i. Higashi, T. Kai, H. Kimura, and T. Masumoto, *Catal. Lett.* 26, 3, 401, (1994)
[67] H. Li and Y. Xu, *Mater. Lett.* 51, 2, 101. (2001)
[68] P. Fouilloux, *Appl. Catal.* 8, 1, 1. (1983)
[69] M. Kawano, N. Takahashi, M. Komuro, and S. Matsui. *Low-cost TSV process using electroless Ni plating for 3D stacked DRAM*. in *Electron.Comp.Technol.Conf.*. 2010: IEEE.

[70] A. R. Rocha, V. M. Garcia-Suarez, S. W. Bailey, C. J. Lambert, J. Ferrer, and S. Sanvito, *Nat. Mater.* 4, 4, 335, (2005)

[71] C. Felser, G. H. Fecher, and B. Balke, *Ange. Chem. Int. Edit.* 46, 5, 668, (2007)

[72] S. K. Dhar, S. R. Ovshínsky, P. R. Gifford, D. A. Corrigan, M. A. Fetcenko, and S. Venkatesan, *J. Pow. Sour.* 65, 1, (1997)

[73] S. Liu, R. A. Dougal, J. W. Weidner, and L. Gao, *J. Pow. Sour.*141, 2, 326, (2005)

[74] M. J. Milden, *Aero.Electron.Sys.Mag., IEEE.* 6, 11, 14. (1991)

[75] W. Xing, F. Li, Z.-f. Yan, and G. Q. Lu, *J. Pow. Sour.* 134, 2, 324 (2004).

[76] M. Jayalakshmi and K. Balasubramanian, *Int. J. Electrochem. Sci.* 3, 1196, (2008).

[77] Q. Huang, X. Wang, J. Li, C. Dai, S. Gamboa, and P. J. Sebastian, *J. Pow. Sour.* 164, 1, 425. (2007)

[78] V. Gupta, S. Gupta, and N. Miura, *J. Pow. Sour.* 175, 1, 680. (2008)

[79] V. Ganesh, S. Pitchumani, and V. Lakshminarayanan, *J. Pow. Sour.* 158, 2, 1523. (2006)

[80] M. L. Wayman and G. C. Weatherly, *Binary alloy phase diagrams*: ASM International.

[81] Y. Shizuku, S. Yamamoto, and Y. Fukai, *J. Alloys. Compd.* 336, 159. (2002)

In: Electrodeposition: Properties, Processes and Applications ISBN: 978-1-61470-826-1
Editor: Udit Surya Mohanty

Chapter 2

DEPOSITION AND PROPERTIES OF ELECTROCHEMICAL COMPOSITE COATINGS

V. N. Tseluikin*
Engels Technological Institute (Branch) of Saratov State Technical University
Engels, Saratov oblast, Russia

ABSTRACT

Results of investigations in the field of electrochemical composite coatings (ECC) are presented. Deposition peculiarities, functional properties and structure of the main types of ECC are also discussed.

INTRODUCTION

An effective approach in modifying and imparting new properties to the metal surfaces lies in the deposition of electrochemical composite coatings (ECC). The technology of ECC is based on the use of electrolyte suspensions from which metals are codeposited with disperse particles of various types and dimensions [1 – 3]. Being incorporated into coatings, dispersed particles substantially improve their operational properties (hardness, wear, and corrosion resistance) and impart them new properties (antifriction, magnetic, catalytic). The first publication reported on the ECC preparation, dates back to 1929 [4]. Nowadays ECC find wide use in machine and instrument building, construction of medicinal tools and chemical industry equipments.

The matrix of ECC usually consists of metals (nickel, chromium, copper, iron, zinc, tin, noble metals), as well as alloys deposited in the absence of external current (Ni–P, Ni–B). As the dispersed phase, solid particles are used. As a rule, they do not exceed 5 μm but in certain cases reach several tens micrometers. This can be oxides (Al_2O_3, TiO_2, ZrO_2, SiO_2, etc.), binary compounds of d-elements (carbides, borides, nitrides, sulfides, etc.), metal and

* E-mail address: tseluikin@mail.ru.

nonmetal powders (Cr, Mo, W, Si, graphite, diamond), high-molecular compounds (polytetrafluoroethylene, caprolactam), etc. Recently, composite coatings with nanosized particles (nanocomposite coatings) were studied most actively. This interest is related to the fact that the structural and functional properties of nanostructured materials significantly differ from those of their coarse-grained analogs. Indeed, as the grain size decreases, the mechanical strength of electrolytic coatings increases (for the same level of plasticity) and their physicochemical properties are improved.

To intensify the ECC deposition, different methods were used, namely, electrolyte stirring, horizontal arrangement of the cathode, the use of concentrated suspensions, nonstationary electrolysis modes, ultra-sonic agitation, magnetic field application, preliminary chemical treatment of the dispersed phase, etc.

The kinetics of ECC formation includes the following stages: delivery of dispersed particles to the cathode, their retention at the cathode surface, and their overgrowing with the metal deposit. Dispersed particles can be delivered to the cathode by agitation, Brownian movement, or gravitation, as well as their adsorption on cations of the deposited metal. The particles retained on the cathode initiate nucleation in sites when placed in contact with the surface, thus stimulating their overgrowthon the metal. Varying the electroplating conditions makes one build such a microrelief surface that would retain dispersed particles of definite size. This, in turn, makes it possible to form coatings with target properties.The present review surveys the significant results on ECC, which were obtained in the recent years.

2.1. Different Types of Electrochemical Composite Coatings (ECC)

2.1.1. Nickel-Based ECC

Coatings with nickel matrices are the most abundant among ECC. Nickel-based ECC are characterized by the high hardness and wear resistance, stability in corrosive media, and good appearance. Dispersed particle of different nature are easily deposited together with nickel. Smaller particles with deformed lattices are captured more readily owing to the heterogeneities of the metal surface. Authors [5 – 9] studied nickel ECC with ultradispersed diamonds (nanodiamonds) as a disperse phase. Nanodiamonds (ND) are prepared by detonation synthesis in a closed volume [9]. ND particles measuring about 4–6 nm, have oval or spherical shape, they lack sharp edges, have well-developed surfaces (up to 450 m^2/g) and high surface energy [8]. The ND structure represents a core (~4–5 nm) of a cubic diamond enclosed into a shell of crystallographically amorphous transient carbon structures 0.4–1.0 nm thick [9]. The ND surface layer is saturated with different preferentially oxygen-containing functional groups that determine the high activity of particles and their trend towards aggregation (coagulation) in electrolyte solutions [9]. That is why ND were preliminarily cleaned from surface impurities-coagulants using ion-exchange resins or the bath containing an ND suspension was subjected to ultrasonic agitation [7]. As a rule, the deposition of nickel–ND ECC involved the use of classical sulfate baths. The ND concentration in a bath reached 30 g/l [7]. The experimental data [5 – 9] point to the positive effect of ND on the quality of nickel–diamond coatings. Composite coatings looked as velours with the color varying from light gray to gray. The content of diamond particles in coatings varied from 0.2 to 1.0 wt % depending on their type and concentration in the electrolyte. Compared with pure nickel coatings, their fretting coefficient decreased from 0.43 to 0.33 and the microhardness increased from 250 to 440 kg m^{-2}. It should be noted that the data of different authors on the properties of nickel–ND ECC differed. Thus in [7], it was shown that for 20 g l^{-1} ND in the bath, the ECC micro-hardness reached 560 kg mm^{-2} and wear resistance increased 5.8-fold. These coatings were deposited at the cathodic current density i_c = 1 A dm^{-2}. However, as i_c increased to 5 A dm^{-2}, the ECC microhardness and wear resistance substantially increased at lower ND concentrations (Table 2.1). The resulting coatings were fine-grained, dense, and of low porosity [9]. According to the data of [10], parts coated with nickel–ND ECC could serve 20-time longer compared with purely nickel coatings. Nickel–nanodiamond ECC were also used for making point contacts on very small areas. In this case, diamond particles incorporated into a nickel coating turned out to be pressed into the contact zone [11]. Deposition of diamond layers together with electroplated binders onto cutting tools produced uniform ECC with the number of particles varying from 20000 to 60000 per cm^2 of the surface [12]. In [5, 13 – 15], the advantages of modifying nickel–diamond deposits with boron were demonstrated. As the boron-containing additive, sodium decahydroborate $Na_3B_{10}H_{10}$ was added to the bath. The microhardness of nickel–B–ND ECC increased from 5 to 12 GPa with an increase of the boron content in the coating. Such an effect was observed for the ND concentration in solution of 5 g l^{-1}. For the higher ND concentrations (10–20 g l^{-1}), the microhardness fell to 4.7–3.7 GPa, which was probably associated with certain nonuniformity of the nanodiamond incorporation into coatings [26]. Structural studies [6] have shown that incorporation of diamond nanoparticles into the nickel matrix leads to a decrease in the grain size, the formation of dislocations looking as tangles and networks along the grain boundaries. A metallographic study of coating cross-sections [5] has revealed that nickel–ND ECC have a columnar structure. The aggregation of sufficiently coarse particles proceeds in the vertical direction. A cross-section of the nickel–ND deposit surface demonstrated in addition to coarse particles, the presence of many fine particles that formed closed chains distributed over the whole volume of the coating. The increase in the microhardness with the addition of boron to nickel–diamond ECC was apparently associated with the changes in the morphology of deposits, namely, the transition from the columnar to the branched chain structure.

Table 2.1. The effect of ND content in the bath on the microhardness and wear resistance of nickel coatings for different cathodic current densities

ND content, g l^{-1}	Microhardness H, kg mm^{-2}			Wear resistance, a.u.
	1.0[a]	1.5[a]	5.0[a]	5.0[a]
0	251	286	308	1.0
0.5	301	312	330	1.8
1	320	331	354	2.3
2	359	380	378	3.8
5	370	408	472	5.7
10	395	467	583	6.4

Note: [a] Values of i_c, A dm^{-2}.

Previous studies [16 – 19] investigated nickel ECC with fullerene C60 and carbon nanotubes (CNT). Fullerene C60 represents a closed shell with unsaturated bonds and can easily and reversibly take up electrons. However, fullerene molecules were found to be hydrophobic and dissolve only in polar solvents. That is why, a method of preparing stable aqueous C60 suspensions containing no organic solvents was developed [20]. To prepare an ECC, the fullerene suspension was added to the nickel-plating sulfate bath [16, 17].

Adding fullerene particles to the bath promotes the cathodic process. ECC nickel–fullerene C_{60} were deposited at less negative potential compared with purely nickel coatings [16]. Being an electron-acceptor, fullerene C_{60} tends to acquire a negative charge in electrolyte solutions. This should in turn favor the adsorption of nickel cations on it, so that dispersed particles move to the cathode and are eventually incorporated into the deposit lattice.In contrast to nickel, ECC has a rough surface with formation of microprotrusions as the dispersed particles were overgrown by metal. The roughness increases as the coating becomes thicker. Hence, being incorporated into the deposit, fullerene particles determine its further growth. The analysis of nickel–UDD ECC by the secondary ion mass spectroscopy has demonstrated the presence of carbon and C–H bonds in the coatings [17]. Obviously, during the electroplating, fullerene particles are hydrogenated by co-deposition of hydrogen.

The sliding friction coefficient of a nickel–C_{60} ECC decreases as compared to pure nickel deposits (Table 2.2).

The nickel–CNT coating deposited with a pulse current source has less porosity, higher hardness and higher wear resistance than that with a direct current source. CNTs greatly improve the coating performance. The wear mechanism is mainly related to the smearing of the nickel–CNTs coatings, instead of the fracture for the nickel coatings [18].

Like the other nickel-based ECC, composite coatings nickel–fluoroplastic (polytetrafluoroethylene) can be prepared from sulfate–chloride baths [21, 22]. However, the formation of an ECC of the given type from the Watt's bath is hindered. A polytetrafluoroethylene (PTFE) molecule is shaped as a spiral that forms a cylinder with the dense outer shell of electronegative fluorine atoms [22]. Hence, PTFE molecules are hydrophobic and tend to coagulate in the bulk of sulfate solutions under the action of attractive van der Waals forces.

This problem can partly be solved by using complexes, e.g., sulfamate [23] or acetate [24] baths in which the adsorption-solvation layer on PTFE particles blockades the aggregation process. The deposition of nickel–PTFE coatings from a sulfamate bath resulted from incorporation up to 20% of the dispersed phase into the coating, whereas the coating deposited from a sulfate bath, included only 4–8% of such a phase [25]. During the deposition of ECC from acetate baths, in a near-electrode layer, the electrocoagulation of PTFE particles occurred, being caused by the appearance of dispersion forces between particles.

Table 2.2. Friction coefficients of pure and fullerene-containing nickel coatings deposited at various cathodic current densities

Coating	Current density i_c, A dm^{-2}				
	6	7	8	9	10
Nickel	0,38	0,34	0,34	0,33	0,30
Nickel–C_{60}	0,20	0,19	0,17	0,15	0,10

This aggregation depended on the pH of the near-electrode layer, the electrolyte composition, and the electrolysis mode. During the ECC deposition, to increase the stability of PTFE in the near-electrode layer, non-ionogenic surfactants were added to acetate baths. The maximum aggregation stability of surfactant-stabilized PTFE particles (diameter 0.3–0.5 μm) was reached at pH ≈ 5. This was caused by the fact that in the baths with pH 5, more than 90% of nickel was bound in acetate complexes, and the discharge of Ni^{2+} ions from complexes proceeded easier by the following scheme [24]:

$$Ni^{2+} + e \rightarrow Ni^{+} + e \rightarrow Ni^{0}.$$

Coatings deposited from the acetate bath were characterized by the uniform distribution of PTFE particles. They manifested a 25–30% higher corrosion resistance, and 2–3 times better tribological behavior compared with pure nickel deposits [24, 25].

Nickel–fluoroplastic coatings can be modified by boron. The synergistic effect of wear-resistance and antifriction properties [26] of ECC nickel–fluoroplastic–boron were investigated. After thermal treatment, the coatings revealed solid phase components Ni, Ni_3B, Ni_2B, and fluoroplastic as the lubricating phase component. After reaching the steady-state mode under the dry friction conditions, the surface layers of ECC revealed the NiB phase.

Using the "concentration wave" model, the linear wear rate and the coefficients of friction were calculated for these coatings. The obtained results adequately agree with experimental data (Table 2.3). The properties of nickel–boron–fluoroplastic ECC are better compared with those calculated in terms of the additive model (positive synergistic effect). Synergism of solid (NiB) and lubricating (PTFE) components of ECC depends on the concentration of the lubricating phases on the friction surface, which enhances the coating resistance.

Recently, scientists have been focused on the functional properties of composite coatings. The ECC properties such as hardness, electric resistance, and stability toward corrosion were dependent on the state of intergranular boundaries between the dispersed phase and the metal matrix [27]. Vickers microhardness testing [28] shows that the nickel–Al_2O_3 nanocomposite coating hardness increases almost 60% in comparison with nickel without disperse phase. Non-oriented nanocomposition nickel coatings with corundum particles exhibited high corrosion resistance [29]. Based on the data [30], the incorporation of corundum particles into the nickel matrix, resulted in formation of Al_2O_3 particles on its surface and inhibited corrosion. An analogous effect was observed when SiC was used as the dispersed phase. Other reports [31, 32], demonstrated that nickel coatings exhibit high wear resistance and hardness if it contain nanosized particles. The use of pulsed current at the deposition of nickel–nanosized SiC ECC made it possible to increase the incorporation rate of dispersed particles into coatings [33].

Table 2.3. Wear and antifriction properties of nickel coatings

Coating	Linear wear rate, $\mu m\ h^{-1}$		Friction coefficient	
	$\langle I_L \rangle^{cal}$	$\langle I_L \rangle^{exp}$	$\langle f \rangle^{cal}$	$\langle f \rangle^{exp}$
Ni–B	1.100	1.100	0.250	0.250
Ni–B–fluoroplastic	0.767	0.740	0.218	0.210

The mechanism of SiC inclusion into a nickel matrix and hence the incorporation rate is determined by the adsorption at low current densities and by the transport of second-phase particles at high current densities [34]. In the presence of SiC particles in the bath, the current density in nickel increased [35]; however, the grain orientation degree of nickel–SiC ECC was 15–90% of those typical of pure nickel coatings. This was probably due to the fact that at the ECC deposition, not only nickel cations and silicon carbide particles but hydrogen [36] was also adsorbed on the cathode.

The structure analysis of the nickel–SiC coatings has shown that the dispersed phase volume content was nonuniform throughout the deposit depth varying from 37% in the lower layers to 14% in the upper ones. Nonetheless, the use of nickel–SiC ECC prolonged the service life of cylinders of combustion engines as regards their permissible wear by a factor of 2–2.5 [37]. The hardness of nickel–SiC coatings increased with an increase in the bath temperature [38]. To reduce the coefficient of friction and to enhance the wear resistance of nickel–SiC and nickel–Al_2O_3 ECC, an additional dispersed phase of molybdenum disulfide can be introduced into the bath. Codeposition of nickel with MoS_2 produced selflubricating coatings [1, 2]. The molybdenum disulfide content in the composition coatings deposited from a fluoroborate bath reached 37% but decreased with an increase in pH and temperature. The dry lubricant effect also manifested itself when nickel coatings included graphite and boron nitride particles.

Nickel-based EC with the dispersed phase of silicon oxide and nitride were always more wear resistant than pure nickel coatings [39]. During the deposition of the aforementioned ECC, dispersed particles were more uniformly incorporated into the metal matrix when the bath circulated in a cell [40]. Incorporation of the dispersed phase of chromium or silicon decreased the wear intensity of nickel coatings 2 to 4-fold [41]. Nickel–Si ECC subjected to diffusive annealing and containing new solid phase Ni_3Si had the best wear resistance. The results of [42] confirmed that nickel–Si ECC exhibit high wear resistance and hardness.

The authors [43] have shown that nickel–phosphorus–ZrO_2 electroplates exhibited enhanced hardness and wear resistance. In turn, the incorporation of dispersed ZrO_2 particles into amorphous nickel–boron–tungsten coatings led to a substantial increase in their heat resistance. They withstood temperatures up to 850°C [44] without oxidation. The tribiocorrosion properties of nickel composite coatings are improved due to dispersed ZrO_2 particles incorporation in the nickel matrix followed by structural modification and an induced dispersion-strengthening effect [45].

2.1.2. Chromium-Based ECC

Chromium electroplates are used to impart good outward appearance to issues, increase the hardness and wear resistance of metal surfaces as well as restoring some worn parts. In the latter two cases, thick chromium layers should be deposited as a rule. However, this is associated with both economical (high cost) and technological (low chromium current efficiency) complications. The synthesis of ECC is one of the ways to decrease the thickness of chromium coatings and thus improve their physical and mechanical properties.

A considerable number of publications [6, 46–50] were devoted to the co-deposition of chromium with ND particles. In [6], it was shown that for chromium–ND ECC deposited from Cr (VI)-based bath the coating microhardness considerably increased, while the coefficients of friction and wear decreased (Table 2.4). The ND content in a coating was 0.3–1.0 mass %. Based on the results of experiments carried out in [48], it was shown that the highest wear resistance of chromium ECC was reached for the ND content of 15–20 g l^{-1} in Cr (VI)-based baths. The increase in the ND concentration to 50 g l^{-1} did not affect this characteristic. A coating formed in the absence of ND, lost 15% of its mass in 20 h of abrasion, whereas chromium–ND ECC lost only 2–3% of the mass.

The authors of [49, 50] studied the structure, physical and some other mechanical properties of chromium– ND ECC deposited from Cr (III)-based baths. Upon the introduction of ND particles to the chrome-plating bath in a concentration of 20 g l^{-1}, the coating microhardness increased from 1000 kg mm^{-2} (pure chromium) to 1365 kg mm^{-2} (ECC). An increase in the ND content in solution to 30 g l^{-1} resulted in the decrease in the microhardness to 940 kg mm^{-2}, which was probably associated with an increase in the coating brittleness upon adding a large number of dispersed particles. For chromium–ND ECC, the optimal dispersed phase content corresponding to the maximum microhardness and the minimum microbrittlness of the deposits [50] was 10.5 vol % (5.6 mass %) ND. The optimal ND concentration in Cr (III)-based baths was 17 g l^{-1}.

The chromium–ND coatings are characterized by incorporating a large number of dispersed particles of different size, which are uniformly distributed in the metal matrix. It is known [51] that ND particles tend to form aggregates of micron sizes. According to [49], the average radius of diamond aggregates in chrome-plating baths was 4530 nm compared with only 204 nm in composition coatings. It was assumed that in the near-electrode layer, ND particles are well dispersed and are incorporated into ECC as finer particles.

To reduce the prime cost of chromium–diamond coatings, it was proposed [52] to use the primary product of nanodiamond synthesis, namely, diamond furnace charge (DFC) that can contain up to 75% ND. The introduction of DFC to the standard chrome-plating bath does not affect the polarizability and the hydrogen evolution overpotential. In the presence of DFC, the current efficiency in chromium increased insignificantly (up to 3%).The microhardness of dispersed-phase-containing ECC increased by 15–20%, the wear resistance increased 2–2.5-fold compared with pure chromium coatings. The optimum concentration of DFC in the bath was 5 g l^{-1}.Further increase in the concentration aggravated the coating wear (Table 2.5).

Chromium–graphite composite coatings can be used in parts operating under the dry friction conditions [53]. According to [50], self-lubricating chromium– graphite ECC can be obtained in a chrome-plating bath containing up to 4 g l^{-1} graphite.

Table 2.4. Physical and mechanical properties of chromium coatings

Coating	ND content in the bath, g l^{-1}	Coating wear, μm	Friction coefficient	Microhardness H, kg mm^{-2}
Cr	0	20.2	0.15	610
Cr–ND	2	8.9	0.14	710
	4	4.8	0.10	920
	10	2.9	0.09	1480
	16	2.0	0.09	2100
	20	5.3	0.13	1900

For the graphite concentrations above 4 g l^{-1}, ECC were characterized by high roughness; for 8 g l^{-1}, the deposit surface was covered with black spots. Coatings formed by the matrix of a chromium–molybdenum alloy and containing graphite as the dispersed phase surpassed pure chromium almost two-fold as regards to the wear resistance and by a factor of six in the corrosion resistance [54].

Good corrosion resistance was also typical of chromium–graphite–phosphorus ECC that could be deposited from both Cr (VI)- and Cr(III)-based baths [55].

As was shown in [56, 57], chromium coatings with cerium dioxide, which were deposited in the presence of tetraethylamine, exhibited very high wear resistance. The optimal range of CeO_2 concentration in a Cr(III)-based bath, which corresponded to ECC with the maximum microhardness (787 kg mm^{-2}), was 5–8 g l^{-1}. As the CeO_2 concentration increased to 20 g l^{-1}, the microhardness decreased to 681 kg mm^{-2}, which was accompanied by cracking and exfoliation of coatings due to high inner strains [86].

The wear resistance and hardness of chromium-based ECC considerably increased with the addition of the dispersed silicon and titanium dioxide [58] particles into the standard chromium-plating bath. When chromium–TiO_2 ECC were deposited in the pulsed mode (the pulse-to-pause current ratio was 1/3), the coating microhardness increased from 4.9–6.5 GPa (pure chromium) to 8.1–8.7 GPa (ECC). In this case, gallic acid was added to the bath [95]. When a dispersed TiN phase was used in the standard chrome-plating bath, it transformed into TiO_2 at higher temperatures.

Later, titanium dioxide was present in the bath as a highly dispersed sediment with the specific surface many times exceeding that of the original TiN powder, which favored dispersion strengthening of the metal matrix. The higher the electrolyte temperature and the finer the powder, the sooner the equilibrium state is reached. The interaction of the ultradispersed titanium nitride powder with chromic acids produced Cr (III) ions without any preliminary treatment of the bath.

Authors of [59] studied the codeposition of chromium with aluminum oxide and silicon carbide particles. The introduction of Al_2O_3 or SiC in a concentration of 0.5 gl^{-1} to the sulfate-oxalate bath increased the hydrogen evolution overpotential. The fraction of current consumed in the deposition of metal chromium, as well as its current efficiency increased.

Table 2.5. The effect of DFC addition on the wear resistance of chromium coatings prepared under different electrolysis conditions

DFC content in the bath, g l^{-1}	Coating mass loss (mass %) at temperature, °C					
	45±1		50±1		55±1	
	Current density, i_c, A dm^{-2}					
	40	50	40	50	40	50
0	3.5	5.4	4.8	5.0	6.2	4.3
1	3.2	3.25	3.1	3.5	3.1	2.0
2.5	2.4	2.55	2.6	2.0	3.0	2.4
5	2.4	2.8	2.2	1.9	2.4	2.6
10	2.5	3.2	2.9	2.0	2.8	2.7

2.1.3. Copper-Based ECC

The main goal of applying copper-based ECC is to impart wear resistance, heat resistance, and antifriction properties to metal surfaces [1, 2]. The deposition of copper-based ECC is most often carried out from sulfate baths. The deposition of copper–ND ECC [6] from the sulfate bath has been investigated. Adding ND to the bath with concentration varying from 0.1 to 0.5 g l^{-1} did not change the nature and mechanism of the eletectrode process. The throwing power of the bath with an addition of ND increased 3-fold as compared with the original solution. The addition of ND to the copper matrix decreased the number of pores from 10 per cm^2 (0.1 g l^{-1} ND) to their total absence (5.0 g l^{-1} ND). The resulting coatings were denser and contained finer grains. Corrosion tests revealed no loss of sample mass. The microhardness of coatings deposited from a bath with the ND concentration of 5.0 g l^{-1} was half as much again the microhardness of coatings formed in the original bath. The wear of copper– ND ECC was 9–10 times smaller than for pure copper.

In [60], different copper-based composition coatings were compared as regards their wear resistance. It was found that Cu–SiC ECC exhibit the best tribiological behavior. The tribological characteristics of copper–C_{60} fullerene ECC revealed [61] that incorporating the fullerene into copper deposits decreases the friction coefficient from 0.50 to 0.22 and the roughness factor from 1.05 to 0.50. Such changes were explained by the dispersed phase effect. In the course of deposition on the cathodic surface, fullerene particles act as the crystallization sites determining the further growth of the deposit. Precisely, they favor the formation of coatings with finer grains compared with pure copper. The presence of a fullerene in coatings determines the changes in the surface state and, correspondingly, the tribological characteristics. A decrease in the friction coefficient of copper coatings is also achieved by the codeposition of copper with molybdenum disulfide [62].

An increase in the microhardness of copper coatings can be achieved by adding the ultrafine powder of titanium nitride or alumina with the grain size of 60–80 nm to the sulfate copper-plating bath [63]. The microhardness increased from 0.85 GPa (pure copper) to 1.27 GPa (copper–TiN) and 1.77 GPa (copper–Al_2O_3). The ECC stabilization was favored by the formation of a large number of dislocations upon the incorporation of ultradispersed particles that prevent the shear strain propagation and partially screen the matrix. At the copper codeposition together with the ASM diamond powder (the grain size of 14–20 μm), the increase in microhardness of coatings up to 1.5–1.8 GPa was also achieved [64].

The effect of the electrolyte nature on the deposition and properties of copper–TiO_2 ECC has been reported [65-67]. Adding polyethylenepolyamide (PEPA) to the copper-plating sulfate bath favored the formation of deposits with the high inner strains and lattice defects [65]. In contrast to copper–TiO_2 ECC, the coherent scattering range was narrower and microtsrains were weaker. This was apparently associated with the changes in the metal-matrix and dispersed-phase properties as a result of adsorption in the bath components. The changes in the matrix structure upon the incorporation of titanium dioxide particles decreased the wear of coatings from 20 g/(m^2 h) (pure copper) to 1.5 g/(m^2 h) (copper–TiO ECC).

The deposition of copper–TiO_2 ECC from a sulfate bath with additions of ammonium molybdate or vanadate decreased the current efficiency in copper and the fraction of dispersed particles in coatings [66]. However, the mentioned additions increased the heat resistance of copper–TiO_2 coatings at 900 °C by a factor of 4–6 [67], while molybdate ions made the deposits brighter [66].

2.1.4. Iron-Based ECC

Electroplated iron coatings are close to steel in hardness and hence are used for restoring the parts of engines and mechanisms [2]. Such coatings are further strengthened if ND is used as the dispersed phase [9]. However, during the deposition of iron-based ECC, it is more reasonable to use cheap diamond furnace charge (DFC) in place of pure nanodiamonds. When the DFC was introduced into the iron-plating bath, the microhardness of ECC doubled compared with the iron–ND coatings. The best results on the wear resistance of iron–diamond ECC were obtained with a partly acidified DFC. Incorporating the B_4C particles into iron matrices reduces the wear of deposits from 9.8 to 4.0 μm/km, while the microhardness increases from 1.8 to 3.2 GPa. At the same time, the deposition rate of iron–B_4C ECC increased as compared with pure iron coatings. The incorporation of B_4C particles into Fe–Co alloys made it possible to substantially enhance their heat stability. Based on the reports [68], for Fe–Co–B_4C ECC, the temperature of oxidation onset was 650 °C and the temperature of the second oxidation step reached 920 °C. In one of the article [69], it was shown that heterogeneous electroplates based on iron and containing V_2O_5 as the dispersed phase exhibited higher catalytic activity in the reduction of nitrogen oxides. Pure iron deposits on steel were catalytically inert in this process.The corrosion behavior of iron–Al_2O_3 ECC in Na_2SO_4 and 5% NaCl solutions has been investigated. The incorporation of the dispersed alumina phase into the iron matrix shifts the corrosion potential in the positive direction and decreases the anodic dissolution current. The ECC corrosion rate decreased as compared with iron coatings. The chemical–thermal treatment (CTT) of ECC, based on Al_2O_3 and CrB_2 by their sulfonitration at 210 °C for 6 h, resulted in strengthening of the bonds of the metal matrix with the dispersed particles and the substrate. As a result, the wear resistance and corrosion stability of deposits increased. The tear tests before sulfonitration revealed the formation of craters due to the solid phase fallout. After the CTT, this phenomenon was not observed. The corrosion rate of iron–CrB_2 ECC decreased from 41.2 (without CTT) to 28.7 g/(m2 h) (after CTT).

2.1.5. Zinc-Based ECC

Zinc-based composition coatings are used in protecting steel surface from corrosion and improving their physical and mechanical properties [6, 9]. In [6, 71, 72], the corrosion properties of zinc–ND ECC deposited from the alkaline zincate and weakly acidic chloride baths were studied. As was shown experimentally, the optimum content of nanodiamonds in the bath was 10 g l^{-1}, which corresponded to 0.7% ND in the resulting coatings. The addition of dispersed ND particles into a zincate bath shifted the potential in the positive direction by 20–25 mV (for the ND concentration of 10 g l^{-1}), which suggested the partial depolarization of the electrical double layer [71]. With an increase in the ND concentration in the bath, the dispersed phase content in it increased and the deposits become more fine grained [6]. The corrosion stability of zinc–ND ECC is substantially enhanced due to the passivation of coatings by the standard chromate and phosphate treatment [72]. The latter should be preferred being environmentally safer [5]. In [73], a method of depositing zinc–silicon dioxide coatings was developed. The zinc–SiO_2 ECC have a better adhesion to the substrate as compared with pure zinc. In place of zinc alloys, it was proposed [74] to deposit ECC containing about 20% of mica or silicon dioxide. As a result, the corrosion and wear resistance of deposits increased.

2.1.6. ECC Based on Noble Metals

Silver-based ECC that contain conductive particles were deposited on electric contacts to improve their conductivity. Moreover, silver ECC were used to enhance the wear resistance of parts in the electric contact [2].The introduction of dispersed nanodiamonds to a cyanide-thiocyanate silver-plating bath resulted in the formation of dense fine-grain semibright deposits containing 1.0 mass % ND [6, 75]. Deposition of silver–ND ECC in the reverse current mode increased the dispersed phase content in the matrix to 2.5 mass % [76].

Table 2.6. Wear resistance of silver coatings

ND content in the bath, g l^{-1}	Coating thickness, μm	Abrasion time, h	Coating mass loss, %
0	5.0	20	33.3
0.2	1.7	20	6.7
0.5	1.3	20	5.0
1.0	1.3	40	2.5
2.0	1.5	25	0

Table 2.7. Wear resistance of gold coatings

ND content in the bath, g l^{-1}	Coating mass loss, 10^{-5} g h^{-1}
0	93.2
0.1	44.0
0.5	4.8
1.0	0.4
2.0	2.5
5.0	5.6

Based on the electrolyte composition and the mode of electrolysis, the microhardness of silver–diamond coatings increased by 200–700 MPa as compared to pure silver [12]. The wear resistance increased by 5−15 fold with an increase in the ND concentration in the bath (Table 2.6) [6]. The soldering ability and the contact resistance of silver–ND ECC were identical to pure silver characteristics [5, 76]. Incorporation of Teflon particles into a silver matrix improves the outward appearance of coatings. Silver– PTFE ECC did not look dull even without special treatment [77]. Gold-plating in electronics is aimed at providing stable junction resistance of contacts. Deposition of gold-based ECC was carried out from potassium cyanide (citrate), alkaline cyanide, and iron cyanide baths. Adding 0.5–1.0 g l^{-1} UDD to the bath allowed deposits containing 0.1–1.0 mass % diamonds to be obtained [6].

Incorporating the diamond nanoparticles into gold coatings substantially improved their physical and mechanical properties [72, 78]. Microhardnes of gold–UDD ECC reached 2.0 GPa and their wear resistance increased 2–200-fold (Table 2.7) [6, 78]. Diamond furnace charge can be used as the dispersed phase for gold-based ECC. The content of DFC particles in gold deposits was 0.5–0.9 mass %, and their microhardness could reach 1.7 GPa [5].

To enhance the wear resistance of jeweleries, they were covered with gold coatings containing the dispersed phases of titanium nitride [79] and boron carbide [80]. However, B_4C particles in ECC synthesized from cyanide baths were nonuniformly distributed.

CONCLUSION

Considering the results on the structure and properties of ECC synthesized in recent years allowed us to state that this field of electroplating continues to be successfully developed. Methods of preparing new coatings with enhanced hardness, wear resistance, corrosion stability, and other operational properties were developed. As a trend, the studies of composite coatings with dispersed nanoparticles may be mentioned. Yet another trend is the fact that most of publications were devoted to the properties and structure of composite coatings, whereas studies dealing with the mechanism and kinetics of ECC formation have been few.

REFERENCES

[1] R. S. Saifullin, *Neorganicheskie kompozitsionnye materialy* (Inorganic Composite Materials), Khimiya, Moscow (1983) [in Russian].

[2] L. I. Antropov and Yu. N. Lebedinskii, *Kompozitsionnye elektrokhimicheskie pokrytiya i materialy* (Composite Electrochemical Coatings and Materials), Tekhnika, Kiev (1986) [in Russian].

[3] R. S. Saifullin, *Physical Chemistry of Inorganic Polymeric and Composite Materials*, Ellis Hornwood Ltd., London (1992).

[4] C. G. Fink and J. D. Prince, *Trans. Am. Electrochem. Soc.* 54, 315 (1929).

[5] L. S. Tsybul'skaya, T. V. Gaevskaya, T. M. Gubarevich, and A. P. Korzhenevskii, *Galvanotekhn. Obrab. Pov-ti.* 4, no. 1, 14 (1996) [in Russian].

[6] Yu. V. Timoshkov, T. M. Gubarevich, T. I. Orekhovskaya, *Galvanotekhn. Obrab. Pov-ti.* 7, no. 2, 20 (1999) [in Russian].

[7] A. D. Toporov, P. Ya. Detkov and S. I. Chukhaeva, *Galvanotekhn. Obrab. Pov-ti.* 7, no. 3, 14 (1999) [in Russian].

[8] G. K. Burkat and V. Yu. Dolmatov, *Physics of Solid State.* 46, 703 (2004).

[9] V. Yu. Dolmatov, *Russ. Chem. Rev.* 76, 339 (2007).

[10] S. I. Chukhaeva, P. Ya. Detkov, A. P. Tkachenko and A. D. Toropov, *Sverkhtverd. Mater.* no. 4, 29 (1998) [in Russian].

[11] T. W. Jelinek, *Galvanotechnik.* 96, 42 (2005).

[12] E. Kranz, *Galvanotechnik.* 89, 2890 (1998).

[13] L. M. Yagodkina, I. D. Loginova and I. E. Savochkina, *Russ. J. Appl. Chem.* 70, 1556 (1997).
[14] L. M. Yagodkina, I. D. Loginova and I. E. Savochkina, *Russ. J. Appl. Chem.* 71, 637 (1998).
[15] G. Novotortseva and T. V. Gaevskaya, *Russ. J. Appl. Chem.* 72, 824 (1999).
[16] V. N. Tseluikin, N. D. Solov'eva and I. F. Gun'kin, *Prot. Met.* 43, 388 (2007).
[17] V. N. Tseluikin, N. D. Solov'eva and I. F. Gun'kin, *Nanotechn. in Russia.* 3, 456 (2008).
[18] Tan, T. Yu, B. Xu and Q. Yao, *Tribology Lett.* 21, 107 (2006).
[19] A. G. Tkachev, Yu. V. Litovka, I. A. D'yakov and O. A. Kuznetsova, *Galvanotekhn. Obrab. Pov-ti.* 18, no. 1, 17 (2010) [in Russian].
[20] V. N. Tseluikin, I. V. Tolstova, I. F. Gun'kin, and A. Yu. Pankst'yanov, *Colloid Journal.* 67, 522 (2005).
[21] E. V. Saksin, A. A. Shevyrev and A. V. Shkurankov, *Zh. Prikl. Khim.* 68, 1822 (1995) [in Russian].
[22] S. V. Devyaterikova, S. V. Khitrin, and S. L. Fuks, *Russ. J. Appl. Chem.* 76, 665 (2003).
[23] E. V. Kuznetsova, *Zh. Prikl. Khim.* 66, 1155 (1993) [in Russian].
[24] N. M. Teterina and G. V. Khaldeev, *Prot. Met.* 34, 276 (1998).
[25] N. M. Teterina and G. V. Khaldeev, *Prot. Met.* 36, 470 (2000).
[26] V. V. Ivanov, V. I. Balakai, A. V. Ivanov, and A. V. Arzumanov, *Russ. J. Appl. Chem.* 79, 610 (2006).
[27] T. Aust, *Metallkunde*. 94, 1066 (2003).
[28] N. A. Badarulzaman, S. Purwadaria, A. A. Mohamad and Z. A. Ahmad, *Ionics.* 15, 603 (2009).
[29] B. Wielage, *Metalloberflaeche.* 57, no. 12, 25 (2003).
[30] V. V. Medyalene, K. K. Leinartas, and E. E. Yuzyalyunas, *Prot. Met.* 31, 89 (1995).
[31] C. Jakob, R. Nutsch and S. Steinhauser, *Metalloberflaeche.* 54, no. 9, 50 (2000).
[32] S. Steinhauser, *Galvanotechnik.* 92, 78 (2001).
[33] G. Heidari, H. Tavakoli and S. M. Mousavi Khoie, *J. Mat. Eng. Perf.* 19, 1183 (2010).
[34] S. H. Yah and C. C. Wan, *Plating Surface Finishing.* 84, no. 2, 54 (1997).
[35] T. W. Jelinek, *Galvanotechnik.* 89, 44 (1998).
[36] B. Szcygel, *Plating Surface Finishing.* 84, no. 2, 62 (1997).
[37] V. V. Zyablintsev, O. V. Zyablintseva and A.M. Velikolug, *Galvanotekhn. Obrab. Pov-ti.* 10, no. 2, 25 (2002) [in Russian].
[38] B. Szcygel, *Galvanotechnik.* 86, 1070 (1995).
[39] H.. Szeptycka, *Galvanotechnik.* 93, 656 (2002).
[40] T. W. Jelinek, *Galvanotechnik* 94, 46 (2003).
[41] G. I. Desyatkova, L. M. Yagodkina, I. E. Savochkina and G. V. Khaldeev, *Prot. Met.* 38, 466 (2002).
[42] D.-H. Cheng, W. Y. Xu and L.Q. Hua, *Plating Surface Finishing.* 85, no. 2, 61 (1998).
[43] B. Wieklage, H. Podlesak, S. Steihauser and D. Nickelmann, *Metalloberflaeche.* 52, no. 5, 386 (1998).
[44] Z. Lingun, Z. Quingpeng and I. Janhua, *Metal Finishing.* 99, no. 7, 28 (2001).
[45] Benea, *J. Appl. Electrochem.* 39, 1671 (2009).

[46] S. V. Vashchenko and Z. A. Solov'eva, *Galvanotekhn. Obrab. Pov-ti.* 1, no. 5-6, 45 (1992) [in Russian].
[47] V. Mandich and J. K. Dennis, *Metal Finishing*, 99, no. 6, 117 (2001).
[48] V. Yu. Dolmatov, T. Fudzhimura, G.K. Burkat and E.A. Orlova, *Sverkhtverd. Mater.* no. 6, 16 (2002) [in Russian].
[49] E. G. Vinokurov, A. M. Arsenkin, K. V. Grigorovich and V. V. Bondar', *Prot. Met.* 42, 204 (2006).
[50] E. G. Vinokurov, A. M. Arsenkin, K. V. Grigorovich and V. V. Bondar', *Prot. Met.* 42, 290 (2006).
[51] V. Yu. Dolmatov, *Russ. Chem. Rev.* 70, 607 (2001).
[52] K. I. Tikhonov, G. K. Burkat, V. Yu. Dolmatov and E. A. Orlova, *Russ. J. Appl. Chem.* 80, 1082 (2007).
[53] D. Jang and C. Jiang, *Metal Finishing*. 97, no. 1, 24 (1999).
[54] R. S. Saifullin, E. P. Zentsova and S. V. Vodop'yanova, *Zashch. Met.* 31, 315 (1995) [in Russian].
[55] W. Plieth, B. Voos, N. Schroder, *Galvanotechnik*. 90, 2425 (1999).
[56] R. Narayanan and S. K. Seshadri, *Plating Surface Finishing*. 87, no. 11, 56 (2000).
[57] R. Narayanan and S. K. Seshadri, *Metal Finishing*. 99, no. 2, 84 (2001).
[58] S. V. Vodop'yanova, E. P. Zentsova and R. S. Saifullin, *Russ. J. Electrochem.* 34, 310 (1998).
[59] E. N. Lubnin, N. A. Polyakov and Yu. M. Polukarov, *Prot. Met.* 43, 186 (2007).
[60] Y. Zhan, *Metallkunde*. 95, no. 2, 91 (2004).
[61] V. N. Tseluikin, N. D. Solov'eva and I. F. Gun'kin, *Perspektivnye Mater.* no. 5, p. 82 (2007).
[62] Y. Z. Wan, Y. L. Wang, H. M. Tao and Y. H. Dong, *Trans. Inst. Metal Finishing*. 77, no. 1, 52 (1999).
[63] T. V. Rezchikova, E. N. Kurkin, V. N. Troitskii, L. S. Kurkin and A. V. Ivanov, *Russ. J. Appl. Chem.* 74, 2035 (2001).
[64] A. Tsisar', G. N. Znamenskii, T. I. Yushchenko and L. V. Paches, *Galvanotekhn. Obrab. Pov-ti.* 4, no. 1, 21 (1996) [in Russian].
[65] A. Abdullin and R. S. Saifullin, *Prot. Met.* 33, 195 (1997).
[66] R. E. Fomina, R. S. Saifullin and G. G. Mingazova, *Russ. J. Electrochem.* 33, 1270 (1997).
[67] R. S. Saifullin, R. E. Fomina, G. G. Mingazova and R. A. Khaidarov, *Prot. Met.* 38, 471 (2002).
[68] Zh. I. Bobanova, N. Yu. Michukova and S. P. Sidel'nikova, *Galvanotekhn. Obrab. Pov-ti.* 8, no. 2, 17 (2000) [in Russian].
[69] I. Akhmerov and G. A. Krinari, *Russ. J. Appl. Chem.* 72, 993 (1999).
[70] V. G. Revenko, T. V. Kozlova, G. A. Astakhov, G.P. Chernova and N. L. Bogdashkina, *Prot. Met.* 39, 77 (2003).
[71] G. K. Burkat and V. Yu. Dolmatov, *Galvanotekhn. Obrab. Pov-ti.* 9, no. 2, 35 (2001) [in Russian].
[72] V. Yu. Dolmatov, *Ultradispersnye almazy detonatsionnogo sinteza* (Ultradispersed Diamonds of Detonation Synthesis), SPbGPU, St. Petrsburg (2003).
[73] Cao and J. Wu, *Mater. Protection*. 33, no. 4, 8 (2000).
[74] M. Azizi, *Galvanotechnik*. 93, 656 (2002).

[75] V. Yu. Dolmatov, G. K. Burkat and V. Yu. Saburbaev, *Sverkhtverd. Mater.* 2, 52 (2001) [in Russian].
[76] A. A. Khmyl', V. A. Emel'yanova, I. I. Mushovets and V. M. Kutsenko, *Galvanotekhn. Obrab. Pov-ti.* 9, no. 3, 26 (2001) [in Russian].
[77] Pagetti, *Oberflachen Polysurfaces*. 42, no. 6, 10 (2001).
[78] E. N. Loubnin, S. M. Pimenov and A. Blatter, *Carbon Technol.* 9, no. 4, 273 (1999).
[79] A. Zielonka and C. J. Raub, *Metalloberflaeche*. 49, no. 6, 349 (1995).
[80] A. Bozzini, *Metal Finishing*. 100, no. 4, 50 (2002).

In: Electrodeposition: Properties, Processes and Applications ISBN: 978-1-61470-826-1
Editor: Udit Surya Mohanty

Chapter 3

ELECTRODEPOSITION OF AU-SN ALLOYS

Anqiang He [*] ***and Douglas G. Ivey*** [†]
Department of Chemical and Materials Engineering, University of Alberta, Edmonton, Canada

ABSTRACT

A review on pulse electrodeposition of Au-Sn alloys from non-toxic and environmentally green electrolytes at ambient temperature is presented based on two main approaches. The first approach is focused on the electrodeposition of Au-Sn alloy films from a single solution. This portion starts with electrolyte preparation and is followed by solution stability and characterization using turbidity measurements, SEM, TEM, XRD, UV/Vis spectroscopy and Cryo-XPS. The microstructures of individual as-deposited films (Au_5Sn and AuSn phases), composite layers formed by alternatively combining Au_5Sn and AuSn phases and reflowed solders are then shown. Factors that affect the compositions of electrodeposited films on patterned wafers are discussed as well. Finally, a process to recover Au from the waste solutions is presented; the Au salt is then used to prepare fresh Au-Sn electrolytes. The second approach is focused on pulse electrodeposition of Au-Sn alloy films using a sequential procedure with pure Au films and pure Sn films from two separate solutions. Within this portion, the development of electrolytes for pure Au deposition and pure Sn deposition are described. The microstructures of sequentially electrodeposited Sn-rich, Au-Sn eutectic solders and reflowed solders are also presented.

INTRODUCTION

Au-Sn eutectic solder (30 at%Sn) has excellent mechanical and thermal properties compared with traditional Pb-Sn solders [1, 2]. It is especially useful for flip-chip and laser bonding in optoelectronics packaging and for MEMS packaging [3-5]. Au–Sn solder has

[*] E-mail: ahe@ualberta.ca.
[†] E-mail: doug.ivey@ualberta.ca.

generally been applied in forms such as preforms, solder paste and through sequential evaporation. The preform method has problems with misalignment and oxidation of the solder prior to bonding. Solder paste also suffers from oxidation prior to bonding, as well as possible solder contamination during bonding from the organic binder in the paste. These two methods are also not easily applicable to patterned circuit boards. Compared with solder preforms and pastes, thin film evaporation techniques can reduce the possibility of oxidation, provide good control over film thickness and can be applied to patterned circuit boards. However, sequential thin film evaporation requires expensive vacuum systems. In contrast to thin film technology, electrodeposition is less expensive, while still providing deposit thickness control. In addition, electrodeposition is applicable to patterned wafers, as long as the electrolyte can reach the regions of interest. Of course, electrodeposition has its own technical challenges.

During the past decade, the research group at the University of Alberta has expended considerable effort to study the Au-Sn system in terms of electrolyte preparation, solution stability, alloy deposition and microstructural characterization of as deposited and reflowed solders. All this work has been done with electrolytes that are non-toxic and environmentally friendly. During electrolyte development, besides using traditional polarization measurements of solutions, several other non-conventional techniques have been employed to characterize solution chemistry and stability. These techniques include turbidity measurements, transmission electron microscopy (TEM), UV-visible spectroscopy and cryo-X-ray photoelectron spectroscopy (cryo-XPS). This Chapter has been divided into two main parts. The first part (Sections 3.1-3.4) is focused on single solution deposition issues, such as solution preparation, solution stability, film deposition and recovery of Au from the waste solutions. The second part (Section 3.5) focuses on sequential deposition issues.

3.1. Electrodeposition of Au-Sn Alloys from a Single Solution

Co-deposition of Au-Sn alloys from a single solution has only been reported in a limited number of scientific publications. Most available literature and patents are related to cyanide based systems [6-10]. Because of the well-known toxicity issues associated with cyanide ions, some researchers have focused development work on non-cyanide solutions for Au-Sn electrodeposition [11-14], including a recent non-aqueous solution [15]. However, no systematic results have been reported; therefore, practical application of electrodeposition for Au-Sn alloys from a single solution has been limited. Here, non-cyanide recipes for Au-Sn electrodeposition are discussed. These recipes have been developed at the University of Alberta and have undergone systematic testing.

3.1.1. Solution Preparation

Two requirements were addressed during the development of non-cyanide solutions for co-deposition of Au-Sn alloys. The first issue was environmental, i.e., no toxic components, including cyanide ions, were to be added to the solution. The second issue was that the developed solutions had to be slightly acidic. High pH, alkaline solutions can cause the photoresist, used in microelectronics/optoelectronics packaging, to delaminate [16]. Based on these two requirements, Sun and Ivey [17] developed a solution which consists of tri-ammonium citrate, potassium tetrachloroaurate (III) ($KAuCl_4{\cdot}xH_2O$ - for simplicity, hereafter referred to as $KAuCl_4$), sodium sulfite (Na_2SO_3), L-ascorbic acid and $SnCl_2{\cdot}2H_2O$. This solution was tested by Doesburg and Ivey [18], Djurfors and Ivey [19], He et al. [20], Zhang and Ivey [21] and Morawej and Ivey [22] under various conditions, with electrodeposition on unpatterned and patterned wafers. Electrodeposition reproducibility was confirmed. Recently, a local MEMS company (Micralyne, Inc.) has applied the recipes to some of their products.

Details about this electrolyte can be found in Ref. [17]; the main points for solution preparation are summarized here. Tri-ammonium citrate is initially dissolved in de-ionized water, followed by (in order) $KAuCl_4$, Na_2SO_3, L-ascorbic acid (after 45 minutes) and finally $SnCl_2{\cdot}2H_2O$. Tri-ammonium citrate was initially added to the solution as a buffer, although it was later found that citrate also acted as a weak reducing agent for the Au ions [24]. The main reducing agent for Au ions in this solution is ascorbic acid, which will be discussed later. Na_2SO_3 is the major complexing agent for the Au ions and is a stabilizer for the solution as well [17]. However, Na_2SO_3 can also act as a reducing agent if excessive amounts of Na_2SO_3 are added to the solution [23-24]. A typical electrolyte recipe for the codeposition of Au-Sn from a single solution is:

Tri-ammonium citrate:	0.411 mol/L (100 g/L)
$KAuCl_4$:	0.026 mol/L (5 g/L)
Na_2SO_3:	0.159 mol/L (20 g/L)
L-ascorbic acid:	0.085 mol/L (15 g/L)
$SnCl_2{\cdot}2H_2O$:	0.022 mol/L (5 g/L)

The as-prepared solution is transparent with a faint yellow color and is stable in terms of electrodeposition for two to three days. To increase the shelf life of the solution, Morawej and Ivey [22] have recently proposed, based on experimental and statistical analysis, changing the amount of Na_2SO_3 and ascorbic acid.

The details are given in Ref. [22]. After careful examination of the roles of the chemical components in the solution, He et al. [24] proposed preparing two separate Au and Sn solutions, and mixing them together as needed. The separate Au and Sn solutions are stable for more than a year and have the following compositions:

Au Solution

Tri-ammonium citrate:	0.411 mol/L (100 g/L)
$KAuCl_4$:	0.026 mol/L (10 g/L)
Na_2SO_3:	0.318 mol/L (40 g/L)

Sn Solution

Triamonium citrate:	0.411 mol/L (100 g/L)
$SnCl_2.2H_2O$:	0.022 mol (10 g/L)

The two solutions can be mixed together as needed; ascorbic acid is added during mixing. The resulting solution is transparent and Au-Sn codeposited films are comparable with those prepared from a fresh solution.

3.2. Solution Stability

3.2.1. Solution Precipitation and Turbidity Measurements

As mentioned in Section 3.1.1, the electrolyte is stable for two to three days. After this period of time, Sn concentrations in codeposited films decrease dramatically [23-24]. There is no apparent change in solution color or appearance after two days and, in some cases, even up to seven days. However, after this time, black or yellow particles are visible. Longer term aging leads to precipitate formation on the container sides and bottoms. Although there is no apparent color change and no precipitates are visually observed within two days after solution preparation, turbidity measurements show that the solution starts to precipitate before two or three days. Turbidity is a measure of the degree to which light is scattered by a liquid sample and, thus, it reflects solution heterogeneity [25]. The higher the intensity of scattered light, the higher the turbidity, which means there is a higher density of suspended particles in the solution - even those that are not visually observable. Figure 3.1 shows a typical plot of solution turbidity as a function of aging time for the Au-Sn electrolyte [23]. The plot can be divided into four major regions, representing different stages during solution aging. The first region shows that the turbidity of a freshly prepared solution decreases from ~4 NTU to ~2 NTU over a period of about one day (here NTU is nephelometric turbidity units). This is followed by a second region with a rapid increase in turbidity, which peaks out about after two days (~12 NTU). The turbidity then decreases gradually in the third region over the next six days before levelling off at almost 0 NTU. The turbidity remains at close to 0 NTU for more than two weeks (fourth region). Similar plots were obtained for all electrolytes measured; the only difference was in the time periods to complete each stage. In some instances the first stage took only about ten hours and in others the maximum turbidity occurred after three days instead of two. During the first and second stages, the solution remained transparent with a faint yellow color and no visible precipitation. However, soon after the peak turbidity was reached, small, suspended particles agglomerated over time and gradually deposited on the bottom and sidewalls of the container.

Clearly, very small solid particles precipitated from the solution long before they became visible to the naked eye. In fact, it is likely that precipitation began as soon as the solution was mixed. The proliferation of precipitates led to more light scattering centers and the observed increase in turbidity during the first 2-3 days. The decrease in turbidity, after peaking out at about two days, was likely due to the agglomeration of the small particles and also possibly due to consumption of the reducing agent.

Both of these effects would reduce the number of scattering centers in solution, resulting in a decrease in turbidity. Analysis by atomic absorption spectroscopy (AAS) showed that there is little Au left in the solution when the solution has a turbidity of 0 NTU, since essentially all the Au in solution has precipitated out and settled to the bottom or on the walls of bottle. It will be demonstrated later that the precipitated Au can be recovered as a salt and reused to prepare fresh electrolyte.

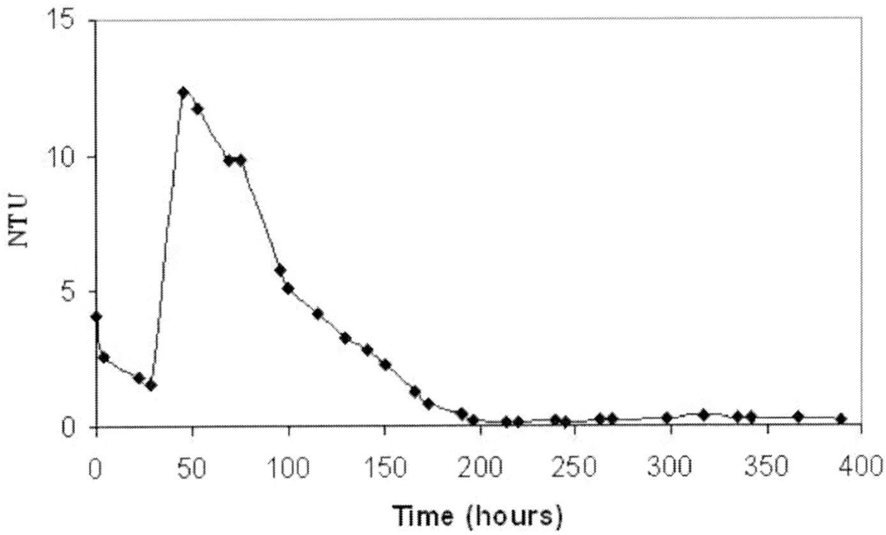

Figure 3.1. Typical turbidity plot as a function of aging time for Au-Sn electroplating solution (He et al. [23]).

3.2.2. Characterization of Precipitates

Transmission electron microscopy (TEM) was employed to confirm the presence of precipitates before they were visible in the solution. TEM samples were prepared by placing a drop of solution onto a C-coated, Cu grid. Once the liquid had evaporated, the grid with remaining residue was imaged in the TEM. Samples were taken from three regions in the turbidity plot, i.e., freshly prepared solution, the solution with the highest turbidity and the solution with almost zero turbidity. TEM bright field (BF) images from two samples, i.e., the freshly prepared solution and the solution with maximum turbidity (turbid solution), are shown in Figure 3.2 [23].

Isolated particles are clearly visible in both cases. The particles are less than 10 nm in size for the freshly prepared solution (Figure 3.2.a) and between 10 and 20 nm for the turbid solution (Figure 3.2.b). The particle concentration is higher in Figure 3.2.b, which is consistent with the higher turbidity. Particle compositions were analyzed using an energy dispersive X-ray (EDX) spectrometer. In both cases, only Au was detected. Convergent beam electron diffraction patterns were obtained from individual particles and confirmed them to be single crystal Au.

The TEM sample taken from fourth region of the turbidity plot (i.e., solution with almost zero turbidity) also contained Au particles (<5 nm in diameter), but the number was much lower than that obtained from the freshly prepared solution. The large particles had already settled at this point. To study the settled particles, a precipitated film was removed from a container with solution that had been aged for 21 days; the film was rinsed and then examined using scanning electron microscopy (SEM), EDX spectroscopy and X-ray diffraction (XRD)

[23]. The low magnification SEM image in Figure 3.3.a shows that the precipitated film is porous, while the higher magnification SE image (Figure 3.3.b) reveals that the film consists of particles, predominately 500 nm or less in diameter. EDX analysis (Figure 3.3.c) of the film shows that it contains only Au, with no Sn or other solution components. XRD analysis (Figure 3.3.d) confirms the SEM results, i.e., only peaks corresponding to Au were detected.

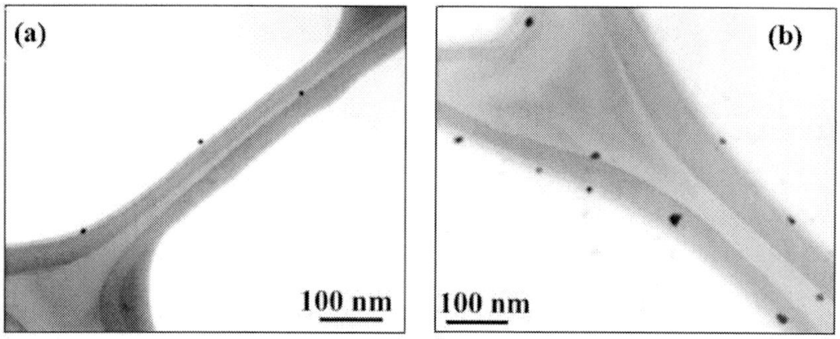

Figure 3.2. a) TEM BF image of a sample obtained from an as prepared Au-Sn solution. b) TEM BF image of a sample obtained from an Au-Sn solution aged for 2 days (He et al. [23]).

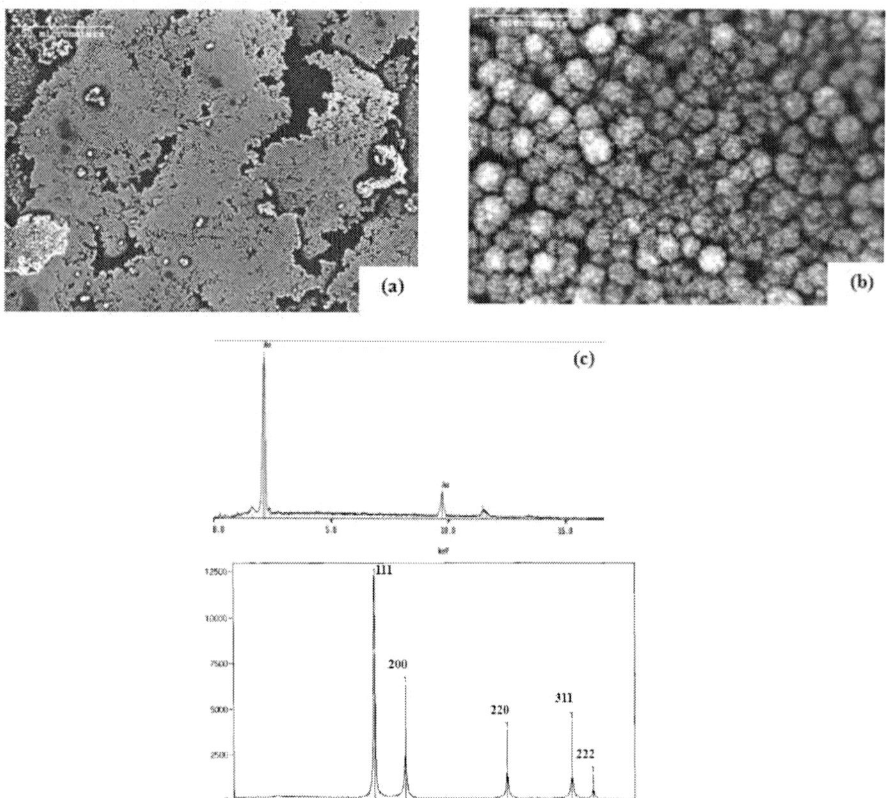

Figure 3.3. a) and b) SEM secondary electron (SE) images of a precipitated film from a solution aged for 21 days. c) EDX spectrum from the region shown in (b). d) XRD pattern from the precipitated film shown in Figure 3a. All diffraction peaks can be indexed to metallic Au (He et al. [23]).

3.2.3. UV/Vis Spectroscopy and X-Ray Photoelectron Spectroscopy Studies

Turbidity measurements are very useful for monitoring precipitation from a solution over time. However, to understand the reasons behind precipitation (such as chemical bonding or electronic states), other techniques, such as UV/Vis spectroscopy and XPS, are needed. The wavelength for UV/Vis light varies from 190 to 1100 nm. The energies associated with photons at these wavelengths have the same magnitude as the energy differences between electronic states of many molecules. Therefore, incident light can be absorbed and shows up in the UV/Vis spectrum as an absorption peak. Electronic transitions, such as $\sigma\rightarrow\sigma^*$, $n\rightarrow\pi^*$, $\pi\rightarrow\pi^*$, etc., for a molecule can be detected by UV/Vis spectroscopic measurements [26]; hence, UV/Vis spectroscopy can be used as a fingerprint to follow changes in solution components. Figure 3.4 shows the UV/Vis spectra as a function aging time for an Au-Sn solution stored in a quartz cell [23]. The peak at ~385 nm was present in all spectra and corresponds to de-ionized water. The peak at ~265 nm is ascribed to ascorbic acid [27]. The intensity of this peak was the highest for the freshly prepared solution (curve **a** in Figure 3.4). The peak decreased with aging time and reached zero after about two days (curve **e** in Figure 3.4). The disappearance of the ascorbic acid peak indicated that it was completely consumed after about two days. The appearance of (reduced) Au particles and disappearance of ascorbic acid indicate that the ascorbic acid acted as a reducing agent in the solution. It should be pointed out that in the literature nanosize Au particles were reported to be detectable with UV/Vis spectroscopy, with a peak appearing in the 500-600 nm range and centred at ~535 nm [28-29]. In this work, no such peak was detected, even though the precipitated Au particles were of similar size. The reasons for the discrepancy between the literature results and the results in this work are unclear.

The stability of Au complexes in the solution is an important issue. As mentioned in Section 3.2.1, the solution color changes when different chemicals are added. This color change is due to a change in the valence of Au. Recently, cryo-XPS was employed in analyzing this change [30]. Since the energy of an X-ray (typically monochromatic Al Kα radiation is used in XPS) is much higher than the energy for UV/Vis light, the chemical shift of an element associated with a chemical reaction can be measured.

The general procedure is to freeze the solution and then analyze it in an ultra-high vacuum system. The experimental details and results are described in the literature [30]. The XPS spectra for Au ions are shown in Figure 3.5. Figure 3.5.a shows spectra from a solution consisting of tri-ammonium citrate and $KAuCl_4$. The major peaks are from Au^+ with a small amount of Au^{3+} in the as-prepared solution. The amount of Au^{3+} gradually increases as the solution is aged from 3 hours to 20 hours. Figure 3.5b shows spectra from a solution with tri-ammonium citrate, $KAuCl_4$ and Na_2SO_3; no extra peaks other than Au^+ appear both for a fresh solution and an aged solution. These results indicate that ammonium citrate is not only acting as a buffer for the solution, but is a weak reducing agent as well [24]. As pointed out by the authors, tri-ammonium citrate can reduce the Au^{3+} in $KAuCl_4$ to form an unstable Au^+-acetonedicarboxylate complex. The XPS work also shows that the addition of Na_2SO_3 to the tri-ammonium citrate-$KAuCl_4$ solution led to the formation of a stable Au^+ complex [30]. Table 1 summarizes the chemical shifts (CS) of Au 4f peaks measured in the cryo-XPS studies.

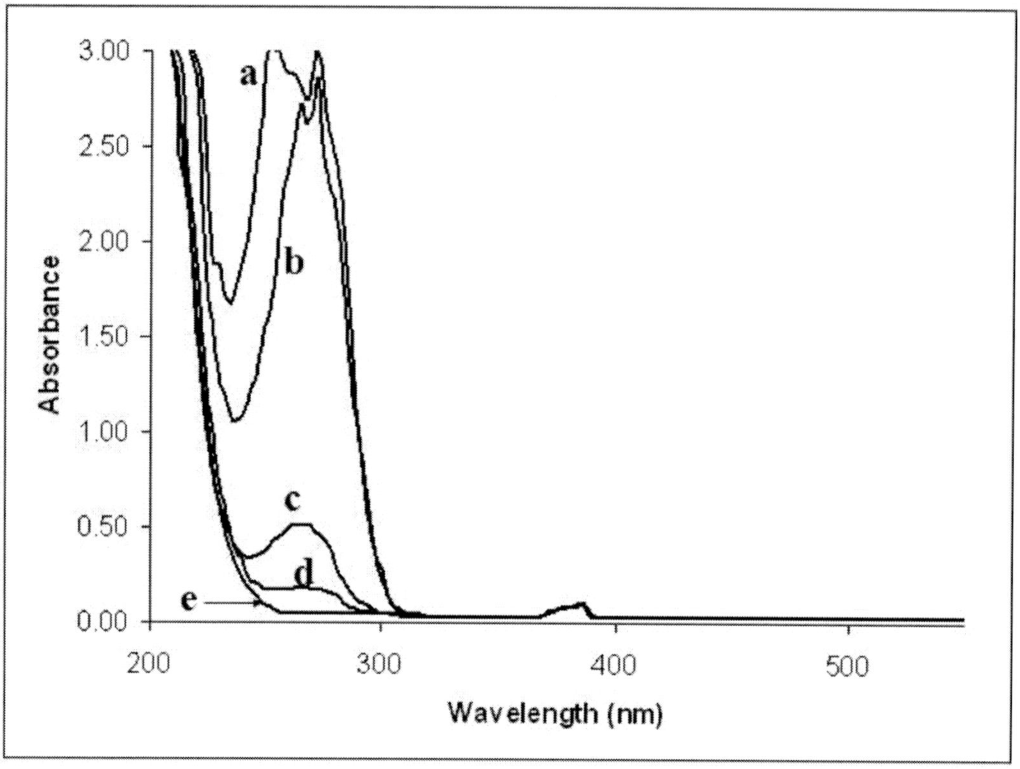

Figure 3.4. UV/Vis spectra obtained from an Au-Sn electroplating solution after aging for: a) 0 h, b) 4.5 h, c) 22 h, d) 28 h and e) 46 h (He et al. [23]).

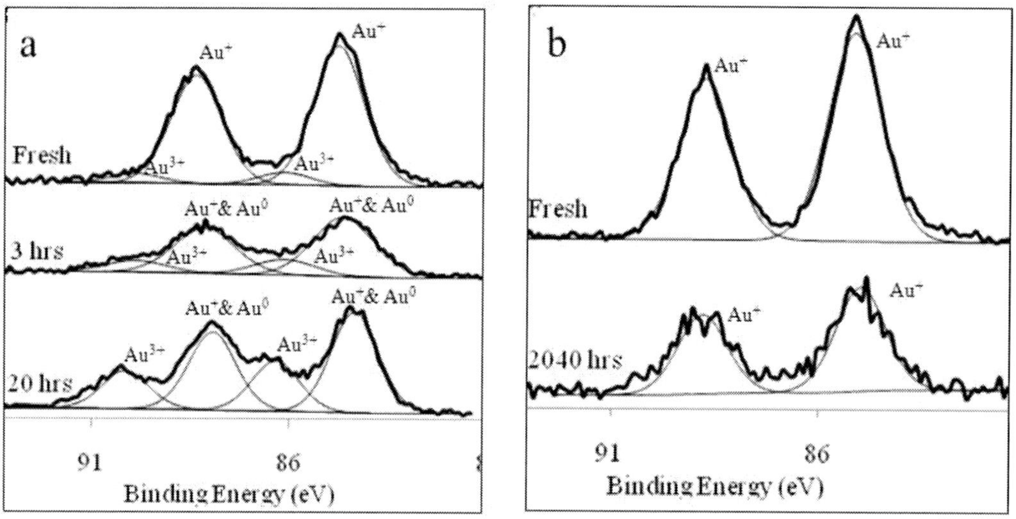

Figure 3.5. XPS spectra of binding energies for Au 4f peaks: (a) from solution with tri-ammonium citrate + $KAuCl_4$, (b) from solution with tri-ammonium citrate+$KAuCl_4$+Na_2SO_3 (Dalili et al. [30]).

Table 3.1. Binding energies (eV) for Au 4f peaks of various standards, freshly prepared solutions and aged solutions (peak indexing was performed with reference to the standards) [30]

Sample	Au $4f_{5/2}$	Au $4f_{7/2}$	CS Au $4f_{7/2}$	Components	pH
Au^o (from pure Au film)	87.95	84.25	----	----	----
Au^{+3} (from $KAuCl_4$)	90.95	87.35	3.1	----	3.8
Au^+ (from ref. [32])	----	----	0.5-2.3	----	----
S1 fresh	88.37	84.77	0.52	Au^+	6.4
	89.90	86.15	1.90	Au^{3+}	
S1 3-h aged	88.15	84.65	0.40	Au^+ & Au^o	6.4
	89.90	86.15	1.90	Au^{3+}	
S1 20-h aged	87.95	84.35	0.10	Au^+ & Au^o	6.4
	90.15	86.35	2.1	Au^{3+}	
S2 fresh	88.75	85.15	0.9	Au^+	7.0
S2 2040-h aged	88.85	85.25	1.0	Au^+	6.9

S1: ammonium citrate + $KAuCl_4$

S2: ammonium citrate + $KAuCl_4$ + Na_2SO_3

3.3. MICROSTRUCTURE OF ELECTRODEPOSITED FILMS

Electrodeposited films, from the Au-Sn solution, can be obtained at average current densities ranging from 0.6 to 4.0 mA/cm^2, using pulse deposition with a 2 ms on time and an 8 ms off time [17-23]. Within this current range, electrodeposited films are continuous and smooth. Two types of Au-Sn phases can be obtained as shown in Figures 3.6-3.8. One phase has a Sn concentration of ~15 at% (corresponds to Au_5Sn as confirmed by XRD) for current densities less than 1.2 mA/cm^2. The other phase has a Sn content of ~50 at% (corresponds to AuSn as confirmed by XRD) for current densities above 2.0 mA/cm^2. Between the two current densities and composition plateaus (Figure 3.8), there is a transition region where deposited films are made up of both phases [19]. SEM images show that deposits consisting of Au_5Sn have fine grains (Figure 3.6), while the grain size is larger for AuSn films deposited at current densities higher than 2.4 mA/cm^2 (Figure 3.7). Cross-section SEM images in Figure 3.6.b and Figure 3.7.b were obtained by cleaving the samples without any further specimen preparation. Deposited films become rough when current densities exceed 4 mA/cm^2.

The deposit formation mechanism is quite different for the two phases: nucleation controlled at low current densities (Au_5Sn) and growth controlled at higher current densities (AuSn) [32]. At low current densities, the driving force for nucleation is small, so the number of Au_5Sn nuclei on the surface is limited. The existing nuclei continue to grow, resulting in a lateral growth. For this reason, the growth rate of Au_5Sn is relatively low, about 0.8 μm per hour, but the surface of the film is very smooth with a mirror-like finish. At high current densities, the driving force is higher and nucleation occurs more readily. A high density of nuclei limits the space for each nucleus to grow laterally, which forces the nuclei to grow along the direction normal to the surface as a columnar structure. Hence, at high current density, the deposition rate is relatively high, about 2 μm per hour if a deposition current of 2.4 mA/cm^2 is applied. Also, due to the relatively high nucleation rate, the surface of the films deposited at high current densities is not as smooth as that for lower current densities.

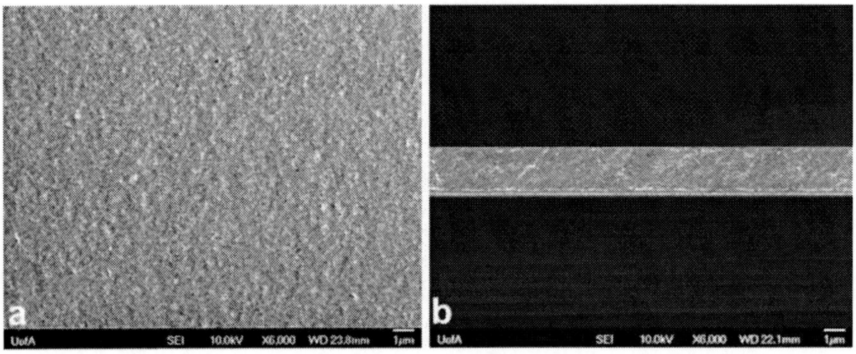

Figure 3.6. SEM SE images of film electrodeposited at 0.8 mA/cm^2 for 120 minutes: (a) plan-view, (b) cross-section. The composition of the film is about 16 at% Sn.

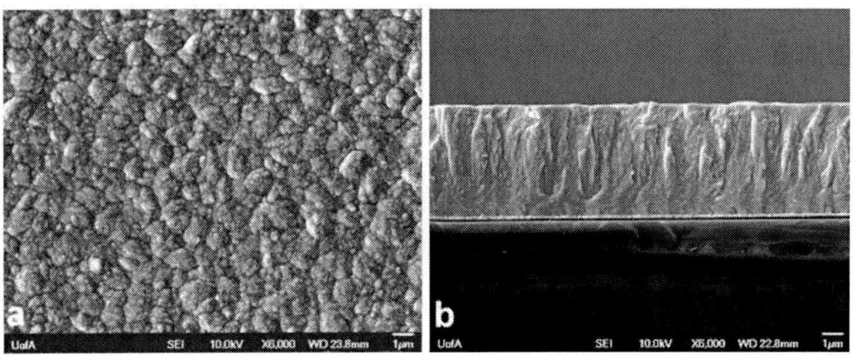

Figure 3.7. SEM SE images of film electrodeposited at 2.4 mA/cm^2 for 90 minutes: (a) plan-view, (b) cross-section. The composition of the film is about 50 at% Sn.

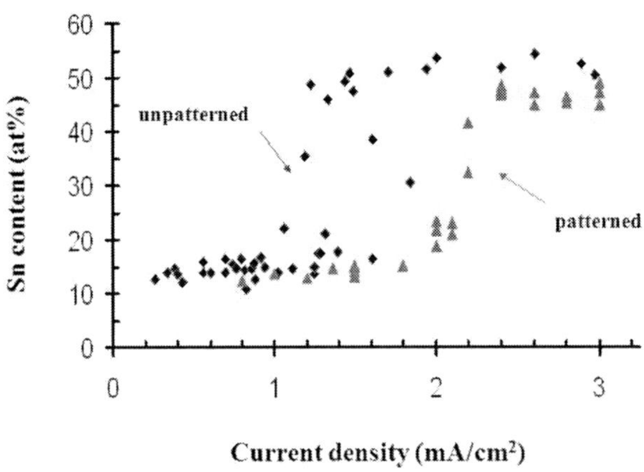

Figure 3.8. Plot of deposit composition vs average current density. Data shown with diamond shapes are from deposits on unpatterned wafers (Djurfors and Ivey [19]), and data with triangle shapes are from deposits on patterned wafers (He et al. [20]).

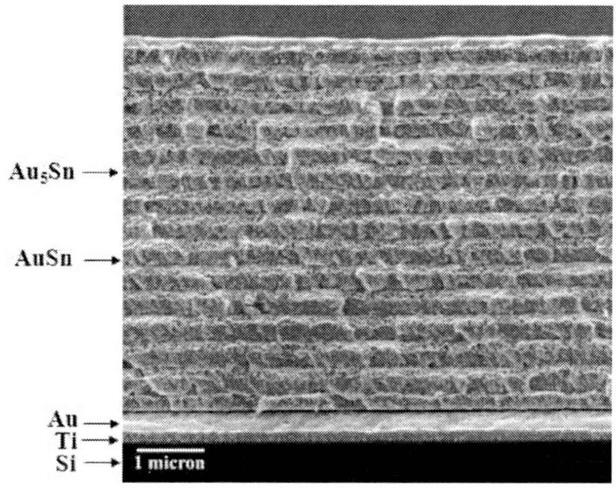

Figure 3.9. SEM SE image of an electrodeposited film consisting of alternating layers. The current density used to deposit AuSn was 2.4 mA/cm^2 (5 minutes); the current density to deposit Au_5Sn was 0.8 mA/cm^2 (21 minutes). The overall composition of the composite film corresponds to 30 at% Sn (He et al. [20]).

3.3.1. Microstructure of Eutectic Deposits and Reflowed Solder

For solder bonding applications, a deposit concentration near the eutectic composition (30 at% Sn) is desired. However, the eutectic composition lies within the transition region of the Sn composition vs deposition current density curve (Figure 3.8), so that the composition is difficult to control. An alternative to using a single deposition current density is to alternate between a lower and higher current density, i.e., within the plateau regions of Figure 3.8, to deposit Au_5Sn and AuSn sequentially. By selecting the appropriate deposition times for each current density, the overall deposit composition can be controlled to between 15 and 50 at% Sn. In addition, by cycling between the two current densities, an artificial eutectic microstructure can be produced (Figure 3.9) [20]. This example has 30 alternating layers of AuSn and Au_5Sn deposited onto an unpatterned Si wafer with an Au seed layer and a Ti adhesion layer. The SEM image of the cleaved sample in Figure 3.9 clearly reveals the ductile nature of the Au_5Sn layers and the brittle nature of the AuSn. Such deposits can be routinely generated from this solution. When electrodeposition is done on patterned wafers, there are issues that arise due to the type of photoresist. For a dry photoresist, there are numerous pin-holes (Figure 3.10.a), which extend through the thickness to the substrate. Electrolyte can seep into the holes and produce Au/Sn deposits within the resist layer (Figure 3.10.b). Current leakage issues associated with dry photoresist can be eliminated through the use of a dense, wet photoresist. Figure 3.11 shows an example of a Au-Sn deposit on a wafer, using wet photoresist for patterning. The openings are smaller than 10 μm and the photoresist thickness is about 3 μm. The coverage and filling of the openings is quite good (Figure 3.11.b) and the multilayered microstructure (alternating Au_5Sn and AuSn layers) is visible (Figure 3.12). There is a difference in the relationship between deposit composition and current density for patterned samples vs unpatterned samples (Figure 3.8). The upper plateau (50 at% Sn or AuSn) and lower plateau (15 at% Sn or Au_5Sn) are shifted to higher current densities. Three possible factors can account for the current density shift.

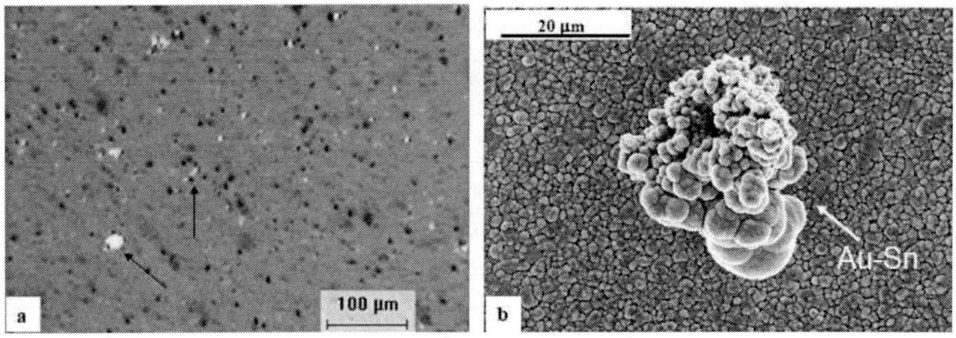

Figure 3.10. (a) Optical microscope image of dry photoresist surface; the bright regions (indicated by arrows) are pinholes that penetrate through the resist layer. (b) SEM SE image showing an Au-Sn particle which formed at the base of a pinhole in the photoresist. The dry photoresist was removed after electrodeposition and prior to imaging (He et al. [20]).

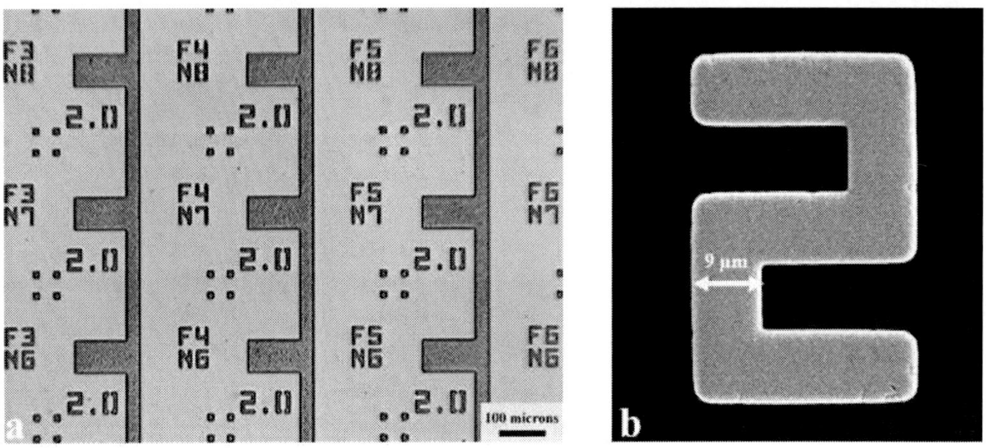

Figure 3.11. (a) Optical microscope image of electrodeposit on a patterned wafer covered with wet photoresist. The patterned areas (grey color and consisting of numbers, letters and lines) are uniformly covered; (b) Higher magnification SEM SE image showing part of the patterned electrodeposit (number "2") (He et al. [20]).

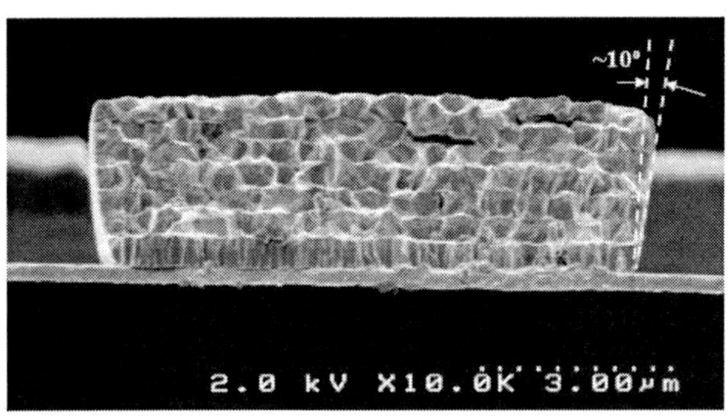

Figure 3.12. SEM SE cross section image through one of openings in Figure 3.11.a. Note that the sides of the deposited film are not perpendicular to the Au seed layer (He et al. [20]).

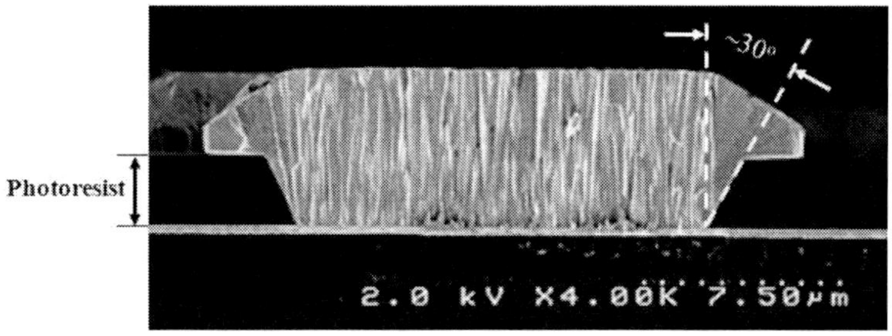

Figure 3.13. SEM SE image from a cross section sample showing edge effects — deposit thickness exceeds resist thickness. Note that the sides of the deposited are not perpendicular to the seed layer (He et al. [20]).

The first factor is related to deposit inclination along the vertical direction (perpendicular to the substrate – Figures 3.12 and Figure 3.13). In these images, it is clear that the deposit sidewalls are not perfectly vertical but are inclined ~10° in Figure 3.12 and ~30° in Figure 3.13. This means that the effective current density decreases, relative to the nominal value, as the deposits thicken, since the deposited areas become larger and larger as electrodeposition proceeds. The second factor is a “muffin-top” shape effect. When the deposited film exceeds the thickness of the photoresist, it will spread laterally as well as grow vertically producing the muffin-top shown in Figure 3.13. This muffin-top shape will also reduce the effective current density. The third factor is related to current leakage at the edge of the photoresist. This is expected to be a minor effect as the edge area is small relative to the entire wafer area. Edge effects can be effectively eliminated by coating wafer edges with lacquer. The second factor, i.e., the muffin-top effect, can be easily avoided if the photoresist is thicker than the deposit. The main factor is sidewall inclination, which results from photoresist openings that are not perfectly vertical and can vary significantly from wafer to wafer (as illustrated in Figures 3.12 and 3.13). Even for the same opening, the inclination angle can vary significantly; e.g., ~25° for the left side of the opening and ~30° for the right side in Figure 3.13. Sidewall inclination is difficult to eliminate completely, as it is related to photoresist etching during lithography. Quantifying inclination effects is also difficult, for the same reasons. It is suggested that higher nominal current densities should be used to offset this effect.

The reflow process for the electrodeposited films occurs rapidly (within seconds) during annealing just above the eutectic temperature (e.g., 300°C). This can be attributed in part to the artificial eutectic structure produced during electrodeposition, which reduces diffusion distances. After reflow, the eutectic microstructure, consisting of Au_5Sn (bright contrast) and AuSn (darker contrast) is easily revealed (Figure 3.14). It should be pointed out that when designing a solder with the eutectic composition, the thickness of the Au seed layer must be accounted for, since the Au seed layer will dissolve into the solder during reflow. Otherwise, the solder will deviate from the eutectic composition, producing a hypoeutectic alloy with a significantly higher melting temperature. From the Au-Sn binary phase diagram [33], as shown in Figure 3.15, it is clear that the slope of the liquidus line on the Au-rich side is quite steep, i.e., a 1 at% Au deviation from the eutectic composition can increase the liquidus temperature by about 30°C. The result may be incomplete melting of the solder during reflow.

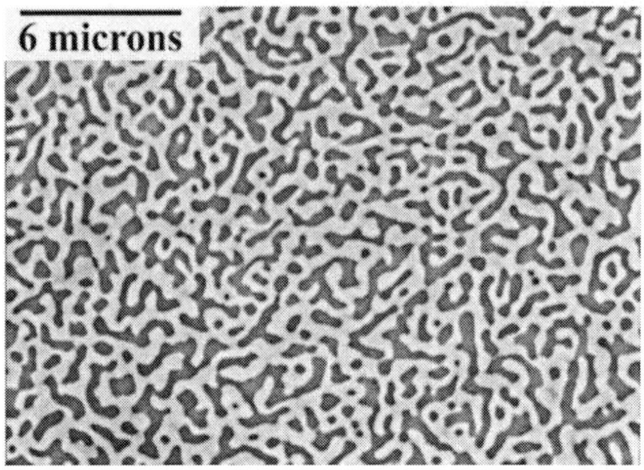

Figure 3.14. SEM back scattered electron (BSE) image showing microstructure of Au-Sn deposit with the eutectic composition, after annealing at 300°C in a forming gas atmosphere (95% N_2 and 5% H_2). The bright and darker contrast regions correspond to Au_5Sn and AuSn, respectively (He et al. [20]).

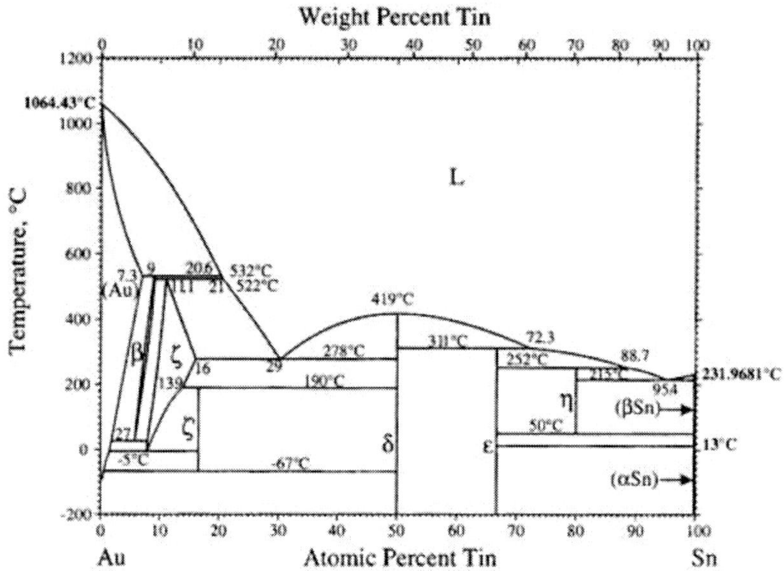

Figure 3.15. Au-Sn binary phase diagram (Okamoto [33], reprinted with permission of ASM International).

Figure 3.16 shows the case where the amount of Au in the seed layer was not accounted for and the annealing temperature had to be increased to 380°C (the eutectic temperature is 280°C) for complete melting. Annealing at such high temperatures is not desirable from a device packaging standpoint. For practical packaging applications, the deposited solder should be slightly hypereutectic so that the composition will become eutectic or near eutectic when the Au seed layer is dissolved. In addition, the liquidus curve has a shallower slope on the Sn-rich side of the eutectic (Figure 3.15 - about 7°C per 1 at% change in composition) [33].

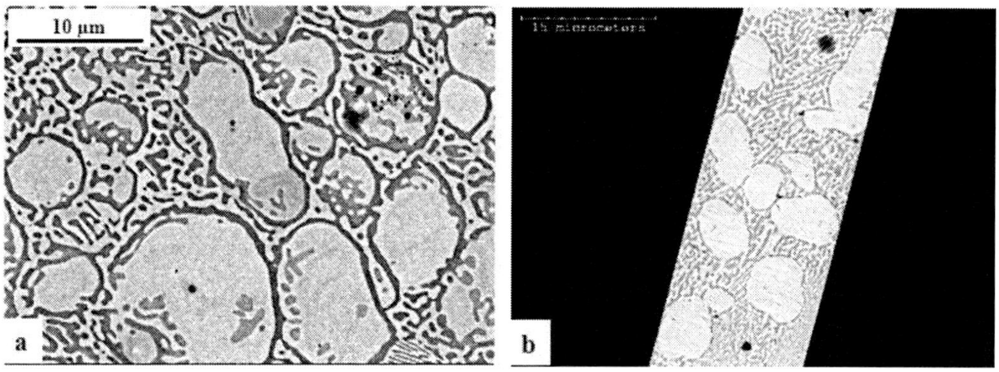

Figure 3.16. SEM BSE images showing the microstructure of an Au-Sn deposit with a hypoeutectic composition, after annealing at 380°C in a forming gas atmosphere (95% N_2 + 5% H_2). The bright and darker contrast regions correspond to Au_5Sn and AuSn, respectively. (a) Plan-view image of annealed deposit; (b) cross section image from two samples, each with Au-Sn deposits, bonded face-to-face.

3.4. Recovery of Au from the Waste Solution

As mentioned above, the developed electrolyte has a useful lifetime of two to three days. After this period, deposits become Au-rich or Sn-deficient. Obviously, the short solution lifetime is not desirable from a practical point of view, as solution costs are increased (particularly for the Au salt). Therefore, in order to reduce cost, a process to recover Au from the used solution was developed within the group at the University of Alberta [34]. The recovery process is initiated by filtering the precipitated residue from the solution, cleaning the extracted particles in HCl to remove impurities and then dissolving the remaining Au in aqua regia (3 parts HCl to 1 part HNO_3 by volume) at 80-85°C for 5 minutes. The resultant Au solution is then evaporated by slowly heating from 85 to 110°C, producing a reddish-brown deposit of $HAuCl_4 \cdot 3H_2O$:

$$Au + HNO_3 + 4HCl \rightarrow HAuCl_4 + NO + 2H_2O \quad (1)$$

$$HAuCl_4 + 3H_2O \rightarrow HAuCl_4.3H_2O \qquad (T = 110°C) \quad (2)$$

Careful temperature control during evaporation is important, as overheating of the solution (120-175°C) results in the formation of a yellow AuCl deposit that is insoluble in water. AuCl then has to be redissolved in aqua regia and the solution re-evaporated. The proposed reaction sequence for the formation of AuCl is as follows:

$$HAuCl_4 \cdot 3H_2O \rightarrow AuCl_3 + HCl + 3H_2O \qquad (T = 110\text{-}120°C) \quad (3)$$

$$AuCl_3 \rightarrow AuCl + Cl_2 \qquad (T = 120\text{-}175°C) \quad (4)$$

The $HAuCl_4$ salt generated through this process is quite susceptible to moisture pick-up at room temperature and, therefore, needs to be sealed and refrigerated during storage. Au-Sn electrolytes were prepared using the procedure documented in Section 3.1.1 (i.e., with Sn chloride, tri-ammonium citrate, sodium sulfite and ascorbic acid); the only difference was that Au was added as $HAuCl_4$ instead of $KAuCl_4$.

The amount of $HAuCl_4$ was adjusted to provide the same amount of Au in solution. Similar deposition behavior was obtained for the electrolytes containing $HAuCl_4$ compared with those containing $KAuCl_4$ [19-20], in terms of deposit concentration, morphology and plating rate, although there were some differences. Deposit concentration, as a function of current density is shown in Figure 3.17 [34]. Deposits obtained from the recovered $HAuCl_4$ salt are shown as open circles, while the diamonds represent deposits obtained using the standard electrolyte with $KAuCl_4$ [19]. In both cases two composition plateaus are obtained, corresponding to Au_5Sn (~15 at% Sn) and AuSn (50 at% Sn). The main difference is a shift in the plateaus to higher current densities for the solution with the recovered $HAuCl_4$ salt. Electrodeposition rates are shown in Table 3.2. The deposition rates are very similar for electrolytes containing $HAuCl_4$ and $KAuCl_4$ at higher current densities (≥2.4 mA/cm^2), but differ somewhat at lower current densities (<1.0 mA/cm^2). This is most likely related to the shift in the deposit composition plateaus to higher current densities for electrolytes with $HAuCl_4$, as shown in Figure 3.17.

Table 3.2. Electrodeposition rates for Au-Sn deposits [34]

Au salt	Current density (mA/cm^2)	Electrodeposition rate (μm/h)
$HAuCl_4 \cdot xH_2O$ (recovered)	0.8	0.2
	1.4	2.2
	2.4	2.6
	2.8	2.9
$KAuCl_4 \cdot xH_2O$	0.8	0.8
	2.4	2.1
	2.8	2.5

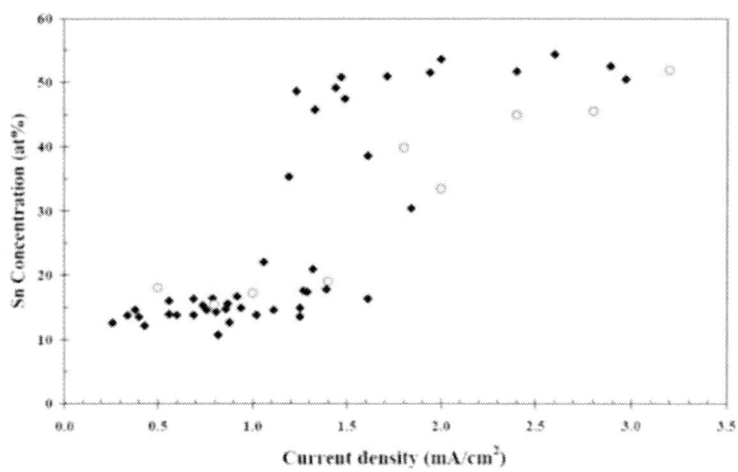

Figure 3.17. Plot of composition vs deposition current density – open circles are from solution prepared with $HAuCl_4$, while the diamonds are from the standard solution prepared with $KAuCl_4$ (Neburchilov et al. [34]).

3.5. SEQUENTIAL ELECTRODEPOSITION FROM TWO SEPARATE SOLUTIONS

Sequentially preparing Au-Sn alloys through electrodeposition has recently been reported by several researchers in the literature [35-40]. These works are mainly focused on the microstructure evolution of sequential Au and Sn films deposited using either electrolytes that are commercially available or recipes developed by the group at the University of Alberta and will be presented in the following sections. Commercial electrolytes are either toxic or have fairly complex chemistries. In contrast, electrolytes for pure Au and pure Sn deposition developed at the University of Alberta are non-toxic and simple to prepare. An example of the application of these two solutions to sequentially prepare Au-Sn solders will also be demonstrated.

3.5.1. Solution Preparation and Electrodeposition of Pure Au Films

Gold deposition has a history of over 200 years and can be traced back to Brugnatelli in 1805 [41]. Until the middle of the 20th century, Au deposition had mainly been applied for decorative purposes, such as jewelry, flatware, hollowware, etc. Since the early 1950s, the electronics industry has required specialized surfaces suitable for electrical contacts or bonding sites for components. Electrodeposited Au films can satisfy many of these demands, so that rapid development of Au deposition ensued. The major deposition baths are cyanide-based, whether they are alkaline, neutral or acidic [42-45]. These baths are relatively mature and well understood. However, one of the problems with cyanide baths is their poor compatibility with positive photoresists, which can result in photoresist cracking [16]. Another major problem is cyanide's toxicity, which can lead to difficulties during deposition operations and waste disposal thereafter. For these reasons, non-cyanide baths have been developed. Of these, sulfite-based baths have been widely studied and utilized [46-49]. Sulfite-based baths can be used to deposit thick films; however, they have stability issues. To increase stability, various stabilizers (such as 2,2'-bipyridine, polyamine or aromatic nitro-compounds) have been added [16, 50-52], which increase the complexity associated with controlling both solution preparation and electrodeposition. Recently, a stable sulfite-thiosulfate system has been reported [53-54]. The preparation procedure is fairly complex, however, not only in terms of controlling the pH value (if the pH exceeds 7.5, thiosulfate can disproportionate) [55] but also in terms of the mixing sequence. In addition to sulfite and sulfite-thiosulfate baths, ammonium-sulfite baths have been studied [56-57], but no details about solution stability and useful lifetime have been reported.

During electrodeposition studies on Au-Sn alloys, the group at the University of Alberta has developed two separate and stable electrolytes for electrodepositing Au and Sn [24]. In this section the development of the electrolyte for soft Au deposition is discussed; the development of the electrolyte for Sn deposition is discussed in the next section.

The procedure to prepare the Au electrolyte has been described in Section 3.11. Following this procedure, five different solution compositions (labeled as solutions I through V in Table 3.3) were prepared and investigated. Among the five, solutions II and III were used to study the roles of tri-ammonium citrate and sodium sulfite in the electrolyte; solutions

I, IV and V were used to determine an optimum electrolyte composition. Figure 3.18 shows SEM plan view and cross-section SE images of films deposited from solution I, which contained tri-ammonium citrate, $KAuCl_4$ and Na_2SO_3. The deposit is dense and relatively smooth when deposited at 1.0 mA/cm^2 (Figure 3.18.a and 3.18.b) and the deposition rate is about 1 μm/h. However, the films become very rough when deposited at higher current densities (e.g., 2.0 mA/cm^2), as shown in Figures 3.18.c and 3.18d. Solution II consists of tri-ammonium citrate and $KAuCl_4$. As mentioned in Section 1.2, the two component solution is not stable, as precipitation occurs within five minutes of mixing. Deposition results from solution II at 1.0 mA/cm^2 and 2.0 mA/cm^2 are shown in Figure 3.19. Comparison of Figure 3.19 with Figure 3.18 indicates that the film deposited from Solution II at low current density (1.0 mA/cm^2) is not as smooth as that obtained from Solution I. Precipitation from solution II is responsible for this behavior, which is evidence that sodium sulfite is a necessary component to stabilize the Au ions.

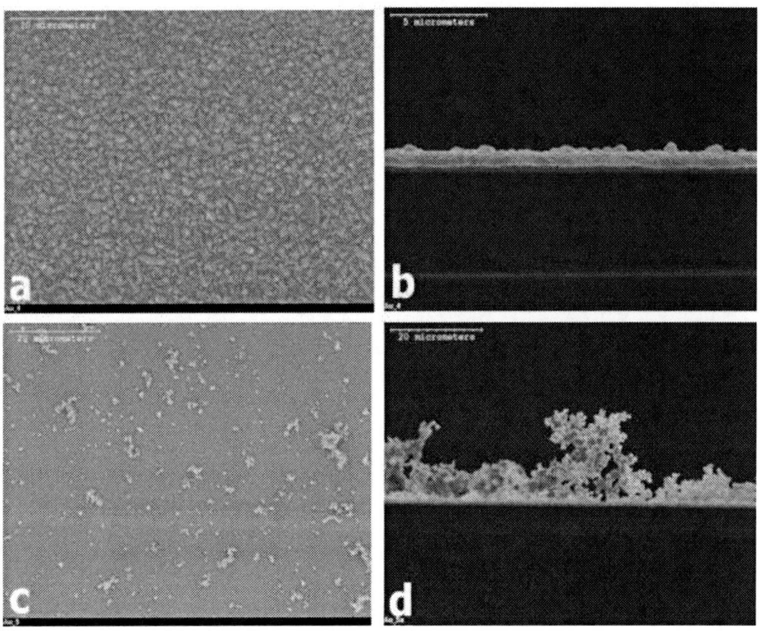

Figure 3.18. SEM plan view and cross-section SE images of electrodeposited films from solution I: (a) and (b) deposited at 1.0 mA/cm^2; (c) and (d) deposited at 2.0 mA/cm^2. The deposition time was 30 minutes (He et al. [58]).

Solution III consists of two components, $KAuCl_4$ plus Na_2SO_3. To prepare this solution, first $KAuCl_4$ and Na_2SO_3 were dissolved separately in deionized water in two separate containers; then the two solutions were mixed together. The resulting solution was initially clear with a low turbidity as shown previously [24]. However, after two months, Au precipitates appeared, indicating that the two-component solution ($KAuCl_4$ and Na_2SO_3) is not as stable as the three-component solution (tri-ammonium citrate, $KAuCl_4$ and Na_2SO_3). Deposition results from Solution III ($KAuCl_4$ and Na_2SO_3) vary. Figures 3.20.a and 3.20.b shows films deposited at a current density of 0.8 mA/cm^2, while Figures 3.20.c and 3.20.d shows films deposited at 1.0 mA/cm^2. These two current densities are quite similar, however, the two films are markedly different. The deposit in Figure 3.20.c and 3.20.d is not as dense as the deposit in Figure 3.20.a and 3.20.b. In addition, deposit morphology is not reproducible even at low current densities (e.g., 0.8 mA/cm^2). To obtain good quality and reproducible films, it appears that tri-ammonium citrate is necessary. Tri-ammonium citrate may play a role as a buffer in the solution, in addition to complexing Au ions [24]. From these simple experiments, it is clear that solutions II and III are not suitable for soft Au deposition; however, the three component solution has the potential to be developed into an efficient Au deposition solution. In a recent study [22] on electrodeposition of Au-Sn alloys from a single chloride-based solution, optimum amounts of tri-ammonium citrate, $KAuCl_4$ and Na_2SO_3 were 0.267 mol/L (65 g/L), 0.013 mol/L (5 g/L) and 0.109 mol/L (13.7 g/L), respectively, combined with 0.022 mol/L (5 g/L) of Sn chloride were used. For solution I, the deposition rate is about 1 μm per hour, which is not practical for application in the electronics industry.

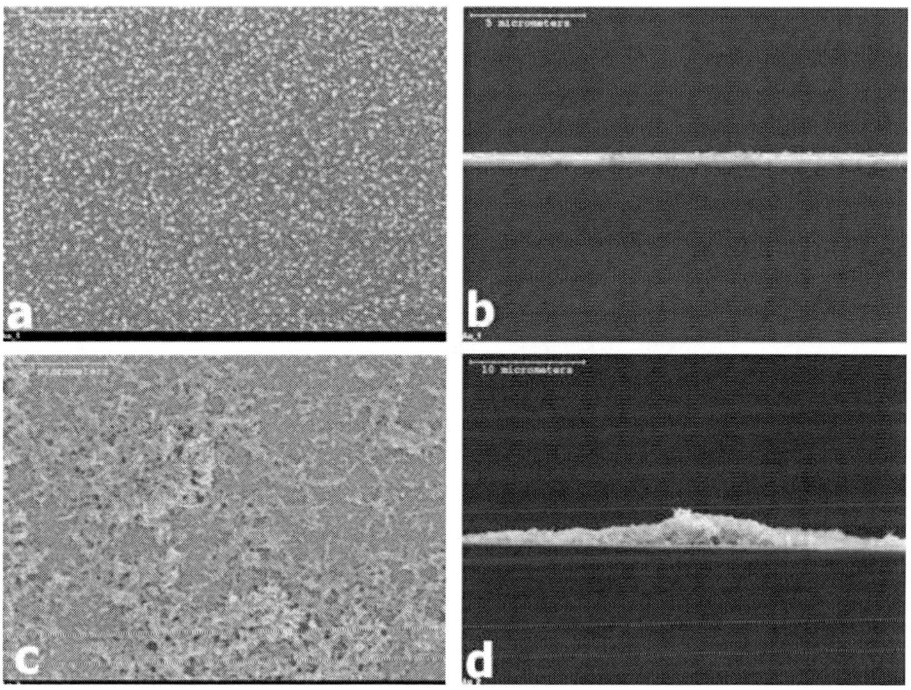

Figure 3.19. SEM SE images of electrodeposited films from solution II: (a) and (b) deposited at 1.0 mA/cm^2; (c) and (d) deposited at 2.0 mA/cm^2. The deposition time was 30 minutes (He et al. [58]).

One way to increase the Au deposition rate is to increase the concentration of $KAuCl_4$. Solution IV was prepared by considering Morawej and Ivey's recipe, in which the amount of $KAuCl_4$ is increased with a proportional increase in the concentrations of tri-ammonium citrate and Na_2SO_3 (previous work has shown that Na_2SO_3 can only complex with Au ions [24]). In this solution, the chemicals, with the exception of tri-ammonium citrate, are four times more concentrated than the solution used by Morawej and Ivey [22]. Films deposited at 1.0 and 2.0 mA/cm^2 are uniform, smooth and dense. However, if the current density is increased beyond 2.0 mA/cm^2, deposit quality degrades as shown in Figure 3.21. In addition, the deposition rate is not increased at current densities higher than 4.0 mA/cm^2. Furthermore, after solution IV was aged for 10 days at room temperature, the films obtained from the aged solutions are suitable only when the current density is $\leq$1.0 mA/cm^2 (Figure 3.22). Figure

3.23 shows films deposited from solution V. In solution V, the amount of $KAuCl_4$ was increased to 0.066 mol/L, The amount of tri-ammonium citrate was reduced to 0.411 mol/L and the amount of Na_2SO_4 was selected to maintain the same ratio for $KAuCl_4/Na_2SO_3$ reported by Morawej and Ivey [22]. The pH value for this solution was 5.5. The solution can be deposited under a wide range of current densities, i.e., from 1-10 mA/cm^2. The deposited films are uniform and dense for current densities from 1.0 to 4.0 mA/cm^2 as shown in Figure 3.23.

Table 3.3. Solutions studied for Au deposition [58]

Solution	Tri-ammonium citrate mol/L (g/L)	$KAuCl_4$ mol/L (g/L)	Na_2SO_3 mol/L (g/L)
I	0.822 (200)	0.013 (5)	0.476 (60)
II	0.822 (200)	0.013 (5)	
III		0.013 (5)	0.467 (60)
IV	1.069 (260)	0.053 (20)	0.435 (54.8)
V	0.411 (100)	0.066 (25)	0.563 (68.5)

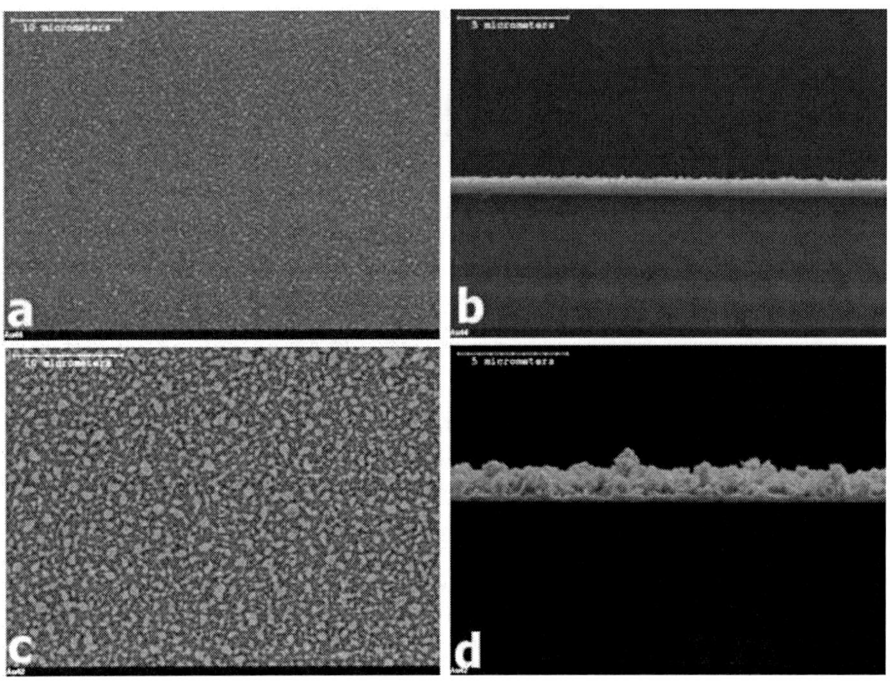

Figure 3.20. SEM SE images of electrodeposited films from solution III: (a) and (b) deposited at 0.8 mA/cm^2; (c) and (d) deposited at 1.0 mA/cm^2. The deposition time was 30 minutes (He et al. [58]).

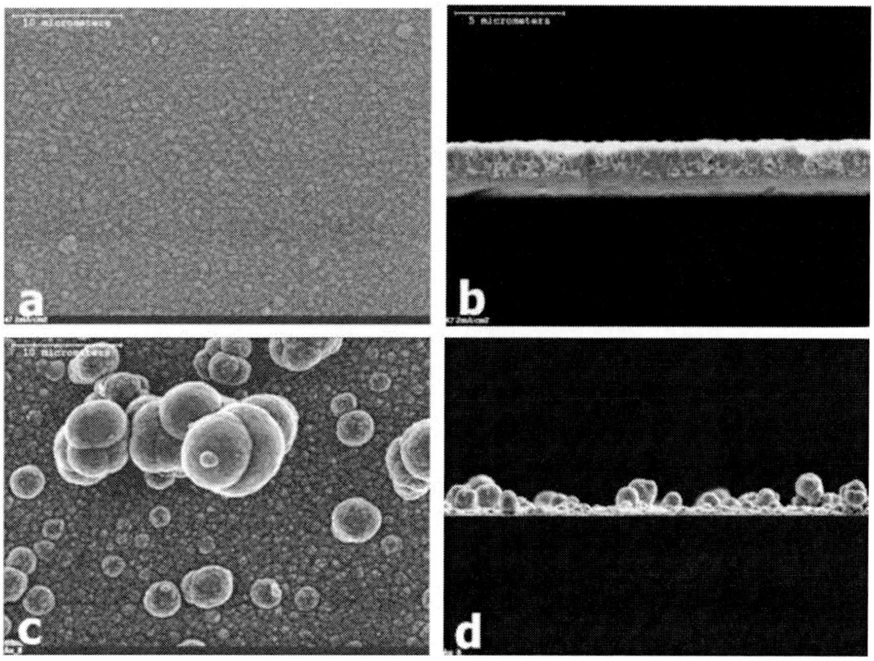

Figure 3.21. SEM SE images of electrodeposited films from solution IV: (a) and (b) deposited at 2.0 mA/cm^2; (c) and (d) deposited at 4.0 mA/cm^2. The deposition time was 30 minutes (He et al. [58]).

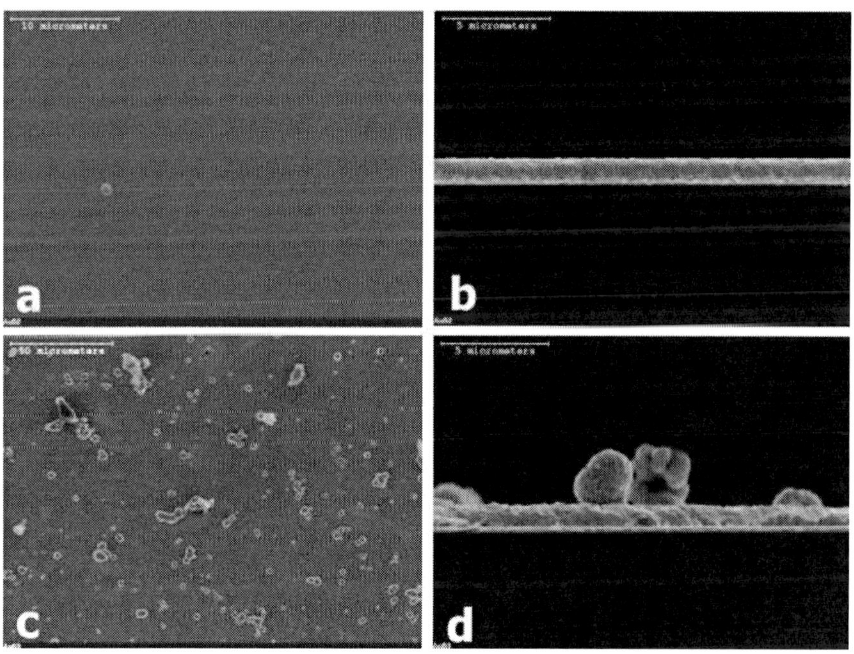

Figure 3.22. SEM SE images of electrodeposited films from solution IV (aged for 10 days): (a) and (b) deposited at 1.0 mA/cm^2; (c) and (d) deposited at 2.0 mA/cm^2. The deposition time was 30 minutes (He et al. [58]).

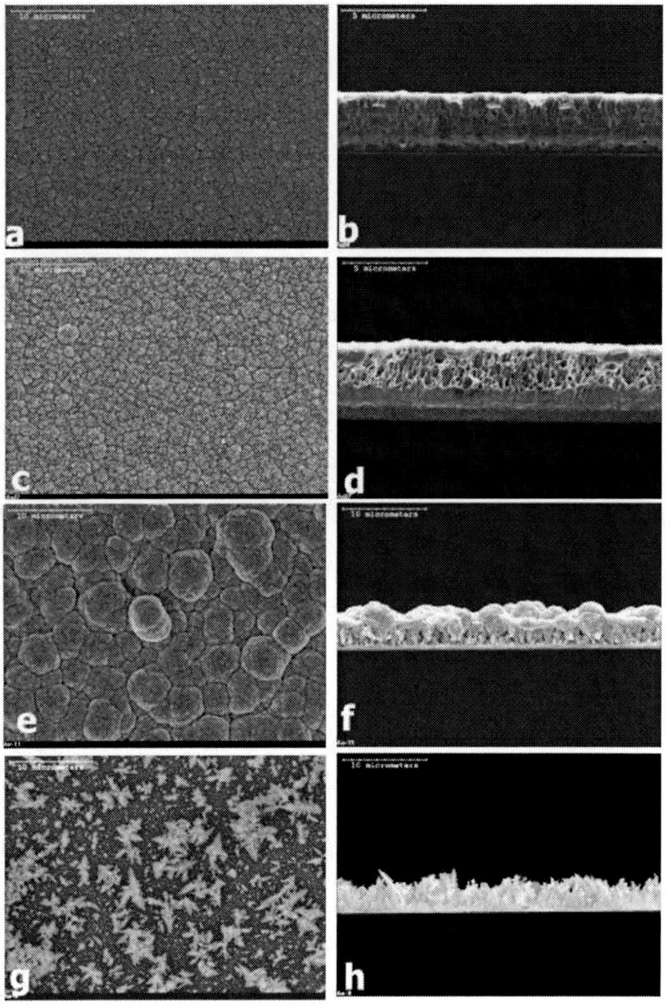

Figure 3.23. SEM SE images of electrodeposited films from solution V: (a) and (b) deposited at 3.0 mA/cm2, (c) and (d) deposited at 4 mA/cm2, (e) and (f) deposited at 7.5 mA/cm2, (g) and (h) deposited at 10 mA/cm^2. The deposition time was 30 minutes (He et al. [58]).

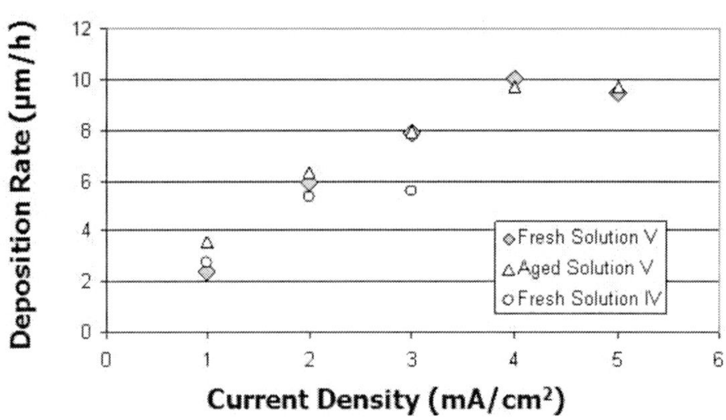

Figure 3.24. Relationship between electrodeposition rate and current density for pure Au deposition (He et al. [58]).

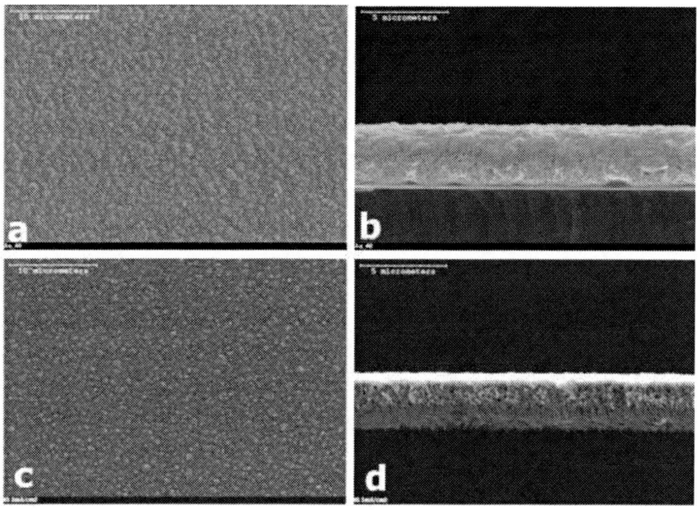

Figure 3.25. SEM SE images of electrodeposited films from aged solution V: (a) and (b) from solution aged for 21 weeks, (c) and (d) from solution aged for 24 weeks. The current density was 3 mA/cm^2 and the deposition time was 30 minutes (He et al. [58]).

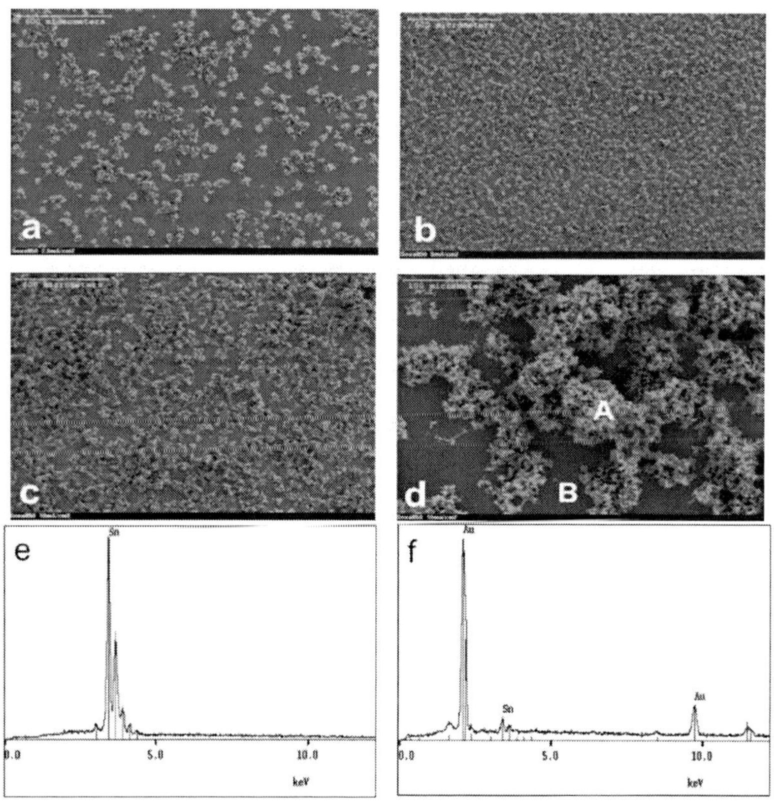

Figure 3.26. SEM SE images and EDX spectra from a deposit obtained from a $SnCl_2{\cdot}2H_2O$ aqueous solution with a concentration of 0.22 mol/L (pulse deposited). The average current densities were: a) 2.5 mA/cm^2; (b) 5 mA/cm^2; (c) 10 mA/cm^2. (d) High magnification image of sample in (c) showing the spongy nature of the deposit. e) EDX spectrum from position A in (d); f) EDX spectrum from position B in (d) (He et al. [80]).

Above a current density of 4.0 mA/cm^2, the quality of the deposits decreases dramatically. The effect of current density on deposition rate, for solution V, is shown in Figure 3.24. The deposition rate increases with increasing current density, up to 4 mA/cm^2, and then slightly decreases at higher values. For comparison, deposition rates for solution IV are also shown in Figure 3.24. Deposition rates for solutions IV and V are similar at current densities below 2.0 mA/cm^2; however, at higher current densities the deposition rate from solution IV is much lower than that from solution V. In addition to electrodeposition on unpatterned wafers, experiments also showed that electrodeposition can be done on patterned wafers with good filling of the patterned structure [58]. The ultimate test of the electrolyte (solution V) is its long term stability and useful lifetime. To test the stability, films were deposited on separate substrates every week from the same solution for the first 3 months, followed by monthly depositions. Films obtained after 21 weeks and 24 weeks of aging are shown in Figure 3.25. The films, in both cases, are smooth and dense and the deposition rates are comparable to those from fresh solutions (Figure 3.23).

3.5.2. Solution Preparation and Electrodeposition of Pure Sn Films

Electrodeposition of Sn has been done for more than one hundred and fifty years. There are two main types of baths for Sn, alkali stannate based baths and acidic stannous salt based baths [59-63]. The alkali stannate baths contain sodium stannate or potassium stannate and the corresponding alkali metal hydroxide. These baths are non-corrosive and do not require other additives. They are, therefore, easy to control and even have some detergent action. Alkali stannate baths must be heated, however, and have a low maximum deposition current density and a restrictive anode current density range. In fact, alkali stannate baths yield satisfactory results only at 65°C and above. At lower temperatures, tin sponge precipitates form in the electrolytes and the anode becomes passivated even at low current densities. In addition, the tin contained in stannate is in the tetravalent state, so twice as much electrical charge is required to deposit the same mass of tin compared with deposition from stannous salt solutions. Electrodeposition of Sn from acidic stannous (divalent) solutions consumes less electricity than alkali stannate baths. In addition, acidic solutions can be used for electrodeposition on circuit boards with patterned photoresist. Alkali baths, on the other hand, can cause the photoresist to delaminate, which limits their application in microelectronic packaging.

The major disadvantage of acidic stannous-based baths is that their composition is complex and satisfactory films cannot be deposited from simple acidic stannous salt solutions. Deposition from simple acidic solutions, containing stannous chlorides, sulfates or fluoroborates, produces loosely adhering acicular growths rather than uniform coatings, so that parts of the cathode remain uncovered [59-63]. To improve deposit morphology and adhesion during acidic Sn deposition, various additives, most of which are organic chemicals, have been tried. These include gelatin [64-65], cresylic acid [66], cresol sulfonic acid [67], an aromatic hydroxyl compound [68-69] or some combination of these. It is, therefore, common for a given acidic stannous Sn bath to contain several components [70-76]. The organic additives are used to improve the quality of deposited films. Their addition, however, leads to more complex solutions, which can be difficult to control. Obviously, acidic stannous baths are advantageous due to their lower power consumption and compatibility with photoresists compared with alkali stannate baths. This is the reason for the focus within the group at the University of Alberta on a simple acidic solution for pure Sn deposition.

The simple acidic solution for pure Sn deposition introduced here consists of only two components, i.e., a Sn (II) chloride salt ($SnCl_2{\cdot}2H_2O$) and a complexing reagent for the Sn (tri-ammonium citrate). The development of this solution at the University of Alberta was actually based on the Sn solution discussed in Section 3.1.1. It should be emphasized that the electrodeposition of Sn from a solution containing stannous and citrate complexes was mentioned in a short communication published in the Transactions of the Kansas Academy of Science as early as 1938 [77]; however, no further work followed. A possible reason for the lack of interest is that corrosion resistance was the main role for Sn coatings at that time. Since chlorine ions are detrimental to steels from a corrosion perspective, other types of deposition solutions for Sn were developed. For microelectronic or MEMS bonding applications, chloride ions are not an issue. As such, this simple solution was pursued.

To prepare the solution, tri-ammonium citrate was dissolved in deionized water first, followed by the addition of $SnCl_2{\cdot}2H_2O$. The solution was stirred until the $SnCl_2{\cdot}2H_2O$ salt was fully dissolved. The tri-ammonium citrate plays an important role in this simple solution, since it is the only complexing agent for the Sn ions in the solution. Without tri-ammonium citrate, Sn can also be directly deposited from a $SnCl_2{\cdot}2H_2O$ aqueous solution (no additives), but the electrodeposits are spongy and cathode coverage is poor, as shown in the SEM images in Figure 3.26. The films were deposited from a solution containing 0.22 mol/L of $SnCl_2{\cdot}2H_2O$ at current densities of 2.5, 5 and 10 mA/cm^2, respectively, with pulsed current (2 ms on and 8 ms off). Platinum was the anode and an Au seed layer was the cathode (the Au seed layer was sputter deposited on a Si wafer with Ti as an adhesion layer). Figure 3.26.d is a higher magnification image of Figure 3.26.c. EDX spectra obtained from the sample in Figure 3.26.d, i.e., regions marked A and B, are shown in Figure 3.26.e and 3.26.f, respectively. The EDX spectra clearly show that the spongy films are pure Sn (position A), while the uncovered regions are the Au seed layer (position B). In addition to the poorly deposited films, the aqueous $SnCl_2{\cdot}2H_2O$ solution quickly turned cloudy upon standing, forming white precipitates. This was likely due to the hydrolysis of Sn^{4+}, which was generated by the oxidation of Sn^{2+} ions.

When $SnCl_2{\cdot}2H_2O$ is added to a tri-ammonium citrate solution, the resulting solution is colorless and transparent if the amount of tri-ammonium citrate is sufficient. Several concentrations of tri-ammonium citrate and $SnCl_2{\cdot}2H_2O$ were tested (Table 3.4). The solution is clear without any precipitation when the molar ratio of tri-ammonium citrate to $SnCl_2{\cdot}2H_2O$ is greater than 1.4. Above this critical ratio, smooth and dense Sn deposits are attainable; the films are matte in appearance and white in color. Figure 3.27.a–d shows examples of such films (cross section images), deposited at current densities of 5, 10, 15 and 20 mA/cm^2, respectively, from a solution with 0.22 mol/L of $SnCl_2{\cdot}2H_2O$ and 0.31 mol/L of tri-ammonium citrate. Corresponding plan view images are shown in Figure 3.27.e–h. The solution pH was 5.0, which is suitable for electrodeposition on photoresist patterned substrates. The images show that the deposited films are dense, uniform and ductile. Cross section samples were prepared by cleaving silicon wafers, which resulted in tearing of the upper portion of the layer for thicker Sn films (Figures 3.27.a-d). Coverage was complete and no acicular regions were observed, unlike earlier simple acidic Sn deposition work [59-63]. Recently, this solution has also served as a starting point for development of simple electrolytes for Sn-Cu-Ag ternary alloy deposition [78].

Table 3.4. Various solution concentrations for Sn electrodeposition [80]

Tri-ammonium citrate mol/L (g/L)	$SnCl_2 \cdot 2H_2O$ mol/L (g/L)
0	0.22 (50)
0.82 (200)	0.044 (10)
0.21 (50)	0.22 (50)
0.31 (75)	0.22 (50)
0.41 (100)	0.22 (50)
0.82 (200)	0.22 (50)
0.41 (100)	0.33 (75)

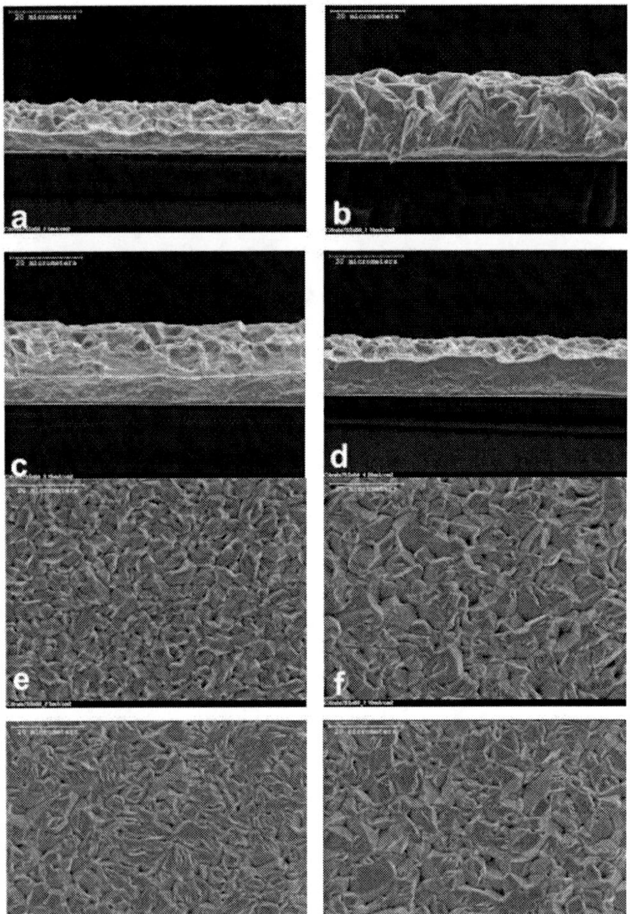

Figure 3.27. SEM SE cross section images (a-d) and plan view images (e-h) from deposits plated from a solution with 0.31 mol/L of tri-ammonium citrate and 0.22 mol/L of $SnCl_2 \cdot 2H_2O$ (pulse deposited). The average current densities were: a), e) 5 mA/cm^2; b), f) 10 mA/cm^2; c), g) 15 mA/cm^2; d), h) 20 mA/cm^2 (He et al. [80]).

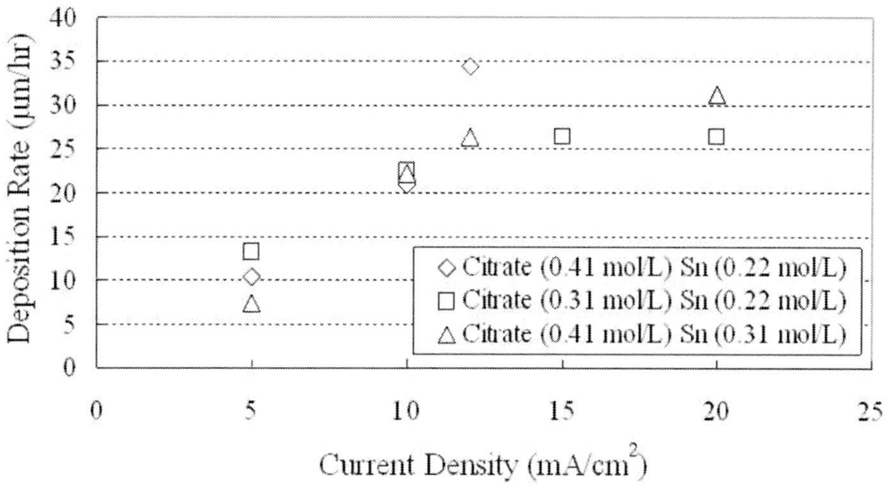

Figure 3.28. Relationship between deposition rate and current density for three different solution concentrations for pure Sn deposition (He et al. [80]).

Figure 3.28 shows the relationship between deposition rate and current density for three solutions containing the following molar concentrations of tri-ammonium citrate/$SnCl_2{\cdot}2H_2O$: 0.31:0.22, 0.41:0.22 and 0.41:0.33. Each data point has an uncertainty of about ±0.5 μm/h. For the 0.31:0.22 molar ratio, the deposition rate increases continuously with increasing current density to about 15 mA/cm^2 and then levels off at ~26 μm/h. The other two solutions show similar behaviour, i.e., higher deposition rate with increasing current density with the deposition rate levelling off at 15–20 mA/cm^2. The current densities corresponding to the start of the deposition rate plateaus likely correspond to the limiting current density for the particular system; higher current densities may lead to hydrogen evolution. It is interesting to note that all three solutions have a similar deposition rate (~21 μm/h) for a current density of 10 mA/cm^2. Pin holes are more prevalent in films deposited at higher current densities, which is evident when Figure 3.27.e and 3.27.f are compared with Figure 3.27.g and 3.27.h.

The ratio of tri-ammonium citrate to $SnCl_2{\cdot}2H_2O$ in the electrolyte has a significant effect on deposit morphology. If a low concentration of tri-ammonium citrate is used, white precipitates form in the solution. These solutions cannot be used to deposit films at high current densities. For example, for a solution with a tri-ammonium citrate/$SnCl_2{\cdot}2H_2O$ ratio of 0.21 mol/L to 0.22 mol/L, films deposited at 5 mA/cm^2 are smooth (Figure 3.29.a and 3.29.b), but the deposit density is dramatically reduced when deposition is performed at 10 mA/cm^2 (Figures 3.29.c and 3.29.d). Similar results are obtained if higher amounts of tri-ammonium citrate are added, as shown in Figure 3.30 where the tri-ammonium citrate to $SnCl_2{\cdot}2H_2O$ ratio is 0.82:0.22.

The films in Figures 3.30.a and 3.30.c. were deposited at 5 and 10 mA/cm^2, respectively. During electrodeposition, hydrogen bubbles are visible on the cathode surface at higher current densities (10 mA/cm^2), and circular features are produced in the films (e.g., Figure 3.29.c and 3.30.c). The surface roughness is increased at higher current densities and deposit density is reduced.

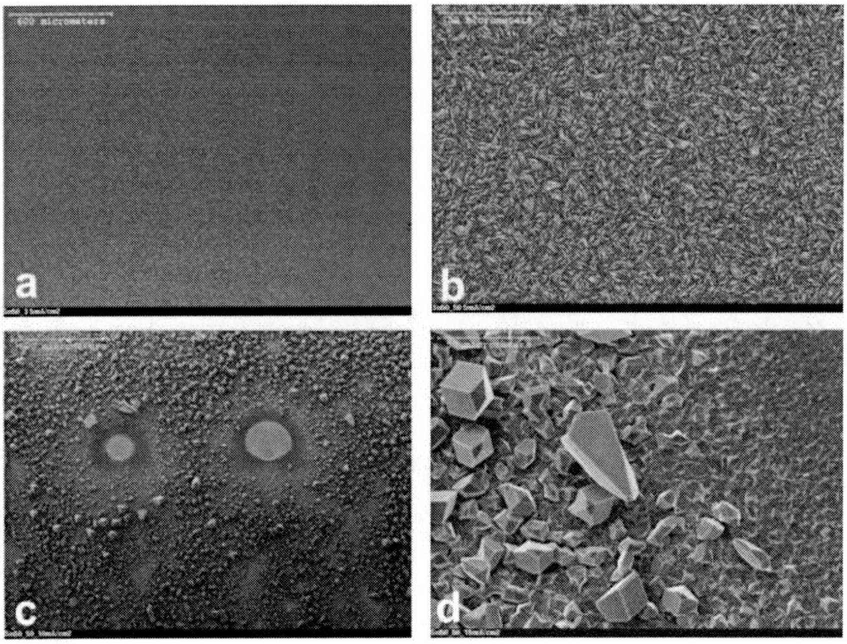

Figure 3.29. SEM SE images of deposits obtained from a solution with 0.21 mol/L of tri-ammonium citrate and 0.22 mol/L of $SnCl_2·2H_2O$ (pulse deposited). The average current densities were: a), b) 5 mA/cm^2; c), d) 10 mA/cm^2 (He et al. [80]).

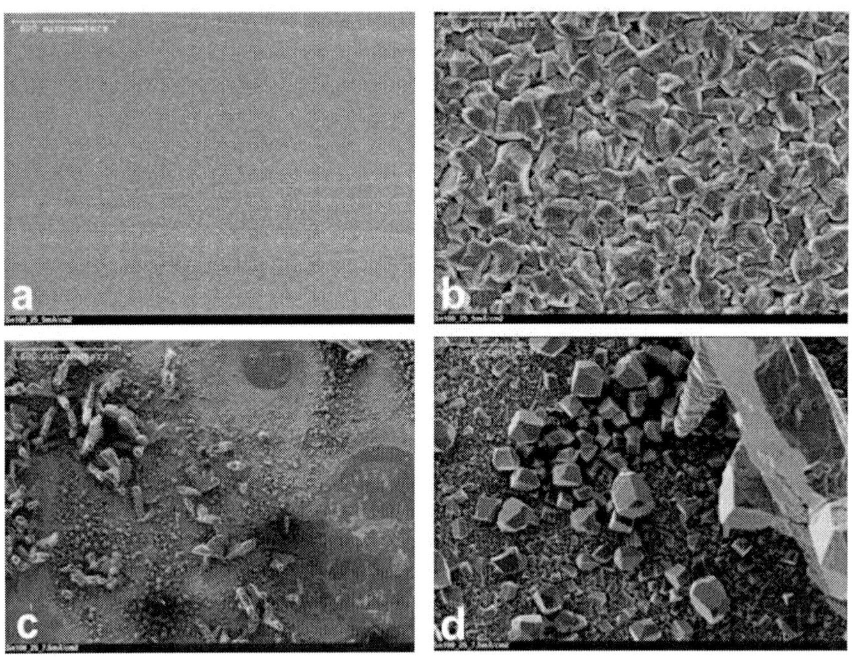

Figure 3.30. SEM SE images of deposits obtained from a solution with 0.82 mol/L of tri-ammonium citrate and 0.22 mol/L of $SnCl_2·2H_2O$ (pulse plated). The average current densities were: a) 5 mA/cm^2; b) 7.5 mA/cm^2; c), d) 10 mA/cm^2 (He et al. [80]).

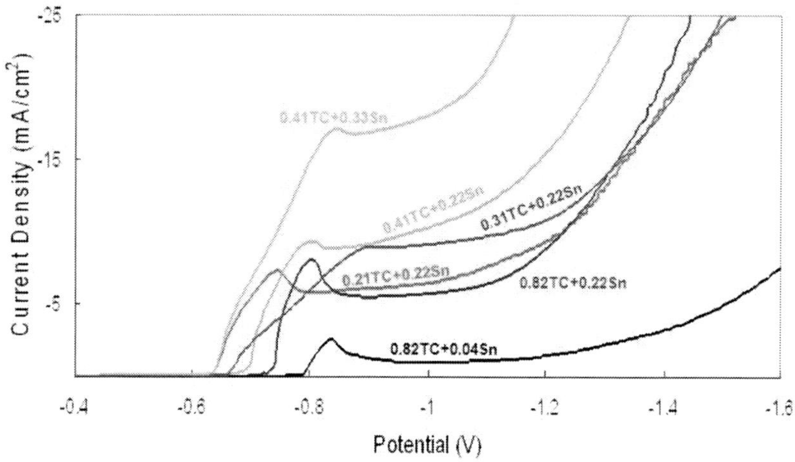

Figure 3.31. Polarization curves for various solution concentrations. Potentials are with respect to SCE (He et al. [80]).

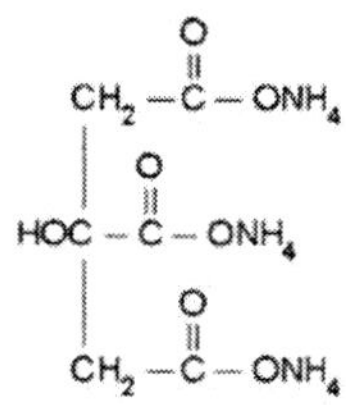

Figure 3.32. Schematic structure of tri-ammonium citrate (He et al. [80]).

In order to better understand the relationship between solution concentration and the current densities used for electrodeposition, polarization measurements were done (Figures 3.31). The curves obtained are quite similar; each one can be divided into four basic stages. For the first stage little or no current flows as the cathodic potential is increased (becomes more negative). The system enters the second stage when a critical potential is reached (between –0.6 and –0.8 V vs SCE). During this stage, current flow starts and the current density increases steadily with increasing cathodic potential. In the third stage, the current density reaches a peak value, then drops slightly and remains at that value as the cathodic potential is further increased. The appearance of a peak and subsequent drop has also been observed by Survila et al. [79], which has been attributed to abrupt changes in the solution composition at the electrode surface under electrolysis conditions. The plateaus in Figure 3.31 span about 0.2–0.4 V and are followed by the fourth stage where the current density increases rapidly with increasing cathodic potential, signalling the reduction of H^+. The critical potential varies from solution to solution and seems to correspond to the onset potential for Sn deposition, because it is more negative at lower Sn^{2+} concentrations or higher tri-ammonium citrate concentrations. The critical potential can be regarded as the open-circuit potential for the specific solution composition. If this is the case, then the peak current density in each case represents the limiting current density for Sn electrodeposition, which is larger for higher tin chloride concentrations (Figure 3.31). Increasing the tri-ammonium citrate concentration in the solution has little effect on the limiting current density for Sn electrodeposition for a given Sn concentration, but causes a shift in the onset potential (beginning of the second stage) to more negative potentials. This is a direct result of the complexing of the Sn^{2+} ions by tri-ammonium citrate.

Based on the deposition experiments and polarization measurements, one can see that solutions containing 0.31~0.41 mol/L of tri-ammonium citrate and 0.22 mol/L of $SnCl_2{\cdot}2H_2O$ (molar ratio of 1.41-1.86) can generate good Sn films and fairly high deposition rates. Films deposited from these solutions have a tetragonal crystal structure and a (110)-type texture [80]. This (110) type texture is considered to be beneficial in preventing Sn whisker growth from Sn films [76]. In this work, Sn whiskers were not observed in electrodeposited films even after aging at room temperature for more than a year. This may be related to the strong (110) texture. The electrolytes are stable, with a shelf life of more than a year.

As mentioned above, the tri-ammonium citrate plays a key role, not only in forming a stable deposition bath, but also in improving the morphology of the deposited films. It is evident that the chemistry of the $SnCl_2{\cdot}2H_2O$ – tri-ammonium citrate system should be addressed. Figure 3.32. displays a schematic diagram of the ammonium citrate molecule. The three ammonium ions can fully or partially separate from the citric ions when ammonium citrate dissolves in an aqueous solution. The citric ions and stannous ions can form complexes as reported by Willey [81], Elbourne and Buchanan [82, 84] and Sherlock and Britton [83]. Britton even pointed out that Sn readily forms complex ions, notably with the organic acids naturally occurring in some fruit and vegetable products [85]. In the study of the corrosion resistance of Sn with food, weak organic acids, such as citric acid, were tested for reaction with tin, and the stability constant of stannous ions with citric ions was estimated to be on the order of 10^7 [81,83]. Obviously, in the electrolyte studied in this work, the Sn ions are present mainly as complexed ions; electrochemically, they are fundamentally different from the simple acid stannous solutions in which Sn may exist as simple cations. Therefore, high quality Sn films are able to be electroplated from the present electrolyte. Recently, Han et al. [86] studied the kinetics of Sn electrodeposition in the tri-ammonium citrate-stannous system and proposed several possible complexes for Sn and citric ions. Readers interested in this topic can find detailed information in [86].

3.5.3. Microstructure of Sequentially Deposited Eutectic Alloys and Reflowed Solder

From the Au-Sn binary phase diagram shown in Figure 3.15, it is clear that there are two eutectic Au-Sn alloy compositions. The first one is an Au–rich eutectic with a Sn concentration of ~30 at% Sn (70 at% Au); the second one is a Sn-rich eutectic with a Sn concentration around 94 at% Sn (6 at% Au). The Sn-rich eutectic contains much less Au than the Au-rich eutectic, which could result in reduced cost for the solder materials. In addition, the Sn-rich solder has a lower eutectic temperature (217°C vs 283°C for the Au-rich eutectic), which is still higher than that for the conventional Pb-Sn eutectic alloy (183°C). Therefore, a Sn-rich, Au-Sn eutectic alloy may be applicable for bonding at relatively high temperatures. The Au-rich, Au-Sn eutectic alloy (30 at% Sn), can be successfully produced by electrodepositing alternative layers of Au_5Sn and AuSn from a single electrolyte. However, the Sn-rich, Au-Sn eutectic alloy has not been successfully electrodeposited from a single solution. Hence, sequential deposition from two separate solutions is an option to prepare the Sn-rich eutectic composition.

There have been reports in the literature of sequential deposition of Sn-rich, Au-Sn eutectic alloys, but the electrolytes are generally cyanide-based Au solutions and complex Sn solutions [34-36, 39-40]. In contrast, the University of Alberta group has successfully prepared Au-rich and Sn-rich eutectic Au-Sn alloys through sequential deposition from the non-toxic solutions described in Sections 3.5.1 and 3.5.2 [87-88]. Since the microstructure of an Au-rich eutectic Au-Sn alloy has already been shown in Figure 3.14, only the microstructure of the Sn-rich eutectic Au-Sn alloy will be presented here. More information related to sequential deposition from these solutions can be found in [37, 87-88].

Figure 3.33.a shows an SEM cross section BSE image of an as-electrodeposited Au/Sn dual-layer film formed by sequentially depositing Sn at 6.1 mA/cm^2 for 60 minutes and Au at 2.0 mA/cm^2 for 5 minutes. These times were chosen to give the Sn-rich eutectic composition. After deposition for 60 minutes, a smooth Sn layer of about 15 μm in thickness is uniformly distributed on the Au coated Si substrate. After subsequent deposition for 5 minutes, the Au layer thickness should be ~0.5 μm. The cross-section image in Figure 3.33.a. is from a polished sample, so that the outer surface is slightly distorted because both Au and Sn are ductile and easily deformed during polishing. The gray zone corresponds to Sn and the white zone is Au. Figure 3.33.b shows an SEM BSE image from the solder after reflow at a temperature of ~290°C for 5 minutes in a quartz tube furnace. A typical eutectic microstructure is apparent, which is similar to that reported by Tang et al. [87] where the white zones correspond to $AuSn_4$ and the dark zones correspond to the Sn. EDX analysis (Figure 3.33.c) of the area in Figure 3.33.b shows that the overall composition of the reflowed solder is about 5 at% Au, which is very close to the Sn-rich Au-Sn eutectic composition.

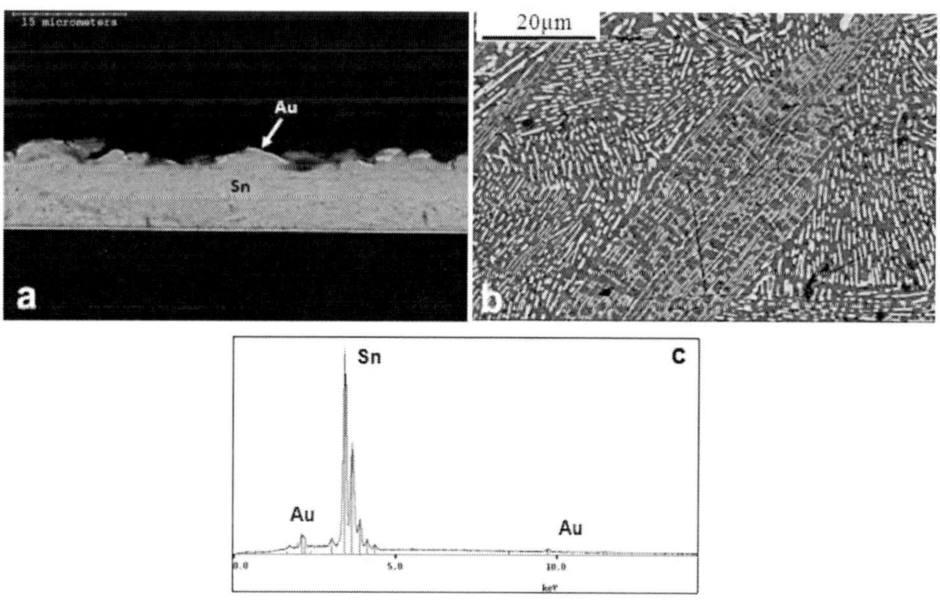

Figure 3.33. (a) SEM BSE image of dual layer Au/Sn film; (b) SEM BSE image of microstructure for reflowed Sn-rich eutectic Au-Sn solder; (c) EDX spectrum from area in (b) .

CONCLUSION

Two methods for electrolyte preparation for electrodeposition of Au-Sn alloys from a single solution have been presented. The electrolyte is a slightly acidic and non-toxic solution, which consists of five components (tri-ammonium citrate, $KAuCl_4$, Na_2SO_3, L-ascorbic acid and $SnCl_2{\cdot}2H_2O$). From this electrolyte, Au_5Sn and AuSn can be deposited, respectively, at different pulsed current densities onto unpatterned and patterned wafers at room temperature. Factors that affect the compositions of deposited films on patterned wafers, covered with dry photoresist or wet photoresist, have been discussed. Au-rich, Au-Sn eutectic solders can be obtained by alternatively depositing Au_5Sn and AuSn phases with an appropriate ratio. The degradation of the electrolyte has been studied using a combination of turbidity measurements, SEM/TEM, XRD, UV/Vis spectroscopy and Cryo-XPS. Degradation begins virtually immediately after the solution is prepared, due to the gradual reduction of ionic complexes of Au by ascorbic acid. The resulting precipitates are initially less than 10 nm in size, but agglomerate after 2 to 3 days to form visible Au particles that settle and coat the container walls. A procedure to recover precipitated Au from the spent solution has been developed.

Non-toxic electrolytes for electrodeposition of pure Au films and pure Sn films have also been developed. The pure Au electrolyte consists of tri-ammonium citrate, $KAuCl_4$ and Na_2SO_3. Deposition rates of ~10 μm/h at 4.0 mA/cm^2 are attainable from solutions containing 0.411 mol/L of tri-ammonium citrate, 0.066 mol/L of $KAuCl_4$ and 0.543 mol/L of Na_2SO_3. The pure Sn electrolyte consists of only two components, i.e., a Sn(II) chloride salt ($SnCl_2{\cdot}2H_2O$) and a complexing reagent for the Sn (tri-ammonium citrate). Deposition rates of 21 μm/h at 10 mA/cm^2 are attainable from solutions with 0.31-0.41 mol/L of tri-ammonium citrate and 0.22 mol/L of $SnCl_2{\cdot}2H_2O$. The two solutions are slightly acidic, which is suitable for deposition on patterned wafers. In addition, the two solutions are stable for more than a year, easy to prepare, environmentally benign and deposition can be done at room temperature. The two solutions can be used to prepare sequential deposits of Au and Sn, producing alloys with both the Au-rich and Sn-rich eutectic compositions.

ACKNOWLEDGMENTS

The authors are grateful to the Natural Sciences and Engineering Research Council (NSERC) of Canada and Micralyne, Inc. for providing funding and Micralyne, Inc. for supplying metallized substrates for electrodeposition. The authors would also like to thank former and current colleagues (N. Dalili, B. Djurfors, J. Doesburg, Q. Liu, N. Morawej, V. Neburchilov, W. Sun, W. Tang and Y. Zhang) for collaboration or useful discussion during this work. The Alberta Centre for Surface Engineering and Science (ACSES) at the University of Alberta is acknowledged for providing access to XPS and SAM facilities.

REFERENCES

[1] G. S. Matijasevic, C.C. Lee and C. Y. Wang, *Thin Solid Films* 223, 276 (1993).

[2] W. Plumbridge, *J. Mater. Sci.* 31, 2501 (1996).

[3] T. Ishii, S. Aoyama and M. Tokumitsu, *J. Elec. Mater.* 30(6), L25 (2001).

[4] Q. Tan, Y.C. Lee and M. Itoh, Chapter 5-Soldering technology for optoelectronic packaging. In: *Passive Micro-Optical Alignment Methods*, R. Boudreau and S. Boudreau (eds.), CRC Press, Boca Raton (2005).

[5] P.P. Maaskanant, M. Akhter, N. Cordero, D.P. Casey, J.F. Rohan, B.J. Roycroft and B.M. Corbett, *Phys. Stat. Sol.*, c2, 2907 (2005).

[6] T. Frey and W. Hempel, DE 4406434 (1995).

[7] W. Kuhn, W. Zilske and A.-G. Degussa, Ger. DE 4,406,434, August 10, 1995.

[8] Y. Tanabe, N. Hasegawa and M. Odaka, *J. Met. Fin. Soc. Jpn.*, 34, 8 (1983).

[9] N. Kubota, T. Horikoshi and E. Sato, *J. Met. Fin. Soc. Jpn.* 34, 37 (1983).

[10] B. Bozini, A. Fanigliulo, G. Giovabbelli, S. Natali and C. Mele, *J. Appl. Electrochem.* 33, 747 (2003).

[11] E. Raub, K. Bihlmaier, *Galvanische Weissgolniederschlage, Mitt. Forschungsinst. Probierants. Edelmetalle Staatl. Hoheren Fachschule Schwab. Gmund* 11, 59 (1937).

[12] R.J. Morrissey and R.I. Cranston, *United States Patent* No. 5,277,790 (1994).

[13] S. Matsumoto, Y. Inomata, *Japan Patent* No. 61 15,992 [86 15.992], January 24, 1986.

[14] J. L. Cardoso and S.G. Santos Filho, *J. Integr. Circ. Sys.* 5(2), 95 (2010).

[15] T.N. Vorobyova and O.N. Vrublevskaya, *Surface and Coatings Tech.* 204, 1314(2010).

[16] H. Honma and K. Hagiwara, *J. Electrochem. Soc.* 142, 81 (1995).

[17] W. Sun and D. G. Ivey, *Mater. Sci. Eng.* B65, 111 (1999).

[18] J. Doesburg and D. G. Ivey, *Plat. Surf. Finish.* 99, 78 (2001).

[19] B. Djurfors and D.G. Ivey, *J. Electron. Mater.* 30, 1249 (2001).

[20] A. He, B. Djurfors, S. Akhlaghi and D.G. Ivey, *Plat. Surf. Finish.* 89, 48 (2002).

[21] Y. Zhang and D. G. Ivey, *Plat. Surf. Finish.* 91, 28 (2004).

[22] N. Morawej and D. G. Ivey, *J. Appl. Surf. Finish.* 2(3), 220 (2007).

[23] A. He, Q. Liu, and D.G. Ivey, *Plat. Surf. Finish.* 92, 30 (2005).

[24] A. He, Q. Liu, and D.G. Ivey, *J. Mater. Sci.: Mater. Electron.* 17, 63 (2006).

[25] D.J. Shaw, *Introduction to Colloid and Surface Chemistry*, 4th Edition, Butterworth-Heinemann, Oxord, UK (1992).

[26] C.N.R. Rao, Ultra violet and Visible Spectroscopy, 3rd Edition., Butterworth, Boston, MA, (1975).

[27] J. V. Paukstells, D.D. Mueller, P.A. Seib and D.W. Lillard, *Ascorbic Acid: Chemistry, Metabolism and Uses*, P.A. Seib and B.M. Tolbert (eds.), ACS Press, Washington DC (1982).

[28] J.A. Creghton and D. G. Eadon, *J. Chem. Soc .Faraday Trans.* 87, 3881 (1991).

[29] S. Mandal, P. Selvakannan, S. Phadtare, R. Pasricha and M. Sastry, *Proc. Indian Acad. Sci. (Chem. Sci.)* 114, 513 (2002).

[30] N. Dalili, A. He, Q. Liu and D.G. Ivey, *J. Electron. Mater.* 39, 1554 (2010).

[31] J.F. Watts, *An Introduction to Surface Analysis by XPS and AES*, New York: Wiley, 2003.

[32] B. Djurfors and D.G. Ivey, *Mater. Sci. Eng.* B90, 309 (2002).

[33] H. Okamoto, *J. Phase Equilib. Diff.* 28, 490 (2007)
[34] V. Neburchilov, A. He, Q. Liu and D.G. Ivey, J. Appl. Surf. Finish. 1(3), 263 (2006).
[35] G.H. Jeong, J.H. Kim, D. Lee and S.J. Suh, *Mater. Res. Soc. Symp. Proc.* 894, LL03-16.1 (2006).
[36] J. Kim, D. Kim and C.C. Lee, *IEEE Trans. Adv. Packag*. 29, 473 (2006).
[37] W. Tang, A. He, Q. Liu and D.G. Ivey, *Acta Mater*. 56, 5818 (2008).
[38] N. Kumar, M.L. Tian, J.G. Wang, W. Watts, J. Kindt, T.E. Mallouk and M.H.W. Chan, *Nanotechnology* 19, 365704 (2008).
[39] K.E. Asamundtveit, K. Wang, N. Hoivik, J.M. Graff and A. Elfving, 2009 Europen Microelectronics and Packaging Conference (EMPC 2009), Vol. 1 and Vol. 2, pp 723.
[40] M.L. Huang, Y. Liu and J.X. Gao, *J. Mater. Sci.: Electron. Mater*. 22, 193 (2011).
[41] R.E. Visco, Chapter 8, in *Modern Electroplating*, ed. F. A. Lowenheim, Jon Wiley and Sons, Inc., New York (1974).
[42] F.H. Reid and W. Goldie, *Gold Plating Technology*, Electrochemical Publications, Ltd., Ayr, Scotland, 1974.
[43] F.R. Lowenheim, *Electroplating*, McGraw-Hill Book Company, New York, (1978).
[44] P. Wilkinson, *Gold Bull.* 19, 75 (1986).
[45] W.S. Rapson and T. Groenewald, *Gold Usage*, Academic Press, London, 1978.
[46] F. Simon and W. Kuhn, *Proceedings of the AESF Technical Conference SUR/FIN, '99*, American Electroplaters and Surface Finishers Society, 373 (1999).
[47] J. Traut, J. Wright and J. Williams, *Plat. Surf. Fin.* 77, 49 (1990).
[48] J. Horkans and L.T. Romankiw, *J. Electrochem. Soc.* 124, 1499 (1977).
[49] H. Honma and K. Kagaya, *J. Electrochem. Soc.* 142, L135 (1993).
[50] R.J. Morrissey, *Plat. Surf. Finish.* 80, 75 (1993).
[51] R.J. Morrissey, *United States Patent* No 5,277,790 (1994).
[52] T. Osaka, A. Kodera, T. Misato, T. Homma, Y. Okinaka and M. Hoshino, *J. Electrochem. Soc.* 144, 3462 (1997).
[53] T. Osaka, M. Kato, J. Sato, K. Yoshizawa, T. Honma, Y. Okinaka and O. Yoshioka, *J. Electrochem. Soc.*, 148, C659 (2001).
[54] M. Kato and Y. Okinaka, *Gold Bull.* 37, 37 (2004).
[55] M.J. Liew, S. Sobri, B. Roy, *Electrochimica Acta* 51, 877 (2005).
[56] E. Smallbrugge, B. Jacobs, S. Falcone, E.J. Geluk and F. Karouta, Proceedings Symposium IEEE/LEOS, Benelux Chapter, Delft, The Netherlands, 143 (2000).
[57] P. Laude, E. Marka and Z. Morrens, *United States Patent* No 4, 192,723 (1980).
[58] A. He, Q. Liu and D.G. Ivey, *J. Mater. Sci.: Electron. Mater.* 20, 543 (2009).
[59] A.G. Gray, (ed.), *Modern Electroplating*, John Willy and Sons, Inc., New York, 387 (1953).
[60] J.W. Cuthberton, *Tin and Its Alloys*, Chapter V, E.S. Hedges (ed.), Edward Arnold Ltd, London, 99 (1960).
[61] F.A. Lowenheim (ed.), *Electroplating*, McGraw-Hill Book Company, New York, 307 (1978).
[62] J.I. Duffy (ed.), *Electroplating Technology-Recent Developments*, Noyes Data Corporartion, Park Ridge, New Jersey, 92 (1981).
[63] J.W. Price, *Tin and Tin Alloy Plating*, Electrochemical Publications Ltd., Ayr, Scotland, 5(1983).
[64] A. Betts, *United States Patent* No. 712,277 (1902).

[65] H.L. Hollis, *United States Patent* No. 916,155 (1909).
[66] F.C. Mathers, *United States Patent* No. 1,397,222 (1921).
[67] A.H. Alexander, J.R. Stack, *Trans Am. Inst. Mining and Met. Eng.* 70, 404 (1924).
[68] M. Schloetter, *British Patent* No. 329,308 (1929).
[69] M. Schloetter, *British Patent* No. 443,429 (1936).
[70] C.A. Disher, F.C. Mathers, *J. Electrochem. Soc.* 102, 387 (1955).
[71] A.M. Harper, A. Mohan, S.C. *Britton, Trans. Inst. Metal Finish.* 34, 237 (1957).
[72] M.E. Roper, *Metal Finish. J.* 12, 255 (1966).
[73] H.K. Wren, W.T. Hobson, *Metal Finish.* 67, 52 (1969).
[74] J.L. Parker, *Electroplat. Metal Finish.* 23, 19 (1970).
[75] H. Nawafune, in *Lead- free Soldering in Electronics: Science, Technology and Environmental Impact*, Chapter 4, K. Suganuma (ed.), Marcel Dekker Inc., New York, Basel, 91 (2004).
[76] K. Whitelaw, J. Crosby, M. Toben, *Circuit World* 32, 23 (2006).
[77] R.H. Hess and R. Taft, *Tans. Kansas Acad. Science* 41, 171 (1938).
[78] C.F. Han, Q. Liu and D.G. Ivey, *Mater. Sci. Eng.* B164, 172 (2009).
[79] A. Survila, Z. Mockus, S. Knapeckaite, *Electrochim. Acta* 46, 571 (2000).
[80] A. He, Q. Liu and D.G. Ivey, *J. Mater. Sci.:Electron. Mater.* 19, 553(2008).
[81] A. R. Willey, *Br. Corros. J.* 7, 29 (1972).
[82] R.G.P. Elbourne, G.S. Buchanan, *J. Inorg. Nucl. Chem.* 32, 493 (1970).
[83] J.C. Sherlock, S.C. Briton, *Br. Corros. J.* 7, 180 (1972).
[84] R.G.P. Elbourne, G.S. Buchanan, *J. Inorg. Nucl. Chem.* 32, 3559 (1970).
[85] S.C. Britton, *Tin Versus Corrosion*, Int. Res. Inst., Pub. No. 510 (1975).
[86] C. F. Han, Q. Liu and D. G. Ivey, *Electrochem. Acta* 53, 8332 (2008).
[87] W. Tang, A. He, Q. Liu and D.G. Ivey, *J. Electron. Mater.* 37, 837(2008).
[88] W. Tang, A. He, Q. Liu and D.G. Ivey, *J. Mater. Sci.: Electron. Mater.* 19, 1176 (2008).

In: Electrodeposition: Properties, Processes and Applications ISBN: 978-1-61470-826-1
Editor: Udit Surya Mohanty

Chapter 4

ELECTROCHEMICAL CORROSION BEHAVIOUR OF LEAD – FREE SOLDER ALLOYS IN 3.5 % NACL SOLUTION

Udit S. Mohanty* and Kwang Lung Lin
Department of Materials Science and Engineering, National Cheng Kung University, Tainan, R.O.C, Taiwan

ABSTRACT

Pb free solders has become increasingly important for industrial applications, since an international effort has been made to develop and characterize lead free solders for replacing traditional Pb-Sn solders as a consequence of governmental regulations and marketing pressure concerning health and environmental hazards of lead. The corrosion behaviour of a set of Pb free solder alloys were investigated in 3.5 % NaCl solution by employing potentiodynamic polarisation techniques. The various Pb free solders used in the present study were Sn-XAg-0.5 cu (X= 1-4 wt %); Sn-8.5Zn-XAg-0.1 Al-0.05 Ga (X=0.1-2 wt %), Sn-8.5Zn-XAl-0.5 Ga (X=0.02-0.5 wt %) and Sn-8.5Zn-XAg-0.1 Al-0.5 Ga where X varies from 0.1 to 2 wt %. Polarisation curves revealed that Sn-XAg-0.5 Cu alloys with higher Ag content (> 2wt %) exhibited a stronger tendency towards passivation and the passivation behaviour was ascribed to the presence of both SnO and SnO_2 on the cathode surface. An increase in the Ag content from 1 to 4 wt % decreased the corrosion current density (Icorr) and shifted the corrosion potential (E_{corr}) towards more negative values. These changes were clearly reflected in the corrosion rate and linear polarization resistance of the solder alloy. Nevertheless for Sn-8.5 Zn-XAg-0.1Al-0.05Ga alloy, an increase in the Ag content from 0.1 to 2 wt % resulted in a progressive increase in the corrosion current density and insignificant improvement in the passivation behaviour of the solder alloy. On the other hand the four element Sn-8.5Zn-XAl-0.5 Ga alloy showed decrease in both the corrosion current density and corrosion rate with increase in Al content from 0.02 to 0.05 wt %. The best results were observed for Sn-8.5Zn-XAg-0.1Al-0.5Ga solder alloy. The corrosion current density underwent a significant decrease and the corrosion potential shifted towards more noble values with increase in Ag content from 0.1 to 1.5 wt %. SE micrographs established that the oxides

*E-mail address: suryaudit@yahoo.com; matkllin@mail.ncku.edu.tw.

and hydroxides of zinc were responsible for the formation of passive film and the presence of Ag atoms in the oxide layer contributed towards the passivation behaviour of solders to a certain extent. SEM and EDX analysis also confirmed than $SnCl_2$ was the major corrosion product formed after the electrochemical experiments in all the investigated alloys.

INTRODUCTION

Electrodeposited and electroless coatings find a wide range of industrial applications for corrosion and wear protection. In particular, zinc and Zn-Ni alloy coatings with about 15-20 % Ni are extensively used in the automobile industry for corrosion protection of steel. Since, nearly all metallic corrosion process involve transfer of electronic charge in aqueous solution there has been a considerable interest among the researchers to examine the electrochemical nature of corrosion in Pb free solder alloys.

The eutectic Sn-37 % Pb alloy has been extensively used in the microelectronic packaging industry for many years mainly due to the unique combination of low melting point, suitable wettability, high strength and low cost. [1-3]. Despite all these useful properties, they are now being replaced with the Pb-free solder alloys because of the high toxicity and the environmental hazards posed by Pb. The European Community, the US EPA (Environmental Protection Agency) and Japan, as well as the electronic industries have launched initiatives for reducing lead usage [4, 5]. Many Sn based alloys have been developed as candidates for the replacement of Pb free solder. These solders are mostly based on Sn containing binary and ternary alloys. The binary systems include Ag-Zn, Au-Sn, Sb-Sn and Sn-Zn alloys [6-11] and some of the ternary alloys include Bi-Sb-Sn [12] and Sn-Zn-In [13]. The Sn-9 Zn and Sn-8Zn-3 Bi solder, with combination of low cost and mechanical attributes, are possible replacements of the Sn-37 Pb solder without markedly increasing the conventional soldering temperature [14]. Although this Sn-Zn system may be an industrial choice, it gives rise to reliability concerns of poor wetting, oxidation and corrosion behaviours because the element Zn is believed to be quite active. They are also susceptible to corrosion when they have long- time contact with corrosive vehicles such as salt or acid matter combined with moisture. In order to improve the corrosion resistance, the galvanizing industry has incorporated Al into the galvanized coatings for steel [15]. The eutectic Zn-5 % Al coating is found to be superior to pure Zn coating as far as atmospheric corrosion is concerned. For the corrosion investigations on Pb-free solders, the presence of Bi in Sn-Bi solders slightly increased the preferential dissolution of Sn in H_2SO_4 solution and dramatically accelerated the dissolution in HNO_3 solution, compared with that of pure tin [16]. An increase of Ag content in Sn-Ag-In solders also increased the corrosion resistance in 3 wt % NaCl solution, whereas the In content [17] did not have a significant effect on the corrosion parameters of the alloy. Sn-Zn-Al alloy [18] underwent more active corrosion than 63 Sn-37 Pb alloy and the alloys 5 In-9 (5Al-Zn) – Y Sn, 10 In-9 (5 Al-Zn)–Sn [19] exhibited passivation behaviour and the polarization behaviour of these two alloys were similar to that of 9 (5Al-Zn) –Sn alloy.

The potentiodynamic polarisation and electrochemical impedance spectroscopy techniques [20-22] are the most commonly used techniques to quantify the corrosion behaviours of materials. One of these methods is described below in details. Generally, when the electric current is applied to an electrochemical cell, the electrode reaction becomes irreversible and the electrode potential departs from equilibrium, which is the so- called polarization. Polarization method refers to changing the potential of the working electrode and monitoring the current, which is obtained as a function of potential. A typical hypothetical polarization diagram in acid solutions for a passivable system with active, passive and transpassive zones is illustrated in Figure 4.1. AB and BF denote cathodic and anodic activation process, while AC and DC correspond to hydrogen evolution reaction ($2H^+ + e = H_2$) and metal dissolution oxidation ($M = M + e^-$) respectively. The intersection i.e. point C shown in Figure 4.1. determines the corrosion current density I_{corr}, which also represent the corrosion rate of the system. It should be noted that at potentials more positive than E_{corr}, an increase in the corrosion rate is seen which reaches maximum at point F. The potential where the current peaks to a maximum is known as the passivation potential E_p and the current at this potential is referred as the critical current density i_{cc}. At transisition region of FG, a sudden decrease in current occurs due to the formation of complex oxides, usually hydrated on the sample surface. Therefore at passive stage of GH, the passive current density i_{pp} is maintained relatively steadily at low level until at point H, breakdown of the passivation begins. Then the curve runs into the transpassive region HJ and the current density tends to rise more and more with the propagation of pits on the surface. The potential at which the breakdown occurs is known as the breakdown potential designated by E_{bd}.

The present chapter discusses the electrochemical corrosion behaviour of a set of Pb free solder alloys namely Sn-XAg-0.5 cu (X= 1-4 wt %); Sn-8.5Zn-XAg-0.1 Al-0.05 Ga (X=0.1-2 wt %), Sn-8.5Zn-XAl-0.5 Ga (X=0.02-0.5 wt %) in 3.5 % NaCl solution. A NaCl solution was used to simulate the sea water condition, which the electronic assemblies may encounter during marine operation.

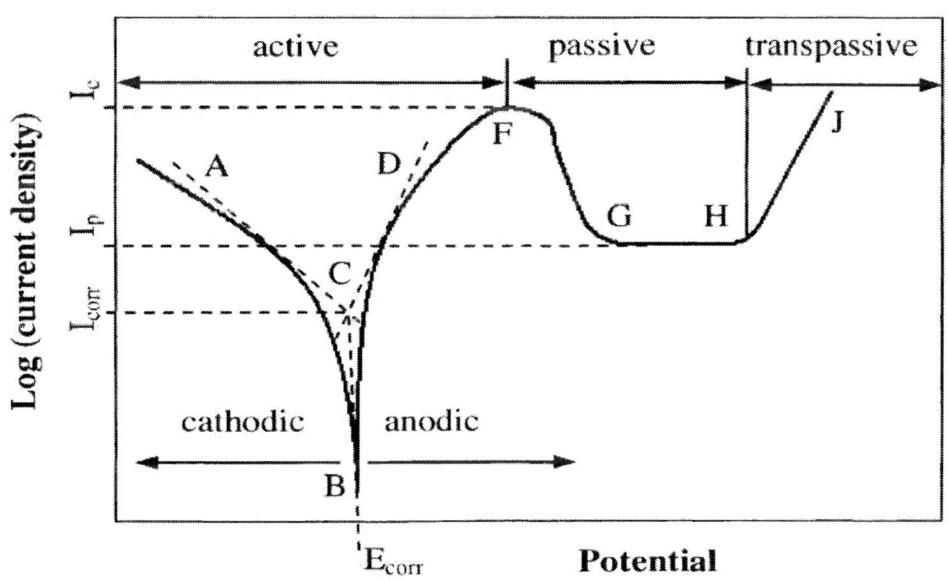

Figure 4.1. Schematic Polarisation curve shown for passivable metals.

4.1. Electrochemical Corrosion Study of Sn-XAg-0.5 Cu Solder Alloys

The eutectic Sn-Ag-Cu (SAC) alloy is considered one of the most promising replacements for the Pb-containing solders in microelectronic packaging and interconnects, since it offers a number of advantageous properties over other candidate alloys [23- 26]. The key advantages of the SAC eutectic alloy include its relatively low melting temperature (~ 217^0 C), reasonable wettability, slower growth of the intermetallics at the interface and superior mechanical properties compared with Sn-Pb and Sn-Ag solder alloys [27, 28]. Among the Sn-Ag-Cu alloys, Sn-3.5 Ag-0.7 Cu and Sn-4.7 Ag-1.7 Cu have been generally recommended as substitutes for Pb-Sn eutectic solder [29]. Several studies have been performed on the phase equilibria, microstructural and mechanical properties of SAC eutectic alloy [30-32] however, very little information is available on the corrosion behaviour of these alloys. In this section we focus on the electrochemical corrosion behaviour of Sn-XAg-0.5C u alloy (X=1 to 4 wt %) by potentiodynamic polarisation techniques performed in a chloride containing environment.

Alloys of Sn, Ag, and Cu were prepared from pure elements (greater than 99.9% pure). The Ag content in the Sn–XAg–0.5Cu alloy varied from 1 to 4 wt%. The constituent elements were weighed and melted in a quartz tube with an inside diameter of 8 mm under an argon atmosphere as reported earlier [28]. The molten alloys were homogenized at 800 °C for 3 h, cooled to 250 °C, and then water quenched to room temperature. The alloy thus formed after cooling was polished with SiC paper of grade ranging from 240 to 2000. Potentiodynamic polarisation studies were carried out in a rectangular cell (Figure 4.2) containing 3.5% NaCl solution. The counter-electrode was a Pt wire, and the reference electrode was a Ag/AgCl electrode saturated with KCl. The effect of Ag addition on the corrosion behavior of Sn-XAg-0.5 Cu alloys were investigated through potentiodynamic polarisation measurements in 3.5 % NaCl solution at a scan rate of 1 mV/s. Polarization studies were carried out in the anodic direction from -2100 to 100 mV at a scan rate of 60 mV/min. The solution used was de-aerated with N_2 for 1 h prior to the start of the polarisation experiments.

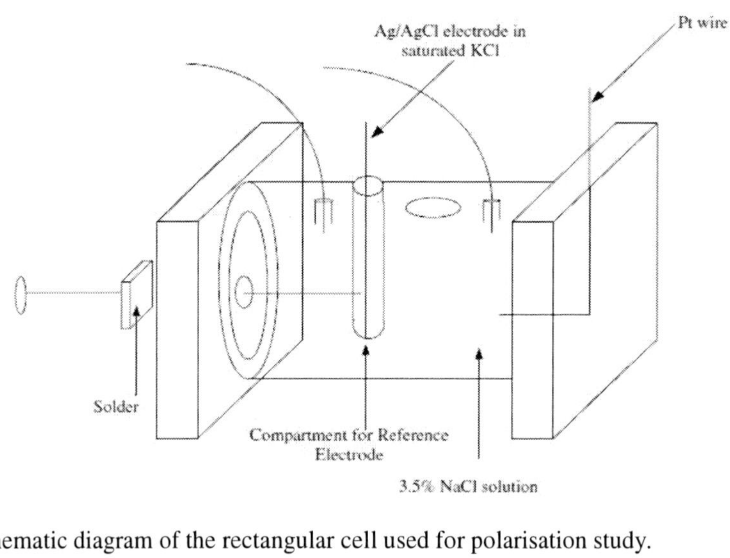

Figure 4.2. Schematic diagram of the rectangular cell used for polarisation study.

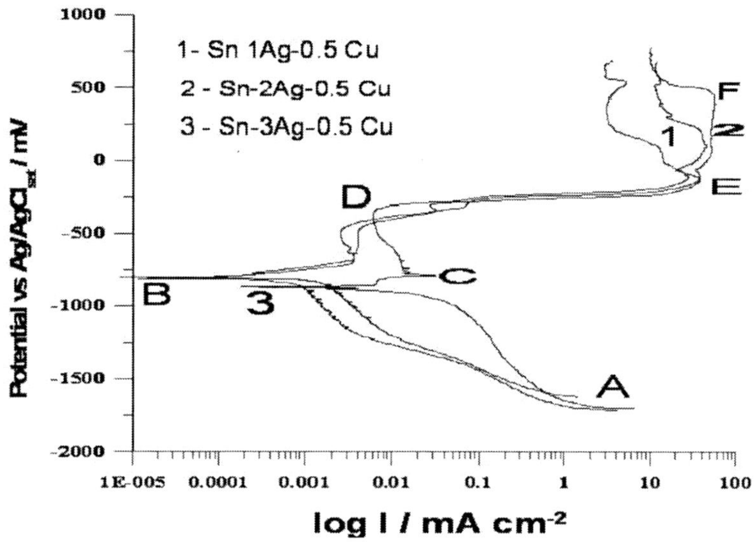

Figure 4.3. Potentiodynamic Polarisation curves for Sn-XAg-0.5 Cu alloy polarized in 3.5 % NaCl solution.

Table 4.1. List of corrosion parameters obtained by the potentiodynamic polarization of Sn-XAg-0.5 Cu alloys in 3.5 % NaCl solution

Ag content in Sn-XAg-0.5 Cu	E_{corr}/ mV	I_{Corr}/ mA cm^{-2}	Corr. Rate/ mm yr^{-1}	LPR/ Ohm cm^{-2}	E_{pp} /mV	E_{bd} /mV	i_{cc} /mA cm^{-2}	i_p / mA cm^{-2}
1Ag	-810	4.6	122.6	21.44	-697	-474	0.040	0.039
2Ag	-814	6.06	161.4	16.29	-677	-473	0.0033	0.0025
3Ag	-804	3.14	84.34	31.17	-788	-370	0.024	0.006
4Ag	-636	2.25	60.1	43.72	-505	-335	0.011	0.009

The surface morphology of the various corrosion products formed on the surface of the specimen was examined by scanning electron microscopy (SEM).Electron probe x-ray microanalyzer (EPMA) was performed on the same specimens used for the SEM observations. Auger Electron Spectroscopy (AES) technique and Auger Depth Profile analysis was used for probing the chemical and compositional surface environment of the corrosion product.

4. 1.1. Polarisation Study

Typical potentiodynamic polarisation curves recorded for Sn–XAg–0.5Cu solder alloy with various Ag content at a scan rate of 1 mV/s and 25 °C are shown in Figure 4.3. Corrosion parameters such as corrosion potential (E_{corr}) and corrosion current (I_{corr}) were estimated from the intersection of the anodic and cathodic Tafel lines. The effects of Ag on the various corrosion parameters obtained by the polarization of Sn–XAg–0.5Cu alloys in 3.5% NaCl solution are demonstrated in Table 4.1. The linear polarisation resistance value calculated in Table 4.1 is also referred to as the polarisation resistance (Rp).

It can be calculated by using the Stern–Geary equation [27].

$$R_p = b_a b_c / 2.303\ I_{corr}\ (b_a + b_c) \quad (1)$$

The data obtained from the polarisation curves were confirmed by repeated polarisation tests. The corrosion potential of Sn–XAg–0.5 Cu alloy is found to shift slightly toward nore negative values (−814 mV) with increase in Ag content from 1 to 2 wt%. Further increase in the Ag content from 2 to 4 wt % makes the corrosion potential (E_{corr}) more noble. These changes are also noted in the corrosion-current density (I_{corr}) and LPR values of the three element alloy in Table 4.1. A progressive decrease in the corrosion rate from 161.4 to 60.1 mm/ year is observed with increase in Ag content from 2 to 4 wt %.The region AB might correspond to the cathodic reduction of H_2O resulting in the formation of hydroxyl ions and hydrogen.

$$2H_2O + 2e^- \rightarrow H_2 + 2OH^- \quad (2)$$

Within the active dissolution region, Sn present in Sn– XAg–0.5Cu alloy takes priority in dissolution. This is also confirmed from the standard electrode potentials of Sn,Ag, and Cu [33].Reports [34] show that Sn–3.5Ag and Sn–0.8Cu alloy exhibits the same dissolution characteristics as Sn, and in these solders the element Sn dissolves faster than others. Since NaCl solution is neutral, hence, the anodic dissolution process might take place by the following scheme [35]

$$Sn + 2OH^- \rightarrow Sn\ (OH)_2 + 2e^- \quad (3)$$

$$Sn(OH)_2 \rightarrow Sn\ O + 2H_2O\ (4)$$

followed by the formation of $Sn(OH)_4$ species

$$SnO + H_2O + 2\ OH^- \rightarrow Sn(OH)_4 + 2e^- \quad (5)$$

$$Sn(OH)_2 + 2OH^- \rightarrow Sn\ (OH)_4 + 2e^- \quad (6)$$

$Sn(OH)_4$ is known to be highly insoluble which precipitates thus giving rise to a passivating film. The stability of the film formed has been reported [36] to increase with dehydration reaction as listed below:

$$Sn\ (OH)_4 \rightarrow SnO_2.\ XH_2\ O + (2\text{-}X\ )\ H_2O \quad (7)$$

Finally when the concentration of tin oxide at the anode surface exceeds its solubility product, the precipitation of solid oxide occurs on the electrode surface, suggesting that the formation of the passive film might have taken place through a dissolution precipitation mechanism.

The potential C in Figure 4.3. is referred as the passivation potential (E_{pp}) [37] where the current density is found to be maximum and the current observed at this potential is the critical current density (i_{cc}). A slight decrease in current density is observed when the potential is scanned from point C to point D indicating an onset of passivation due to the formation of oxides of Sn namely SnO and SnO_2. From Table 4.1 it is noted that Sn–1Ag–0.5Cu alloy possesses the highest i_{cc} value as compared to other solders. It is well known [38] that for promoting the metal passivity a large value of i_{cc} is required to produce enough metal cations to form passive film. Sn–1Ag–0.5Cu solder, despite having a higher value of i_{cc} (0.040 mA/cm^2) also shows an almost similar value of i_p (0.039 mA/cm^2) indicating a decrease in the tendency towards passivation.

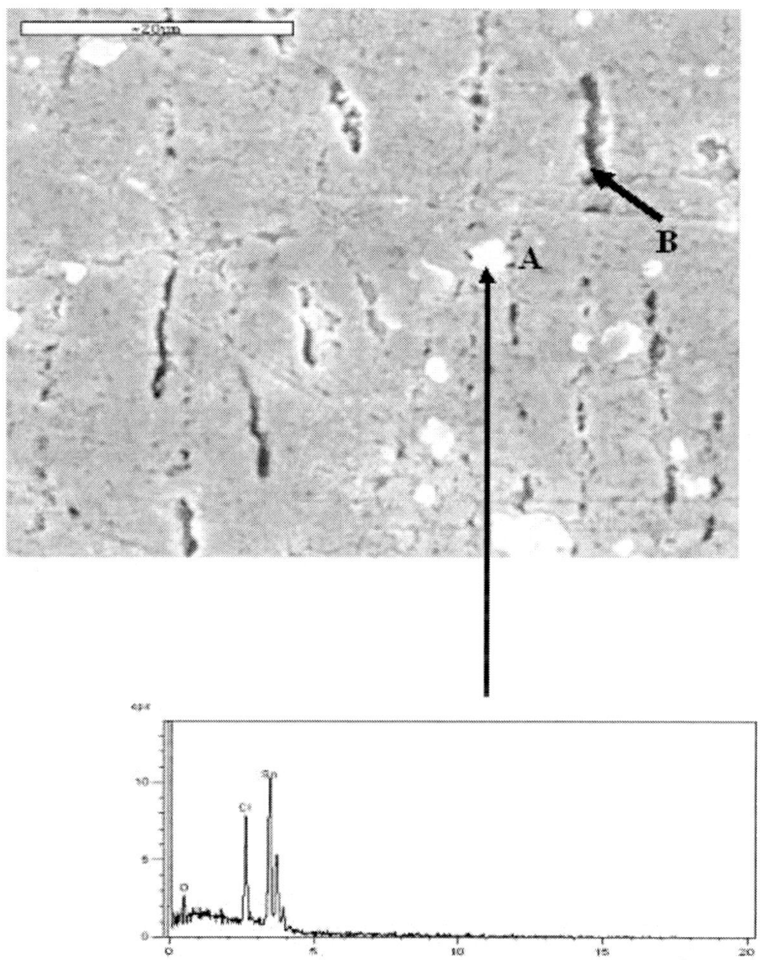

Figure 4.4. SE micrograph and EDX spectrum for Sn-1Ag-0.5 Cu polarized to -474 mV in 3.5 % NaCl solution. (Reprinted from [50], © 2007, with permission from Cambridge University Press).

Nevertheless, a broad potential is observed for Sn-3Ag-0.5 Cu alloy and a significant difference is noticed between the i_{cc} (0.024 mA/cm^2) and i_p (0.006 mA/cm^2) values. Lower values of i_p and i_{cc} correspond to the stability and protective nature of the passive film [39]. In the present study Sn–3Ag–0.5Cu and Sn–4Ag–0.5 Cu solder alloys provides greater stability and passivation ability than other solders. The passivation behaviour of Sn–XAg–0.5Cu solder alloy is attributed to the formation of thin film of oxides of both SnO and SnO_2 on the anode surface.

As reported [34] SnO_2 contributes towards passivity during electromigration of Sn–3.5Ag solder. Though addition of Ag to the alloy increases the passivation ability of Sn–3Ag–0.5Cu and Sn–4Ag–0.5Cu alloys, neither Ag nor its oxide contributes towards the formation of the passive film because the noble element Sn undergoes preferential dissolution and forms an oxide film near the passive region.

4.1.2. Surface Characterisation Studies

Surface analytical techniques like SEM, EPMA and AES were used for the characterisation of the corroded surface formed after polarization studies. SEM micrograph in Figure 4.4. represents the microstructure of Sn–1Ag–0.5Cu polarized to −474 mV (point D, Figure 4.3). It is noted that tin oxide is formed non-uniformly on the surface of the alloy along with chloride ions (Cl^-) in region A. A pit or a crack is observed at region B, and point D is referred to as the pitting potential or the breakdown potential (Epit or Ebd) where the current increases rapidly with increase in potential in the anodic direction to point E. The rapid rise in anodic current density at E_{pit} has been attributed to the initiation and growth of pitting attack. Breakdown of the passive layer and initiation of pitting attack is ascribed to the competitive adsorption between Cl^- and oxygenated species at adsorption sites on an oxide-covered layer [40]. The adsorbed Cl^- ions can penetrate through the passive layer especially at its point defects and flaws with the assistance of high electric field across the passive film to reach the base metal surface and initiate pitting [41]. However, the number of pitting sites reduced for Sn-4Ag-0.5 Cu alloy polarized to the same potential around -474 mV.

EPMA (electron micro probe analysis) performed on the surface of the Sn–4Ag–0.5Cu alloy (Figures 4.5.a and 4.5.b) polarized to the end of the electrochemical experiment (point E) revealed that the surface of the alloy is mainly composed of Sn and Cl. Thus the corrosion product formed at point E would be $SnCl_2$. A similar sheet-shaped corroded product has also been reported [42] for Sn–3.5Ag solder alloy. $SnCl_2$ can absorb O_2 from air to form insoluble oxychloride. The EPMA results however show that Cu and Ag are present in insignificant amounts on the surface of the Sn–4Ag–0.5 Cu alloy.

AES (Auger Electron Spectroscopy) was used to investigate the corrosion behavior of Sn–XAg–0.5 Cu alloys at specified potentials (potentials of interest) during the polarization studies. A typical AES depth profile results for Sn–4Ag–0.5Cu alloy polarized to the corrosion potential (−636 mV) is depicted in Figure 4.6.a. The atomic concentration of Sn is found to be around 40%, which slowly increases to 90% and further remains constant with increase in sputtering time to 600 s which indicated that only a small amount of Sn has oxidized, and rest remains in the sputtered layers and forms the bulk of the alloy.

Nevertheless, the atomic concentration of O is observed around 60%, which decreases with an increase in sputtering time to 600 s. No Ag or Cu appears on the surface of the alloy. However, at higher sputtering times some Ag is noticed in the layers. These results confirm that oxides of tin are only present on the surface of Sn–4Ag–0.5 Cu alloy at the corrosion potential. Figure 4.6.b. shows the Auger depth profile result for Sn–4Ag–0.5 Cu alloy polarized to −335 mV (point D, Figure 4.3).

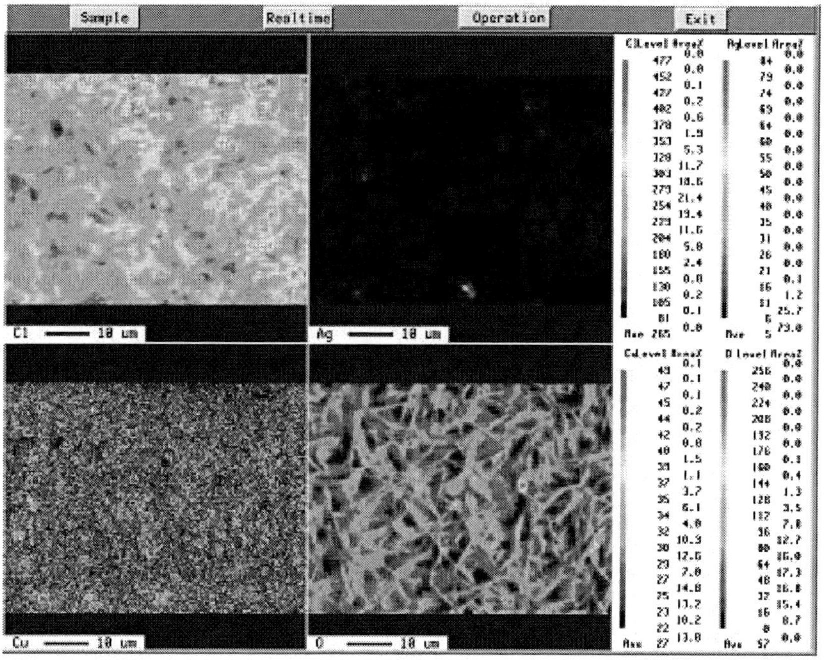

a

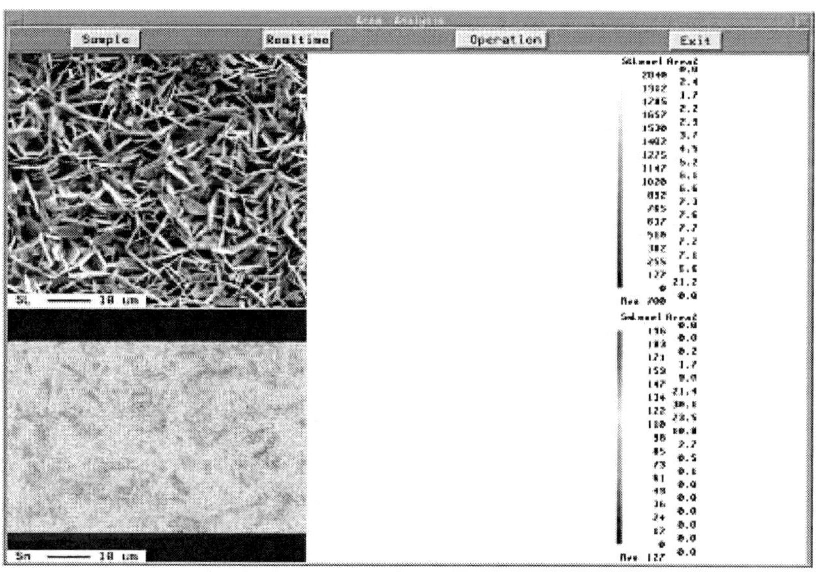

b

Figure 4.5. EPMA results for Sn-4Ag-0.5 Cu alloy potentiodynamically polarized to + 110 mV in 3.5 % NaCl solution. (a) Analysis for elements Cl, Ag, Cu and O (b) Analysis for the element Sn. ((Reprinted from [50], © 2007, with permission from Cambridge University Press).

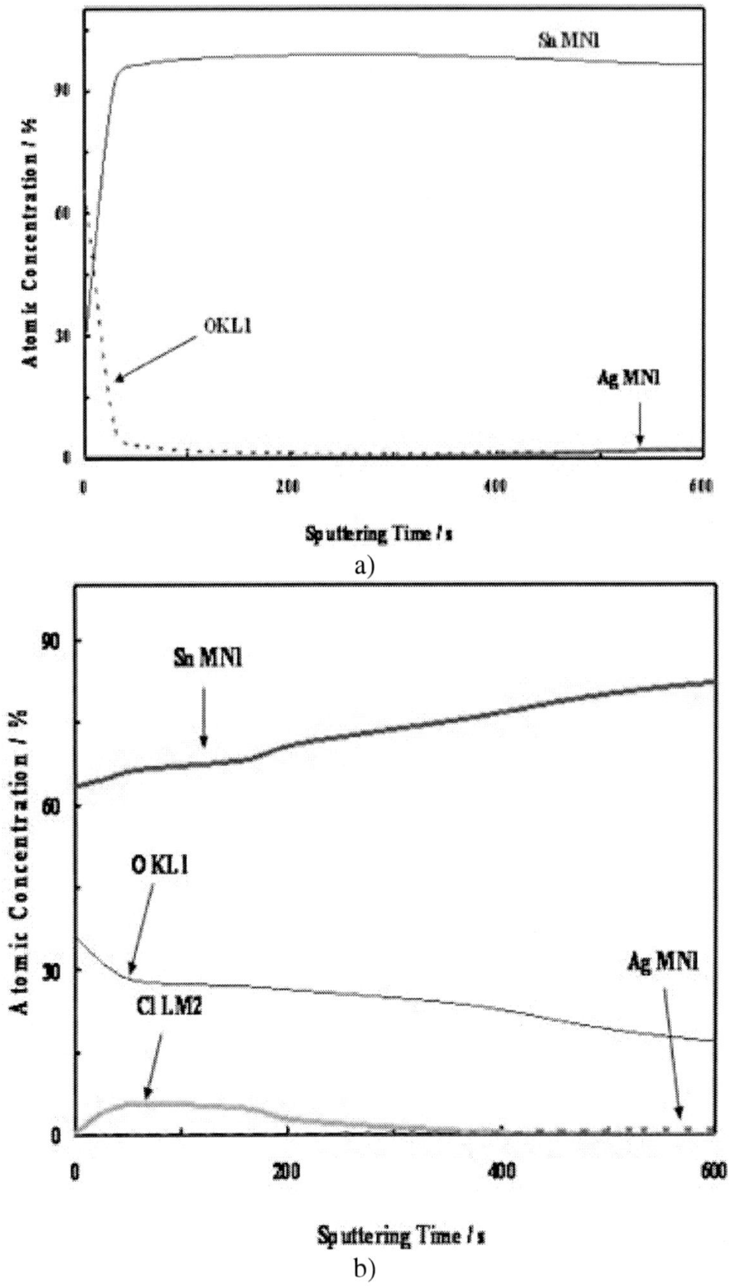

Figure 4.6. (a) Auger depth profile analysis for Sn-4Ag-0.5 Cu alloy potentiodynamically polarised to -636 mV, (b) Polarized to -335 mV (Reprinted from [50], © 2007, with permission from Cambridge University Press).

It is clearly noticed from the figure that the surface of the alloy is composed of 60% tin and around 40% oxygen in atomic concentration units. The atomic concentration of tin shows an initial increase followed by a constant concentration with increase in sputtering time to 600 s. The atomic concentration of oxygen, on the other hand underwent a progressive decrease with gradual increase in sputtering time, indicating the presence of high concentration of native oxide on the surface than in the bulk. The atomic concentration of Cl^- is found to be the lowest, which undergoes an initial rise and then forms a hump before decreasing. This low amount of Cl^- is responsible for the slow pitting behavior on the surface of Sn–4Ag–0.5 Cu alloy at point D.

4.2. ELECTROCHEMICAL CORROSION STUDY OF SN-8.5ZN-XAG-0.1AL-0.05GA SOLDER ALLOYS

Sn–Zn alloy has been reported [43] to be one of the best substitutes for a Sn–Pb solder because its melting temperature i.e. 198^{0}C for Sn–9 wt% Zn eutectic composition is close to that of a Sn–Pb eutectic alloy. The advantages of the Sn–Zn alloy system are its cheaper price, its relative friendliness in terms of health and the environment. Sn–Zn alloys are also known to provide mechanical integrity in electronic packaging but are susceptible to oxidation and corrosion [44–46]. Several authors [47, 48] have put forth their views on the addition of Ag to the Sn-Zn solder system. Small alloying additions of Ag dramatically improved the mechanical properties of the ternary Sn–8.5Zn–5In alloy [47]. Also, addition of Ag to the Sn–9Zn solder alloy inhibited the anodic dissolution of zinc and enhanced the wettability of solder alloy on a Cu substrate [48]. The alloying element Al used in this study is usually employed in the galvanizing industry for improving the corrosion resistance of galvanized coatings for steel [49]. The solder element Ga because of its inherent characteristics of wetting to most metals and oxidizing without the use of flux also plays an important role in the microelectronic packaging industry. This section deals with the corrosion study of Sn-8.5 Zn-XAg-0.1 Al-0.05 Ga by potentiodynamic polarisation methods where the Ag content varies from 0.1 to 2 wt %. The experimental methods used in this study are similar as described in our previous study [50] .

4.2.1. Polarization Study

Potentiodynamic polarisation techniques were used to study the electrochemical corrosion behaviour of Pb-free Sn–8.5Zn–XAg–0.1Al–0.05Ga solder alloy in 3.5% NaCl solution with varying Ag content. A typical potentiodynamic polarisation curve for the above alloy immersed in 3.5% NaCl solution is shown in Figure 4.7. It is noted that increasing the potential upto Point B results in a sharp peak which might be attributed to the dissolution of zinc according to the following reaction

$$Zn + 2\,OH^- \rightarrow Zn(OH)_2 + 2e^- \qquad (8)$$

Scanning Electron (SE) micrograph for the Sn–8.5Zn–1Ag–0.1Al–0.05Ga solder polarized up to the corrosion potential (-1190 mV) along with EDX analysis are shown in Figure 4.8. The oxidation of zinc at Point B exposes the underlying metal Sn present in the alloy. The active dissolution of zinc increases with increasing potential until the zincate concentration reaches a critical value and supersaturates the surface of the alloy (Point C1 in Figure 4.7). A very small peak 'C1' close to the dissolution peak of Zn is noticed in the potential range of -1110 to -1150 mV for 1 wt% Ag which might be attributed to the dissolution of Al from the surface of the alloy. Furthermore, the dissolution of Al takes place at a potential more noble than the dissolution of Zn which is in conformity with the theoretical value [51]. The current density at Point C1 shows a decrease and remains constant with increase in potential up to D for the alloy containing 0.1 wt% Ag. This region is described as the passive region and the lowest current density noted in this region is known as the passivation current density (i_p) which is around 0.02 mA cm^{-2}.

The current density is found to increase with increase in potential beyond D which might be attributed to the active dissolution of Sn to Sn (II) species [52, 53] according to the reactions

$$Sn + 2OH^- \rightarrow Sn(OH)_2 + 2e^- \tag{9}$$

$$Sn(OH)_2 \rightarrow SnO + H_2O \tag{10}$$

A steady state region is formed between E to F for 0.1 wt % Ag where the oxide film attains stability over a certain period of time. However increase in the Ag content to 1 wt%, changes the passivation region to EF where the current density undergoes a decrease at around 0.22 mA cm^{-2}. Further increase in the potential beyond 'F' for 1 wt% Ag resulted in a sharp increase in current density up to G. This increase in current might be due to oxygen evolution [39] or breakdown of the oxide film due to the incorporation of Cl^- in the oxide layer. As reported [54] the penetration of Cl^- ions occurs into the oxide film and causes the mechanical breakdown of the film. Figure 4.9 depicts that an increase in the Ag content from 0.1 to 1 wt% increases the E_{corr} value towards more active values i.e the corrosion potential shifts towards more negative values. These results suggest an increase in anodic dissolution and a higher rate of corrosion of the alloy evident in Table 4.2.

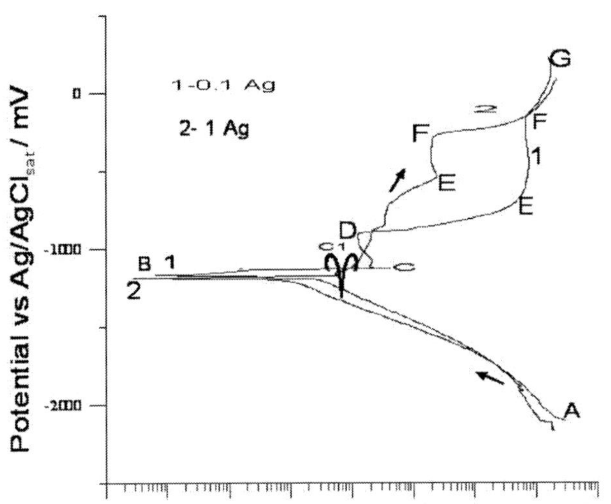

Figure 4.7. Potentiodynamic polarisation curves obtained for Sn-8.5 Zn-XAg-0.1Al-0.05 Ga solder alloy in 3.5 % NaCl solution with varying Ag content.

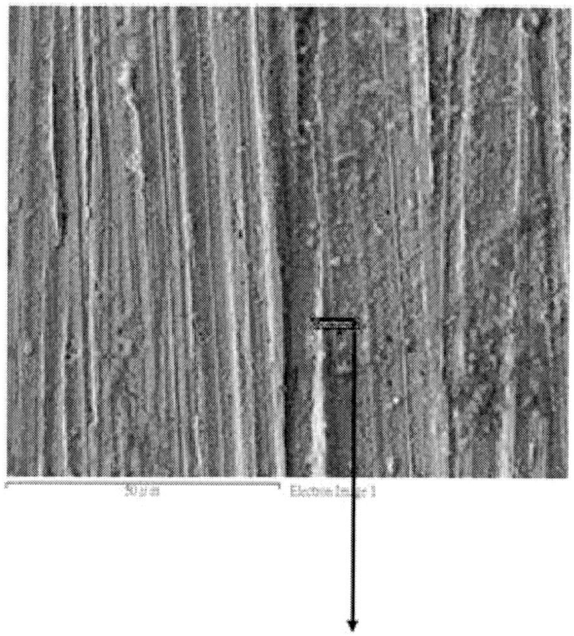

Element	Weight %	Atomic %
O K	13.08	45.10
SnL	53.47	24.85
Zn L	30.90	26.07

Figure 4.8. Scanning electron micrograph for Sn-8.5Zn-1Ag-0.1Al-0.05 Ga solder alloy polarized to -1190 mV.

However the linear polarisation resistance of the Sn–8.5Zn–XAg–0.1Al–0.05Ga solders alloy decreases with increase in Ag content in the solder alloy. Also the corrosion resistance of the alloy does not improve significantly with increase in Ag content from 0.1 to 2 wt %.This can be explained from the dissolution potentials of Ag. The dissolution of Ag starts around +390 mV as derived from the literature studies [55] while the dissolution of Sn–Zn alloy occur around -1000 mV vs. $Ag/AgCl_{sat}$ electrode. Hence the dissolution of Sn–Zn alloy remains unaffected by the presence of alloying element Ag in its composition. Hence, in our present study the decrease in LPR value as noted in Table 4.2 might be attributed to the presence of Ga in the investigated five element alloy.

Table 4.2. Effect of Ag content (wt %) on the electrochemical corrosion parameters of Pb free Sn-8.5Zn-XAg-0.1 Al-0.05 Ga solder in 3.5 % NaCl solution parameters of Pb free Sn-8.5Zn-XAg-0.1 Al-0.05 Ga solder in 3.5 % NaCl solution

Ag / wt %	E_{corr} / mV	I_{corr} / mA $cm^{-2)}$	Corrosion Rate / mm yr^{-1}	LPR/ Ohm Cm^{-2}	i_p / mA cm^{-2}	i_{cc} /mA cm^{-2}
0.1	-1160	1.02	27.36	96.1	1.76	0.69
0.25	-1168	1.16	31.01	84.7	7.35	6.63
0.5	-1183	1.25	33.3	78.8	4.05	4.47
1	-1190	1.3	34.83	75.4	0.22	0.20
2	-1207	1.59	42.5	61.4	2.01	2.75

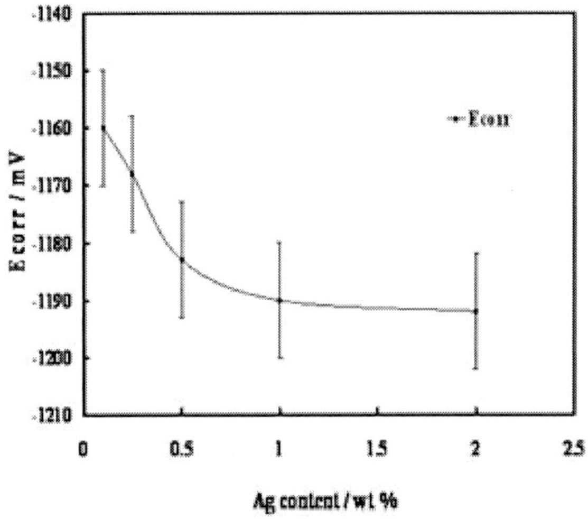

Figure 4.9. Effect of varying Ag content on the E_{corr} value obtained by the potentiodynamic polarisation of Sn-8.5Zn-XAg-0.1 Al-0.05 Ga alloy.

Addition of Ga to alloying element Al has been found [56, 57] to facilitate chloride adsorption and produced localized attack. The adsorption of Cl^- ion was responsible for the breakdown of the oxide film formed on the surface of the alloy.

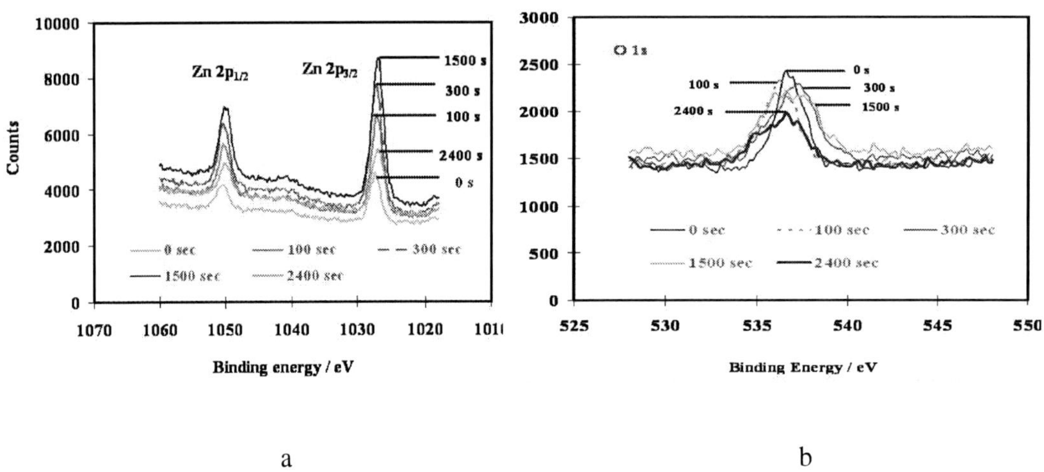

Figure 4.10. XPS spectra for Zn and O obtained by argon ion sputtering of Sn-8.5Zn-1Ag-0.1 Al-0.5Ga alloy polarized to -290 mV (a) Zn 2p (b) O1s.

4.2.2. Surface Characterisation Studies

Surface analytical techniques such as XPS (X-ray photo electron spectroscopy) and SIMS (Secondary Ion Mass Spectrometry) were used to investigate the characteristics of Sn–8.5Zn–XAg–0.1Al–0.05Ga solder alloy polarized to different potentials. For XPS measurements, the surface energy of the adsorbed C on the surface of the specimen is compared with the standard binding energy, 248.8 eV. XPS spectra acquired on the outer surface of the typical Sn–8.5Zn–1Ag–0.1Al–0.5Ga alloy polarized to -1190 mV are shown in Figure 4.10. XPS spectra for Zn 2p revealed the presence of 2 peaks (Figure 4.10.a), one major peak at binding energy (BE) value of 1021.57 eV and another peak at BE value of 1044.8 eV. Similarly one major peak for oxygen with BE value of 530.97 is shown in 4.10 b. The XPS spectra of Zn 2p shown in Figure 4.10.a. consists of two peaks corresponding to a binding energy of 1027.2 (Zn $2p_{3/2}$) and 1050 (Zn $2p_{1/2}$) eV, respectively. The binding energy of major peak Zn $2p_{3/2}$ decreases to a value of 1026.8 eV and a decrease in the peak intensity is observed with increase in sputtering time to 2400 s. The decrease in the intensity of Zn $2p_{3/2}$ peak with increase in sputtering time to 2400 s could be due to the fact that the thickness of the ZnO layer might have decreased with sputtering time. Nevertheless, the XPS spectra of O1s peak revealed a major peak at 536.6 eV denoted by A. The figure shows that the BE value of peak A shifts towards higher energy values with increase in sputtering time to 300 s, further increase in sputtering time to 2400 s did not change the binding energy value. These changes in the binding energy may be attributed to the modification in the oxidation state of metals present at various sputtering depths. XPS studies suggested that Sn and Al did not play a significant role in the formation of oxides as Zn. SIMS concentration profiles of the various elements in the Sn–8.5Zn–XAg–0.1Al–0.05Ga solders polarised to -290 mV are depicted in Figure 4.11. The SIMS depth profile results (Figure 4.11.a) reveal that the the oxygen concentration is high on the surface indicating the spontaneous formation of a thin and adherent oxide layer. It is also clearly noticed that the oxide layer is rich in Zn and Sn but low in Al, Ag and Ga. The results indicate that the alloying element Ag does not form any oxides in the region EF.

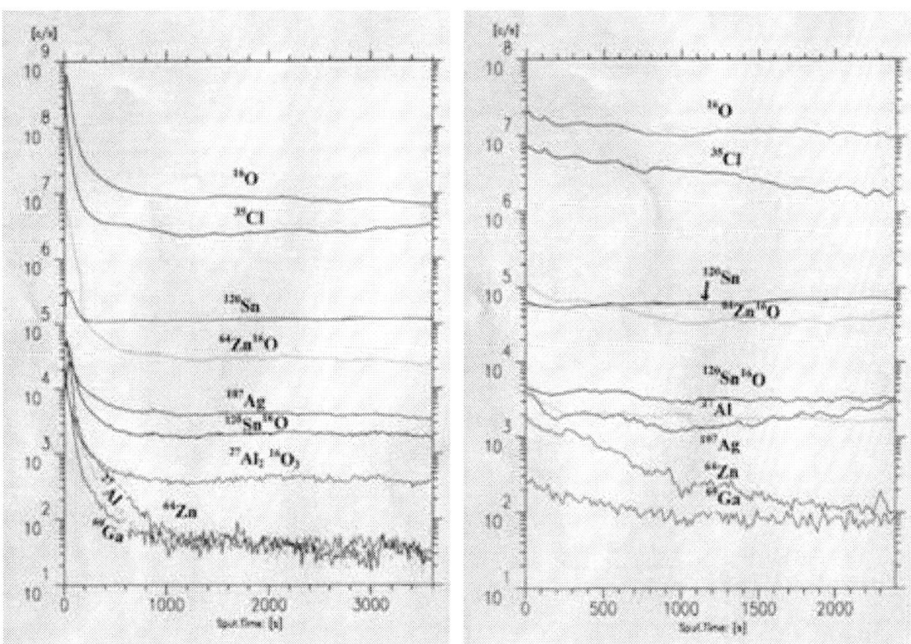

a b

Figure 4.11.a.SIMS depth profile results on Sn-8.5Zn-0.5Ag-0.1 Al-0.05 Ga alloy polarized to -290 mV; b. -329 mV.

The SIMS depth profile for 2 w% Ag shows (Figure 4.11.b) that there is a significant reduction of oxygen, ZnO and Sn as compared to the alloys containing 0.5 and 1 wt% Ag. These results suggested that increase in Ag content reduces the concentration of oxides thereby affecting the stability of the oxide film. SIMS depth profile could establish the presence of Ag, Ga in the polarized solder alloy, though present in low concentration. Surface characterisation by SIMS demonstrated that alloying elements like Ag, Ga might be present in the subsequent layers but does not form any oxides on the outer surface of the alloy.

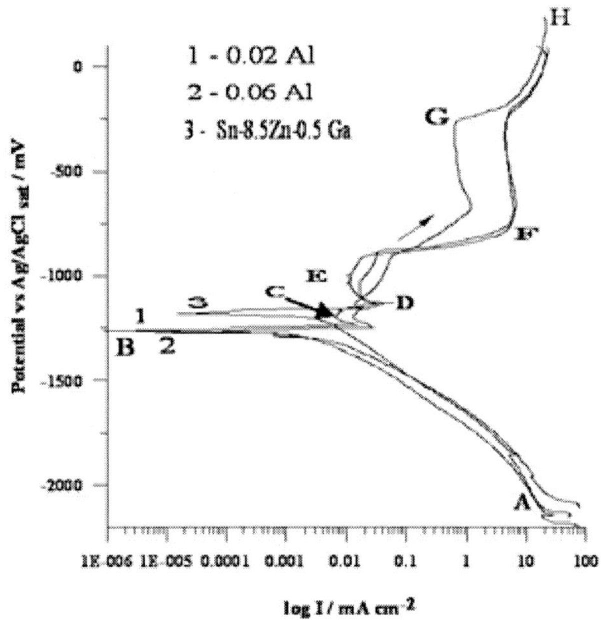

Figure 4.12. Potentiodynamic polarization curves obtained for Sn-8.5Zn-0.5 Ga and Sn-8.5Zn-XAl-0.5Ga alloys in 3.5 % NaCl solution.

4.3. Electrochemical Corrosion Study of Sn-8.5Zn-X Al-0.5 Ga Solder Alloys

Several investigations have been carried out on lead-free solder alloys such as Sn-Zn, Sn-Ag alloys. Among them, Sn-Zn eutectic alloy has emerged as a promising lead-free solder alloy in the recent years [58]. It has attracted much attention because of its low cost. More importantly, it has a melting point (198.5°C) very close to conventional Sn-37wt.% Pb alloy (183°C) which is much lower than that of other Sn-based alloys e.g. Sn-Cu (227°C),Sn-Ag (221°C) and Sn-Ag-Cu (217°C) [59].And its tensile stress and creep resistance are better than that of Sn-37 Pb alloys[57].Several investigations have shown that addition of third element such as Al [60], In [61] improved the deficiencies in the Sn-9Zn binary alloys. Polarisation studies on Sn-8Zn-3Bi-XCr [62] solder revealed that addition of small amount of Cr improved the corrosion resistance of Sn-Zn-Bi alloy. It shifted the corrosion potential towards more noble values and these changes were also reflected in the current density of solder alloy in passivation, corrosion rate, respectively. Corrosion studies on the Sn-Zn-Al [18] and Sn-Zn-Al-In [19] alloys have established that Sn–Zn–Al alloy underwent more active corrosion than 63Sn–37Pb alloy. Nevertheless, 5In–9 (5Al–Zn–Sn) alloy was slightly more susceptible to corrosion than 10In–9 (5Al–Zn–Sn) alloy. The polarisation behavior of these two alloys was similar to that of 9 (5Al–Zn-Sn) alloy.This section chapter describes the potentiodymic polarisation measurement of Sn-8.5 Zn-XAl-0.5Ga alloy in 3.5% NaCl solution.

4.3.1. Polarisation Study

Figure 4.12 displays the potential–current density profiles for Sn–8.5Zn–0.5Ga alloy and Sn–8.5Zn–XAl–0.05Ga alloy potentiodynamically polarised to +100 mV. Point B is referred to as E_{corr} or corrosion potential as mentioned before, where the extrapolated anodic and cathodic Tafel slopes intersect and the current becomes zero. A sharp rise in current is seen at point B which is attributed to the dissolution of zinc as governed by the following reactions [63, 64]:

$$Zn + OH^{-} \rightarrow Zn\,(OH)_{ads} + \_2e^{-} \quad (11)$$

$$ZnOH_{ads} + 2OH- \rightarrow Zn\,(OH)_3^{-} + e^{-} \quad (12)$$

$$Zn\,(OH)_3^{-} + OH^{-} \rightarrow Zn\,(OH)_4^{2-} \quad (13)$$

$$Zn\,(OH)_4^{2-} \rightarrow Zn(OH)_2 + 2OH^{-} \quad (14)$$

$$Zn + 2\,OH^{-} \rightarrow ZnO + H2O + 2e- \quad (15)$$

A very small current density peak C is noticed very close to the end of the dissolution peak for the polarization curves 1 and 2. This peak might be attributed to the dissolution of Al [65] which occurs by the following steps

$$Al\,(s) + OH^{-} \rightarrow Al\,(OH\,)_{ads} + e- \quad (16)$$

$$Al\,(OH)_{ads} + OH- \rightarrow Al\,(OH)_{2\,ads} + e- \quad (17)$$

It can be seen that the active dissolution of Al continues with increasing potential until the hydroxide concentration reaches a critical value and supersaturates the surface of the alloy at point D (Figure 4.12). Further increase in potential in the anodic direction results in the formation of very broad current density peak E, which decreases with the increase in Al content in Sn–8.5Zn–XAl–0.5Ga alloy (Figure 4.12). This peak might be attributed to the dissolution of Ga and formation of gallium hydroxide on the surface of the alloy. The possibility of formation of Ga $(OH)_3$ is further supported by the high hydration energy of gallium (-4.6×10^3 kJ/ mol) [66], hydrophilicity [67] and the susceptibility of Ga^{3+} aqueous ions to hydrolysis [68]. The transition from active dissolution occurs at the passivation potential (point F, Figure 4.12) where the solid species, oxides/hydroxides of Zn, Ga, and Al become more stable than the parent metal ion. These oxides/hydroxides cover the surface of the electrode and passivation thus begins to set in, thereby decreasing the current density with further increase in potential to point G .The region FG is not really a passive region; however, a slight decrease in current density is observed which remains constant with increase in potential in the anodic direction, indicating slight passivation behavior. The ip and i_{cc} values obtained for Sn–8.5Zn–0.5Ga alloy are found to be lower than the values obtained for Sn–8.5Zn–XAl–0.5Ga alloys. The four-element alloys (Table 4.3) containing Al in the range of 0.02–0.05 wt % show lower i_p and i_{cc} values, thus indicating the stability of the passive film at this content. It has been established [69] that lower values of ip and i_{cc} correspond to the stability and protective nature of the passive film. However, increase in Al content beyond 0.05 wt % increases the passivation current density (ip), thereby destabilizing the passive film and resulting in a higher rate of corrosion of the alloy (low LPR value, Table 4.3). The passivation behavior thus observed at low content of Al might be attributed to the formation of oxides of gallium and zinc on the outer surface of the alloy. Metallic gallium [70] present in the outermost layers of the deposit is usually readily oxidized during air atmosphere. The alloying element Al and Sn contributes a little to the formation of passive film on the surface. A large increase in current density with small increase in the anodic potential is observed beyond FG which indicates the transpassive behavior of the alloy and the potential is referred as the breakdown potential. The transpassive region is usually associated with the evolution of a gas, particularly oxygen, or metal dissolution or breakdown of the passive film, leading to localized corrosion. The breakdown of the passive film occurs in presence of aggressive chloride Cl^- ion. In general, aggressive halide ions [71] penetrate the oxide layer and get incorporated as an impurity thereby altering their physical and electrical properties. The E_{bd} value in Table 4.3 shows that Sn–8.5Zn–0.5Ga alloy is found to be almost similar to the value obtained for Sn–8.5Zn–0.02Al–0.5Ga alloy. The values are found to shift toward more negative values with further increase in Al content to 0.5 wt % in the four element alloy.

Pure Al can be activated [57] in solutions containing gallium ions and can shift the operating potential toward more negative values, thus facilitating the increased adsorption of Cl^- ions. Hence, presence of a high content of Al in the four-element alloy causes faster breakdown of the passive film due to the increased adsorption of Cl^- ions. The corrosion products formed after the end of the electrochemical experiment (+100 mV) up to H (Figure 4.13) were examined by SEM which reveals a mixture of needle and sheet like structures with numerous pores and openings on the surface. These results were confirmed by EDX analysis thereby indicating the presence of Sn, Cl, and O in the analyzed specimen. Tin chloride formed in our study reacted with oxygen to form tin oxychlorides, which is consistent with the previous results [72].

Table 4.3. Effect of Al content (wt %) on the corrosion parameters obtained by polarization of Sn-8.5Zn-XAl-0.5Ga solder in 3.5 % NaCl solution

Al content / wt %	E_{corr} / mV	I_{corr} / mA cm^{-2}	Corrosion rate / mm $year^{-1}$	LPR / Ω cm^{-2}	i_{cc} / mA cm^{-2}	i_p / mA cm^{-2}	E_{bd} /mV
0.02	-1254	1.13	30.3	86.7	1.79	1.21	-271
0.05	-1260	1.23	32.8	80.0	1.97	1.31	-276
0.06.	-1268	1.41	37.6	69.8	2.82	1.79	-294
0.1	-1272	1.48	39.5	66.4	5.59	4.37	-321
0.5	-1296	1.88	50.1	52.4	5.92	4.6	-359

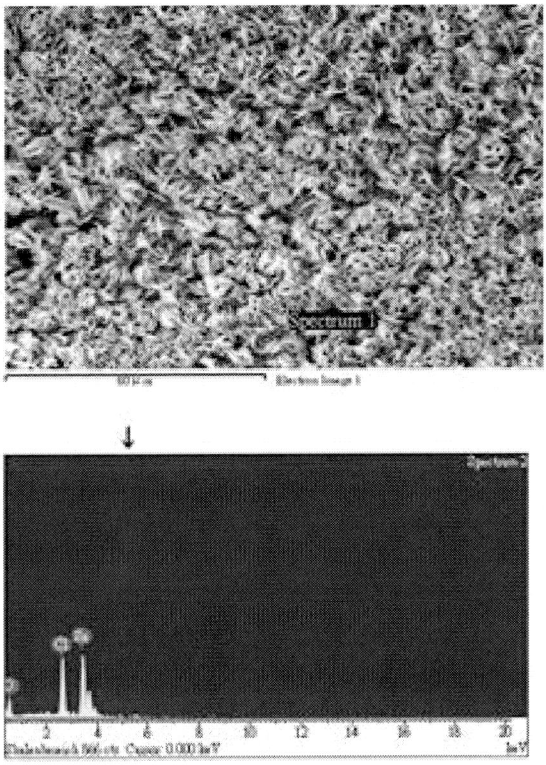

Figure 4.13. SE micrograph and EDX spectrum for Sn-8.5 Zn-0.02Al-0.5Ga solder polarised upto +100 mV.

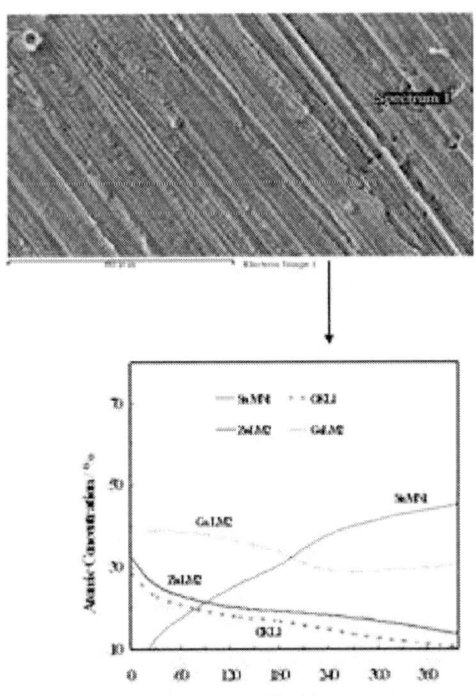

Figure 4.14. Auger depth profile for Sn-8.5 Zn-0.02Al-0.5Ga solder polarised upto -1020 mV.

The corrosion potential (E_{corr}) and the corrosion current density (I_{corr}) for Sn–8.5Zn–0.5Ga alloy are found to be −1174 mV and 1.01 mA cm^{-2}, respectively in Table 4.3. Nevertheless, an increase in the Al content from 0.02 to 0.5 wt % in the Sn–8.5Zn–XAl–0.5Ga alloy progressively increases the corrosion current density (Icorr) and shifts the corrosion potential toward more negative values. Similar shift in the corrosion potential was observed [73] while studying the electrochemical behavior of aluminium alloys in the presence of small amounts of gallium and indium. Hence an enhancement in the corrosion rate of the alloy was observed with increase in the Al content in Sn–8.5Zn–XAl–0.5Ga alloy.

4.3.2. Surface Characterisation Studies:

Auger electron spectroscopy (AES) was employed to examine the corrosion behavior of the alloy at specified potentials of interest during the polarization studies.Thc atomic concentration depth profiles of elements, i.e., Sn, Zn, O, and Ga plotted vs the sputtering time, for the Sn–8.5Zn–0.02Al–0.5Ga alloy polarised to point E (Figure 12) i.e −1020 mV is depicted in Figure 4.14. The Auger depth profile was performed at the point referred as spectrum 1 in the SE micrograph. Auger depth profile analysis revealed that the atomic concentration of Ga is around 40% on the outer surface and the atomic concentration of Zn and O are around 30%. The profile also demonstrates that very little Sn has segregated to the surface along with Zn and Ga. Since the Ga concentration is highest among other elements, it is clear that gallium oxide must be the dominant oxidized species on the surface of the alloy, followed by zinc. These results confirm the formation of Ga_2O_3 on the surface of the four element alloy potentiodynamically polarised to point E. Nevertheless when the sputtering time reaches 480 s there is a gradual increase in the concentration of Sn which finally reaches around 90% with increase in sputtering time to 2400 s. These results suggest that oxidation of Ga and other elements from the outer surface exposes the underlying metal Sn present in the bulk of the alloy.

Figure 4.15. displays the Auger depth profile analysis for Sn–8.5Zn–0.02Al–0.5Ga alloy, polarized to point F −604 mV. It is clearly noticed from the figure that the atomic concentration of Zn and O are very high on the surface but subsequently decreases with increasein sputtering time to 2400 s. The atomic concentration of Al undergoes a slight decrease upto sputtering time of 100 s followed by a continuous increase with increase in sputtering time to 300 s, then a gradual decrease. These results indicate that Al has not completely oxidized from the surface; it is still present in the sputtered layers. On the other hand, the atomic concentration of Ga increases sharply up to a sputtering time of 480 s followed by a sharp decrease in the peak, indicating that all the Ga has not reached the outer surface of the alloy; some are still present in the sputtered layers. The atomic concentration of Sn is very low on the outer surface which increases progressively to around 90% with increase in sputtering time to 2400 s. These findings establish that the oxides of Zn, Al, Ga, and Sn tend to accumulate near the passivation potential F before the formation of the passive film.

Figure 4.16 reveals the Auger depth profile for Sn–8.5Zn–0.02Al–0.5Ga alloy polarized to the end of the passive region G (−271 mV). The atomic concentration of Ga and O are found to highest, i.e., around 17–20%, on the outer surface of the alloy, indicating the possible existence of Ga_2O_3. It is also observed that the concentration.of Zn decreases with an increase in sputtering time to 2400 s, indicating that all the Zn might have segregated to the surface and was oxidized as ZnO. However, Al does not get completely oxidized from the outer surface; it is still present in the sputtered layers. The stability of the oxides of Ga, Zn, Al, and Sn, can be determined from their corresponding Gibbs free-energy values $\Delta G^0{}_F$ [74] for $Ga_2O_3 = -998.3$ kJ/mol, $\Delta G^0{}_F$ for $Al_2O_3 = -1582.3$ kJ/mol, and $\Delta G^0{}_F$ for $ZnO = -320.5$ kJ/mol.

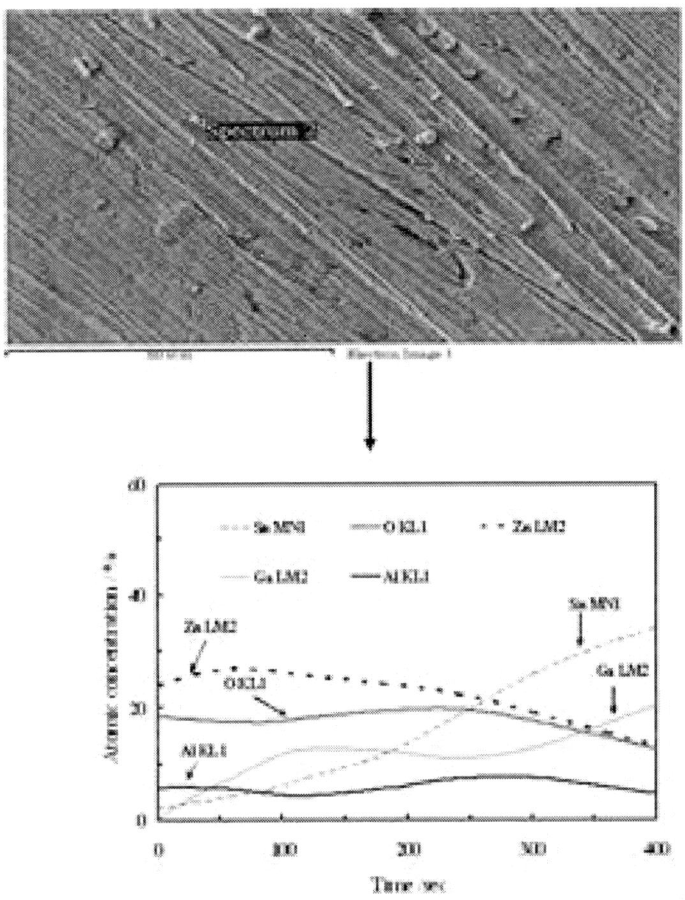

Figure 4.15. Auger depth profile for Sn 8.5Zn-0.02Al-0.5Ga solder polarized upto -604 mV.

It can be opined from the free-energy values that the formation of ZnO is more thermodynamically unfavorable and could not be formed as easily on the surface as the other two oxides cited above. Ga_2O_3, being more stabler than ZnO, does not get completely oxidized and still remains in the sputtered layers.

Although Al_2O_3 corresponds to the most stable oxide with highest $\Delta G^0{}_F$, its atomic concentration on the surface is found to be low as compared to Zn and Ga. This confirms that the entire Al has not oxidized from the outer surface; some are present in the sputtered layers. The outer surface of the passive film is thus primarily composed of a high content of Ga_2O_3 and a small amount of ZnO. Tin oxide, because of its highest Gibbs free energy ΔG^0 F for $SnO = -251.9$ kJ/ mol is unstable and is formed in very low concentration on the surface. It is, however, present in the sputtered layers and represents the base metal.

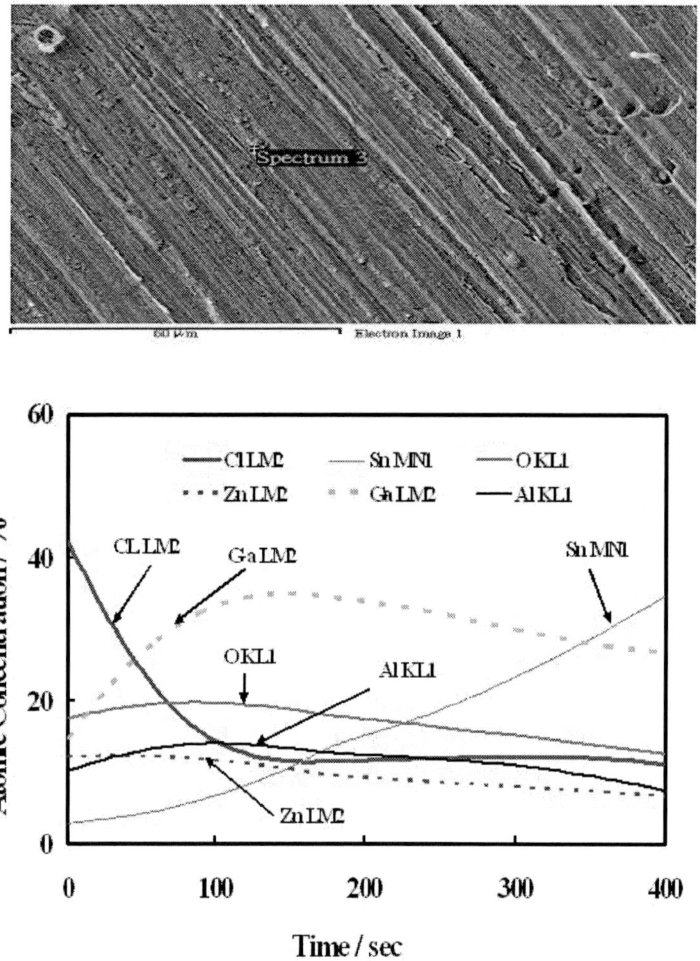

Figure 4.16. Auger depth profile for Sn-8.5 Zn-0.02Al-0.5Ga solder polarized upto -271 mV.

The presence of a high concentration of Cl^{-} observed on the outer surface (Figure 4.16) is responsible for the breakdown of the passive film and lowering of corrosion resistance of the alloy, especially for higher Al content. Secondary ion mass spectrometry technique was used to determine the formation of different oxides in the passive film. SIMS depth profile analysis (Figure 4.17) for the Sn–8.5Zn– 0.02Al–0.5Ga alloy polarized to the end of passive region G revealed that chloride ions (Cl^{-})is present in very high concentration on the surface of the alloy, which can destroy the passivity and result in breakdown of the passive film.

However, zinc oxide though present in high concentration on the outer surface decreases with increase in sputtering time. These results indicate that most of the Zn might have segregated to the surface and have oxidized. Ga_2O_3 is also formed on the outer surface of the alloy; nevertheless its atomic concentration remains constant with increase in sputtering time to 3500 s, indicating that some Sn is also present in the sputtered layers.

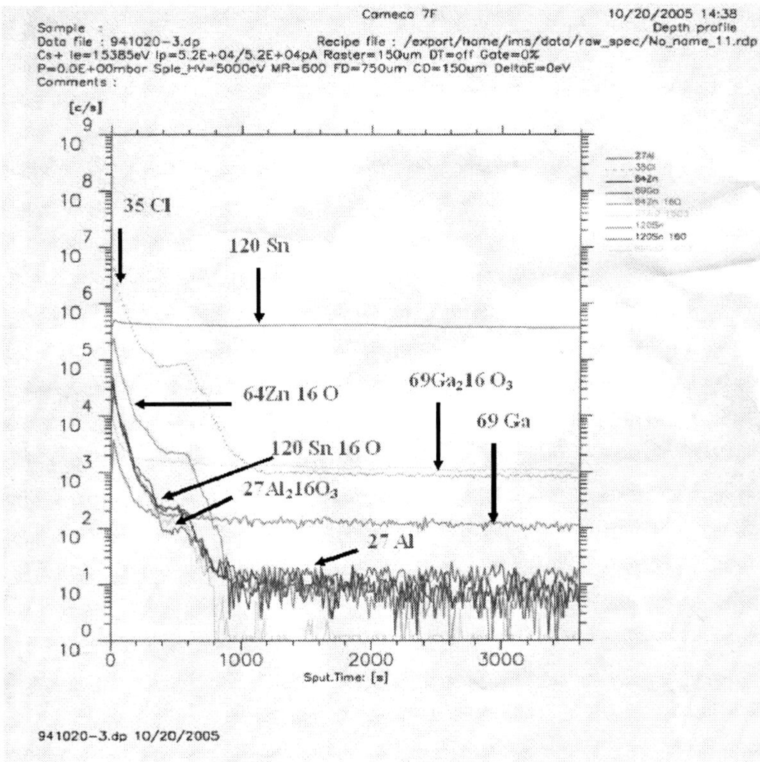

Figure 4.17. SIMS depth profile for Sn-8.5Zn-0.02Al-0.5Ga solder polarized to the end of passive region G.

CONCLUSION

The following conclusions can be made from the study.

(a) Potentiodynamic polarisation studies on Sn–XAg–0.5 Cu alloys in 3.5% NaCl solution revealed that an increase in Ag content in the alloy beyond2 wt % increases the ability towards passivation. Further increase in the Ag content to 4 wt % also results in a decrease in the corrosion current density and corrosion rate of the alloy. However, the linear polarisation resistance (LPR) increases and the corrosion potential (E_{corr}) shifts toward more noble values. Surface characterization studies by XPS and AES revealed that the outer surface of the passive layer is composed of a mixture of both SnO and SnO_2. The presence of chloride ion (Cl^-) on the surface of the alloy initiates pitting and is also responsible for the breakdown of the passive film. The breakdown potential (Ebd) also underwent a shift towards more positive values with increase in Ag content in the alloy.

(b) Electrochemical corrosion studies on Sn–8.5Zn–XAg–0.1Al–0.05Ga solder demonstrated that an increase in the Ag content from 0.1 to 2 wt% progressively increases the corrosion current density and shifts the corrosion potential towards more negative values. The linear polarisation resistance (LPR) values showed a decrease with increase in the weight content of Ag. Increase in the Ag content from

0.1 to 2 wt % did not have a significant effect on the passivation behaviour of the five element alloy. XPS and SIMS depth profile results reveal that the oxides of Zn and Sn are present on the outer surface of the solder alloy with very low concentration of Al_2O_3.

(c) Polarisation studies made on Sn–8.5Zn–XAl–0.5Ga solder alloy established that the corrosion current density (I_{corr}) and corrosion rate of the alloy are found to be lower for Al content in the range of 0.02–0.05 wt %, however, it undergoes a significant decrease with increase in Al content beyond 0.05 wt %. Auger depth profile and SIMS analysis results show that the outer surface of the passive film is primarily composed of a high concentration of gallium oxide and a low concentration of zinc oxide. The other oxides aluminum oxide and tin oxide contributes a little to the formation of the passive film on the surface. Increase in the Al content to 0.5 wt % shifts the breakdown potential toward more negative values and facilitates the increased adsorption of Cl^- ions in the passive layer.

REFERENCES

[1] W.H.Zhong, Y.C. Chan, M.O.Alam, B.Y.Wu and J.F.Guan, *J. Alloys Compd.* 414, 123 (2006).
[2] R.A. Islam, B.Y. Yu, M.O.Alam, Y.C.Chan and W.Jillek, *J. Alloys Compd.* 392, 149 (2005).
[3] K. K. Mohan, V. Kripesh, L. Shen, K.Zeng and A.A.O, Tay, *Mat. Sci. Eng A*. 423, 57 (2006).
[4] US Environmental Protection Agency, 1991, advanced Notice of Rulemaking.
[5] US Environmental Protection Agency, 1991, Comprehensive Review of lead in the Environmental under TSCA, 56 FR 22096-98.
[6] P.T. Vianco and D.R. Frear, *J. Met.* 47, 14 (1993).
[7] C.H.Raeder, L.E. Felton, V.A. Tanzi, and D. B. Knorr, *J. Electron. Mater.* 23 (7) (1994) 611.
[8] L. E. Felton,C. H. Raeder and D. B. Knorr, *J. Met.* 47, 62 (1993).
[9] W. L. Winterbottom, *J. Met.* 47, 20 (1993).
[10] W. J. Tomlinson and I. Collier, *J. Mater. Sci.* 22, 1835 (1987).
[11] M. Harada and R. Satoh, IEEE Trans. *Comon. Hybr. Manuf. Technol.* 13 (1990) 736.
[12] G. Ghosh, M. Loomans and M. E. Fine, *J. Electron. Mater.* 23, 619 (1994).
[13] M. McCormack, S. Jin, H. S. Chen and D. A. Machusak, *J. Electron. Mater.* 23, 687 (1994).
[14] M. Abtew and G. Selvaduray, *Mater. Sci. Eng. Rep.* 27, 95 (2000).
[15] L. Allegra and J. C. Zoccola, *Mater. Perform.* 22, 5 (1993).
[16] M. Mori, K. Miura, T. Sasaki and T. Ohtsuka, *Corros. Sci.* 44, 887 (2002).
[17] H. Oulfajrite, a. Sabbar, M. Boulghallat, A. Jouaiti, R. Lbibb and A. Zrineh, *Mater. Lett.* 57,4368 (2003).
[18] K. L. Lin and T. P. Liu, *Mat. Chem. Phys.* 56, 171 (1998).
[19] K. L. Lin, F. C. Chung and T. P. Liu, *Mat. Chem. Phys.* 53, 55 (1998).

[20] H.Qiu,Polarisation Curves of Passivating Metals, www. Ntu.edu.sg/home/asjqiu/me/ME303 L11. HTM (Accessed June, 2004).
[21] Gamry Instruments, PA: Electrochemical Impedance Spectroscopy Primer, www.gamry.com/ App-Notes/EIS_ Primer/ EIS Primer.pdf (accesed July, 2005).
[22] Corrosion Doctors' Site Map: Polarization Diagram of a Passivable System, www.Corrosion-doctors.org/index.htm (accesed February, 2004).
[23] W.C. Luo, C. E. Ho, J. Y. Tsai, Y. L. Lin and C. R. Kao, *Mat. Sci. Eng. A* 396, 385 (2005).
[24] L.C. Shiau, C. E. Ho and C. R. Kao, solder. *Surf. Mount.Technol.* 14, 25 (2002).
[25] J. Y. Tsai, Y. C. Hu, C. M. Tsai and C. R. Kao, *J. Electron. Mater.*32, 1203 (2003).
[26] I. E. Anderson, *J. Mater. Sci. Mater. Electron.* 18, 55 (2007).
[27] D. Z. Li, C. Q. Liu and P. P. Conway, *Mat. Sci. Eng. A.* 391, 95 (2005).
[28] J. W. Yoon, S. W. Kim and S. B. Jung, *J. Alloys Compd.* 392, 247 (2005).
[29] K. W. Moon, W. J. Boetinger, U. R. Kattner, F. S. Biancaniello and C. A. Handwerker, *J. Electron. Mater.* 29,1122 (2000).
[30] Y. J. Park, C. U. Kim, T. Carper and V. Puligandla, *J. Electron. Mater.* 32, 1297 (2003).
[31] K. S. Kim, S. H. Huh and K. Suganuma, *J. alloys Compd.* 352, 226 (2003).
[32] F. Zhu, H. Zhang, R. Guan and S. Liu, *J. Alloys Compd.* 438, 100 (2007).
[33] Corrosion, Vol. 13, 9th Ed., Metals Handbook (ASM International, Metals Park, OH, (1987).
[34] H. Tanaka, F. Ueta, S. Yoshihara and T. Shirakashi, *Mater. Trans.*42 (9), 3401 (2001).
[35] S. A.M. Refaey and S. S. Abd El-Rehim, *Electrochim.Acta.* 42, 667 (1997).
[36] J. Kirstein; Atlas of metal ligand equilibrium in aqueous solutions. Chester, p.623 (1978).
[37] R. Mishra and R. Balasubramanium: *Corros. Sci.* 46, 3019 (2004).
[38] L. Zhong, H. Zhu, J. Hu, S. Xiao, and F. Gau. *Electrochim. Acta* 51, 5494 (2006).
[39] V. B. Singh and A.Gupta, *Mater. Chem. Phys.*85 (1), 12 (2004).
[40] A. Kolics, J.C. Polkinghorne, and A. Wieckowski, *Electrochim. Acta* 43, 2605 (1998).
[41] J.A. Ricardson and G.C. Wood: *Corros. Sci.* 10, 313 (1970).
[42] T.C. Chang, M.H. Hou, M.C. Wang, and D.Y. Lin: Electrochemical behaviors of the Sn–9Zn–XAg lead-free solders in a 3.5 wt % NaCl solution. *J. Electrochem. Soc.* 151(7), C484 (2004).
[43] M. Hansen, *Construction of Binary Alloys,* Mc-Graw Hill Book Co. Inc, New York (1985).
[44] Y. Chonau, T. Komiyama, J. Muki, R. Unao, T. Kimura and T. Nagano, *Mater. Trans.* 43 (8) 1887 (2002).
[45] S.P. Yu, H.J. Liu, M.H. Hou and M.C. Wang, *J. Mater. Sci.* 11 (6), 461 (2000).
[46] *NCMS, Lead-free solder Project Final Report, NCMS Report 0401 RE 96,* National Centre for Manufacturing Sciences, Michigan (1997).
[47] M. McCormack, *J. Electron. Mater.* 23, 715 (1994).
[48] T. Takemoto, T. Funaki and A. Matsunawta, *J. Jpn. Welding Soc.* 17, 251 (1994).
[49] L. Allegra, J.C. Zoccola, *Mater. Perform.* 18 (10), 13 (1979).
[50] U. S. Mohanty and K.L. Lin, *J. Mater. Res.* 22 (9), 2573 (2007).
[51] X. Zhang, S. Lorusso, A. Miotello, L. Guzman, E. Cattaruza, P.L. Bouora and L. Benedetti, *Surf. Coat.Technol.* 141, 187 (2001).

[52] B.N. Stirrup and N.A. Hampson, *Surf. Technol.* 5, 429 (1977).
[53] M. Drogowska, H. Menard and L. Brossard, *J. Appl. Electrochem.* 21, 84 (1991).
[54] P.C. Pistorius and G.T. Burstein, *Corros. Sci.* 38, 1525 (1994).
[55] H. Tanaka, *Espec Technology Report*, 14, pp. 1–9 (2002).
[56] C.B. Breslin and W.M. Carroll, *Corros. Sci.* 33 1735 (1992).
[57] H.A. El Shayeb, F.M. Abd El Wahab, S. Zein El Abedin, *Corros. Sci.* 43, 643 (2001).
[58] C. M. Chen, Y. M Hunga and C. H. Lina,J. All and Compd.475, 238 (2009).
[59] T. Ichitsubo, E. Matsubara, K. Fujiwara, M. Yamaguchi, H. Irie, S. Kumamoto and T. Anada, *J. Alloys Compd.* 392 , 200, (2005),
[60] K.I. Chen, S.C. Cheng, S. Wu and K.L. Lin, *J. Alloys Comp.* 416, 98 (2006).
[61] K.L. Lin, T.P. Liu, *Oxid.Met.* 50 , 255 (1998),.
[62] M. McCormack, S. Jin, H.S. Chen, D.A. Machusak, J. Electron. Mater. 23 , 687 (1994),.
[63] J. Hu, A. Hu, M. Li, D. Mao , *International Conference on Electronic Packaging Technology and High Density Packaging (ICEPT-HDP)* 701 (2009)..
[64] W. Furbeth and M. Stratmann, *Corros. Sci.*, 43, 207 _2001_.
[65] M. Pourbaix, *Atlas of Electrochemical Equilibria*, Permagon Press, Toronto, Canada p.404 (1966).
[66] W. A. Badawy, F. M. Al-Kharafi, and A. S. El-Azab, *Corros. Sci.*, 41, 709 (1999).
[67] N. V. Federovich, *Elektrokhimiya*, 14, 5 (1979)
[68] V. P. Vasilev, E. K. Zolotarev, and A. F. Kapustinskii, *Zh. Fiz. Khim.*, 34, 1762 (1960).
[69] C. F. Baes and R. E. Mesmer, *The Hydrolysis of Cations*, Wiley, New York (1976).
[70] D. P. Butt, Y. S. Park, and T. N. Taylor, *J. Nucl. Mater.*, 264, 71 (1999).
[71] S. M. Abd El Haleem, *Br. Corros. J., London*, 11, 215 (1976).
[72] M. Windholz, S. Budavali, R. F. Blumeti, and E. S. Otterbin, Merck Index, 10th ed., *Merck and Company*, Inc., Rahway, NJ, p. 1257 (1983).
[73] A. R. Despic, D. M. Drazic, M. M. Purenovic, and N. Cikovic, *J. Appl.Electrochem.*,6, 527 (1976).
[74] David R. Lide, *Handbook of Chemistry and Physics*, CRC press, Boca Raton, FL 73rd ed., pp. 5–75 (1993).

In: Electrodeposition: Properties, Processes and Applications ISBN: 978-1-61470-826-1
Editor: Udit Surya Mohanty

Chapter 5

PROPERTIES AND APPLICATIONS OF NICKEL COATINGS SYNTHESIZED BY PULSE ELECTRODEPOSITION

Guozhe Meng,[1, 2,*] Feilong Sun[1], Yang Li[1] and Fuhui Wang[1, 2]
[1]Corrosion and Protection Laboratory, Key Laboratory of Superlight Materials and Surface Technology (Harbin Engineering University)
Ministry of Education, Harbin, China
[2]State Key Laboratory for Corrosion and Protection, Institute of Metal Research, Chinese Academy of Sciences, Shenyang, China

ABSTRACT

Coatings with less defects and high corrosion resistance can be obtained by pulse electrodeposition, which can be divided into two categories according to the applied current waveform. (1) Pulse current electrodeposition (PC) employed the pulses in one direction; (2) Pulse reverse current electrodeposition (PRC) employed the anodic and cathodic pulses. The results on the pulse electrodeposition studies carried in our lab in recent years has been summarized in this chapter. In one of the investigations, a nickel coating with high density nano-scale twins (NT) was synthesized by using pulsed electrodeposition technique. The passive films formed on nickel coatings showed higher pitting corrosion resistance and a bi-layer semi conducting structure distribution, compared with that of industrial electrodeposited nickel. The passive films behaved like p-type semiconductors at lower potentials, but they behaved like n-type semiconductors at higher potentials. The coating exhibits high potential as a substitute material in the processing industry, power plant. In another study the inhibition of hydrogen blister on Cu–Sn alloy coatings was demonstrated. The Cu–Sn alloy coatings were synthesized on 27SiMn steel by direct current (DC), pulse current (PC) and pulse reverse current (PRC) electrodeposition techniques from pyrophosphate-based electrolyte. The hydrogen permeation amount of different electrodeposition techniques decreased in the order: DC > PC > PRC. Frequency and duty cycle have significant effect on the sub-surface concentration of atomic hydrogen and hydrogen permeation amount. The appropriate

* E-mail address: mengguozhe@hrbeu.edu.cn.

PRC technique can prevent or greatly decrease the occurrence of hydrogen blister during the industrial electrodeposition process.

INTRODUCTION

Pulse electrodeposition can be divided into two categories according to the applied current waveform. (1) Pulse current electrodeposition (PC) employed the pulses in one direction; (2) Pulse reverse current electrodeposition (PRC) employed the anodic and cathodic pulses. Typical waveforms include: sine-wave pulses, square-wave pulses, Duplex pulses, Pulse-on-Pulse and cathodic pulses followed by anodic pulses etc. Among these, the square-wave pulses are used widely [1].

Pulse and pulse reverse electrodeposition have many advantages compared with direct current electrodeposition (DC):

(1) PC raises the limiting current density [2].
(2) Coatings with desired structure, composition and property can be obtained by modifying pulse parameters [3-5].
(3) PC can reduce the additive requirement by 50-60%. PRC can enhance bath stability and efficiency with negligible additive consumption [6].
(4) PRC can eliminate thickness and porosity of coating due to the existence of reverse current.The application of PC and PRC electrodeposition is extensively used because of its advantages. It can be used to prepare coatings of metals, alloys, composites and semiconductors etc. The two examples in this chapter are just preparation of coatings with desired structure and property using PC and PRC.

5.1. THE SYNTHESIS OF NICKEL COATING WITH HIGH DENSITY NT

5.1.1. The Synthesis of Nickel Coatings with High Density NT

Pure Ni coatings were synthesized on Q235 steel by using reverse pulsed electrodeposition technique in sulphate-based baths with 0 and 0.2 g/L phytic acid addictive. The phytic acid addictive exhibited significant effect on the microstructure of coatings. TEM observations of the typical microstructure in as-deposited Ni specimens are shown in Figure 5.1 [7]. Figure 5.1. a shows the bright-field TEM image of Ni coating obtained in electrolyte without phytic acid. It can be seen from the image that it consists of irregular shaped grains, most of which are roughly equiaxed in three dimensions and the grain size is approximately 500 nm. The corresponding electron diffraction pattern is shown in Figure 5.1.b, which indicates that this Ni coating has the typical face-centered cubic (fcc) crystal structure (Crystal plane index are marked in the figure. A and B denotes interplanar spacing, where $A/B=\sqrt{2}$). Figure 5.1.c. shows the bright-field TEM image of Ni coating obtained in electrolyte with 0.2 g/L phytic acid and Figure 5.1.d. shows the corresponding electron diffraction pattern. It indicates that the as-deposited Ni coating obtained in electrolyte with 0.2 g/L phytic acid consists of irregular equiaxed shaped grains in three dimensions with a grain size of about 500 nm, which is similar with that obtained in electrolyte without phytic acid. However, the noticeable difference is that high densities of growth twins are present in the interior of each grain, which is of the {111} / [112] type (Figure 5.1.d). The diffraction spots of twins and matrix form mirror symmetry along the crystal plane of (111). Measurements of the lamella thickness show a wide distribution ranging from about 50 nm to 220 nm (Figure 5.2. [7]). The high density growth twins separate submicron-sized grains into nanometer-thick twin lamellar structures. Based on the discussion above, it can be concluded that the addition of organic additive phytic acid changes the microstructure of pure Ni coating. The addition of phytic acid favors the growth of twins in Ni sample, which contradicts the previous report that it is difficult to form twins due to its high SFEs [8-10]. Therefore, the SFEs of Ni must be lowered during the deposition process in the presence of additive phytic acid. This may be the result of modification of the state of adsorption when phytic acid was added into the bath with respect to the direction of preferential growth which was determined by a competitive adsorption of various species (especially hydrogen species) on the crystal faces of the growing nickel crystals [11-14].

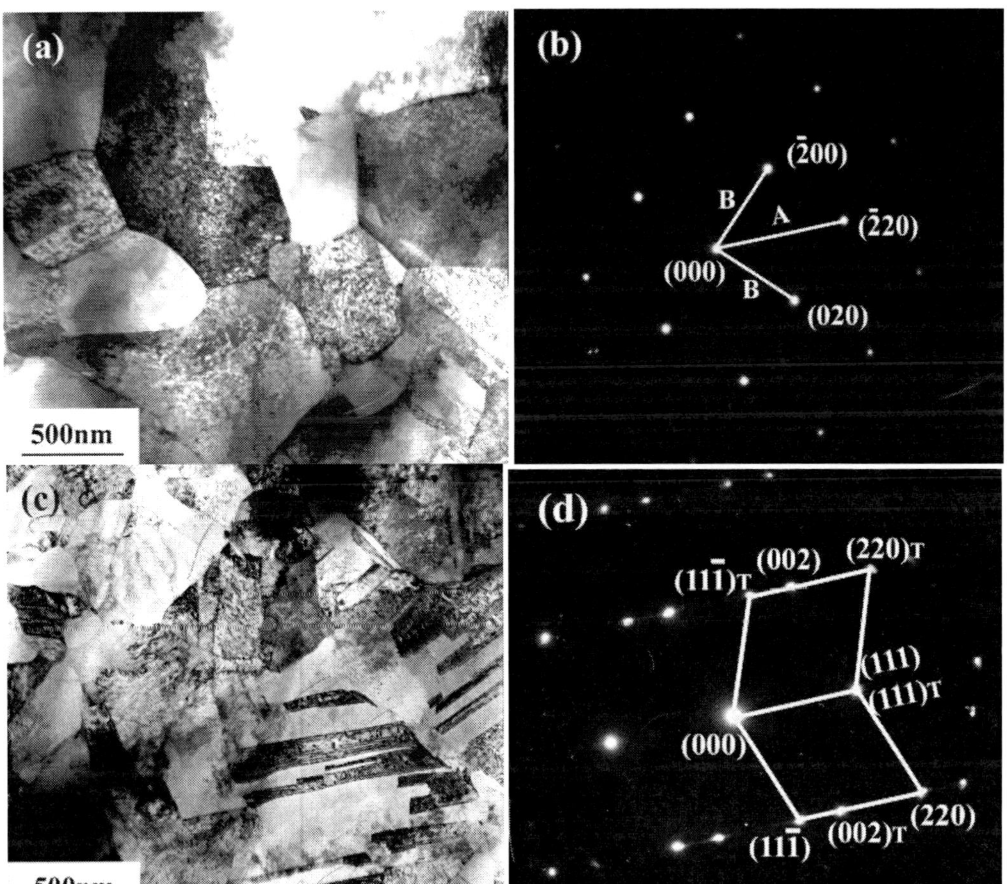

Figure 5.1. TEM observations of the typical microstructure in as-deposited Ni samples. A bright-field TEM image (a) and the electron diffraction pattern (b) of Ni coating obtained from the bath without phytic acid; a bright-field TEM image (c) and the electron diffraction pattern (d) of Ni coating obtained from the bath with 0.2 g/L phytic.

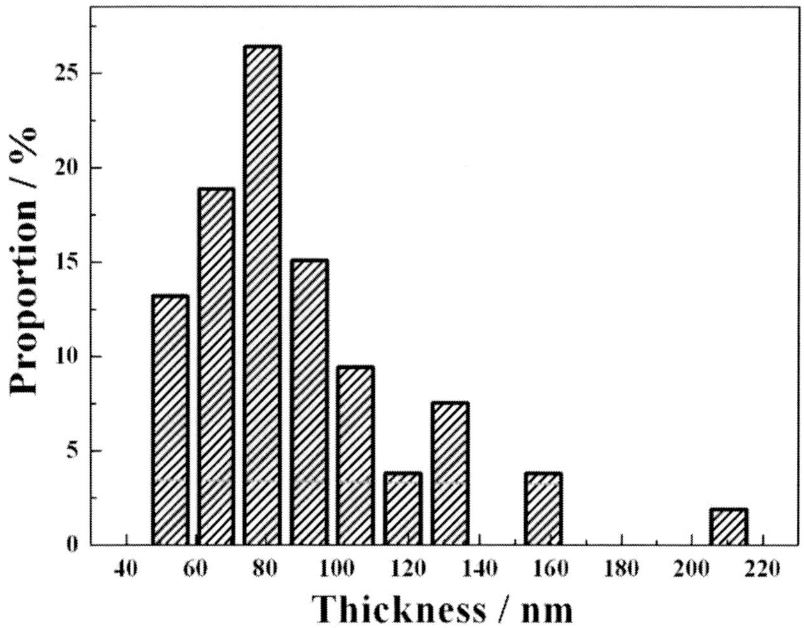

Figure 5.2. The statistical distributions for thickness of the twin lamellae.

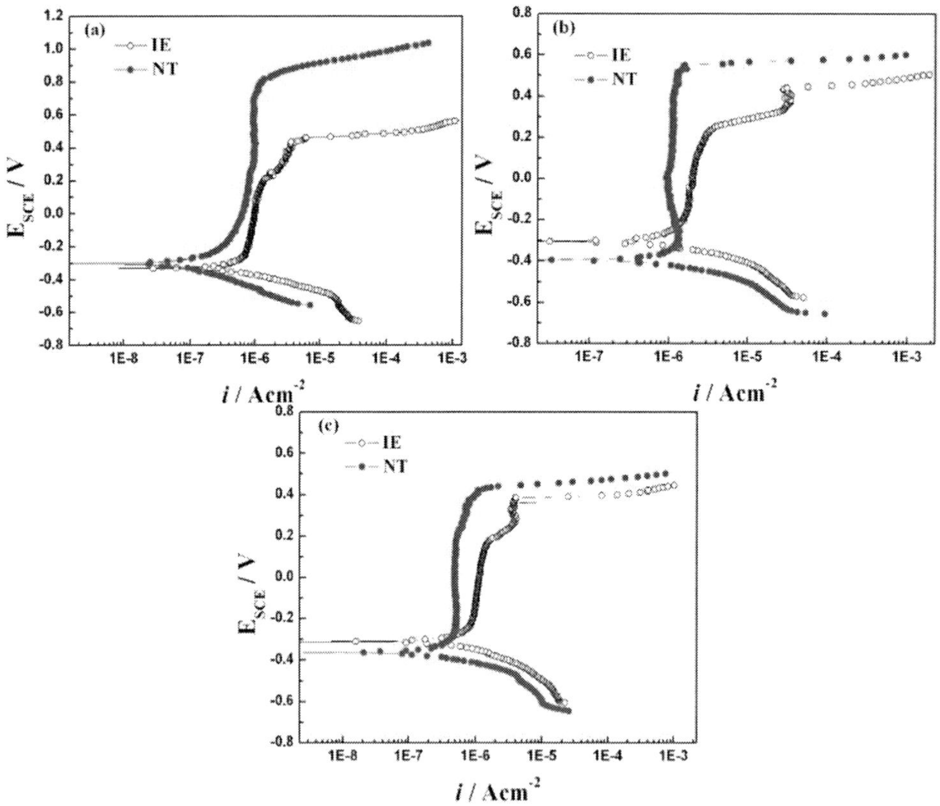

Figure 5.3. Potentiodynamic polarization curves of IE and NT nickel in borate buffer solution with different concentrations of NaCl [(a) 0.02M, (b) 0.04M, (c) 0.06M].

5.1.2. The Corrosion Property of Nickel Coating with High Density NT

The corrosion properties of nickel coating with NT were studied, comparing with that of industrial electrodeposited (IE) nickel. A comparison of corrosion behavior between IE nickel and NT nickel in borate buffer solution with [Cl^-] of 0.02, 0.04 and 0.06M are shown in Figure 5.3, respectively [15]. It can be observed that IE nickel and NT nickel both show self-passive behavior in all the three kinds of solutions. The corrosion resistance of NT nickel is much higher than that of IE nickel. The main characters of NT nickel, which are different from that of IE one, are summarized as follows: (1) With decrease in Cl^- concentration [Cl^-], the pitting potential of NT nickel greatly shifts in the positive direction and reaches up to 900mV in the solution containing [Cl^-] of 0.02M, and the passive area broadens correspondingly. But, it only reaches about 500mV in the same solution for IE one. (2) The passive currents of NT nickel are lower and more stable than that of IE one. These differences may result from the change of crystalline structure. Nanocrystallization reduces size of grains and increases density of lattice defects. It may reduce [Cl^-] in single lattice defect and decrease the sensitivity to pitting corrosion [16]. And twin formation generally reduces the grain boundary energy during grain growth [17], the grain boundary energy of the part where the twins are emitted is likely to be lower than that of the initial random boundary. Formation of a twin can introduce a low energy segment into the random high angle grain boundary and can sometimes result in low-CSL structures. The lower energy may decrease the adsorption amount of [Cl^-] at the grain boundary, where the pits always preferably initiate and develop. It led to the decrease of susceptibility to pitting corrosion for NT nickel.

5.1.3. The Semiconducting Behavior of Passive Film of NT Nickel

The relationship between capacitance and the applied potential is given by the well known M–S equation [18]. For n-type semiconductor, M–S equation is shown in (1), for p-type semiconductor, M–S equation is shown in (2).

$$\frac{1}{C^2} = \frac{2}{\varepsilon\varepsilon_0 q N_D}\left(E - E_{FB} - \frac{KT}{q}\right) \tag{1}$$

$$\frac{1}{C^2} = -\frac{2}{\varepsilon\varepsilon_0 q N_A}\left(E - E_{FB} - \frac{KT}{q}\right) \tag{2}$$

Where C is the capacitance of passive film semiconductor, ε is the dielectric constant of the passive film, ε_0 is the permittivity of vacuum (8.854×10^{-14} F/cm), q is the electron charge (1.602×10^{-19} C), ND and NA are the donor and acceptor densities, respectively. E is applied potential, EFB is the flat band potential, K is the Boltzmann constant (1.38×10^{-23} J/K), and T is the absolute temperature. Assuming the dielectric constant of the passive film on nickel as 12 [19].

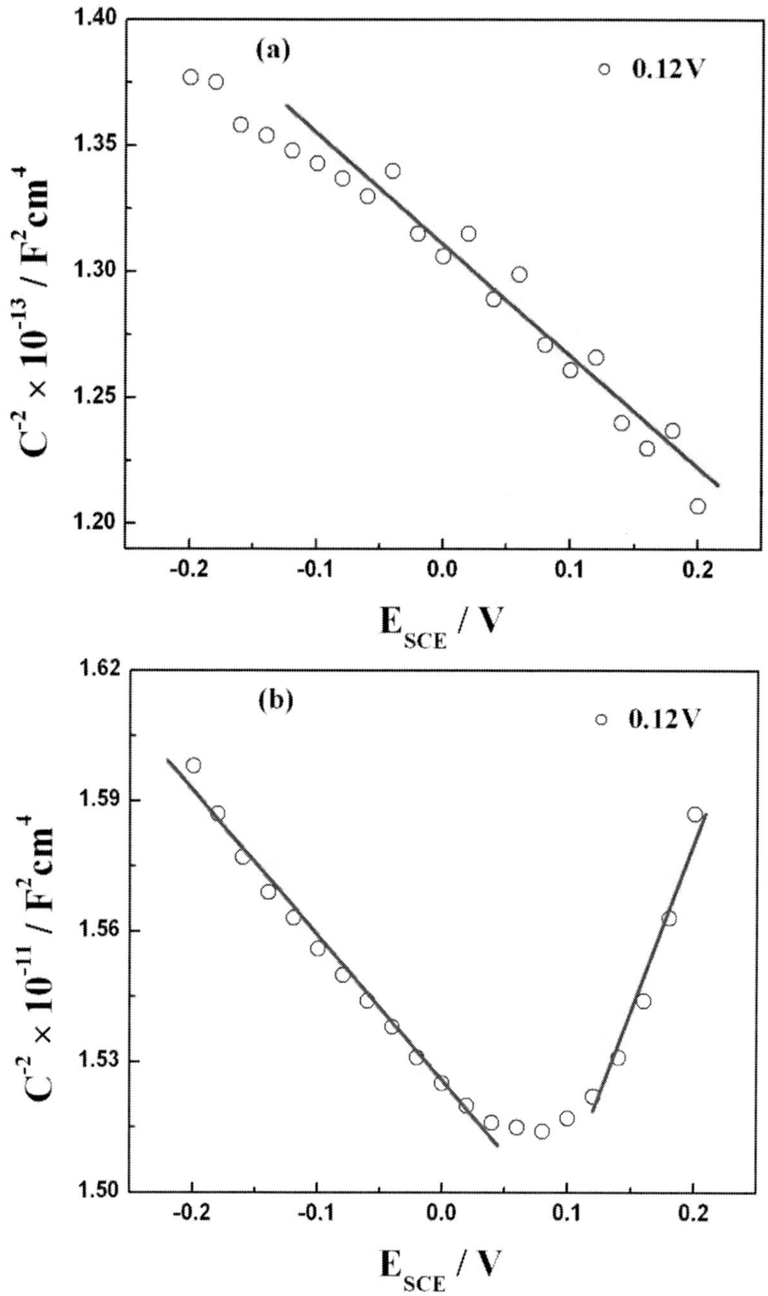

Figure 5.4. The M–S plots of IE nickel (a) and NT nickel (b) in borate suffer solution with 0.06M NaCl.

The M–S plots of the passive films for IE nickel and NT nickel in borate buffer solution with 0.06M NaCl are shown in Figure 5.4. [15]. The passive films formed on IE nickel in borate buffer solution with 0.02M, 0.04M, 0.06M and 0.10M NaCl have a p-type semi conducting behavior. However, the passive films of NT nickel show a bi-layer structure distribution in the same solution. The passive films behave like p-type semiconductors at low potentials, but they behave like n-type semiconductors at higher potentials.The semi conducting properties of Ni oxides have been studied by many authors [20, 21]. These studies show that Ni (II) oxides in both hydrated and dehydrated forms behave like p-type semiconductors, while phases containing Ni at higher oxidation states behave like n-type semiconductors. Therefore, the n-type character of NT nickel at high potentials is probably due to oxidation of Ni (II) to Ni(III) or injection of Ni(II) interstitials into the barrier layer.

Figure 5.5. The 27SiMn steel rod as cantilevers of coal hydraulic support: (a) with Cu-Sn electrodeposition as an undercoat; (b) Hydrogen blisters occurred in the Cu-Sn plus Cr combination coating after plating for 7days.

5.2. Inhibition of Hydrogen Blister on Cu–Sn Alloy Coatings

27SiMn steel is always used to make hydraulic supports for coal. In order to enhance its corrosion resistance and wear resistance, a steel cantilever rod of a hydraulic support is electrodeposited with Cu-Sn alloy as an undercoat (Figure 5.5 a [22]) and chromium as the uppercoat. However, a large amount of blisters began to occur on the Cu-Sn plus Cr combination coatings after plating for 7 days during the industrial manufacture of the solid cantilevers of hydraulic supports (Figure 5.5 b). It greatly influences the quality and safety of the hydraulic supports. During plating, hydrogen evolution occurred and a part of the adsorbed hydrogen atoms permeated into the coating and substrate steel by diffusion; others electrochemically recombined to molecular hydrogen and left the surface in gas form [20, 21, 23]. The diffusion of hydrogen can markedly deteriorate various properties of steels, which has attracted the interest of several researchers [24–30]. It was proved that the diffused hydrogen would react with the metal atoms to form brittle metal hydride, leading to the failure of structure even under the applied stresses far below the yield strength [24]. The diffused hydrogen may also combine into hydrogen molecules and accumulate in the microvoids which is located close to the surface of the material. When the pressure exerted by the accumulated molecular hydrogen was high enough, hydrogen blisters would occur on the surface [31]. This process would take a long time due to the lower diffusivity of hydrogen at the room temperature. For the electrodeposited coatings, the molecular hydrogen may accumulate at the boundary between coatings and the substrate, which will cause the hydrogen blisters in the coatings as shown in Figure 5.5 b. Our goal is to investigate the effect of electrodeposition parameters, such as current density, pulse frequency and duty cycle, on the hydrogen permeation behavior during the electrodeposition of Cu-Sn alloy. This may give important implications on preventing or greatly decreasing the occurrence of hydrogen blister during the industrial electrodeposition.Cu–Sn alloy coatings were synthesized from pyrophosphate solutions on 27SiMn steel by DC, PC and PRC electrodeposition techniques, respectively. The hydrogen permeation behavior during the plating was studied by Devanathan-Stachurski method [32]. The hydrogen permeation measurements were performed in a two cell system based on the Devanathan-Stachurski techique [33] with some modification of the original setup as shown in Figure 5.6 [22].

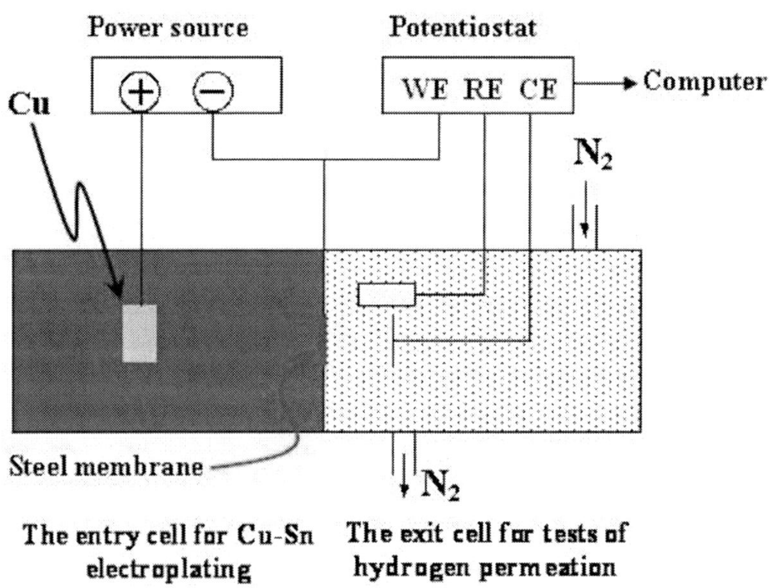

Figure 5.6. Schematic diagram of apparatus for electrochemical investigation of hydrogen permeation.

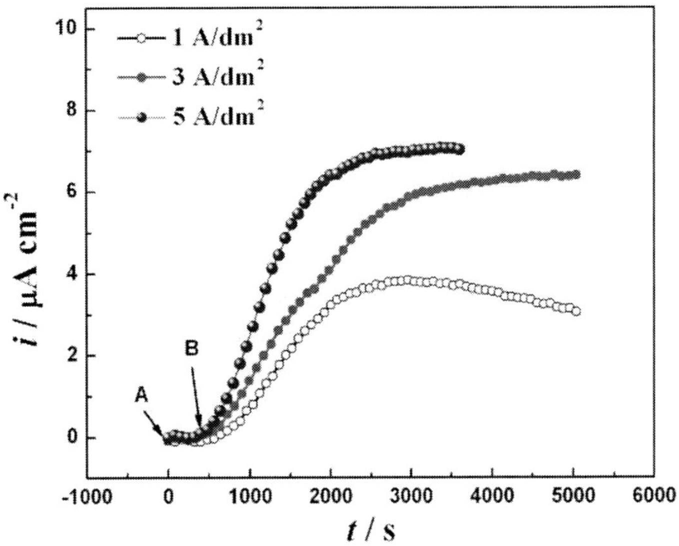

Figure 5.7. The hydrogen permeation curves obtained during DC plating at different current density. Point A indicated the initiation of electrodeposition; point B indicated the detection time of first hydrogen passed through membrane.

The set-up consists of two symmetrical electrochemical cells mounted on both sides of the vertical 27SiMn steel membrane. The exposed area was 1 cm^2 on both sides. The hydrogen entry side of the membrane was subjected to cathodic Cu–Sn plating. The hydrogen extraction side (the side with a thin Ni layer) was constantly polarized at 0.24 $V_{vsAg/AgCl}$ in order to oxidize the hydrogen atoms which diffused through the substrate. Ag/AgCl reference electrode and Pt counter electrode were used in the hydrogen extraction cell. Prior to the electrodeposition of Cu-Sn, 0.2 M NaOH solution was introduced into the hydrogen extraction cell and a 0.24 $V_{vsAg/AgCl}$ potential was applied for sufficient time to get a residual anodic current density of about 0.1 $\mu A/cm^2$.

5.2.1. Influence of Current Density on Hydrogen Permeation

The hydrogen permeation curves are recorded during Cu–Sn alloy plating at different current density, as shown in Figure 5.7 [22]. The electrodepositing method applied is DC electrodeposition. In Figure 5.7, point A indicates the initiation of electrodeposition; point B indicates the detection time of first hydrogen passed through the steel membrane. It can be seen that the current densities continue to remain at approximatly 0.1 $\mu A\ cm^{-2}$ from point A to B. The time span between point A and B represents breakthrough time. This time is due to finite rate constant for the transfer of hydrogen atoms from the surface of steel membrane into the steel phase. After point B, the hydrogen permeation current densities abruptly increase and finally reach a steady value.

It clearly shows the breakthrough time gradually decreases with the increase of current density (Figure 5.7), which indicates that the increase in current density can accelerate the process of entry of hydrogen to the bulk and consequently accelerate permeation of hydrogen. The permeation rate of hydrogen passing through the steel membranes can be indicated by the steady value of the hydrogen permeation current density. Therefore, Figure 5.7 also represents that the permeation rate of hydrogen passing through the steel membranes markedly increases (from 3.2 μA cm^{-2} to 7 μA cm^{-2}) with the increase of deposition current density (from 1 A dm^{-2} to 5 A dm^{-2}).

5.2.2. Influence of PC Frequency and Duty Cycle on Hydrogen Permeation

Figure 5.8. and 5.9 show the effect of various PC deposition parameters on the hydrogen permeation behavior [22]. The average pulse current density applied in PC electrodeposition is 5 A/dm^2. The features of these curves are similar to those obtained during DC electrodeposition. However, the steady values of the hydrogen permeation current density obtained during PC deposition are markedly less than that obtained under DC deposition. Figure 5.8 depicts the effects of PC frequency (from 200 Hz to 2500 Hz) on the hydrogen permeation transients for Cu-Sn PC electrodeposition at constant duty cycle (25%) [22]. It can be seen that the steady value of hydrogen permeation current density firstly decreases with the increase of frequency (reach the lowest value 2.7 μA/cm^2 at 1000 Hz). Then, it increases when the frequency is higher than 1000 Hz, and finally it falls down again at 2500 Hz. The results indicate that the hydrogen uptake is minimum (2.7 μA/cm^2) for the PC plating at 1000 Hz and 2500 Hz. The effects of duty cycles (from 10% to 70%) on the hydrogen permeation transients at constant frequency 1000 Hz are compared in Figure 5.9. It shows that the steady value of hydrogen permeation current density firstly decreases (reaches the lowest value 1.8 μA/cm^2 at 40%), and then increases with the increase of duty cycle. This implies that the hydrogen uptake is also greatly affected by the duty cycle and its minimum value occurs when the duty cycle equals to 40%. Therefore, the results demonstrate that the minimum hydrogen uptake during PC plating could be obtained with the deposition parameters: frequency 1000 or 2500 Hz, duty cycle 40%.

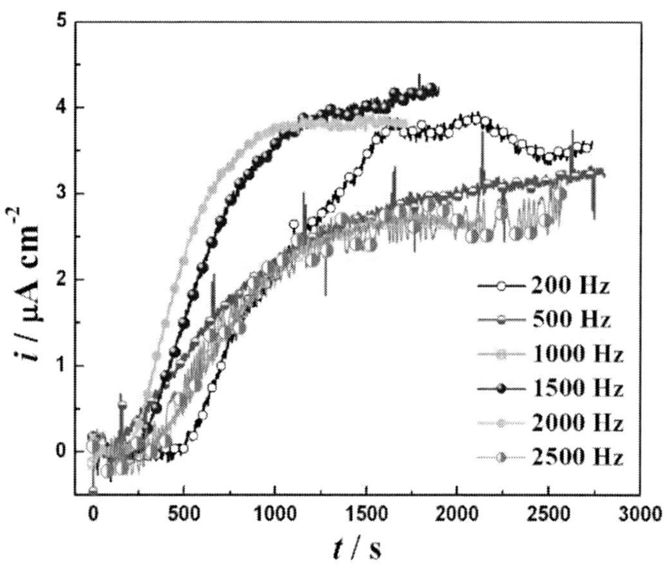

Figure 5.8. The hydrogen permeation curves obtained during PC electrodeposition at different frequencies.

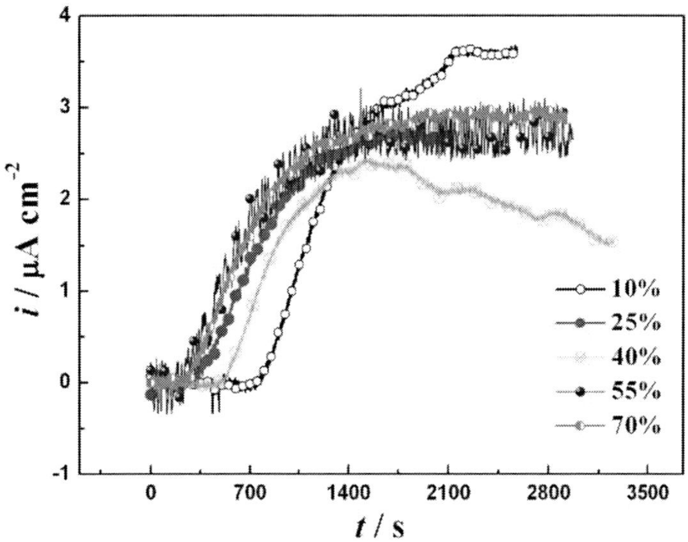

Figure 5.9. The hydrogen permeation curves obtained during PC electrodeposition at different duty cycle.

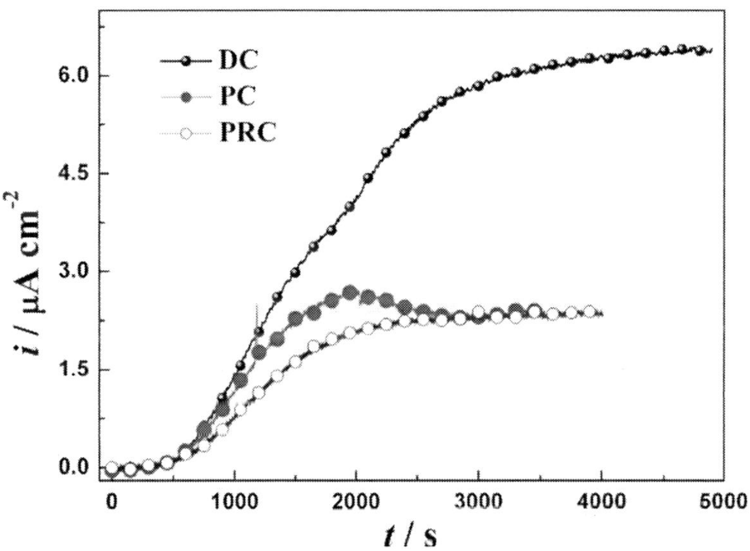

Figure 5.10. The hydrogen permeation curves obtained during DC, PC and PRC electrodeposition.

5.2.3. Influence of Reverse Pluse on Hydrogen Permeation

The PRC electrodeposition is carried out based on PC electrodeposition. The normal pulse parameters of PRC electrodeposition are the same with those of PC electrodeposition. The reverse pulse parameters are: frequency 1000 HZ, duty cycle 10%. And the reverse average pulse current density is one fifth of the normal average pulse current density.The hydrogen permeation currents density during DC, PC and PRC electrodeposition (frequency is 1000HZ and duty cycle is 40%) are compared in Figure 5.10. The average current density applied in all electrodeposition techniques used is 3 A/dm^2. It can be seen that the values of steady hydrogen permeation current density varies greatly. Apparently, the maximum value of steady hydrogen current density (6.4 μA/cm^2) is observed during the DC plating, while the minimum (2.2 μA/cm^2) is observed during the PRC plating and an intermediate value (2.6 μA/cm^2) is observed during the PC plating. All of these results indicate that the PC and PRC plating can effectively reduce the hydrogen uptake compared with the DC plating. Although the PRC deposition can further reduce the permeation rate of hydrogen permeation, the reduced extent is very small. It means that introduction of pulse current can effectively retard the hydrogen permeation, and reverse current has a little effect on the permeation.

5.2.4. The Amount of Hydrogen Permeation during Various Deposition Processes

It has been well accepted that hydrogen evolution reaction proceeds through three steps in neutral or alkaline solutions [20, 21, 23]: (1) electrochemical reduction of water, (2) electrochemical recombination of hydrogen atoms, and/or (3) chemical desorption of hydrogen atoms. Furthermore, hydrogen evolution is always accompanied with the process of hydrogen absorption and penetration into the electrode material followed by diffusion of hydrogen into the bulk metal. In the present work, a part of current was used for the electrolysis of water during the electrodeposition of Cu-Sn alloy:

$$H_2O + e \rightarrow H + OH^- \tag{3}$$

$$H + H \rightarrow H_2 \tag{4}$$

Part of the atomic hydrogen electrochemical recombined into hydrogen molecular and leaves the surface in gas form, the other part permeated into the steel in the form of hydrogen atom. The atomic hydrogen permeated into the steel can be divided into two kinds according to the activation energy of hydrogen trappings [34]. The first kind of atomic hydrogen is those trapped by irreversible traps in the steel, which do not diffuse and combine into hydrogen molecule in steel and it would not lead to hydrogen blisters. The second kind of atomic hydrogen is those trapped by reversible traps in steel, which will congregate locally in material and combine into hydrogen molecules at the defect sites, leading to hydrogen blisters in material when stress reaches to a certain value.The exit side of Devanathan-Stachurski apparatus was polarized anodically ($0.24V_{vsAg/AgCl}$) in order to detect the hydrogen which diffused through the steel membrane. And the permeation current was recorded against time until it reached steady state. The detection of hydrogen is according to the electrochemical reaction:

$$H \rightarrow H^+ + e \tag{5}$$

The hydrogen atom, which enters the steel membrane in the permeation experiment, will escape from the other side and been oxide. This does not take place in the electroplated solid hydraulic supports except that these hydraulic supports are hollow in the centre. In this case, hydrogen is unable to escape like the hydrogen permeation experiment due to its high concentration in the coating. Therefore, the amount of the hydrogen uptake in the steel during the plating can be approximately estimated in terms of the hydrogen permeation curves by calculating the areas (Q_H) beneath the permeation transients of Figure 5.5, 5.6 and 5.7. According to Faraday's Laws, the amount of M_{H_2} (mol) permeated into the steel is governed by the following equation:

$$M_{H_2} = \frac{Q_H}{2qN_A} \tag{6}$$

where q is the electron charge (1.602×10^{-19} C), N_A is Avogadro constant (6.02×10^{23} mol^{-1}). According to equation (6), the amount of MH_2 (mol) permeated into the steel during the PC plating at different frequencies and duty cycles are shown in Figure 5.11. It can be seen that the amount of permeated hydrogen MH_2 was minimum with the deposition parameters: frequency 1000 or 2500 Hz, duty cycle 40%.

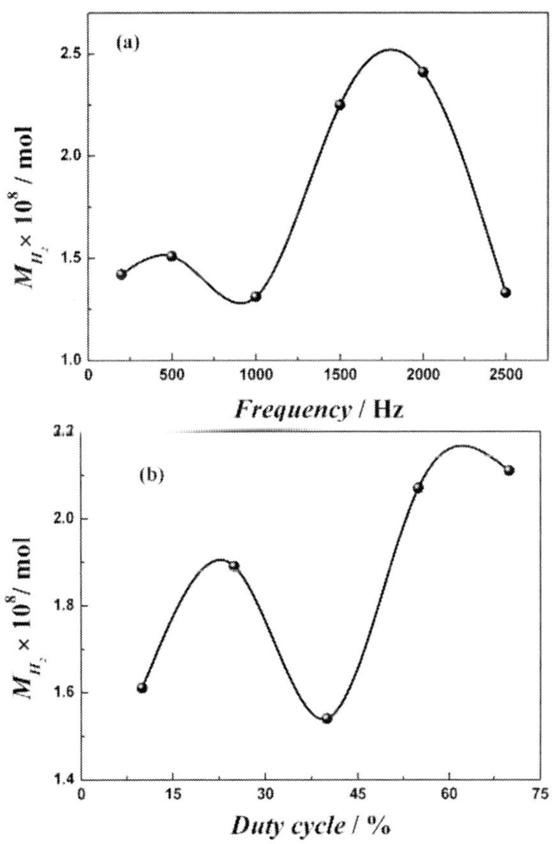

Figure 5.11. The amount of hydrogen molecule M_{H2} (mol) obtained during PC electrodeposition at different frequency (a) and duty cycle (b).

CONCLUSION

The additive phytic acid had significant effect on microstructure of coatings synthesized by pulse electrodeposition. The addition of phytic acid favored the growth of NT in the interior of grains due to the lowered SFEs of Ni during the electrodeposition proces.The passive films formed on NT coatings showed higher pitting corrosion resistance and a bi-layer semi conducting structure distribution, comparing with those formed on industrial electrodeposited nickel. The passive films behaved like p-type semiconductors at lower potentials, but they behaved like n-type semiconductors at higher potentials.The hydrogen permeation rate during electrodeposition increased with the increase of average deposition current density (5 Adm^{-2} > 3 Adm^{-2} >1 Adm^{-2}) for all kinds of electrodeposition. Frequency and duty cycle had significant effect on the hydrogen permeation behaviors. The optimization values of frequency was 1000 or 2500 Hz and duty cycle remained around 40%.The amount of permeated hydrogen M_{H_2} in steel decreased in the order: DC electrodeposition (6.03×10^{-8} mol) > PC electrodeposition (3.17×10^{-8} mol) > PRC electrodeposition (2.70×10^{-8} mol).

REFERENCES

[1] J. Puippe, F. Leaman, *Theory and Practice of Pulse Plating*, AESF Publication, Orlando, (1986).

[2] S. Roy, A. Connell, M. Ludwig, N. Wang, T.O. Donnell, M. Brunet, P.M Closkey, C.O. Mathuna, A. Barman, R.J. Hicken, *J. Magn. Magn. Mater.* 290, 1524(2005).

[3] H. Liu et al., *Electrochimica Acta* 47, 671(2001).

[4] Elizabeth J.Podlaha., *Nano Letters* 1, 413(2001).

[5] Allen Bai, Chi-Chang Hu., *Electrochimica Acta* 50, 1335(2005).

[6] P.T. Tang, *Trans. Inst. Met. Finish.* 85, 51(2007).

[7] G. Meng, F. Sun, Y. Shao, T. Zhang, F. Wang, *Electrochimica Acta* 55, 5990(2010).

[8] G.T. Gray III, *Acta Metall.* 36, 1745 (1988).

[9] Y. Zhang, N.R. Tao, K. Lu, *Scripta Materialia* 60, 211(2009).

[10] K. Lu, L. Lu, S. Suresh, *Science* 324, 349(2009).

[11] J. Amblard, M. Froment, N. Spyrellis, *Surf. Technol.* 5, 205(1977).

[12] J. Amblard, I. Epelboin, M. Froment, G. Maurin, *J. Appl. Electrochem.* 9, 233(1979).

[13] O. Devos, A. Olivier, J. P. Chopart, Aaboubi; G. Maurin, *J. Electrochem. Soc.* 145, 401(1998).

[14] O. Devos, O. Aaboubi, J. P. Chopart, E. Merienne, A. Olivier, J. Amblard, *J. Electrochem. Soc.* 145, 4135(1998).

[15] F. Sun, G. Meng, T. Zhang, Y. Shao, F. Wang, C. Dong, X. Li, *Electrochimica Acta* 54, 1578(2009).

[16] R. B. Intrui, Z. Szklarska-Smialowska, *Corrosion* 48, 398 (1992).

[17] Fullman R.L., Fisher J.C., *J. Appl. Phys.* 22, 1350(1951).

[18] S. R. Morrison, *Electrochemistry at Semiconductors and Oxidized Metal Electrodes.* Plenum Press, New York (1980)

[19] Barral, S. Maximovitch, F. Njanjo-Eyoke, *Electrochim. Acta* 41, 1305 (1996).

[20] Lasia, *Curr. Topics Electrochem.* 2, 239(1993).

[21] E.R. Gonzalez, G. Tremiliosi-Filho, M.J. de Giz, *Curr. Topics Electrochem.* 2, 167(1993).

[22] G. Meng, F. Sun, S. Wang, Y. Shao, T. Zhang, F. Wang, *Electrochimica Acta* 55, 2238(2010).

[23] G.Z. Meng, C. Zhang, Y.F. Cheng, *Corrosion Science* 50, 3116(2008).

[24] A.M. Brass, J.R. Collet-Lacoste, *Acta Mater.* 46, 869(1998).

[25] H. Addach, P. Berçot, M. Rezrazi, M. Wery, *Mater. Lett.* 59, 1347(2005).

[26] P.W. Liu, J.K. Wu, *Mater. Lett.* 57, 1224(2003).

[27] M. Monev, L. Mirkova, I. Krastev, Hr. Tsvetkova, St. Rashkov, W. Richtering, *J. Appl. Electrochem.* 28, 1107(1998).

[28] L. Mirkova, G. Maurin, M. Monev, Chr. Tsvetkova, *J. Appl. Electrochem.* 33, 93(2003).

[29] Hansung Kim, Branko N. Popov, Ken S. Chen, *Corrosion Science* 45, 1505(2003).

[30] H. Addach, P. Berçot, M. Rezrazi, J. Takadoum, *Corrosion Science* 51, 263(2009).

[31] H.C. Rogers, *Science* 159, 1057(1968).

[32] ASTM International, *Designation: G 148 - 97* (Reapproved 2003), United States.

[33] M. A. V. Devanathan, Z. Stachurski, *Proc. R. Soc.* 270, 90(1962).

[34] R. N. Iyer, H. W. Pickering, *Scripta Metallurgica* 22, 911(1988).

In: Electrodeposition: Properties, Processes and Applications
Editor: Udit Surya Mohanty
ISBN: 978-1-61470-826-1

Chapter 6

SYNTHESIS OF Cu_2O AND ZNO NANOWIRE ARRAYS BY ELECTROCHEMICAL DEPOSITION PROCESS

***Yu-Min Shen*[1], *Pramoda K. Nayak*[1], *Sheng-Chang Wang*[2] and *Jow-Lay Huang*[1,3,4,*]**

[1]Department of Materials Science and Engineering, National Cheng Kung University, Tainan, Taiwan
[2]Department of Mechanical Engineering, Southern Taiwan University, Tainan County, Taiwan
[3]Center for Micro/Nano Science and Technology, National Cheng Kung University, Tainan, Taiwan
[4]Research Center for Energy Technology and Strategy, National Cheng Kung University, Tainan, Taiwan

ABSTRACT

Polycrystalline Cu_2O nanowire arrays have been grown via porous alumina membranes using three-electrode electrochemical deposition. The effect of electrolyte, pH value, deposition potential, annealing temperature, and annealing atmosphere on the growth of Cu_2O nanowire arrays has been investigated. On the other hand, ordered ZnO/AZO/PAM nanowire arrays have been prepared by seed layer assisted electrochemical deposition. The comprehensive of electrochemical process adopted (e.g. electrolytes, deposition parameters); various characterization techniques and the outcome of the results have been summarized in details. It is observed that the ZnO nanowire arrays are assembled into the nanochannel of porous alumina template with diameter of 120~140 nm. The crystalline structure of single ZnO nanowire is dependent on AZO seed layer. The nucleation and growth process of ZnO/AZO/PAM nanowires are interpreted by seed layer assisted growth mechanism.

---

[*] Corresponding author: Jow-Lay Huang, E-mail: jlh888@mail.ncku.edu.tw.

Keywords: Electrochemical deposition, Cu_2O nanowire, AZO seed layer, Seed layer assisted growth mechanism, ZnO nanowire

INTRODUCTION

Electrochemical deposition is a process by which a film of solid metal is deposited from a solution of ions onto an electrically conducting surface. Generally, the deposited film has dimensions within the nanoscale and hence the resulting product has gone through a process of nanomanufacturing. People have been using electrochemical deposition for centuries. Its practice has evolved from an art to a science and accordingly there have been many discoveries that pushed its use towards depositing nanoscale films. The nanostructured thin films exhibit novel properties due to their small size and thickness, which largely differ from the bulk materials. Due to their significant properties, it can be used as microelectronic materials, bacteriostatic materials, catalytic materials or magneto recording materials, antibacterial materials, cryogenic superconducting materials and biosensor materials. Generally the shape, size, and size distribution of particulates and grains can be controlled by adjusting the reaction condition such as external and internal parameters like temperature, electrolyte concentration, current density and pH of the solution. Electrochemical deposition of metals and alloys involves the reduction of metal ions from aqueous and organic electrolyte. The reduction of metal ions M^{z+} in aqueous solution can be represented by

$$M^{z+} + ze \rightarrow M_{lattice} \quad (1)$$

where, the reducing agent in the solution is the electron source.

There are different aspects involved in deposition process represented in eq^n.1 such as (1) metal-solution interface as the locus of the deposition process, (2) kinetics and mechanism of the deposition process, (3) nucleation and growth process of the metal lattice ($M_{lattice}$), and (4) structure and properties of the deposits [1].

Electrochemical (EC) technologies attract constantly rising interest because it is the way for controlled modifications of the surface even with atomic precision. EC deposition is a widely used approach for creation of thin metal films with unique properties. In recent years, more attention has been paid to electrochemical deposition technique for manufacturing thin films and devices of carbon based materials due to its simplicity, its low capital equipment cost, and its ability to be scaled up for large production [2-4].

6.1. ELECTROCHEMICAL CELLS AND REACTIONS

In electrochemical systems, we are concerned with the processes and factors that affect the transport of charge between the chemical phases, i.e. between the electrode and electrolyte. When an electric potential is applied and current passes, the charge is transported through the electrode by the movement of electrons and holes. Typical electrode materials include solid metals (e.g., Pt, Au), liquid metals (Hg, amalgams), Carbon (graphite) and semiconductors (indium-tin oxide, Si). In the electrolyte phase, charge is carried by the movement of ions. The most frequently used electrolytes are liquid solutions containing ionic species, such as H^+, Na^+, Cl^- in either water or a non-aqueous solvent.

Mostly, in an electrochemical cell, the total system is defined as two electrodes separated by at least one electrolyte phase. A difference in electric potential can be measured between the electrodes in an electrochemical cell. The magnitude of the potential difference at an interface affects the rate of charge transfer. Therefore, the control of the cell potential is one of the important aspects, which can be taken into consideration. In an electrochemical process, the overall chemical reaction taking place in a cell is made up of two independent half reactions, which describe the real chemical changes at the two electrodes. Each half reaction responds to the interfacial potential difference at the corresponding electrode. The electrode at which this reaction occurs is called working electrode. The other half of the reaction occurs at reference electrode made up of phases having constant composition [5].

The internationally accepted primary reference is the *normal hydrogen electrode* (NHE), which has all components at unit activity:

$$Pt/H_2\ (a = 1)/H+\ (a = 1,\ \text{aqueous}) \qquad (2)$$

Potentials are often measured with respect to reference electrodes other than NHE. A common reference is the *saturated calomel electrode* (SCE), which is

$$Hg/Hg_2Cl_2/KCl\ (\text{saturated in water}) \qquad (3)$$

its potential is 0.242V vs. NHE. Another saturated electrode is *silver-silver chloride electrode*,

$$Ag/AgCl/KCl\ (\text{saturated in water}) \qquad (4)$$

With a potential of 0.197V vs. NHE. It is common to see potentials identified in the literature as "vs. Ag/AgCl" when this electrode is used.

6.2. NANOWIRE ARRAYS

Nanostructured materials have received considerable growing interest because of their fascinating properties and various applications as compared to their bulk or microsized counterparts. Recently, one-dimensional (1D) nanostructures such as wires, belts, rods, and tubes have become the focus of intensive research as a result of their unique properties and potential usages [6]. Nanowires have attracted extensive interest in view of their interesting electronic and optical properties and for their potential applications as building blocks for nanoelectronic and nanophotonic devices. To control the transport of electric carriers and phonons through nanowires and thereby enhance their functionalities, great efforts have been made to create nanowires with controllable structural complexity. The synthesis of dense, conductive nanowire arrays is critical for ultimately realizing the potential benefits of nanowire components in electronic devices. Uniform nanowire lengths will allow for contacting the largest number of wires possible in an electronic device, leading to maximum device performance. Among the various methods used for fabrication of nanoarrays, template synthesis [7, 8] using electrodeposition has proved to be a low cost and high yield technique for producing large arrays of nanowires. The template synthesis method has been playing an important role in the fabrication of many kinds of nanowires for its interesting and useful features. It is particularly useful for producing nanowires from different materials, with diameters less than 100 nm [9-11].

It is well known that the anodic alumina membranes (AAM)/porous alumina membranes (PAM) grown in acid electrolytes possess hexagonally ordered porous structures [12] with channel diameters ranging from below 10 to 200 nm, channel lengths from 1 to over 100 mm, and channel density in the range 10^{10}–10^{12} cm^{-2}. These unique structure features and its thermal and chemical stability make AAM ideal templates for the fabrication of ordered nanostructures. However, one of the most pervasive challenges for using PAM to template devices on prefunctionalized substrates is the presence of an intrinsic alumina barrier at pore bottoms .This oxide layer, typically 20–60 nm thick, hinders the creation of electrical contact to either the remaining aluminum (Al) beneath the pore bottoms or to a buried conductive layer that is designed to serve as an electrode in a subsequent electrochemical deposition process. Because of the structural features of the anodic alumina membranes, such as nanometer-sized channels with adjustable structural parameters (diameter, spacing, and length) and ordered pore arrays, the alumina membrane is an excellent template to fabricate ordered arrays of nanowires and nanotubes. So far many kinds of nanowire and nanotube arrays have been prepared using anodic alumina membranes, which include metals [13-15], semiconductors [16, 17], carbon [18, 19], polymers [20] and other types of materials. After the deposition of the nanowires or nanotubes into the pores of the alumina membranes, the nanowires or nanotubes can be released from the template by simply removing the alumina membranes using acid or alkali solutions, and finally resulting in free-standing nanowire or nanotube arrays.

6.2.1. Cu_2O Nanowire Arrays

Cuprous oxide (Cu_2O), which crystallizes into a cubic structure, exhibits direct band gap energy (2.1 eV) and high exciton energy (~150 meV) [21]. These excellent properties have been widely applied in the electrode of Li batteries and in high temperature superconductors. The synthesis of good quality cuprous oxide (Cu_2O) is important because of its usefulness for several semiconductor devices [22, 23]. The use of Cu_2O in optoelectronic devices has recently gained impulse because it posses natural p-type conductivity between the great variety of metallic oxide semi- conductors, for instance ZnO. Studies on the electrical properties demonstrate that it has an acceptor level at +0.4 eV, and two donor levels at 1.1 and 1.3eV below the conduction band [24].

Electrodeposition is a simple and inexpensive technique to deposit Cu_2O films on many substrates. Studies have shown that the electrical properties of Cu_2O are mainly controlled by the intrinsic defects such as copper and oxygen vacancies [25]. Most of the preparation techniques produce p-type Cu_2O due to the presence of Cu vacancies. It is well understood that in p-type Cu_2O, the dominant defect of Cu vacancies creates an acceptor level of 0.4 eV above the valence band. In the thermally grown Cu_2O in addition to Cu vacancies, the existence of oxygen vacancies has been observed [26]. Yet the conductivity of Cu_2O is observed to be p-type due to the lower density of O_2 vacancies as compared to that of Cu vacancies.

The polycrystalline Cu_2O nanowire arrays growth via porous alumina membranes was studied by three electrode electrochemical deposition [27-30]. Oh et al. [28] reported the Cu_2O based nanowires on PAM using aqueous $CuSO_4$ and lactic acid solutions with pH = 9. Choi et al. [29] observed that the nanowires arrays of Cu were transferred to Cu_2O due to the change in pH value (6 to 10). Inguanta et al. [27] found that the Cu_2O nanowire arrays was synthesized at 55^0C with pH=6.5 and -0.2V vs. SCE in copper acetate and sodium acetate. They also explored the different potential form on pulse to avoid the co-deposition of Cu and Cu_2O [30]. Although many studies have been published concerning the effect of electrolyte, temperature, pH value, and potential form, there is less work reported about the heat treatment of the Cu_2O/PAM nanowire arrays. In this study, Cu nanowire arrays have been synthesized via PAM template using three electrode electrochemical deposition process. The Copper nanowires have been transformed to Cu_2O phase with the space limitation of PAM template. The microstructures of Cu_2O nanowire arrays have been characterized by applying various spectroscopic techniques. The effect of electrolyte, pH value, deposition potential, annealing temperature, and annealing atmosphere on the growth of Cu_2O nanowire arrays has been investigated in sufficient details.For the synthesis of porous alumina membranes, high purity Al foils (99.9995%) were first degreased in ethanol. Before anodic process, the Al foils were annealed in argon of 10^{-2} torr at 500°C for 1 hour and etched by HF-HNO_3-HCl-H_2O (1:10:20:69 vol%) solutions. The Al foils were degreased in acetone by ultrasonic bath for 10 min. The samples were electro polished in a mixture of $HClO_4$-C_2H_5OH (1:4 vol %) at 10^oC, by applying 100 mA/cm^2 for 1 min. The anodization was carried out in 0.3M oxalic acid at constant voltage of 80V using Pt foil as a counter electrode. The electrolyte was rigorously stirred, and its temperature was kept at 3^oC during anodization. After first anodization for 5 h, the alumina film was selectively etched away in a mixture of H_3PO_4- CrO_3-H_2O (2g-3.5ml-100ml) at 70 $^{\circ}$C for 40 min. After second anodization under the same conditions for 18 h, the remaining Al was dissolved by saturated $HgCl_2$ solution. The released membranes were etched in 5 vol% H_3PO_4 solution at 60^oC for 30 min to dissolve the barrier layer on the bottom side of the PAM and room temperature for 20 min to widen the pores (shown is Figure 6.2.(a)). Before the deposition of nanowire arrays, a layer of Ti (10 nm) and Ag (100 nm) was coated onto one side of membrane and was sticked on the Cu substrate to be served as the working electrode (cathode). Pt foil (anode) and saturated calomel electrode (SCE, 0.241V) were served as the counter and reference electrode, respectively, in a three-electrode electrochemical deposition system. The PAMs/Ti/Ag/Cu were degassed using the mixture of 0.3M $CuSO_3$, 0.15M H_3BO_3 solution with pH =2 (controlled by H_2SO_4) and deposited at 230, 180, 80, -180, -230 mV for 2 hours. After electrodeposition, the Cu nanowire arrays embedded in PAM were heated in air at 400^oC for different time (4 to 12 hrs) and different partial oxygen pressure (0.02 to 0.12 atm). The samples were immersed in 3M NaOH solution for 15 min to dissolve the PAM template and characterized using field emission electron microscopy (FESEM, Hitachi S4800-I and Phillips 6700). The crystal structure of the samples was investigated by XRD (30kV, 20mA, Cu Kα, λ=1.54Å). The single nanowire was dispersed in ethanol and investigated using FETEM (Philips Tecnai G2 F20).

6.3. Synthesis of Cu/PAM Nanowires

Figure 6.1 shows the X-ray diffraction (XRD) spectra of nanowire arrays with different deposition potential. The applied potentials from A to E are 230, 180, 80, -180, and -230 mV, respectively. No Cu peak was found for the potentials of 230, 180, and 80 mV except -180 and -230 mV. It can be seen clearly that the diffraction patterns of Cu has no preferred orientation, indicating that the Cu nanowires are polycrystalline. In addition, the trace amount of Cu_2O was observed in all conditions. The relationship of the deposition potential can be described from the following equations: [30].

$$Cu^{2+} + 2e^{-} \rightarrow Cu \ E^0 = 0.099V \text{ (vs. SCE)} \tag{5}$$

$$2Cu^{2+} + 2e^{-} + H_2O \rightarrow Cu_2O + 2H^{+} \tag{6}$$

$$Cu_2O + 2e^{-} + 2H^{+} \rightarrow 2Cu + H_2O \tag{7}$$

Where E^0 is standard potential of an electrode. The reduction potential of Cu is 0.099V, so that the Cu was appeared at -180 and -230 mV. There are no Cu peaks found at 80 mV, as this potential is close to the reduction potential. Less amount of Cu may be present there, which could not be detected by XRD. Equations (6) and (7) indicate that the Cu was synthesized due to the reduction in the presence of H^+ ions. In the previous study [30], it has been demonstrated that the Cu and Cu_2O nanowires were co-deposited in continuous potential. Therefore, it is obvious that the Cu_2O is likely to be formed under all conditions.

Figure 6.2(a) shows SEM observations of the PAM template and Figure 6.2(b, c, and d) display the SEM images of Cu deposited in the PAM template. In Figure 6.2(a), the channel diameter of the two-step anodized PAM was about 110–140 nm. Figure 6.2(b) demonstrates the cross sectional image of a Cu/PAM nanowire. It is very clear that the Cu nanowires were deposited via the PAM template, and that the nanowire diameter depends on the template pore size. Figure 6.2(c) and (d) shows SEM images for different deposition potentials, with the PAM template dissolved in 3 M NaOH. The figures show that the nanowires were more densely distributed for the −180 mV sample than for the −230 mV sample. So, −180 mV was chosen as the working potential for the present study. This result indicates that the Cu nanowires were synthesized due to diffusion of Cu^{2+} ions into the PAM channel. The reaction rate of Cu under −230 mV was faster than that under −180 mV, which caused the depletion area of Cu^{2+} ions in the PAM channel. This depletion area may prohibit the growth of Cu/Cu_2O inside the channel. Thus, nanowire density and uniformity increased when the potential was changed from −230 mV to −180 mV.

As depicted in the TEM image (Figure 6.3(a)), the single nanowire was formed by nanocrystalline particles. It appears that the particles aggregated in the PAM channels to form nanowires. In the electron diffraction pattern (Figure 6.3(b)), the broad rings can be identified as Cu (111) and Cu_2O (220), respectively. This implies that the Cu/Cu_2O nanowires are polycrystalline in nature and consist of nano-size grains, which is consistent with XRD analysis results. Inguanta et al. [30] also indicated the polycrystalline nature of Cu/Cu_2O nanowires synthesized at a −200 mV potential, which is very close to −180 mV (in this case).

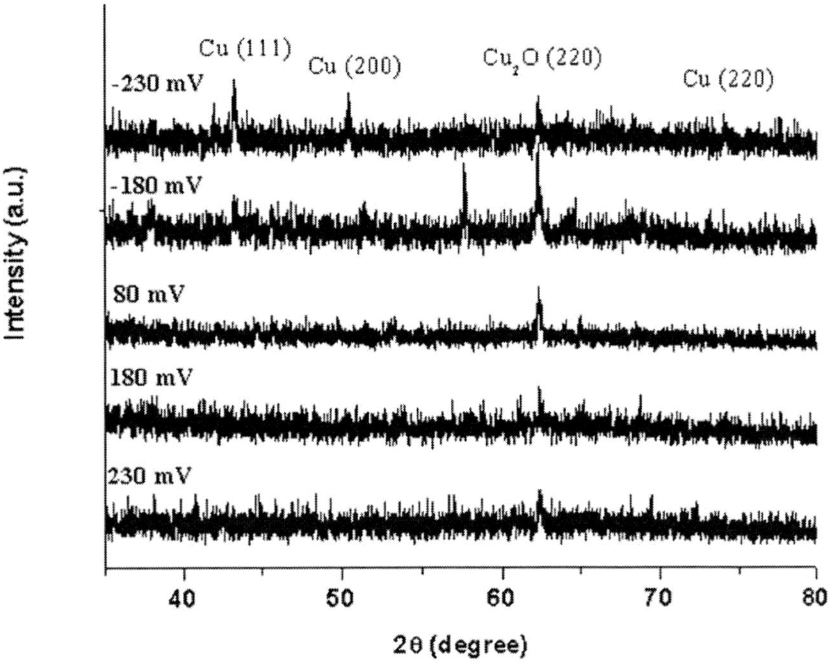

Figure 6.1.XRD patterns of Cu deposited into PAM templates at various potentials.

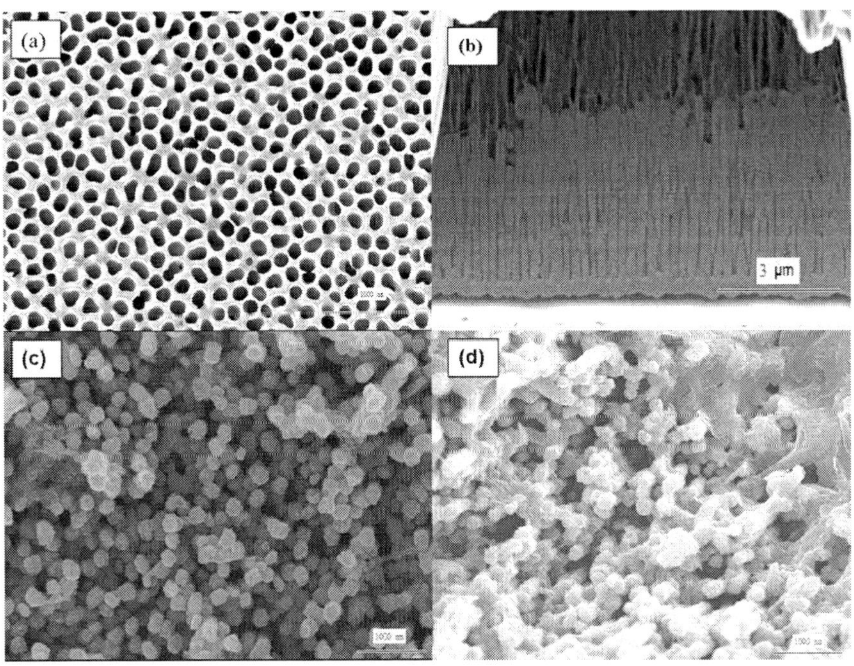

Figure 6.2. Plane-view SEM images of (a) PAM template and Cu nanowire arrays within a PAM at (b) -180mV (cross sectional), (c) -180mV, and (d) -230mV.

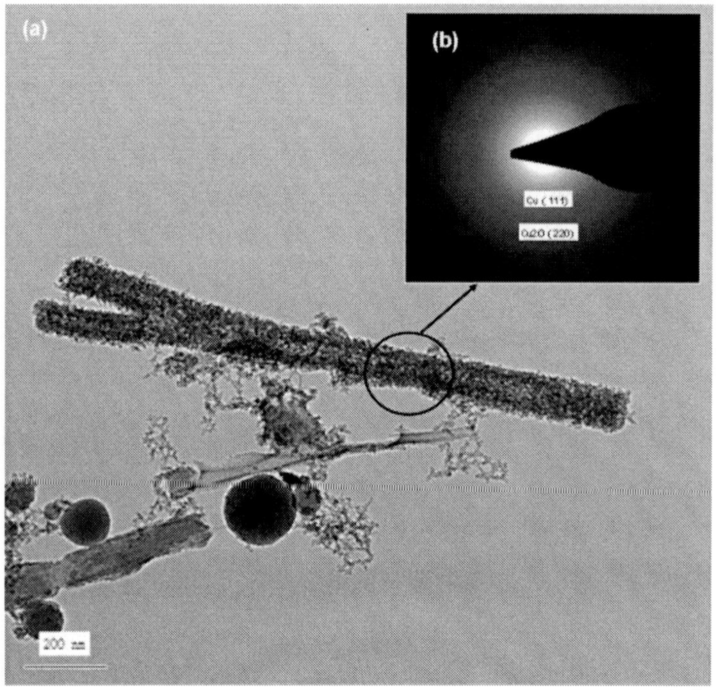

Figure 6.3. TEM image of (a) a single Cu nanowire and (b) electron diffraction pattern. The deposition potential is -180 mV.

6.3.1. Effect of Annealing Time on the Growth of Cu_2O Nanowire Arrays

The above results indicate that highly concentrated Cu/Cu_2O nanowires were synthesized at −180 mV owing to co-deposition of metal and oxide. Before the annealing process, the PAM template was not removed. Figure 6.4. shows SEM images of Cu_2O/PAM nanowire arrays after 400 °C heat treatment for 4, 8, and 12 h in air (P_{O2}= 0.2 atm). It was found that the contrast of the nanowire from inner to outer side was different for 4 h heat treatment (shown in Figure 6.4(a)). The core–shell structure indicates that the outer layer of Cu nanowires was oxidized to Cu_2O. The core–shell structure disappeared with the increase of annealing time. To further investigate the effect of annealing time on structural change, TEM images are shown in Figure 6.5. The results show that perfect crystalline nanowires were obtained after 4 and 8 h of heat treatment. Figure 6.5(c) indicates that the nanowires are polycrystalline with aggregated by nanocrystalline particles. The diffraction patterns shown in Figure 6.5(d), 6.5 (e), and 6.5 (f) indicate that a cubic Cu_2O structure was observed at 4 and 8 h. However, monoclinic CuO and cubic Cu_2O structures were found to be seen at 12 h. This result can be explained by assuming that after 4 h of heat treatment, less amount of oxygen entered the Cu/PAM nanowires. Only the outer layer of Cu was oxidized, forming a core–shell structure. At 8 h condition, most of the Cu was oxidized to form Cu_2O. With an increase in annealing time, CuO was found due to increase in the reaction between oxygen and copper, which gives rise to fractional Cu_2O converted to CuO.

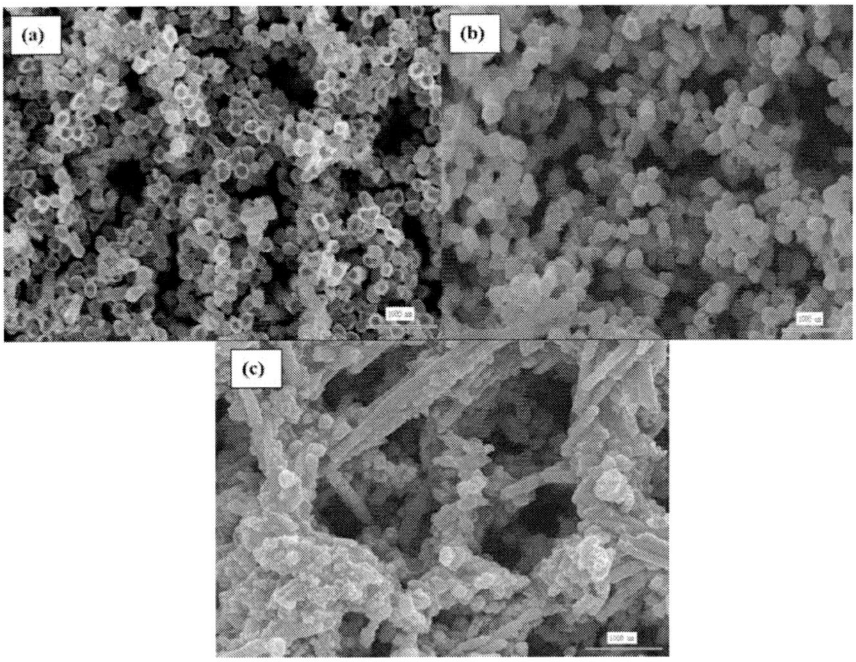

Figure 6.4. Plane-view SEM image of Cu_2O nanowire arrays within PAM after heat treatment at 400°C for (a) 4, (b) 8, and (c) 12 h.

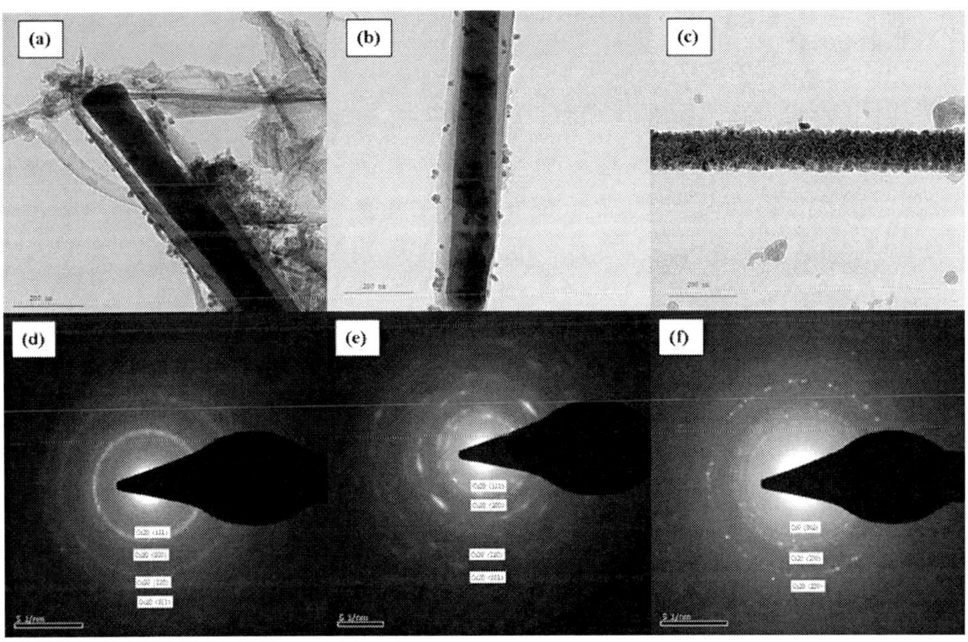

Figure 6.5. TEM images and SAED patterns of a single Cu_2O nanowire after 400°C heat treatment in air for (a,d) 4, (b,e) 8, (c,f) 12 h. The upper figures are electron diffraction patterns.

Figure 6.6. Plane-view SEM images of Cu_2O nanowire arrays within PAM after heat treatment at 400°C for 8 h. The partial oxygen pressure is (a) 0.12, (b) 0.06, and (c) 0.02 atm.

6.3.2. Effect of Annealing Atmosphere on the Growth of Cu_2O Nanowire Arrays

The annealing atmosphere affected the Cu_2O nanowire arrays in the present work. The results indicate that a pure Cu_2O structure was synthesized after 8 h of annealing. This time was thus chosen for further experiments. Figure 6.6 illustrates the SEM images of Cu_2O/PAM nanowire arrays after 400 °C heat treatment for 8 h under various partial oxygen pressures (P_{O2} is 0.12, 0.06, and 0.02 atm). The figure shows that ordered nanowire arrays were obtained. Figure 6.6(a) and (b) shows that the diameter of each nanowire is found to in the order of (150 to 160 nm), which is larger than that (130 to 140 nm) as observed in Figure 6.6(c). Figure 6.7 shows TEM images and diffraction patterns for the above three samples. The variation of nanowire diameter was identical to that shown in SEM results. The diffraction patterns (Figure 6.7(d), (e), and (f)) imply that with a decrease of oxygen, the nanowire changes from polycrystalline to single-crystalline. The polycrystalline cubic-Cu_2O structure is shown in Figure 6.7(d), Figure 6.7(e) and 6.7(f) illustrates the diffraction patterns taken from the main zone axes of [1 1 0]. The planetary zone is observed from these patterns. From the distribution of diffraction spots in Figure 6.7(e), it was found that $d_{Cu2O(200)}$, $d_{Cu(200)}$ are 2.11 Å and 1.78 Å, respectively, and that the main structure is Cu_2O. This result indicates that Cu_2O precipitates from the Cu surface, and that Cu_2O (200) and Cu (200) are parallel to each other.

In Figure 6.7(f), the d-spacing of Cu_2O (220) and Cu(220) are 1.31 Å and 1.15 Å, respectively. Furthermore, most of the Cu single crystal structure has been investigated in this condition. This finding seems to show that the amount of oxygen affects the main structure of nanowires. It is known that the copper may form diverse oxides as it is a transitional metal. Oxides such as Cu_4O, Cu_8O, and $Cu_{64}O$ may also form due to the oxygen diffusion from the surface into the copper lattice [31, 32]. Our results established that the compound of nanowire is mainly Cu_2O and not CuO, with controlled annealed temperature and atmosphere. Two possible mechanisms have been presented by PAM template space limitation. During the oxidation process, less amount of oxygen diffused from the surface into the Cu lattice due to the PAM channel limitation effect, which caused the transfer of the transitional metal Cu to Cu_2O phase.

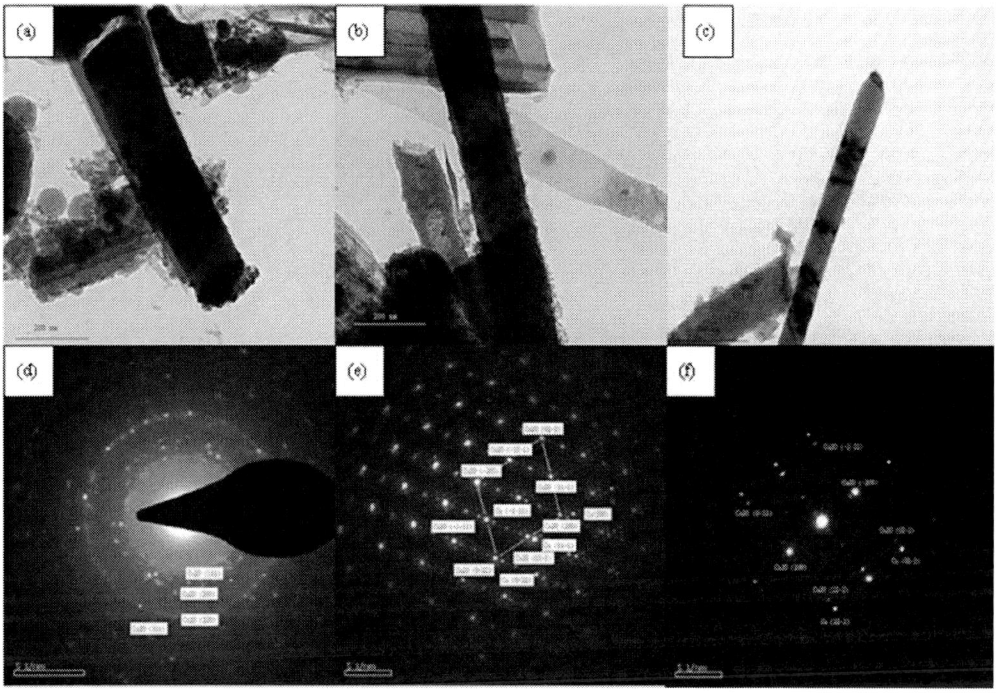

Figure 6.7. TEM images and SAED patterns of a single Cu_2O nanowire after 400°C heat treatment for 8 h in (a, d) 0.12, (b, e) 0.06, and (c, f) 0.02 atm of partial oxygen pressure. The upper figures are electron diffraction patterns.

On the other hand, the volume expansion due to stress-relaxation is also responsible for this conversion. Similar observation has also been reported by Zhou [33]. During the Cu oxidation process in the PAM channel, the compressive stress occurred due to volume expansion. A comparison of the lattice constant of CuO (monoclinic, lattice constant: a=4.662 Å, b=3.416 Å, c=5.118 Å, β=99.4°) and that of Cu_2O (cubic, lattice constant: a=b=c=4.26 Å) shows that the volume of CuO is greater than that of Cu_2O, i.e. ($CuO>Cu_2O$). In order to release the compressive stress, Cu_2O was preferentially formed. Therefore, the results indicate that the formation of Cu_2O is preferred in the PAM channel [34].

6.4. ZnO Nanowire Arrays

ZnO is a wide band gap (3.37 eV) semiconductor with a large exciton binding energy (~ 60 meV) and it crystallizes in wurzite hexagonal structure. It is one of the most promising oxide semiconductor materials because of its good optical, electrical, and piezoelectrical properties. ZnO is thus widely applied in the transparent conductivity films [35], short wave emission materials [36], UV-lasers [37], and sensors [38]. Numerous studies on the preparations and physical properties of zinc oxide films and nanoparticles have been reported in the literature. Besides this, one-dimensional ZnO nanostructures have been attracting intense scientific interest as these materials exhibit near-UV emission [39], UV absorption [40] and field-emission capabilities [41].

A number of physical and chemical techniques have been developed to fabricate one-dimensional ZnO nanostructures. High-temperature thermal evaporation [42,43], CVD [44,45] and pulsed laser deposition [46-48] have been used to produce high-quality ZnO nanorods and nanowires. The properties of high aspect ratios and small sizes of ZnO nanowires are expected to improve the luminescence efficiency of the electro-optical devices and the sensibility of the chemical sensors.

The uniform order of the ZnO nanowire arrays and the quasi-continuous controllability of lengths and diameters of the ZnO nanowires make it more practical to apply ZnO in electrodevices and chemical sensors. Ordered ZnO nanowire arrays embedded into PAMs by tuning the electrochemical deposition parameters, such as annealing time, electrolyte solution, deposition potential, temperature, and additives, have already been reported in many studies [49-52]. Using template assisted method, although nanowire's size, shape and uniformity are very easy to control, but the nanowire shows polycrystalline in nature [51]. On the other hand, seed layer assisted method is more preferable over template method for obtaining single crystalline ZnO nanowire. ZnO nanowire by seed layers assisted method has been reported in [53, 54].

However, the nanowire's size, shape, and uniformity aren't easy to control by seed layers assisted method. In order to overcome such difficulties, few researchers have focused on the growth of ZnO nanowire by combining both templates and seed layers assisted method. In this study, ZnO nanowire arrays were embedded into PAMs, which was assisted by AZO seed layers by electrochemical deposition. The microstucture of ZnO nanowire arrays embedded in AZO/PAMs was characterized by employing various spectroscopic techniques.

6.4.1. Preparation of Porous Alumina Membranes (PAMs)

The porous alumina membranes (PAM) template was prepared using two step anodization processes, which is exactly similar to that mentioned above. The flow chart showing the synthesis of PAM is given in Figure 6.8.

6.4.1.1. Preparation of Al-Doped Zinc Oxide (AZO) Seed Layers

For synthesis of AZO seed layers, the 0.3 M ZnO solution was prepared by mixing 9.877g zinc acetate, 150 mL ethanol, and 2.7 mL ethanolamine. The mixing solution was heated to 60°C and stirred for 30 min followed by stirring under room temperature for 24 h. The AZO solution was obtained by mixing ZnO solution and 0.1688 g aluminum nitrate and heated to 60°C for 30 min. A glass substrate was taken and cleaned by H_3PO_4-H_2SO_4 (3:1 vol%) solution under 60°C for 20 min. Then, it was degreased in DI water in an ultrasonic bath for 1 min, and finally dried up by N_2 gas. Initially, the AZO seed layers was coated on the glass substrate by spin coating and annealing in vacuum at various temperatures (400,

600°C) for various time (0, 2 hr). Finally, the PAMs were degreased in 0.05M H_2SO_4 solution in an acid bath for 1 min and AZO seed layers were coated by spin coating.

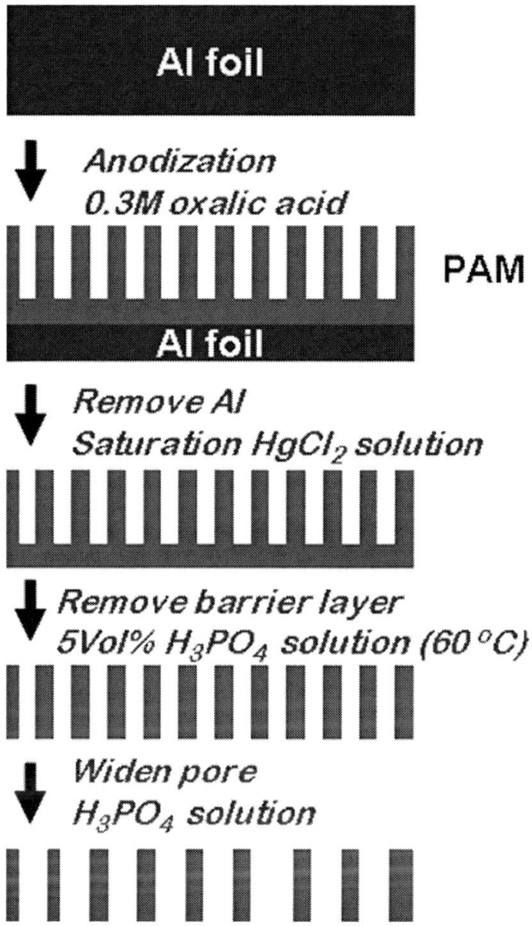

Figure 6.8. The flow chart showing the synthesis of PAM.

6.4.1.2 Preparation of ZnO/AZO/PAMs Nanowire Arrays

Before the deposition of nanowire arrays, layers of Ti (10 nm) and Ag (100 nm) were coated onto one side of the membrane and was sticked on the Cu substrate to serve as the working electrode (cathode). Pt foil (anode) and a saturated calomel electrode (SCE, 0.241V) served as the counter and reference electrode, respectively, in a three-electrode electrochemical deposition system. The PAMs/AZO/Ti/Ag/Cu were degassed using a mixture of 0.15 M $ZnSO_4$ and 1.5 M H_2O_2 solution deposited under -1 V at 65 and 80°C for 3 h, respectively. After electrodeposition, the composites were heated in air at 400°C for various durations (4 and 8 h) to obtain ZnO/AZO/PAMs nanowire arrays. The schematic diagram of three electrode electrochemical deposition process along with the flow chat showing the synthesis of ZnO/AZO/PAMs nanowire arrays are given in Figure 6.9.

6.4.2. Characterization of ZnO/AZO/PAMs Nanowire Arrays

The surface and cross sectional morphology of ZnO/AZO/PAMs nanowire arrays was observed using field emission electron scanning microscopy (FE-SEM). For surface morphology, the samples were prepared by immersing in 3 M NaOH solution for 15 min to dissolve the PAM template and for cross sectional morphology, samples were cut by using focus ion beam (FIB). The crystal structure of the single nanowire and ZnO/AZO/PAMs composites were investigated by HR-TEM, for which, the samples were prepared by dispersing in ethanol and cutting by FIB, respectively.

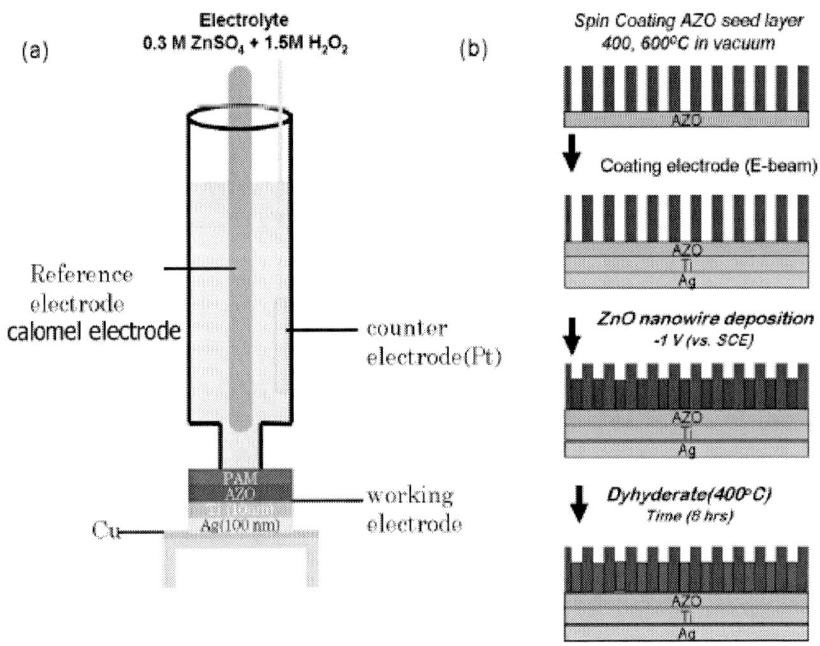

Figure 6.9. (a) The schematic diagram of three electrode electrochemical deposition process, (b) the flow chart showing the synthesis of ZnO/AZO/PAMs nanowire arrays.

6.4.3. Characterization of AZO Seed Layers

Before synthesis of ZnO/AZO/PAMs nanowire arrays, the AZO films and PAMs were initially characterized. Figure 6.10 shows the SEM images of AZO films coated on the glass with annealing at 400, 600°C for 0 and 2 hr in vacuum. The average sheet resistance of AZO films annealed at 400°C and 600°C is found to be 129.37 and 326.34 Ω/sq, respectively. From figure 6.10(a) and (b), the average grain size is calculated to be 21.8 and 44.2 nm, respectively, which indicates that an increase in the annealing temperature promotes the grain growth. The loose structure under 600°C restrained the conductivity of electrons which caused the increase in average sheet resistance. In addition, with the increase of annealing time to 2 hr, the average sheet resistance was raised to 1023 Ω/sq. As shown in Figure 6.10. (c), the porosity of AZO films was increased. It indicates that the increase in resistance is consistent with increase in annealing temperature and duration of annealing.

Figure 6.11 shows the average sheet resistance with various layers of AZO seed, which indicates that the resistance was decreased with increasing the number of layers. The relationship between layers of AZO seed and resistance can be expressed as following equation:

$$R_s=1/\sigma t \tag{8}$$

where R_s is the average sheet resistance, σ is conductance of thin films, and t is the grain size. The conductivity of thin films can also be described as

$$\sigma=nq\mu \tag{9}$$

where n is thickness of thin films, q is electric charge, and μ is mobility of electron.

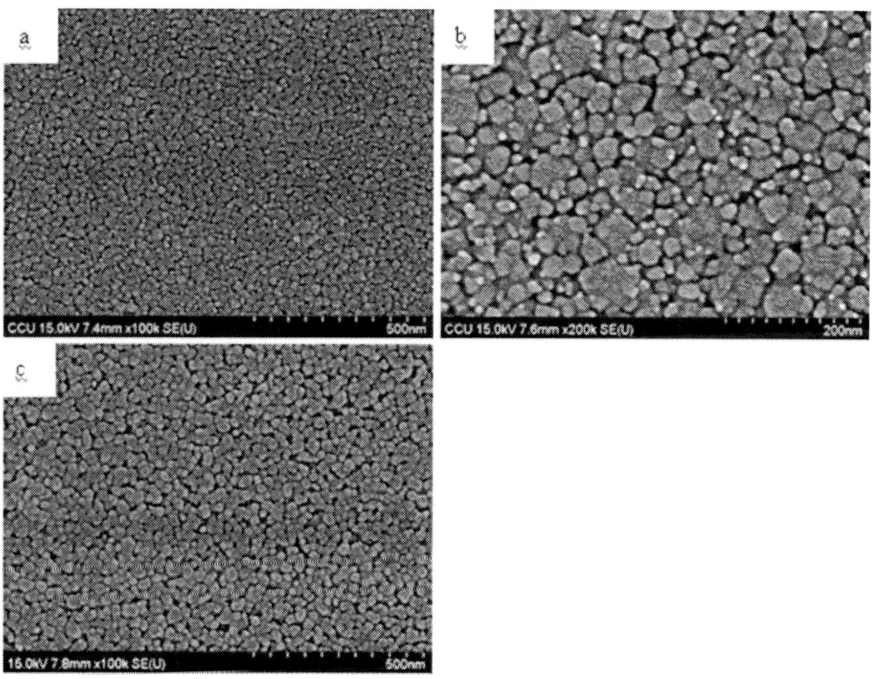

Figure 6.10. Plane-view SEM images of AZO seed layers at annealing temperature and time (a) 400°C, 0 h (b) 600°C, 0 h, and (c) 400°C, 2h in vacuum.

These equations suggest that the conductivity of thin films can be improved with the increase in thickness of thin films. However, the AZO films was underdone heat treatment after each spin coating process and it could be destroyed the PAMs under many times heat treatment. In our further work, 5-layers AZO seed was chosen to coat at the back side of PAMs for electrochemical deposition electrode.

Figure 6.12 shows the TEM images of (a, b) AZO/PAMs composite, (c) high resolution image, and (d) electron diffraction pattern. The AZO/PAMs TEM sample was prepared by focus ion beams (FIB). In Figure 6.12(a), point 1 shows the carbon films, which protected the sample during cutting process, point 2 shows the AZO films, and point 3 shows the PAMs.

It can be found that there was an interface (shown as B layer) between point A and C in Figure 6.12(b). The high resolution image indicates that point A was polycrystalline; point B and C are amorphous structure. The results of TEM image also imply that AZO films are well-adhesion on PAMs. The EDX results (shown as Figure 6.12(e)) indicate that the Zn composition is decreased and Al composition is increased from A to C. In addition, the electron diffraction pattern implies that point A exhibits ZnO structure, in which the planes are (100), (101), (102) and (110).

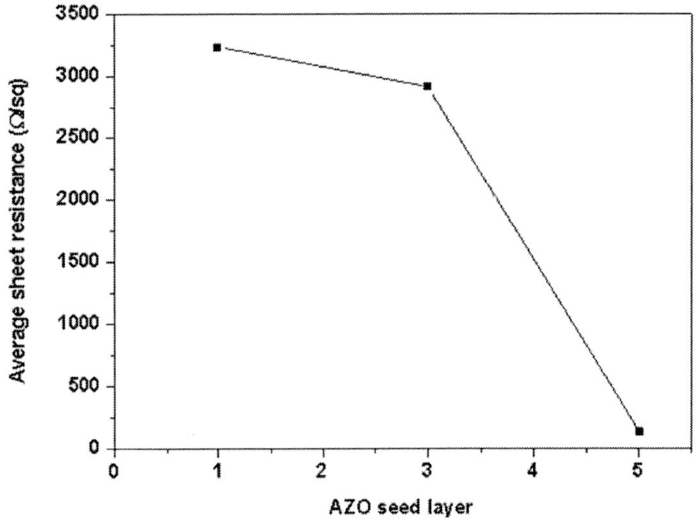

Figure 6.11. Plot of average sheet resistance versus various layers of AZO seed films.

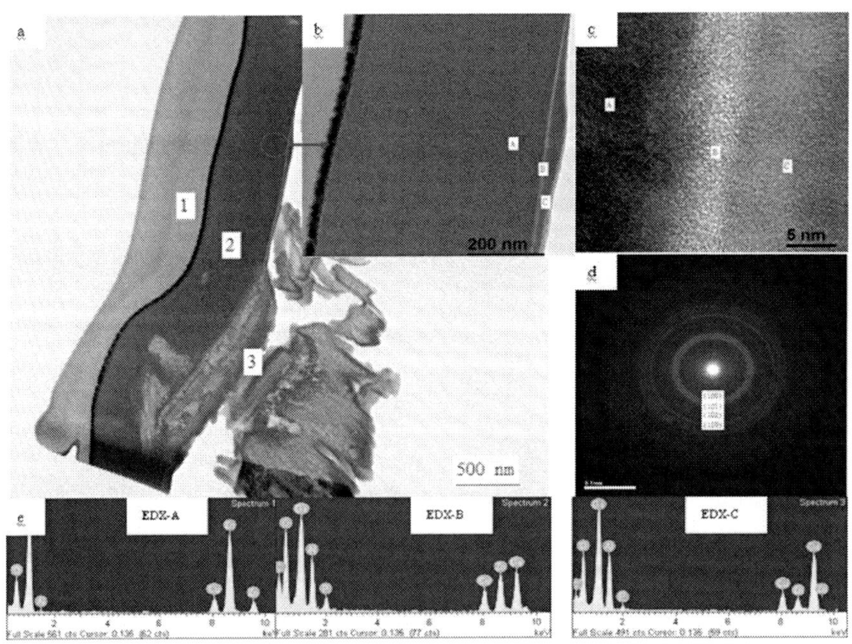

Figure 6.12. TEM images of AZO/AAO composite (a) low magnification, (b) high magnification, (c) high resolution image, (d) electron diffraction pattern, and (e) EDX spectra.

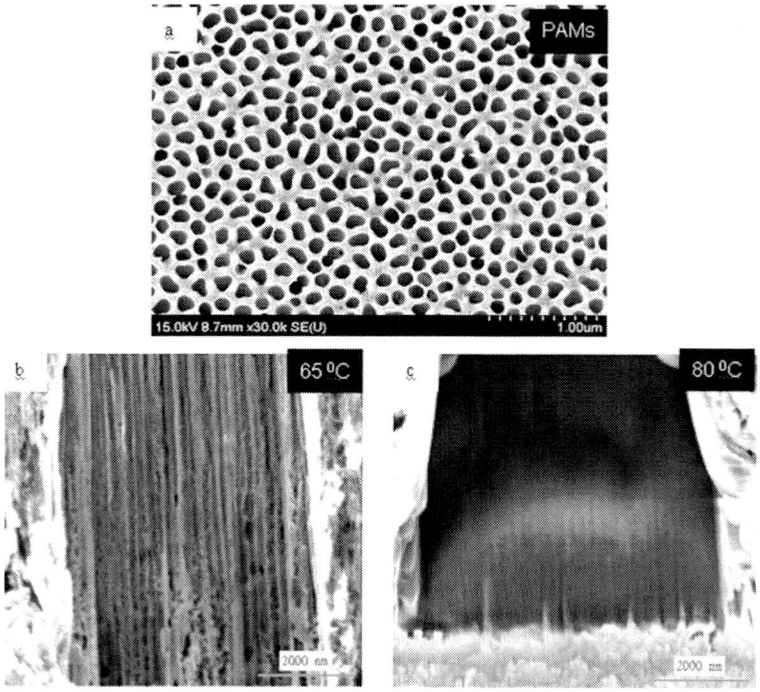

Figure 6.13. Plane-view SEM images of (a) PAMs and cross sectional images of nanowire arrays deposited in the PAMs under -1 V/SCE at (b) 65 and (c) 80°C.

6.5. Synthesis of Nanowire Arrays

Figure 6.13 shows SEM observations of the (a) PAMs and (b, c) cross sectional images, in which nanowire arrays were deposited in the PAMs under -1 V/SCE at 65 and 80°C. In Figure 6.13(a), the channel diameter of the two-step anodized PAMs is about 110~140 nm. For the deposition temperature of 65°C (Figure 6.13(b)), there are many particles exist in PAMs channel, and the obvious nanowires are formed at 80°C (Figure 6.13(c)). Thus, the size and shape is controlled by PAMs channel. The crystallinty of nanowire was investigated from TEM observation as shown in Figure 6.14(a) and 6.14(b) for the deposition temperature of 65 and 80°C, respectively. We can see that individual nanowires are formed by particles and the particles aggregate in the PAMs channel to form nanowires. The high resolution image results indicate the amorphous structure of the nanowire. In the other ways, compared the TEM results between 65 and 80°C, a modicum fibrillar structure was observed in 80°C expect for 65°C. The contrast of 65 and 80°C TEM images also implies that the crystallinity of 80°C is greater than 65°C. The formation of nanowire can be described by the following equations:

$$Zn^{2+} + 2e^- \rightarrow Zn \quad E^0 = -1.01 \text{ V (vs SCE)} \tag{10}$$

$$H_2O_2 + 2e^- \rightarrow 2OH^- \quad E^0 = 0.694 \text{ V (vs SCE)} \tag{11}$$

$$Zn^{2+} + 2OH^- \rightarrow Zn(OH)_2 \tag{12}$$

$$Zn(OH)_2 \rightarrow ZnO + H_2O \quad (13)$$

Eq^n. (10) and (11) show the reduction potential of Zn and OH^-. In this work, the deposition potential was -1 V, so that the Zn^{2+} **wasn't reduced to Zn metal instead of** Zn^{2+} assembling on the surface of work electrode. The OH^- was formed on the electrode and reacted with Zn^{2+} to form the $Zn(OH)_2$ hydrate (as shown in Eqn.(12)). The fibrillar structure was the transitional product during crystallization. This result implies that the fibrillar structure in 80°C was obtained due to partial dehydration of Zn $(OH)_2$ to ZnO.

6.5.1. Effect of Annealing Time

The above results indicate that the well-crystallized and highly concentrated nanowire arrays were synthesized at 80°C. So, this temperature was selected as suitable electrochemical deposition temperature for further work. Before annealing process, the PAMs were not removed. Figure 6. 15 (a, b) illustrates cross sectional SEM images of nanowire arrays after 400°C heat treatment for 4 and 8 h in air.

It was found that in PAMs channel, the whiskery structure was observed at 4 h and the nanowire was filled entirely at 8 h. This whiskery structure was formed due to under dehydration course, i.e. there was not enough time for diffusion of H_2O from PAMs channel to the surface.

On the contrary, dense nanowire was obtained at 8h heat treated specimen. In order to comprehend the surface morphology of nanowire arrays, 3M NaOH was used to remove the template. Figure 6.15(c) shows the top-view image of 8 h heat treated specimen, which indicates that the size and distribution of nanowire are well incorporated onto the PAMs. Figure 6.16 demonstrate TEM and electron diffraction results of single nanowire for 4 and 8 h heat treatment samples.

As shown in TEM images, the nanocrystallinity was decreased gradually to form the fully crystalline with respect to increase in the annealing time. The ZnO crystalline ring was found in electron diffraction pattern at 4 and 8 h heat treated samples. It is very clear that the ring pattern was more remarkable under 8 h and the spot pattern was also observed. This result can be evidenced that during the dehydration process, ZnO was formed and the crystallinity was improved.

Figure 6.17 shows the TEM image of ZnO/AZO/PAMs nanowire arrays with heat treatment at 400°C for 8 h. Points A, B and C correspond to ZnO/PAMs, interfacial of ZnO-AZO, and AZO seed layers, respectively. As shown in electron diffraction pattern, the zone pattern with hexagonal structure was found in AZO seed layers (point C), which was formed by submicron grains.

The electron diffraction patterns (point A and B) indicate that the crystallinity of ZnO nanowire was improved when this nanowire was closed to the AZO seed layers. It was found that the d-spacing of ZnO nanowire was 2.5Å, and the growth direction was [001], which was closed to AZO seed layers (d=2.7Å) as shown in high resolution and FFT images. This result implies that under the electrochemical deposition and annealing process, the ZnO aggregated at the seed layers. Thus, the heterogeneous nuclei of AZO seed layers improved the growth of ZnO nanowire. The ZnO nanowire approached to weak crystallinity, when the nanowire was farther seed layers. Therefore, the results indicate that the formation of ZnO nanowire is preferred in the AZO seed layers.

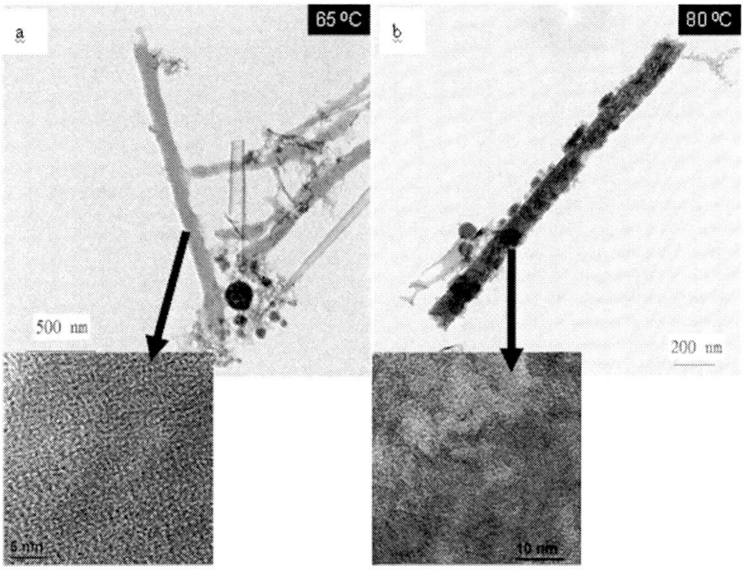

Figure 6.14. TEM and high resolution images of a single nanowire at deposition temperature (a) 65°C and (b) 80°C.

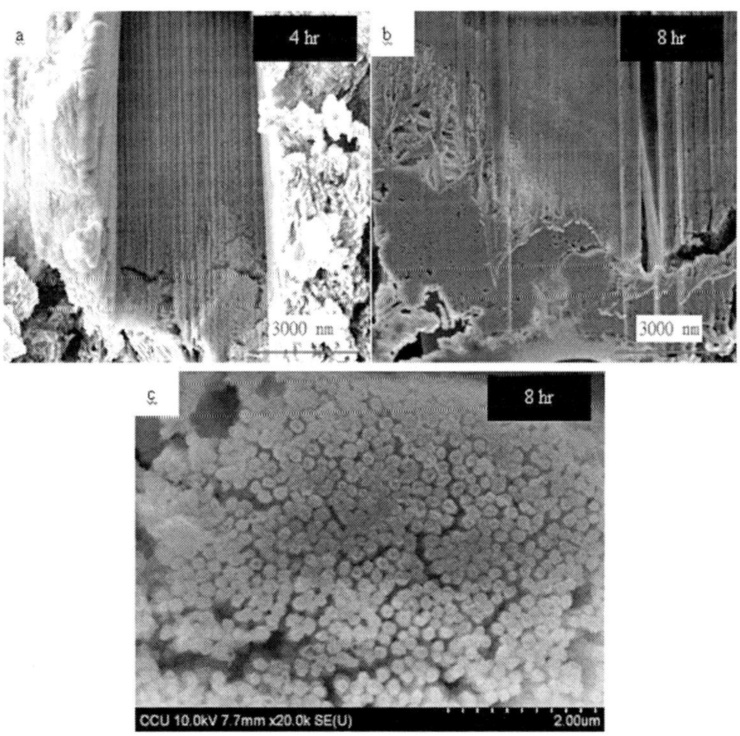

Figure 6.15. Cross sectional images of ZnO nanowire after 400°C heat treatment in air for (a) 4, (b) 8 h, and (c) 8 h (top-view SEM image).

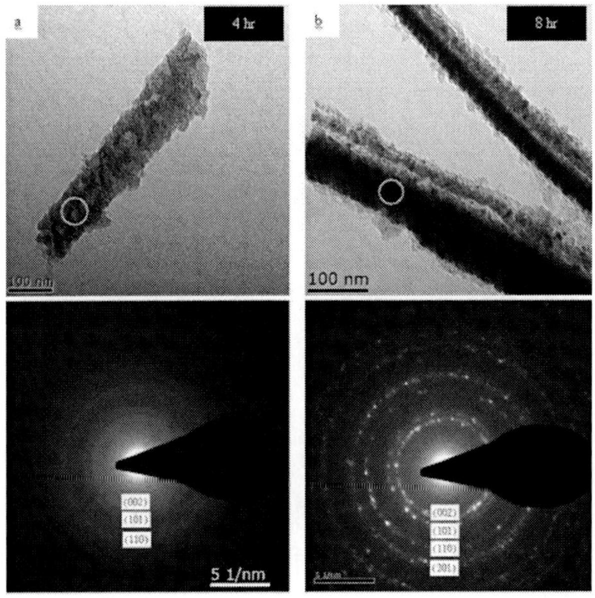

Figure 6.16. TEM image of single ZnO nanowire with 400°C heat treatment for (a) 4 and (b) 8 hr. The upper figure is electron diffraction pattern.

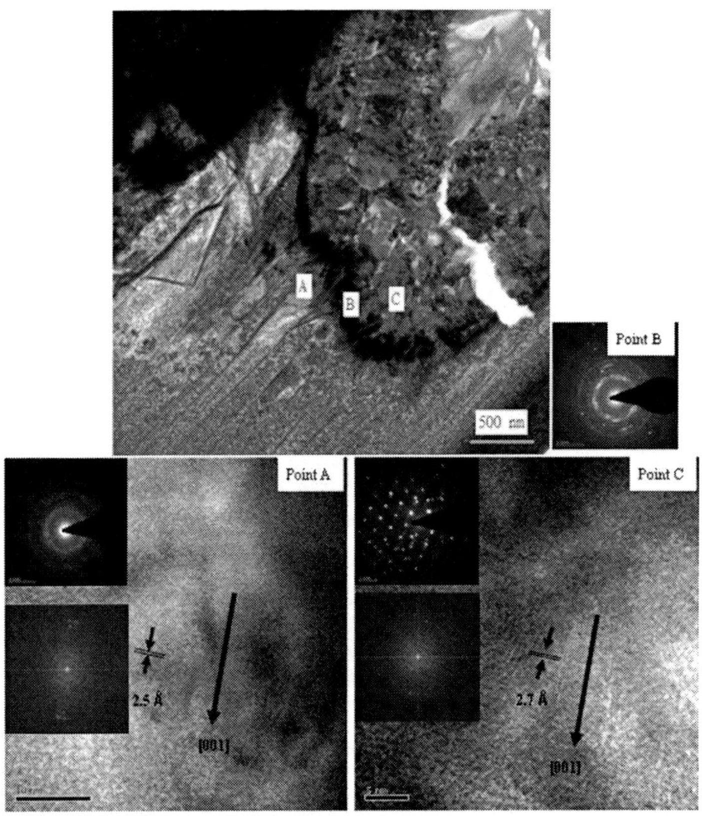

Figure 6. 17. TEM image of ZnO/AZO/PAMs nanowire arrays with heat treatment at 400°C for 8 hr. The sample was prepared by focus ion beam (FIB). Points A, B and C correspond to ZnO/AAO, interface between ZnO and AZO, and AZO, respectively.

CONCLUSION

Uniform and ordered Cu_2O oxide nanowire arrays were synthesized via PAMs template using three-electrode electrochemical deposition process. The structure of nanowires for controlling the annealing time from 4 to 12 hours is polycrystalline Cu_2O/Cu core-shell structure, fully Cu_2O, and Cu_2O/CuO. Under various annealing atmosphere conditions (from 0.02 to 0.12 atm), the structure varies from polycrystalline Cu_2O, to single-crystal rich-Cu_2O/Cu, and then single-crystal rich-Cu/Cu_2O. It was found that the template space limitation promotes Cu_2O nanowire growth by oxygen diffusion and the release of compressive stress.

Uniform and ordered polycrystalline ZnO/AZO/PAM nanowire arrays were synthesized using seed layer assisted electrochemical deposition. The average sheet resistance of 5-layers AZO seed was 129.37 Ω/sq under 400°C annealed in vacuum. The AZO seed layers were polycrystalline, which demonstrated greater adhesion on PAMs. The $Zn(OH)_2$ hydrate was obtained during electrochemical deposition under -1V at 65 and 80°C. Compared with 65 and 80°C, partially ZnO compound was formed at 80°C during the dehydrate process, which improved the crystallinity. Shape and size of ZnO nanowire arrays were controlled by PAMs of diameter around 110-140 nm. The crystallinity of ZnO was dependent on the AZO seed layers, which grew along the PAMs channel. The heterogeneous nuclei of AZO seed layers also improved the growth of ZnO nanowire.

ACKNOWLEDGMENTS

Authors are thankful to National Science Council of Taiwan for its financial support under the contract No: NSC98-2221-E-006-008 to carryout the present work.

REFERENCES

[1] M. Pounovic and M. Schlesinger, *Fundamental of Electrochemical deposition*, 2nd edition (2006), The Electrochemical Society Series, John Wiley and Sons, Inc., Hoboken, New Jersey, pp.1-6.

[2] Y. Namba, *J. Vac. Sci. Technol.* A 10, 3368 (1992).

[3] S. E. Kwiatek, V. Desai, P. J. Moran, P. M. Natashan, *J. Mater. Sci.* 32, 3123 (1997).

[4] D. Zhou, A. R. Krauss, D. M. Gruen, *J. Appl. Phys.* 82, 4051 (1997).

[5] Allen J. Bard and Larry R. Faulkner, *Electrochemical Methods: Fundamentals and Applications*, JOHN WILEY and SONS, INC. 2nd edition (2001), pp. 1-43.

[6] Y. Xia, P. Yang, Y. Sun, Y. Wu, B. Mayers, B. Gates, Y. Yin, F. Kim and H. Yan, *Adv. Mater.*15 353 (2003)

[7] C. R. Martin, *Science* 266, 1961 (1994).

[8] C. Scho¨nenberger, B. M. I. van der Zande, L. G. J. Fokkink, M. Henny, C. Schmid, M. Kru¨ger, A. Bachtold, R. Huber, H. Birk, and U. Staufer, *J. Phys. Chem. B* 101, 5497 (1997).

[9] M. Lederman, R. O'Barr, and S. Schultz, *IEEE Trans. Magn.* 31, 3793 (1995).

[10] Blondel, J. P. Meier, B. Doudin, and J.-Ph. Ansermet, *Appl. Phys. Lett.* 65, 3019 (1994).
[11] K. Ounadjela, R. Ferre´, L. Louail, J.-M. George, J. L. Manrice, L. Piraux, and S. Duboix, *J. Appl. Phys.* 81, 5455 (1997).
[12] H. Masuda and K. Fukuda, *Science* 268, 1466 (1995).
[13] K. Liu, C. L. Chien, P. C. Searson, and K. Y. Zhang, Appl. Phys. Lett. 73, 1436 (1998).
[14] T. M. Whitney, J. S. Jiang, P. C. Searson, and C. L. Chien, Science 261, 1316 (1993).
[15] K. Nielsch, F. Muller, A. P. Li, U. Gösele, *Adv. Mater.* 12, 582 (2000).
[16] K. B. Shelimov, M. Moskovits, *Chem. Mater.* 12, 250 (2000).
[17] Y. Lei, L. D. Zhang, G. W. Meng, G. H. Li, X. Y. Zhang, C. H. Liang, W. Chen, S. X Wang, *Appl. Phys. Lett.* 78, 1125 (2001)
[18] G. L. Che, B. B. Lakshmi, E. R. Fisher, C. R. Martin CR, *Nature*, 393, 346 (1998).
[19] J. Li, C. Papadopoulos, J. M. Xu, *Nature* 402, 253 (1999).
[20] Q. Lu, F. Gao, S. Komarneni, T. E. Mallouk, *J. Am. Chem. Soc.* 126, 8650 (2004)
[21] B. Balamurunga, B.R. Mehta, Thin Solid Films 396, 90 (2001).
[22] K. Akimoto, S. Ishizuka, M. Yanagita, Y. Nawa, Goutam K. Paul, T. Sakurai, *Sol. Energy* 80, 715 (2006).
[23] D. K. Zang, Y. C. Liu, Y. L. Liu, H. Yang, *Phys. B* 351, 178 (2004).
[24] E. Rakshani, Solid State Electron. 29, 7 (1986).
[25] G.P. Pollack, D. Trivich, *J. Appl. Phys.* 46, 163 (1975).
[26] J. Bloem, *Philips Res. Rep.* 13, 167 (1958).
[27] R. Inguanta, C. Sunseri, S. Piazza, *Electrochem. Solid State Lett.* 10, K63 (2007).
[28] J. Oh, Y. Tak, and J. Lee, *Electrochem. Solid-State Lett.* 7, C27 (2004).
[29] J. S. Choi, E. S. Ko, J. W. Kang, Y. S. Tak, and J. Lee, *J. Ind. Eng. Chem.*, 13, 305 (2007).
[30] R. Inguanta, S. Piazza, and C. Sunseri, *Electrochim. Acta*, 53, 6504 (2008).
[31] R. Guan, H. Hashimoto, K. H. Kuo, *Acta Crystallogr. B*, 40, 560 (1984).
[32] R. Guan, K. H. Kuo, H. Hashimoto, *Proceedings of the 8th European Congress on Electron Microscopy*, Budapest, 1984, p.1173
[33] G. Zhou, *J. Appl. Phys.,* 105, 104302 (2009).
[34] Y. M. Shen, Y. T. Shih, S. C. Wang, P. K. Nayak and J. L. Huang, *Thin Solid Films,* 519, 1687 (2010).
[35] M. Chen, Z. L. Pei, X. Wang, C. Sun, and L. S. Wen, *J. Vac. Sci. Technol. A*, 19, 963 (2001).
[36] Y. Chen, D. M. Bagnall, H. Koh, K. Park, Z. Zhu, and T. Yao, *J. Appl. Phys.* 84, 3912 (1998).
[37] Z. K. Tang, G. K. L. Wang, P. Yu, M. Kawasaki, A. Ohtomo, H. Koinuma. and Y. Segawa, *Appl. Phys. Lett.* 72, 3270 (1998).
[38] E. A. Meulenkamp, *J. Phys. Chem. B*, 102, 5566 (1998).
[39] R. E. Service, *Science*, 276, 895 (1997).
[40] E. A. Meulenkamp, *J. Phys. Chem. B*, 102, 5566 (1998).
[41] D. Banerjee, S. H. Jo, Z. F. Ren, *AdV. Mater.* 16, 2028 (2004).
[42] M. H. Huang, S. Mao, H. Feick, H. Q. Yan, Y. Y. Wu, H. Kind, E. Weber, R. Russo, P. D. Yang, *Science 292*, 1897 (2001).
[43] X. D. Wang, C. J. Summers, Z. L. Wang, *Nano Lett.* 4, 423 (2004).
[44] W. I. Park, D. H. Kim, S. W. Jung, G. C. Yi, *Appl. Phys. Lett.* 80, 4232 (2002).

[45] M. Monge, M. L. Kahn, A. Maisonnat, B. Chaudret, *Angew. Chem., Int. Ed.* 42, 5321 (2003).
[46] J. H. Choi, H. Tabata, T. Kawai, *J. Cryst. Growth* 226, 493 (2001).
[47] T. Nobis, E. M. Kaidashev, A. Rahm, M. Lorenz, J. Lenzner, M. Grundmann, *Nano Lett.* 4, 797 (2004).
[48] Y. Sun, G. M. Fuge, N. A. Fox, D. J. Riley, M. N. R. Ashfold, *AdV. Mater.* 17, 2477 (2005).
[49] Y. Li, G. W. Meng, and L. D. Zhang, *Appl. Phys. Lett.*, 76, 2011 (2000).
[50] J. G. Wang, M. L. Tain, N. Kumar, and T. E. Mallouk, *Nano Lett.*, 5, 1247 (2005).
[51] Q. Wang, G. Wan, B. Xu, J. Jie, and X. Han, Mater. Lett. 59, 1378 (2005).
[52] J. B. Yi, H. Pan, J. Y. Lin, J. Ding, Y. P. Feng, S. Thongmee, T. Liu, H. Gong, and L. Wang, *Adv. Mater*, 20, 1170 (2008).
[53] C. H. Ku, H. H. Yang, G. R. Chen, and J. J. Wu, *Crystal Growth and Design* 8, 283 (2008).
[54] C. L. Kuo, R. C. Wang, C. P. Liu, and J. L. Huang, *Nanotechnology,* 19, 035605 (2008).

In: Electrodeposition: Properties, Processes and Applications ISBN: 978-1-61470-826-1
Editor: Udit Surya Mohanty

Chapter 7

TAILOR-DESIGNED ELECTRODEPOSITED METALLIC THIN FILMS, NANOSTRUCTURES AND NANOWIRES TOWARDS TARGETED APPLICATIONS

Mohamed S. El-Deab, * ***Ahmad M. Mohammad and Bahgat E. El-Anadouli***
Department of Chemistry, Faculty of Science, Cairo University, Cairo, Egypt

ABSTRACT

This chapter describes our findings over the last few years, concerning the use of electrodeposition as a facile technique for the fabrication of several thin films of noble metals, nanostructures and nanorods onto various substrates. This includes three tasks. Firstly: the electrodeposition of metallic Ni or Cu thin films on reticulated vitreous carbon (RVC), and Pd or black Ni onto copper screens for use as electro-catalytically active cathodes for the hydrogen evolution reaction (HER). Moreover, the effects of some hydrodynamic and solution parameters concerning the performance of the electrodeposition process of heavy metal ions (e.g., lead ions) from flowing wastewater are briefly discussed. Secondly: the fabrication (via electrodeposition) of metallic (e.g., Au) and metal oxide (e.g., manganese oxide) nanostructures for application in fuel cell catalysis including the oxygen reduction and evolution reactions in addition to formic acid oxidation. Details of the morphological and electrochemical characterizations of the thus-electrodeposited nanostructures are outlined. This includes scanning electron microscope imaging (SEM), electron back scatter diffraction (EBSD) and X-ray diffraction (XRD) patterns. Thirdly: the fabrication of one-dimensional nanostructures, i.e., nanowires, for vital electronic, chemical, biological, and medical applications. Several methodologies are outlined for this sake; however, the electrodeposition in templates seems more interesting in terms of cost and potential for high volume production. The recent work combining the electrodeposition and the template growth approach for nanofabrication is introduced. Special emphasis is dedicated to assembling metal-semiconductor nanowire contacts in which the contact interface is formed along the cross-section of the wire.

* E-mail address: msaada68@yahoo.com, ahmad0873@yahoo.com, bahgat30@yahoo.com.

INTRODUCTION

Electrodeposition of thin solid films and nanoparticles onto solid substrates is among the many approaches for modifying electrode materials and is considered the most familiar binder-free technique. It is a facile and flexible technique which results in the direct attachment of the nanoparticles to the substrate without the need for immobilization (by anchoring agent) after the preparation step in addition to the facile control of the characteristics of the metal (or the metal oxide) (e.g., size, crystallographic orientation, mass, thickness and morphology of the nanostructured nanoparticles) by adjusting the current density, bath chemistry and temperature [1-4]. This technique, advantageously, allows to cover surfaces with metal (or metal oxide) nanoparticles on a macroscopic scale (for example 1 cm x 1 cm) within seconds or few minutes, in contrast to other techniques which require several steps and consume relatively longer time, such as sol-gel [5,6], micelle-based cluster generation [7,8] and metal vapor synthesis routes [9,10].

In this regard, catalysis and electrocatalysis by nanoparticles have been a subject of continuously growing interest and have been applied for diverse applications in which the metal (or metal oxide) nanoparticles are dispersed onto a relatively inert substrate. The incentive behind this interest is driven by the unusual and fascinating properties of nanoparticles compared to those of the corresponding bulk materials, such as the high effective surface area, catalytic activity, quantum confinement, etc [11-16]. The stimulus for this growth can be traced to new and improved approaches of making and assembling, positioning and connecting, imaging and measuring the properties of nanostructured materials with controlled size and shape, composition and surface microtopography, charge and functionality, for use in the macroscopic real world [17]. The catalytic performance of the nanoparticle-based electrodes is inherently connected with particle size, crystallographic orientation, the nature of the support as well as the method of preparation [18-22]. Thus, the preparation of catalysts with a tailored design in terms of size and crystallographic structure is a subject of unequivocal importance. These parameters are crucial for the kinetics of electrode reactions that involve adsorption of intermediates [23, 24].

The aim of this chapter is to shed some light on our recent results regarding the preparation (via electrodeposition technique) of metallic Ni or Cu thin films on reticulated vitreous carbon (RVC), and Pd or black Ni onto copper screens for use as electro-catalytically active cathodes for the hydrogen evolution reaction (HER). This also includes the electrodeposition of Au nanoparticles and/or metal oxides (e.g., manganese oxide and nickel oxide) nanoparticles onto various substrates. Another part deals with the fabrication of nanowires, for vital importance in electronic, chemical, biological, and medical applications. Thus, this chapter is divided into five parts, i.e.:

1. Electrodeposition of metallic Ni or Cu thin films on reticulated vitreous carbon (RVC) for use as electro-catalytically active cathodes for the hydrogen evolution reaction (HER).
2. Investigation of the effect of the solution parameters (e.g., viscosity and conductivity) on the overall performance of a porous electrode during the electrodeposition of metallic Pb from flowing alkaline solutions.

3. Electrodeposition of Au nanoparticles (nano-Au) onto carbon substrate, e.g., glassy carbon (GC) electrodes. Also investigation of the various parameters which affect the particle size, and the crystallographic orientation of the electrodeposited nano-Au for application as cathodes for the oxygen reduction reaction.
4. Electrodeposition of metal oxide nanoparticles (e.g., manganese oxide nanorods) onto GC and Pt electrodes and their use as efficient anodes for water electrolysis and for formic acid oxidation.
5. Fabrication of one-dimensional nanostructures, i.e., nanowires, via the electrodeposition in templates.

7.1. Electrodeposition of Metallic Thin Films

The production of pure hydrogen is important both for potential industrial and energy storage applications. The development of reliable and cost effective technologies for the production of pure hydrogen is a challenging issue from electrochemical engineering point of view. For hydrogen production from alkaline solutions, the cathodes that are used must have high stability, high specific surface area, and good electrocatalytic properties [25-27].

The choice of RVC as a porous cathode in the present study is based on the unique characterization of this material. These include its high area/volume ratio, high porosity, good electrical conductivity, high liquid permeability and its chemical and electrochemical inertness over a wide range of potential and chemicals. RVC is also an easily machinable material that could be formulated in different geometries. RVC porous electrodes have been used extensively as flow- through porous cathodes for the recovery of heavy metal ions from dilute streams [26-29] and also for the destruction of organic wastes from dilute streams [30,31]. The RVC cylinders were electroplated by black nickel prior to the insertion into the electrolysis cell. The electroplating bath composition was $(NH_4)_2MoO_4$ (35 g l^{-1}), $NiSO_4$ (140 g l^{-1}) and H_3BO_3 (20 g l^{-1}) [32]. Figure 7.1. is a schematic illustration of the cell and experimental arrangements of the porous-bed electrode. Electroplating of the RVC porous matrix was performed in situ with copper or nickel using flowing electroplating solutions. The electrode arrangement was described above (see Figure 7.1). The electroplating solutions and conditions are listed in Table 7.1 [33]. The electroplating solutions were circulated for 1 h at a linear flow rate of 1.0 cm s^{-1} [34]. Figure 7.2. shows i–E relations for HER on both RVC and black nickel-coated RVC (RVC/Ni) from 1 MKOH at 30°C and at an electrolyte flow rate of 0.6 cm s^{-1} [31]. A mathematical model helped to understand the effects of the electrolyte flow rate on the overall behavior of the porous electrode and on the potential and current distributions of the HER within the electrode [31]. It was also possible to extract the exchange current density of the HER using different electrodes. Fitting of the theoretical results with the polarization curve for HER on RVC/Ni was done using a value of $i_{0;H}$ equals 2×10^{-4} A cm^{-2}. This value of $i_{0;H}$ on RVC/Ni is comparable with those reported on the Ni electrode surfaces [35]._Figure 7.3. shows SEM micrographs for RVC cathodes coated with (a) Cu and (b) Ni. The coatings remain adhered for long times thus have the required performance and stability [34]. The decrease in the cathodic polarization when RVC was coated with Cu or Ni, leads to a corresponding energy saving at the cathode at a particular current, *i*. This is reflected on the overall energy consumption of the HER process.

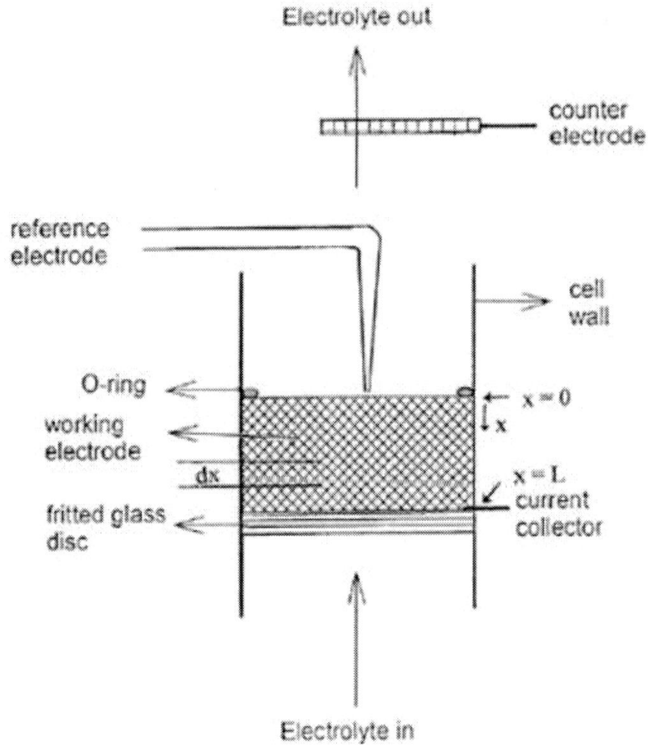

Figure 7.1. Schematic illustration and the experimental setup of a porous flow through electrode.

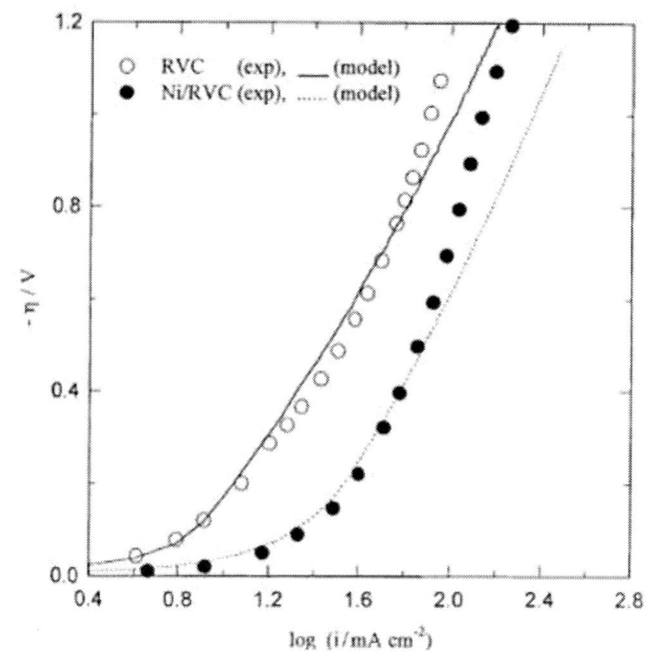

Figure 7.2. Comparison of the i–E relations for HER between RVC and RVC/Ni from 1 M KOH. Symbols show the experimental data and lines show the model predictions. Q = 0:6 cm s^{-1}, S = 70 cm^{-1} and T = 30°C.

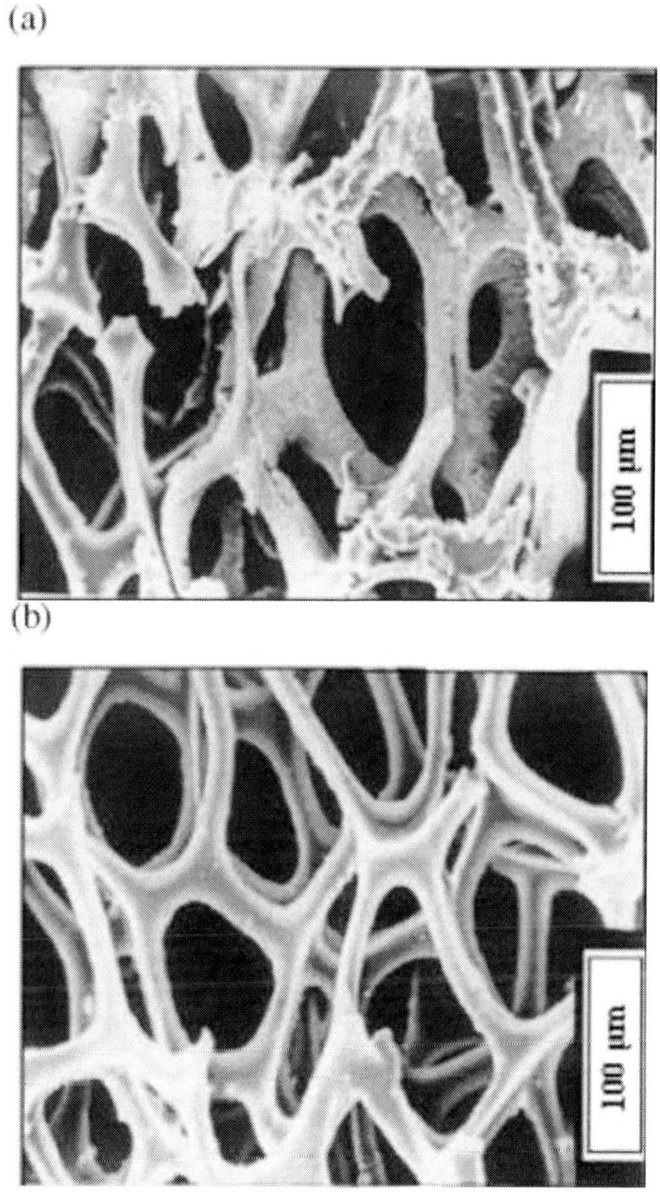

Figure 7.3. SEM micrographs of the different electrodes: (a) RVC/Cu and (b) RVC/Ni after an operation period of 30 h of HER at _2:0 V. The white zones are the matrix.

7.1.1. Effect of the Solution Parameters on the Performance of a Porous Electrode

The rate of the electrochemical processes taking place at porous flow-through electrodes is critically dependent on the flow rate, resistivity, diffusivity and viscosity of the electrolyte. The effects of the electrolyte viscosity on the behaviour of porous flow-through electrodes were studied during the electrodeposition of lead from alkaline solutions. The viscosity of an electrolyte is an important transport property which affects the rate of momentum transfer

[36] within the electrolyte inasmuch as the diffusivity, thermal conductivity and electrical resistivity affect the rates of mass, heat and charge transfer, respectively. As such, the viscosity, η, appears in expressions of the Reynolds number [37], which determines whether the flow regime is laminar or turbulent, i.e.

$$Re = 2\, v\theta rd/\eta$$
$$Re = 2vr\theta/\upsilon \ (1) \quad (1)$$

where *Re* is the Reynolds number, *v* is the superficial electrolyte flow rate, cm s^{-1}, *r* is the equivalent pore radius of the porous medium, cm, θ is the porosity of the porous electrode, and υ is the kinematic viscosity (viscosity(η)/density(*d*)) of the electrolyte, $cm^2\ s^{-1}$.

In view of the mechanism of ionic transport within electrolytes, the viscosity, diffusivity and electrical conductivity (ionic mobility) are inherently related [38-40]. Thus the Walden rule relates the ionic mobility (μ) to the viscosity of the medium (η) within which the ion migrates under the effect of a potential field [40], i.e.,

$$\mu\eta = \text{constant} \quad (2)$$

Furthermore, the Nernst-Einstein equation relates the diffusivity, *D*, and ionic mobility, μ, [41], i.e.

$$D = (RT/F)\,\mu \quad (3)$$

For this reason, the Walden rule has been frequently cited in the form [41-43]

$$D\eta = \text{constant} \quad (4)$$

In view of this, a change in the viscosity of the electrolyte must affect the behaviour of the electrode by affecting the rate of mass transfer of the electroactive species. Figure 7.4. shows the effect of the electrolyte viscosity on the current-potential (*i*-*E*) relations for the electrodeposition of lead from 1 M NaOH containing 19.3×10^{-5} M (40 ppm) lead ions at electrolyte flow rate of 1.25 cm s-1 [44].

Table 7.1. Bath composition and operating conditions for Cu and Ni electroplating on RVC and GC electrodes

Metal	Bath composition / g L^{-1}	c.d. / mA cm^{-2}	Duration / h
RVC/Cu	$CuSO_4$ - 150 H_2SO_4 - 50	100	1
RVC/Ni (Watts nickel)	$NiCl_2$ - 25 $NiSO_4$ - 90 H_3BO_3 - 60	100	1

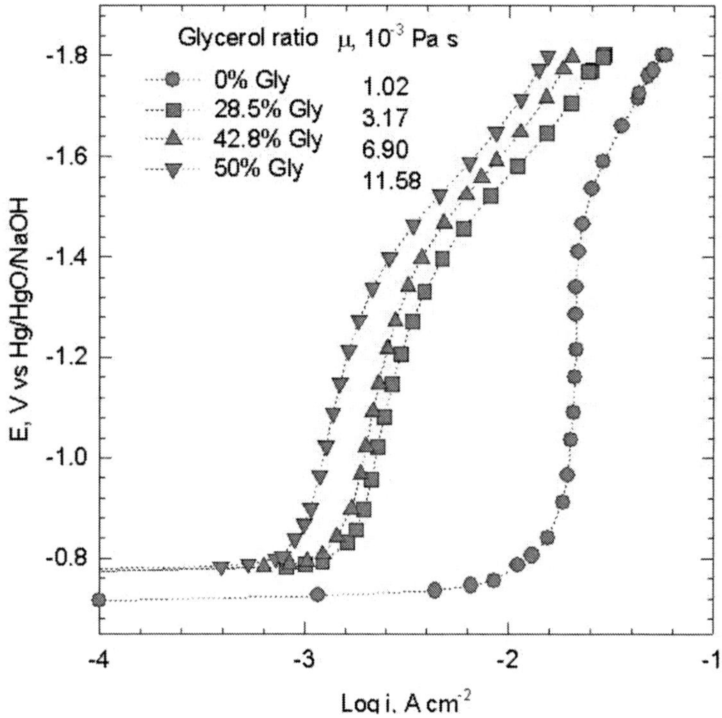

Figure 7.4. Effect of electrolyte viscosity (η) on the *i-E* relations for the lead deposition from 1 M NaOH solution containing 193 μM lead ions (40 ppm) and different ratios of glycerol. The working porous electrode was 0.33 cm thick (6 screens of 60 mesh) and the electrolyte flow rate was 1.25 cm s^{-1}.

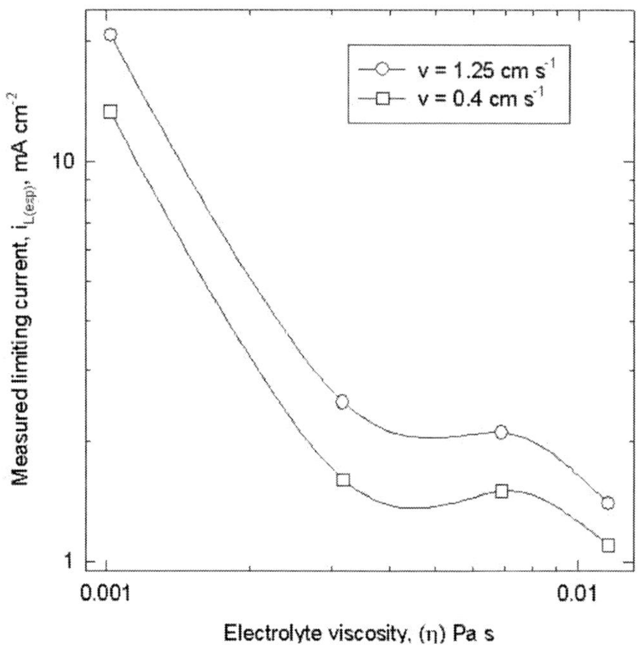

Figure 7.5. Effect of electrolyte viscosity (η) on the experimentally measured limiting current ($i_{L(exp)}$) at v equals (□) 0.4 and (O) 1.25 cm s^{-1}.

Table 7.2. Variation of the specific conductivity (κ), viscosity (η), density (d), and kinematic viscosity (υ) of the various electrolytes used in this study

Physical property → Electrolyte ↓	glycerol vol. %	κ / ohm^{-1} cm^{-1}	η / (10^{-3}) Pa s	d / g cm^{-3}	υ / (10^{-2}) cm^2 s^{-1}
1.00 M NaOH	0.0	0.13	1.14	1.03	1.10
1.00 M NaOH	28.5	0.04	3.17	1.09	2.90
1.00 M NaOH	42.8	0.03	6.91	1.13	6.79
1.00 M NaOH	50.0	0.02	11.58	1.14	10.16
2.00 M NaOH	0.0	0.32	1.35	1.07	1.25
3.00 M NaOH	0.0	0.39	1.75	1.11	1.57
5.26 M NaOH	0.0	0.39	3.12	1.19	2.62
0.10 M KOH	0.0	0.03	0.91	1.01	0.90
0.20 M KOH	0.0	0.06	0.91	1.01	0.90
0.30 M KOH	0.0	0.08	0.91	1.01	0.90
0.40 M KOH	0.0	0.12	0.91	1.01	0.90

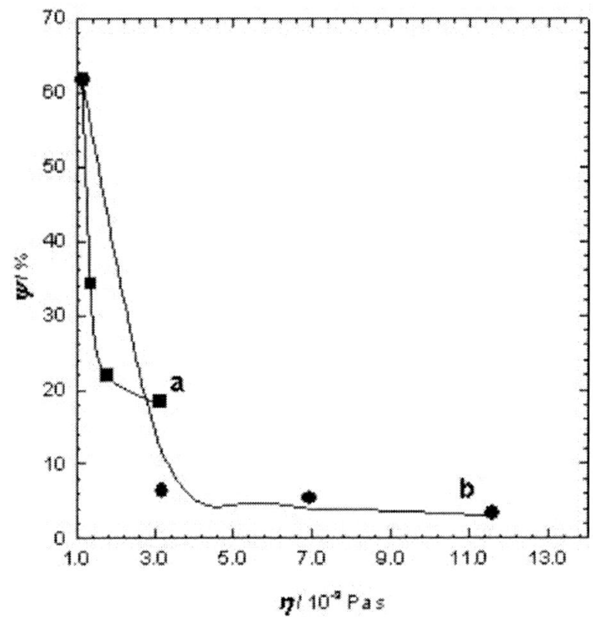

Figure 7.6. Effect of electrolyte viscosity (η) on the collection efficiency for the lead electrodeposition reaction. All electrolytes contained 193 μM lead ions (40 ppm) and are flowing at rate of 1.25 cm s^{-1}. The electrode was 0.33 cm thick (6 screens of 60 mesh). (a) Different NaOH concentration and (b) same concentration of NaOH (1 M) with different volume ratios of glycerol (see Table 7.2).

The electrodes were 0.33 cm thick (6 screens of 60 mesh) having a specific surface area, S, = 74 cm^{-1} and a porosity, θ, of 0.6. Inspection of this figure reveals that for all electrolytes a limiting current, $i_{L(exp)}$, is reached as the potential becomes sufficiently cathodic the value of which depends on the ratio of glycerol in the electrolyte. Figure 7.5. shows the effect of electrolyte viscosity on $i_{L(exp)}$. Table 7.2. shows the physical properties (specific conductivity, viscosity, density and kinematic viscosity) of the electrolytes used for the various tests, as measured during this work. This table shows that as the viscosity increases the conductivity of the electrolyte decreases[44].

The increase in electrolyte viscosity (which is accompanied by a decrease in its conductivity, see Table 7.2) leads to three pronounced effects [44]:

1. An increase in the potential at a particular current. This can be readily explained by the decrease of electrolyte conductivity that accompanies the increase in its viscosity.
2. A decrease in the measured limiting current, $i_{L(exp)}$, (see Figure 7.5) and
3. A less well defined limiting current. This effect is also observed with the more viscous, glycerol-free NaOH electrolytes.

Figure 7.6. shows the effect of the electrolyte viscosity (η) on the collection efficiency, Ψ. This figure shows that the increase of η decreasesΨ. The increase of diffusion coefficient (D) and/or decrease of the kinematic viscosity (υ) results in an increase in the mass transfer coefficient (km) and hence iL(exp) increases. This figure shows that ψ is much more sensitive to changes in η at low than at high values of η. [44]

7.2. Electrodeposition of Au Nanoparticles (Nano-Au) onto Carbon Substrate

Au nanoparticles-based catalysts, among others e.g., Pt, Ru, Pd, Ag, etc, are the most intensively used nanoparticles and are widely applicable in many vital processes, e.g., reduction of NO with propene, CO or H_2, removal of CO from H_2 streams, selective oxidation, e.g., epoxidation of olefins as well as a selective hydrogenation of CO and CO_2. In addition to their extraordinary catalytic activity for the oxygen reduction [45-51] Au nanoparticle-based substrates have been efficiently utilized for the hydrogenation of unsaturated organics [52,53] as well as low-temperature oxidation of CO [54-56].

Because gold tends to bond with some organic molecules, the attachment of gold nanoparticles on several substrates usually uses peculiar binder molecules such as aminopropyl siloxane, mercaptopropyl siloxane and other analogues utilizing the bonding ability of the silanol group to the substrate and the affinity of the –SH or $-NH_2$ group toward the gold particles [57]. However, from catalytic point of view, the binder layer may affect the catalytic reactivity of the gold nanoparticles and reduce the surface conductivity. In addition, the oxidative or reductive desorption of the binder layer may shorten the applicable potential window of the electrode. Thus, direct attachment of nanoparticles on the electrode substrates without using organic binder molecules may overcome these disadvantages.Among the many approaches, the most familiar binder-free technique is the electrodeposition of nanoparticles. Figure 7.7. shows a linear sweep voltammetry (LSV) for the electrodeposition of Au nanoparticles onto GC (nano-Au/GC) electrode (diameter = 3.0 mm) from acidic 0.5 M H_2SO_4 solution containing 1.0 mM $Na[AuCl_4]$ at a potential scan rate of 100 mV s^{-1}. The reduction peak of Au ions to Au metal is centered at ca. 0.5 V vs. Ag/AgCl/KCl(sat). Thus, a potential step electrolysis from an initial potential of 1.1 V to a final potential of 0.0 V (vs. Ag/AgCl/KCl(sat)) is proved to be sufficient to electrodeposit the nano-Au from this solution for different electrolysis time (Figure 7.8).

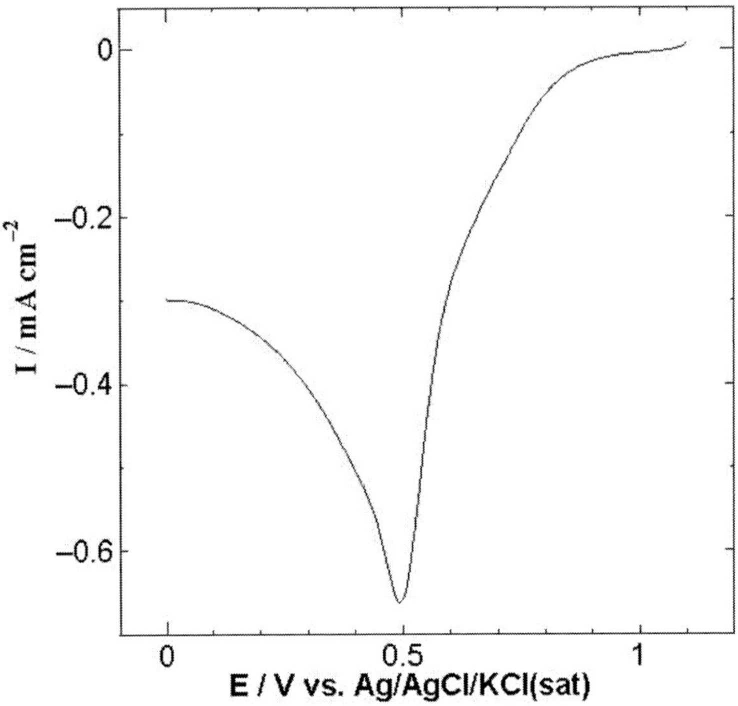

Figure 7.7. LSV for electrodeposition of Au nanoparticles at GC electrode (d = 3.0 mm) from 0.5 M H_2SO_4 solution containing 1.0 mM $Na[AuCl_4]$. Potential scan rate: 100 mV s^{-1}.

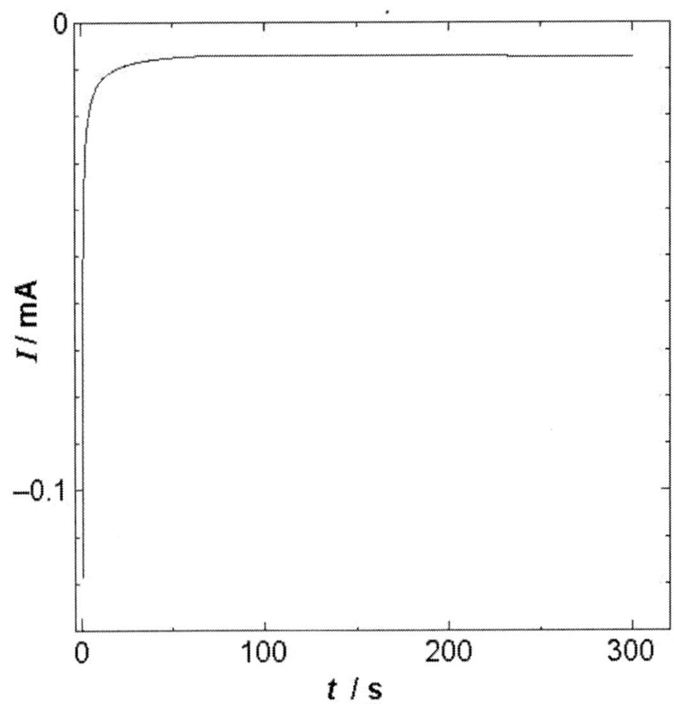

Figure 7.8. I-t relation obtained at GC electrode (d = 3.0 mm) during the potential step electrolysis experiment in 0.5 M H_2SO_4 containing 1.0 mM $Na[AuCl_4]$.

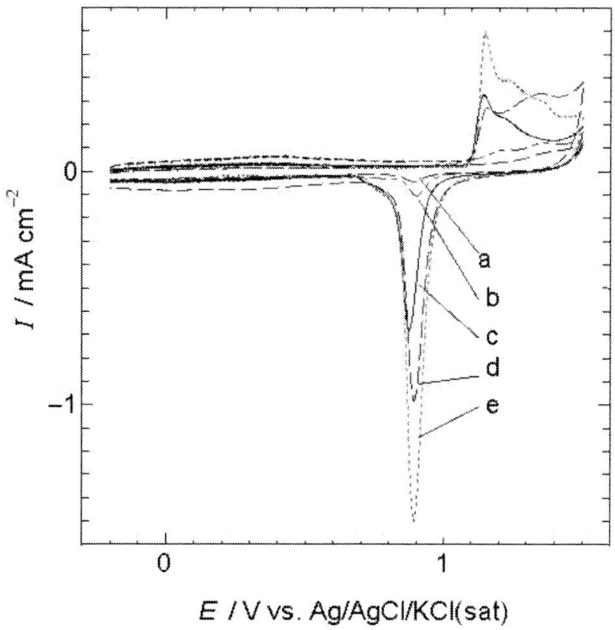

Figure 7.9. CV response, in N_2-saturated 0.05 M H_2SO_4, for Au nanoparticles-electrodeposited GCEs prepared from 0.5 M H_2SO_4 solution containing 0.11 mM Na[$AuCl_4$] (a) or 1.0 mM Na[$AuCl_4$] (b-d) via (a,b) 5, (c) 300 and (d) 600 s potential step from 1.1 to 0.0 V vs. Ag/AgCl/KCl (sat). Potential scan rate: 100 mV s^{-1}. Curve (e) represents the CV response of the bulk Au electrode.

Integration of the amount of the Faradaic charge passed during this potential step electrolysis helped in the estimation of the loading of the nano-Au at the electrode. The successfulness of the electrodeposition of the nano-Au onto the GC electrode is verified by the measurement of the characteristic cyclic voltammetric (CV) response of the nano-Au/GC electrode in N_2-saturated 0.05 M H_2SO_4 (Figure 7.9) [47]. The appearance of the two characteristic peaks corresponding to the formation of the Au surface oxide monolayer formation (in the potential range of ca. 1.1 to 1.5 V) and its reduction (at ca. 0.9 V) revealed the successful electrodeposition of the nano-Au over the GC substrate. The increase of the reduction peak current at ca. 0.9 V reflects the increase of the Au loading on the GC electrode [47]. The different shape of the oxidation peak (in the potential range 1.1 to 1.5 V) might originate from the different ratios of the single crystalline facets constituting the electrodeposited Au nanoparticles by changing the electrodeposition conditions. Furthermore, the morphological characterization is carried out by taking the scanning electron microscope (SEM) imaging of the nano-Au/GC electrodes (Figure 7.10) [47]. It is clear from this figure that the increase of either the time of deposition or the concentration of the Na [$AuCl_4$] resulted in an increase in the average particle size of the Au nanoparticles. That is, the Au particles deposited from the 0.11 mM Na[$AuCl_4$] solution are in the nano-scale size (typically ≈ 20 nm) and homogeneously distributed throughout the glassy carbon surface (Figure 7.10a), while in the case of the electrodeposition from the 1.0 mM Na[$AuCl_4$] solution, in addition to the 20 nm-scale particles, bigger particles of about 80-100 nm size (clusters) are formed (Figures 7.10b-d). The real surface area of the Au loading was estimated by calculating the amount of charge consumed during the reduction of the Au surface oxide monolayer using a reported value of 400 μC cm^{-2} [58-60].

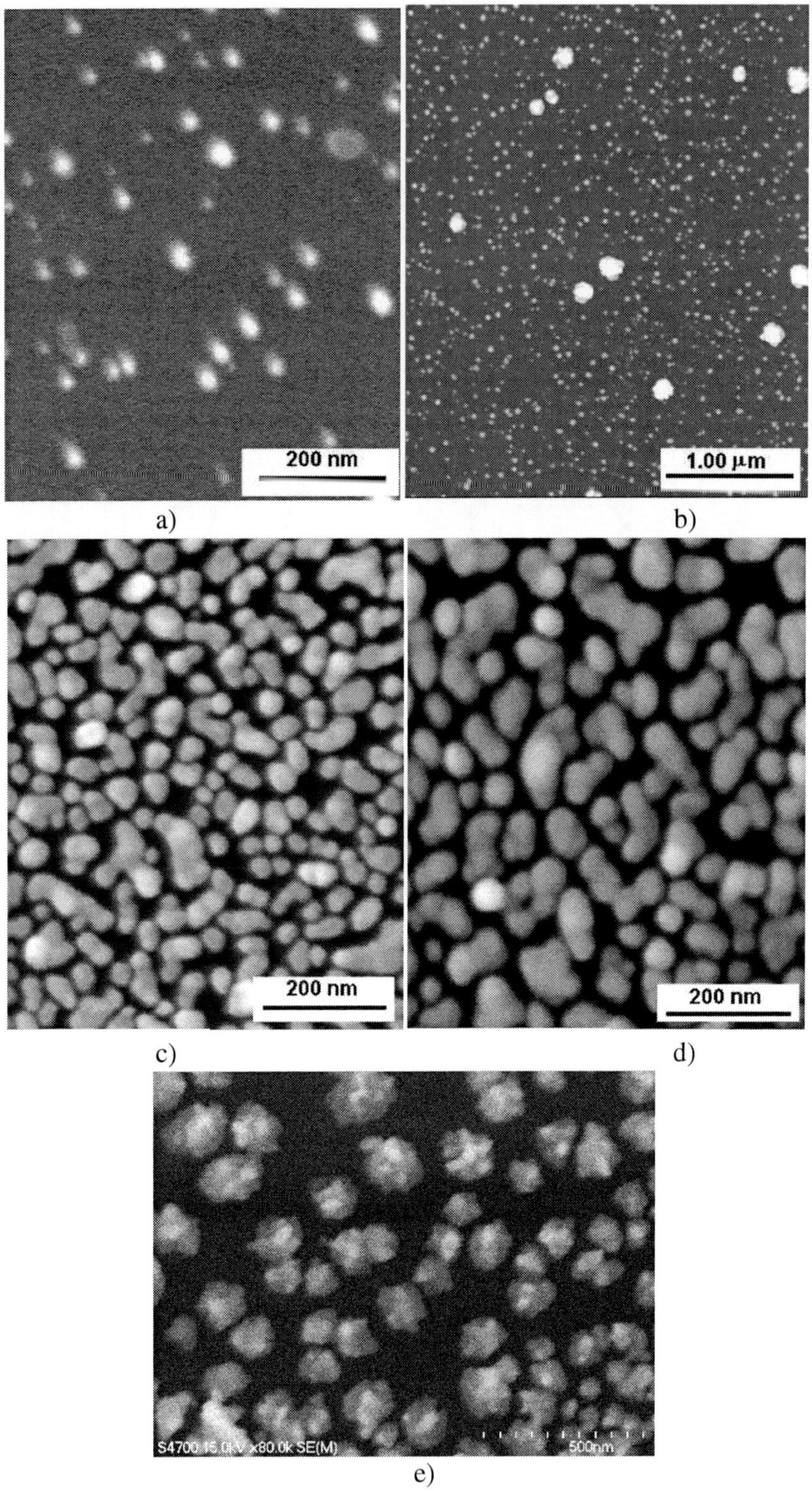

Figure 7.10. SEM micrographs of Au nanoparticles electrodeposited onto (a-d) GC electrodes and (e) CF microelectrode prepared from 0.5 M H_2SO_4 solution containing (a) 0.11 mM $Na[AuCl_4]$ and (b-e) 1.0 mM $Na[AuCl_4]$ via (a, b) 5, (c) 300, (d) 600, and (e) 60 s potential step from 1.1 to 0.0 V vs. Ag/AgCl/KCl (sat).

Table 7.3. Characterization of the Au loadings on GC electrodes under different conditions of electrodeposition

Electrode No.	Deposition time/s	Bath composition	Au loading[a] / g cm^{-2}	Average surface area of Au loading[b] / cm^2	Equivalent film thickness[c] / nm
1	5	0.11 mM $Na[AuCl_4]$ in 0.5 M H_2SO_4	$2.78 x 10^{-7}$	5.40×10^{-3}	0.14
2	5	1.0 mM $Na[AuCl_4]$ in 0.5 M H_2SO_4	$1.50 x 10^{-6}$	1.68×10^{-2}	0.77
3	300	As above	$1.98 x 10^{-5}$	1.07×10^{-1}	10.3
4	600	As above	$4.16 x 10^{-5}$	1.46×10^{-1}	21.6

[a]as calculated from the i-t curve during the potential step electrodeposition,

bas estimated from the charge consumed for the reduction peak of the surface oxide monolayer of Au (the peak at 900 mV in Figure 9) using a reported value of 400 μC cm-2.

[c] it is the thickness of a homogeneous Au film that covers the entire surface area of the GCE and has the same Au loading as that of nanoparticles.

Table 7.4. The particle size distribution and the particle density of the Au nanoparticles electrodeposited from 0.5 M H_2SO_4 containing 1.0 mM $Na[AuCl_4]$ in the presence of different additives. The potential was stepped from 1.1 V to 0 V for 60 s

Additives	Particle size distribution / nm	*Particle density[a] / particle μm^{-2}*
none	20 - 200	~ 72
10 μM cysteine	50 - 250	~ 64
100 μM cysteine	50 - 300	~ 32
100 μM I^- ions	10 - 40	~ 300

[a]The number of particles per 1 μm^2.

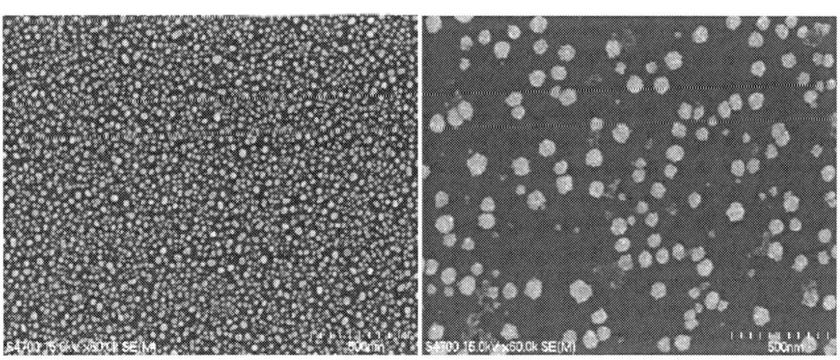

a) b)

Figure 7.11. SEM images for the Au nanoparticles electrodeposited on GC electrodes from 0.5 M H2SO4 containing 1.0 mM Na[AuCl4] in the presence of (a) 0.1 mM I– ions and (b) 0.1 mM cysteine. The potential was stepped from 1.1 V to 0 V for 60 s.

Table 7.3. summarizes the loading characteristics of the Au nanoparticles electrodeposited on GC electrodes under different conditions of electrodeposition. Similarly, Au nanoparticles have been successfully electrodeposited onto carbon fiber (CF) electrodes (10 μm in diameter) applying the same procedure as mentioned above for the GC substrate, and the corresponding SEM images are shown as Figure 7.10.e [61].

In a trial to control the particle size and/or its crystallographic orientation, the electrodeposition of Au nanoparticles has been performed in the presence of some additives, typically cysteine or iodide ions. The Au nanoparticles prepared in the presence of I^- ions are of the smallest particle size in addition to the relatively narrow range of particle size distribution (10-40 nm) (Figure 7.11.a), and the particle density of this nano-Au/GC electrode is highest among the nano-Au/GC electrodes examined here (Table 7.4) [3]. This fact may be attributed to the strong specific adsorptivity of I^- ions to the freshly prepared Au nanoparticles. That is, the adsorption of I^- ions may carry a negative charge on the Au nanoparticles and consequently the coalescence of the neighboring Au nanoparticles can be prevented effectively. On the contrary, the Au nanoparticles prepared in the presence of cysteine are bigger (aggregates) with a particle size reaching 300 nm (Figure 7.11b), suggesting that the adsorption of cysteine molecules is not effective for preventing the coalescence of the neighboring Au nanoparticles.

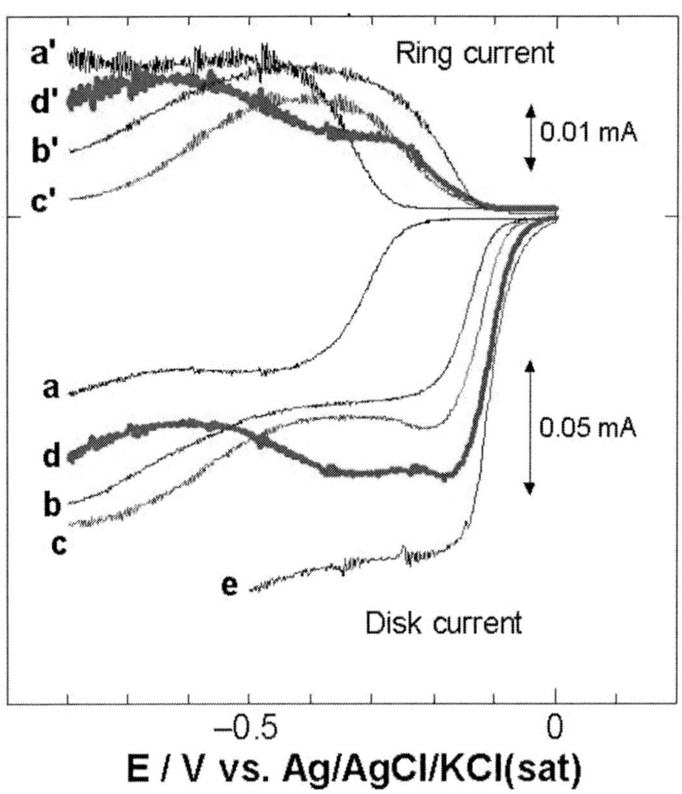

Figure 7.12. RRDE voltammograms for the ORR, in O_2-saturated 0.5 M KOH, at (a) bare GC, (b-d) nano Au/GC disk electrodes (d = 3.0 mm), electrodeposited in (b) the absence of additives, and the presence of (c) 0.1 mM iodide ions and (d) 0.1 mM cysteine, respectively. Curve (e) measured at Pt-disk. Curves a'-d' represent the corresponding Pt-ring currents (polarized at 0.5 V). Rotation rate: 200 rpm. Potential scan rate of the disk electrode: 20 mV s^{-1}.

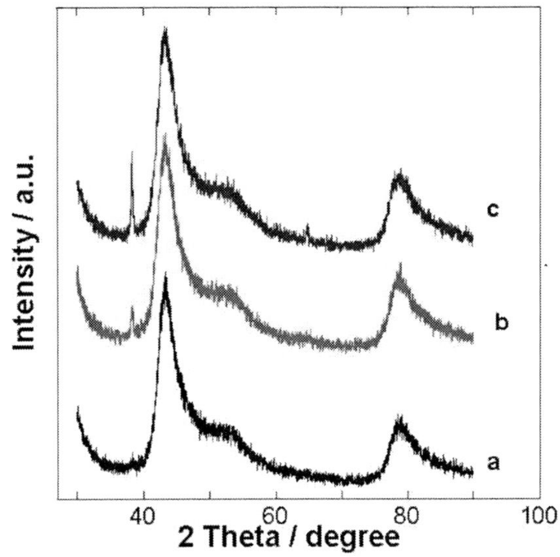

Figure 7.13. XRD patterns of nano-Au/GC electrodes prepared in the same way as in Figure 1 with a step width of (a) 60, (b) 300 and (c) 900 s.

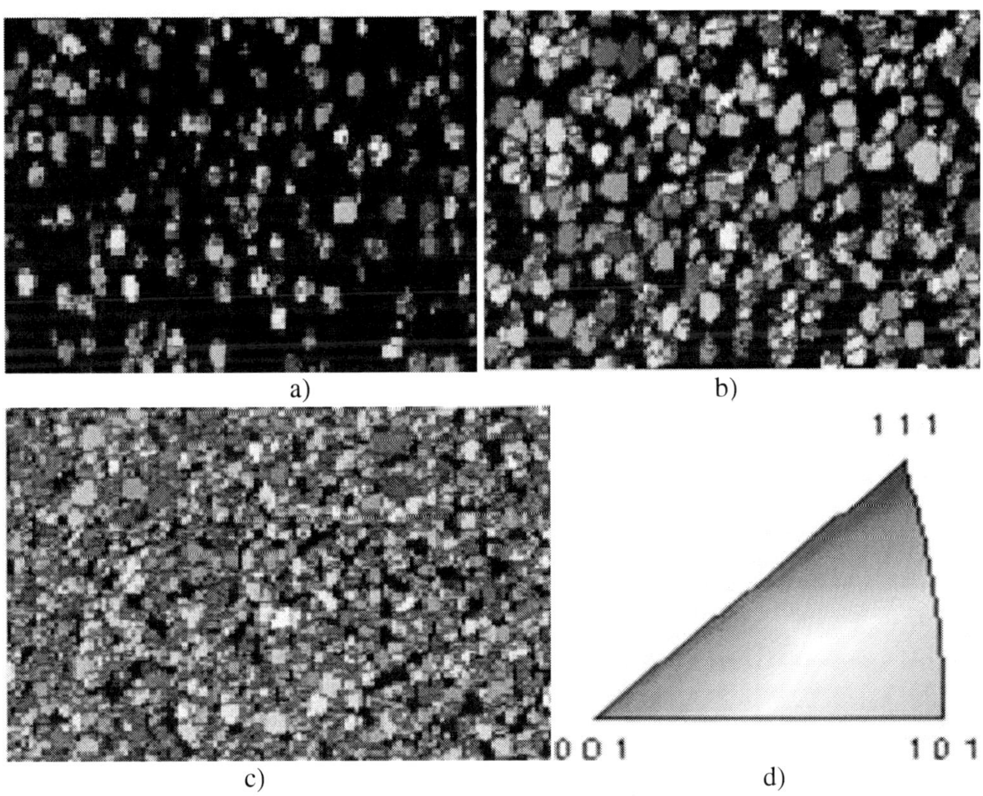

Figure 7.14. Electron backscatter diffraction (EBSD) patterns of Au nanoparticles electrodeposited onto GC electrodes (at magnification factor of 80k, $S = 500$ nm). Electrodeposition time: (A) 60, (B) 300 and (C) 900 s. (D) is a color coding to identify the crystallographic orientation of Au nanoparticles.

The electrocatalytic behaviour of the nano-Au/GC electrodes towards the ORR has been assessed by the measurements of the steady-state hydrodynamic voltammetric using a rotating ring disk electrode (RRDE). Figure 7.12. shows the steady-state hydrodynamic voltammograms for the ORR at (a) bare GC disk and (b-d) different nano-Au/GC disk electrodes in O_2-saturated 0.5 M KOH solution at a rotation rate of 200 rpm [45]. The current of the Pt ring (potentiostated at 0.5 V, corresponding to the oxidation of hydrogen peroxide produced at the disk electrode) is shown at the upper side of Figure 7.12. (curves marked a'-d'). The Pt disk current (curve e) is shown for comparison at which the direct 4-electron reduction of O_2 to OH^- takes place. As can be readily seen from this figure that the nano Au/GC disk electrode (marked d, prepared in the presence of 0.1 mM cysteine) supports the highest current among the examined nano-Au/GC electrodes up to a disk potential of ca. –0.5 V vs. Ag/AgCl/KCl(sat). This high value (maxima) of the disk current (I_D) is associated with a low value of the corresponding Pt ring current (I_R) (see curves d and d'), indicating a significant contribution of the direct 4-electron reduction of O_2 to OH^- at this electrode compared with the others. Furthermore, this electrode showed the highest positive shift (almost close to that obtained at the Pt disk, curve e) of the onset potential of the ORR current flow. The enrichment of the electrocatalytically active Au (100) (and Au(110)) facet domains (prepared in the presence of cysteine) at the expense of the less active Au(111) facets is the primary factor which results in the high electrocatalytic activity towards the ORR in this alkaline medium. The determination of microstructure of Au nanoparticles is important because it determines many of their physical and electrocatalytic properties. Figure 7.13. shows the XRD patterns of the Au nanoparticles electrodeposited onto GC electrodes for various durations. The diffraction peaks located at 2θ values of ca. 45 and 80° are attributed to the (011) and (110) planes of the carbon substrate, respectively [62]. This figure reveals the existence of two peaks at 2θ of ≈ 38° and 64° corresponding to the Au(111) and Au(220) facet domains of the Au nanoparticles, respectively [63]. For the present case, the substrate is glassy carbon, which is of less ordered surface structure and the quantitative estimation of the preferential crystallographic orientation of Au nanoparticles is difficult due to the noisy background of the XRD patterns caused by the interference from the GC substrate. EBSD is a powerful technique which advantageously allows for obtaining crystallographic information (i.e., crystal orientation mapping) of samples (e.g., particles, clusters, aggregates). Figure 7.14. shows typical crystal orientation maps obtained for Au nanoparticles electrodeposited for various durations, i.e., (a) 60, (b) 300 and (c) 900 s. Note that particles of the similar orientations are in similar colors. Black regions of the mapping image represent areas undetected by electron backscattering due to roughness of the GC substrate [64]. This figure reveals the following points, (i) Au nanoparticles electrodeposited at short time (60 s, image A) are rich in the (111) orientation (as indicated from the predominance of the blue color) and (ii) longer electrodeposition time (≥ 300 s) leads to enrichment of Au nanoparticles with the (100) and (110) orientations as reflected from the high intensity of the red and green colors of images B and C [65]. This observation is consistent with the theoretical studies of the equilibrium and growth forms of three-dimensional crystals, in which the growth forms of three-dimensional crystals contain closely packed crystallographic orientation (the most thermodynamically stable facet i.e., the (111) facet for Au) at the early stage of electrodeposition [66]. On the other hand, at longer electrodeposition times the percentage of (100) and (110) increases probably due to the more complex mechanism of growth and coalescence of the neighboring Au nanoclusters [66]. The fractions for the (001), (101) and (111) orientations corresponding to the three samples are given in Table 7.5. The total fraction of a specific facet is the percentage of a specific orientation relative to the total area of the analyzed region of the surface [67]. This table shows that the increase of t_d resulted in (i) increase in the total fraction of the three low index facets of Au nanoparticles and (ii) a decrease in the percent of the (111) orientation. It should be mentioned here that the analysis of the EBSD images (given in Table 7.5) provided the relative percentage of the three low index facet orientations of the electrodeposited Au nanoparticles and cannot be used to estimate the total surface coverage of GC.

Table 7.5. Variation of the total fraction of the (001), (101) and (111) facet domains for the Au nanoparticles electrodeposited for various durations

t_d / s	Total fraction			% of (111)
	(001)	(101)	(111)	
60	0.010	0.026	0.030	45
300	0.055	0.097	0.069	31
900	0.089	0.210	0.108	26

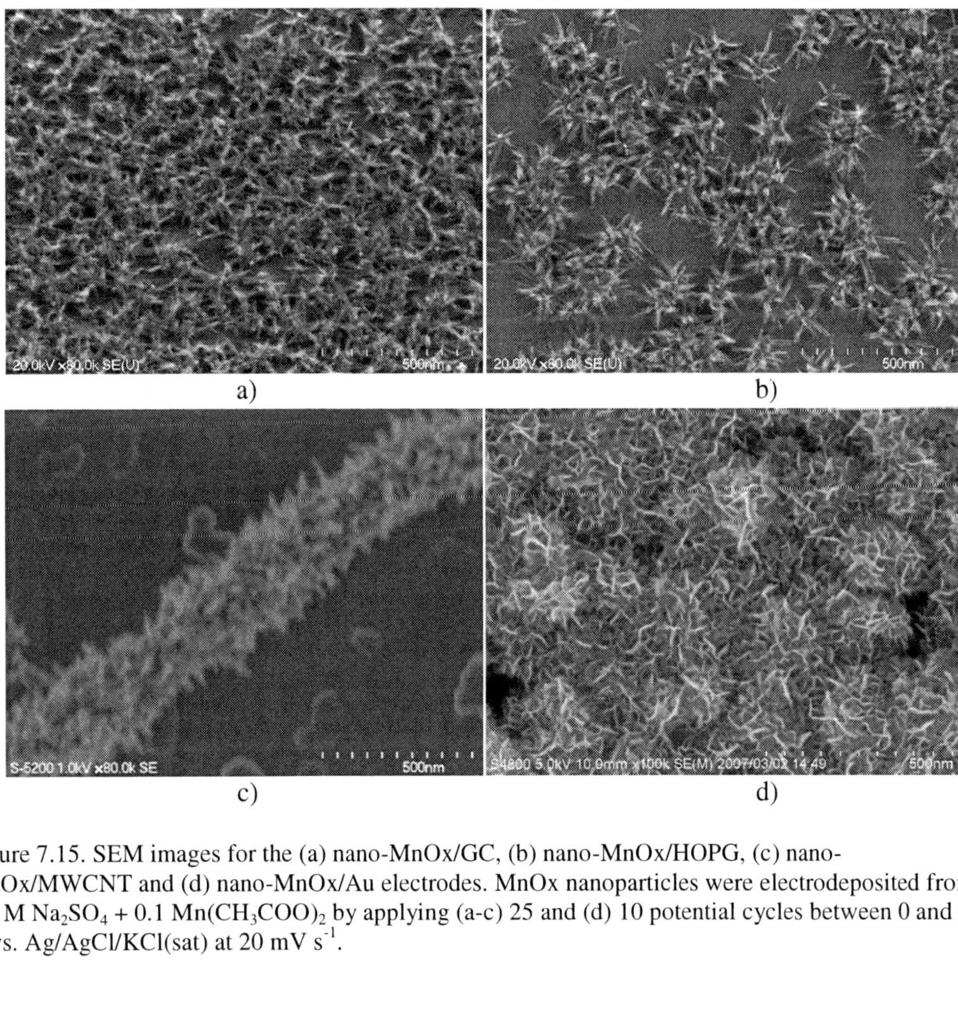

Figure 7.15. SEM images for the (a) nano-MnOx/GC, (b) nano-MnOx/HOPG, (c) nano-MnOx/MWCNT and (d) nano-MnOx/Au electrodes. MnOx nanoparticles were electrodeposited from 0.1 M Na_2SO_4 + 0.1 $Mn(CH_3COO)_2$ by applying (a-c) 25 and (d) 10 potential cycles between 0 and 0.4 V vs. Ag/AgCl/KCl(sat) at 20 mV s^{-1}.

7.3. Electrodeposition of Metal Oxide Nanoparticles

Metal oxide nanostructures represent an important category of nanoscale materials which attracts increasing technological and industrial applications in view of their innovative properties, e.g., optical, magnetic, electrical, and catalytic properties. Additionally, their general characteristics such as mechanical hardness, thermal stability and/or chemical passivity are well recorded [68-88]. The application of the metal oxide nanoparticles includes: (i) materials with a very high specific surface area (used in the manufacturing of supercapacitors) and (ii) materials with size-dependent physical properties [89]. Crystallographic orientation as well as size and surface topography of the metal oxide nanostructures can be controlled during the preparation procedure [90]. This is a crucial point in the fabrication of efficient catalysts [91-93].

Manganese oxide (MnO_x) is among the most widely studied metal oxides in view of its excellent catalytic activity toward the chemical disproportionation of hydrogen peroxide into water and molecular oxygen [93-96]. This material can be prepared chemically as reported in the literature [97] in different phases. Thus, it is very interesting to study the feasibility of its direct electrodeposition onto the electrode substrate in a trial to control its amount, phase, and structural properties. MnO_x nanoparticles (nano-MnOx) were electrodeposited onto different substrates from a solution containing 0.1 M Na_2SO_4 + 0.1 M $Mn(CH_3COO)_2$ via cycling the potential between 0.0 and 0.4 V vs. Ag/AgCl/KCl(sat) for different numbers of cycles at 20 mV s^{-1} [98].

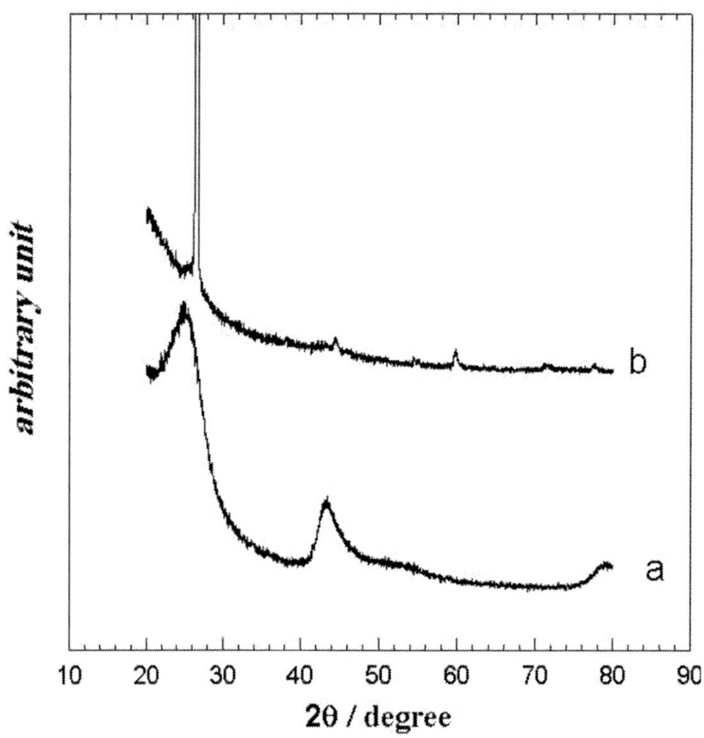

Figure 7.16. XRD patterns for the (a) nano-MnOx/GC and (b) nano-MnOx/HOPG electrodes. Nano-MnOx were electrodeposited from an aqueous solution of 0.1 M Na_2SO_4 containing 0.1 M $Mn(CH_3COO)_2$ by applying 25 potential cycles between 0 and 0.4 V at 20 mV s^{-1}.

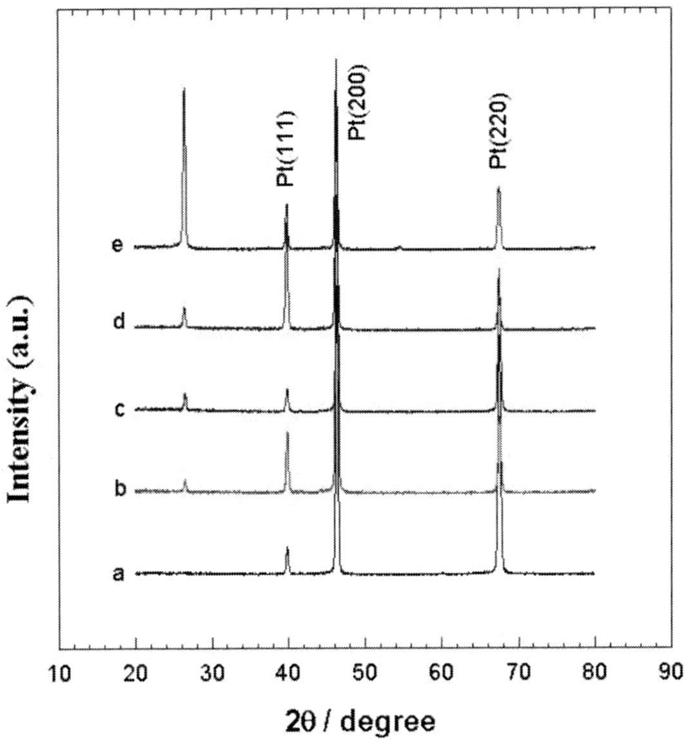

Figure 7.17. XRD patterns of the bare Pt (a) and Pt electrodes modified with electrodeposition of nano-MnOx for (b) 5, (c) 25, (d) 50 and (e) 100 potential cycles between 0 and 0.4 V vs. Ag/AgCl/KCl(sat) at 20 mV s^{-1}.

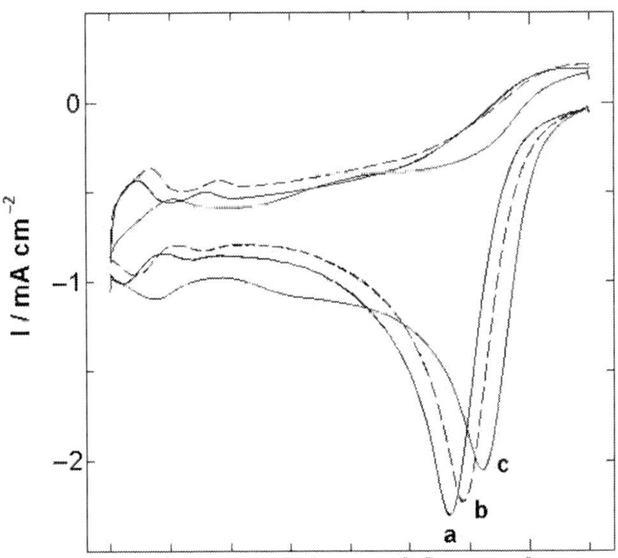

Figure 7.18. CVs for the ORR at (a) bare Pt and (b, c) nano-MnOx/Pt electrodes measured in O_2-saturated 0.1 M KOH solution. Potential scan rate: 100 mV s^{-1}. Nano-MnOx was electrodeposited by employing (b) 5 and (c) 25 potential cycles.

Figures 7.15. a-d show typical SEM micrographs obtained for (a) nano-MnOx/GC, (b) nano-MnOx/HOPG, (c) nano-MnOx/MWCNTs, and (d) nano-MnOx/Au electrodes. The MnOx was electrodeposited in a porous texture composed of nanorods onto the various electrodes. This texture covers homogeneously the entire surface of the GC (HOPG, MWCNTs and Au) electrode in a rather porous texture form, which enables the accessibility of the solution species to the underlying substrate through nano channels across the MnOx nanotexture. The beneficial electrocatalytic effect of this structure will be demonstrated later towards the oxygen reduction reaction (ORR) as well as the oxygen evolution reaction (OER). Figure 7.16. shows the XRD patterns of the nano-MnOx electrodeposited onto GC and HOPG. In contrary to the amorphous GC, the highly crystalline nature of the HOPG substrate advantageously enables the monitoring of the XRD pattern of the electrodeposited MnOx. This figure shows that manganese oxide is electrodeposited in a crystalline phase as a sharp peak located at 2θ of ca. 26° emerged upon its electrodeposition (which is overlapped with the C(002) peak of the HOPG substrate).Figure 7.17. shows the typical XRD patterns of Pt substrates loaded by different amounts of nano-MnOx (curves b-e) in comparison with the unloaded (bare) Pt substrate (curve a). The peaks located at 2θ of ca. 40, 46.5 and 78.5° correspond to the Pt(111), Pt(200) and Pt(220) single crystalline facets of the underlying polycrystalline Pt substrate. This figure reveals that MnOx is electrodeposited in a crystalline phase as a sharp peak, located at 2θ of ca. 26.5°, emerged upon its electrodeposition onto the Pt substrate. This peak corresponds to the $(11\bar{1})$ crystallographic plane of the manganite phase [62] (manganese oxide hydroxide, MnOOH [99]). Figure 7.18. shows the CV response for the ORR at (a) bare Pt and (b,c) nano-MnOx/Pt electrodes measured in O_2-saturated 0.1 M KOH at 100 mV s^{-1}. This figure shows that the electrodeposition of nano-MnOx resulted in two significant observations. Firstly, a positive shift of the onset potential of the ORR, that is, the potential of the peak current shifts positively from −140 mV at the bare Pt to −110 and −70 mV at the nano-MnOx/Pt electrodes prepared by applying 5 and 25 potential cycles, respectively.

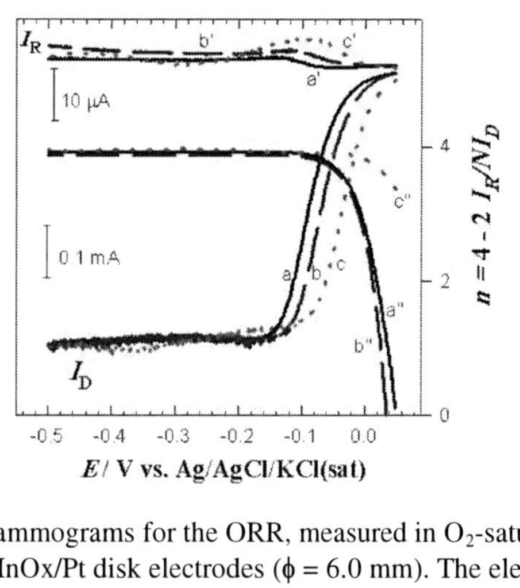

Figure 7.19. Steady-state voltammograms for the ORR, measured in O_2-saturated 0.1 M KOH, at (a) bare Pt disk and (b, c) nano-MnOx/Pt disk electrodes (ϕ = 6.0 mm). The electrodeposition conditions used for the MnOx preparation were the same as in Figure 8 ((b) 5 and (c) 25 potential cycles). Rotation rate: 200 rpm. Potential scan rate of the disk electrode: 10 mV s^{-1}. Curves a'-c' represent the corresponding Pt-ring currents (polarized at 0.5 V). Curves a''-c'' represent the number of electrons (n) involved in the ORR at the different disk electrodes.

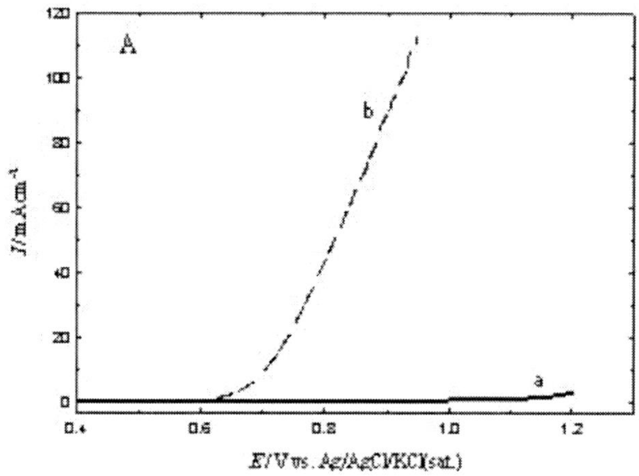

a

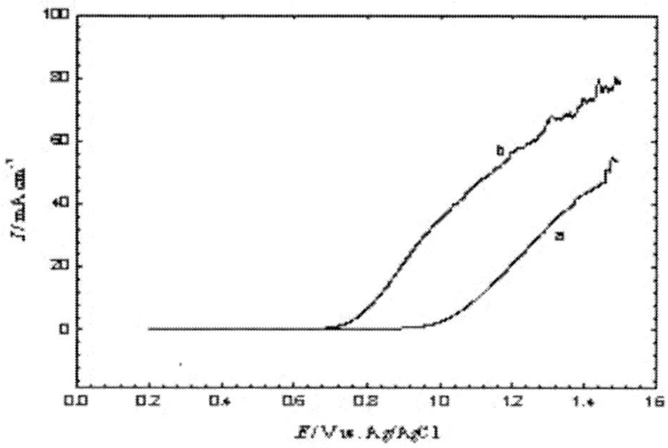

b

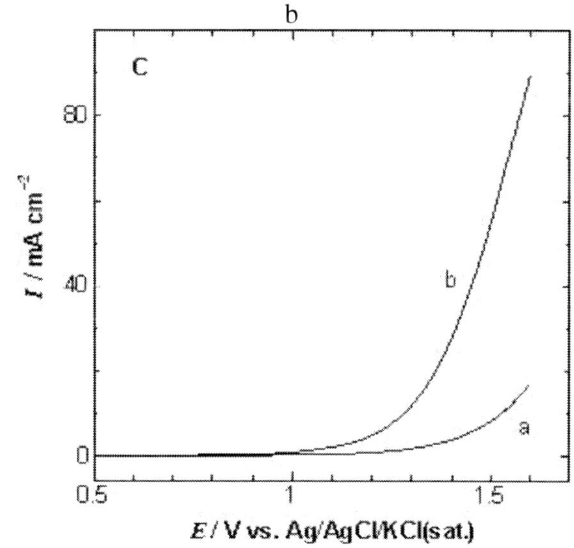

c

Figure 7.20. LSV response for the OER in N_2-saturated 0.5 M KOH at (A) Au, (B) Pt and (C) GC electrodes modified with electrodeposited nano-MnOx (curves labeled b) at potential scan rate of 20 mV s^{-1}. Curves labeled a correspond to the LSV response of the unmodified (i.e., bare) electrodes. The nano-MnOx has been electrodeposited by applying 10 potential cycles between 0 and 0.4 V vs. Ag/AgCl/KCl(sat.) at potential scan rate of 20 mV s^{-1}.

And secondly, a slight decrease in the peak current intensity is noted. The first fact reflects an enhancement of the electrocatalytic activity of the nano-MnOx/Pt electrodes towards the ORR. The second one is due to the retarded O_2 diffusion through the porous nano-MnOx matrix. The similarity in the CV response of the ORR at the nano-MnOx/Pt (curves b and c) and the unmodified Pt (curve a) electrodes can be reasonably attributed to the formation of a micro-disk type Pt electrode (in the nano-MnOx/Pt) at which the diffusion layers at individual bare Pt spots overlap each other to form a linearly expanding diffusion region. This assumption would be verified by comparison of the diffusion layer thickness ($\delta = \sqrt{(\pi D_{O2}\, t)} \approx 0.0224$ cm; D_{O2} is the diffusion coefficient of O_2 in alkaline aqueous medium ($\approx 10^{-5}$ cm^2 s^{-1}) and t is the electrolysis time (s)) and the average distance between bare Pt spots $\approx$ 50 nm. Thus δ overweighs the average distance between the bare spots of Pt at the nano-MnOx/Pt electrode. The corresponding steady-state hydrodynamic measurements for the ORR at the same electrodes are shown in Figure 7.19. [100]. This figure reveals two important points, i.e., (i) A positive shift of the onset potential of the ORR to different extent at the different nano-MnOx/Pt disk electrodes (curves b and c) compared to that obtained at the bare Pt disk (curve a) and (ii) A noticeable increase of the corresponding ring current (curve c'), which indicates a slight contribution of the two-electron reduction pathway of molecular oxygen as a result of the electrodeposition of nano-MnOx.

Table 7.6. Values of the anodic potential (E_{50}) of the various electrodes (unmodified and nano-MnOx-modified Au, Pt and GC) during the OER in N_2-saturated 0.5 M KOH operating at a current density of 50 mA cm^{-2}

	E_{50} / V vs. Ag/AgCl/KCl(sat)	
Substrate	Bare	Modified
Au	>> 1.2	0.8
Pt	1.45	1.16
GC	> 1.6	1.48

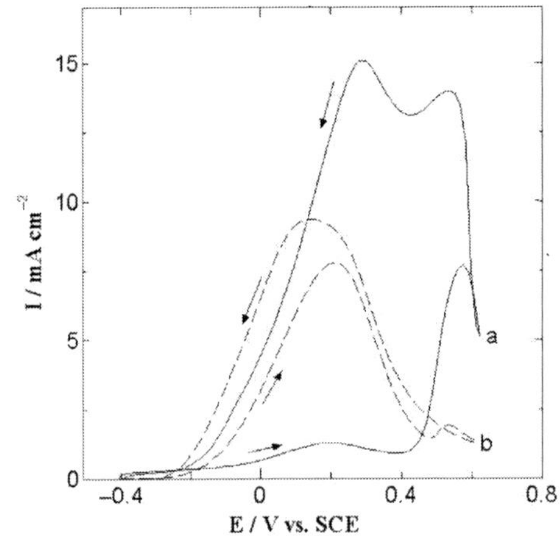

Figure 7.21. CVs for formic acid oxidation at (a) unmodified Pt and (b) nano-MnOx/Pt ($\theta \approx 30\%$) electrodes in 0.3 M HCOOH (pH 3.45) at 50 mV s^{-1}. The inset shows the current transients obtained during formic acid oxidation at the same electrodes at +100 mV vs. SCE in the same solution.

In further application of GC, Au and Pt electrodes modified with electrochemically deposited crystallographically oriented manganese oxide nanorods (nano-MnOx) have been examined as novel anodes for the electrocatalytic evolution of oxygen gas from alkaline medium (N_2-saturated 0.5 M KOH).

Figure 7.20. shows linear scan voltammograms (LSV) for the oxygen evolution reaction (OER) measured in N_2-stuarated 0.5 M KOH solution at bare (curves labeled a) and nano-MnOx modified (curves labeled b) (A) Au, (B) Pt and (C) GC electrodes. This figure shows a significant enhancement of the polarization behaviour of the three substrates to different extent depending on the nature of the substrate. That is, a 550 mV negative shift of the onset potential of the current flow has been obtained upon the modification of the Au electrode by nano-MnOx (Figure 7.20A). Similarly, a 300 mV negative shift of the onset potential of the OER has been observed upon the modification of Pt (Figure 7.20.B) and GC (Figure 7.20.C) electrodes. Table 7.6. lists the electrode potentials (E_{50}) of the individual electrodes corresponding to a current density of 50 mA cm^{-2}. This table shows a significant decrease in the E_{50} values at the modified electrodes compared to the unmodified electrodes. This corresponds to an equivalent decrease in the energy consumption accompanying the overall electrolysis process. The nano-MnOx modified Au electrode, advantageously, shows a lower anodic overpotential "η" towards the OER than that observed at Ni electrodes, which are the most common electrodes for the OER, in similar media [101]. That is, a value of η of 0.3 V is required to support a current density of 10 mA cm^{-2} at the nano-MnOx modified Au electrode, whereas the same current density is supported by the Ni electrode with η of 0.5 V in similar alkaline media [101]. Thus, a potential saving of about 200 mV is achieved at the proposed electrocatalyst.

Nano-MnOx modified Pt electrodes have been suggested as a novel electrocatalysts for the oxidation of formic acid. CV measurements (shown in Figure 7.21) have been performed in a deoxygenated 0.3 M formic acid (pH 3.45) at (a) bare Pt and (b) nano-MnOx/Pt ($\theta \approx$ 30%) [102]. The first peak (I_p^d) is assigned to the direct oxidation of formic acid to CO_2 which is retarded indicating that poisoning of the surface takes place during the course of the formic acid oxidation. The oxidation of the poisoning intermediate CO at the surface accounts for the observation of the second peak [103-108]. The modification of Pt with nano-MnOx (curve b of Figure 7.21) resulted in a significant increase of the first peak, I_p^d, concurrently with a noticeable depression of the second oxidation peak, I_p^{ind}, indicating that the direct pathway is enhanced and less poisoning intermediate (CO) is produced. The stability of nano-MnOx/Pt electrode has been tested by recording the current transients (*I-t* curves, inset of Figure 7.21) at +100 mV vs. SCE for formic acid oxidation. The *I-t* curves show that nano-MnOx/Pt (curve b) supports a higher oxidation current (about 4 times) than bare Pt (curve a). This demonstrates the preferred oxidation of formic acid via the direct path at the modified electrode and its high CO tolerance.

7.4. Fabrication of One-Dimensional Nanostructures

In microelectronics, the modern ultra-large scale integrated circuits (ULSI) consist of tens to hundreds of millions of devices to establish a complex functionality. In order to build more sophisticated circuits, more complex devices will be required. This should lead to a continuous scaling down of the feature sizes in microelectronic devices and circuits. Nanofabrication which deals with manipulating materials on an atomic or molecular scale appears interesting in this regard. The fabrication of one-dimensional nanostructures (nanowires, nanorods, and nanotubes) is, in particular, an exciting research area as they promisingly participate in vital electronic, chemical, biological, and medical applications. Several methodologies have been reported for this sake; however, the electrodeposition in templates seemed more interesting in terms of cost and potential for high volume production. In addition, nanostructures with small monodispersed diameters similar to that of the pores of the membrane can be prepared and moreover be freed from the template and collected for further use. Finally, segmented multi-metal rods of controllable length can be easily prepared by sequential electrodeposition. Herein, the recent work combining the electrodeposition and the template growth approach for nanofabrication is introduced. A special emphasis will be dedicated to assembling nanowire contacts in which the contact interface is formed along the cross-section of the wire. The recent advanced revolution in nanoscience has promoted a structural and electronic improvement in integrated circuits for the sake of establishing a sophisticated functionality. This enhancement has targeted both of the circuits (which consist of tens to hundreds of millions of devices), and interconnects (the wires that provide the power and other signal to these devices). To achieve this, a continuous scaling down of the feature sizes in microelectronic devices and circuits was required.

Shrinking of the feature size has a great impact on the physical and chemical properties of materials. In reality, nanomaterials possess an immense surface area per unit volume, a high proportion of atoms in the surface and near surface layers, and the ability to exhibit quantum effects. The resulting unique properties of nanoparticles cannot be anticipated from a simple extrapolation of the properties of bulk materials. For example, carbon-sheathed Ge nanowires (55 nm in diameter and a micron in length) started to melt at 650 ^{o}C , which is almost 280 ^{o}C lower than the regular melting point of bulk Ge (930 ^{o}C)[109]. Theories also predict a dramatic reduction in the thermal conductivity of semiconductors with shrinkage of the size due to the enhanced surface and grain boundary scattering [110]. The impact of this enhancement in the thermal properties of nanomaterials will be reflected in the decrease in the annealing temperature required to prepare high-quality, defect-free electrical components. Another implication may appear in the capability of cutting, linking, and welding nanomaterials at relatively modest temperatures, facilitating the integration of these nanostructures into functional devices and circuitry. One more interesting property is the electrical conductivity of metal-semiconductor contacts, which sometimes increased four order of magnitudes as a consequence of shrinking of the features sizes down to the nanometer-scale [111]. The optical properties are also influenced, and nanostructures acquire different colors based on size [112, 113].

Among the nanostructures, nanowires received substantial attention as they have the potential to answer fundamental questions about one-dimensional systems and are expected to play a vital role in electronic technologies. They also offer a great potential as building blocks to realize nanodevices and molecular electronics. Interestingly, semiconductor nanowires can easily be doped and approach a metallic regime [114]. Therefore, one may not get surprised when seeing that great deal of effort dedicated to realize the applications of nanowires in electronic, magnetic, optical, and mechanical nanodevices [115-124].To realize electronic nanodevices, a controlled assembling of metallic and semiconductor nanowires as well as metal-semiconductor nanowire contacts, where the contact interface is formed along the cross-section of the nanowire will be of great importance. Several metals are of interest in this regard, however, silicon, which is the long dominant semiconductor material in microelectronic industry, stands as one of the few preferred semiconductor material in nanoelectronics. Preliminary investigations indicated that silicon nanowires, SiNWs, exhibited fascinating electrical conductivity [125, 126], thermal conductivity [110], field emission [127, 128], and visible photoluminescence due to quantum confinement effect [129, 130].

Electrodeposition, particularly when proceed in nanoporous templates, is considered among the simplest, most convenient and cheapest tools for synthesizing metallic nanowires. On the other hand, many successful strategies for the fabrication of SiNWs have been proposed including electron beam lithography[131-135], reactive ion etching[115, 131, 133, 136, 137], solution-phase self-assembly[129], thermal evaporation and sublimation[138-140], and laser ablation (which may be oxide[141-144] or metal[114, 145, 146] assisted). The expensive cost and the sophistication accompanied with these approaches have renewed again the interest in the classical metal catalyzed vapor-liquid-solid (VLS) growth mechanism introduced by Wagner and Ellis[147-150] and detailed by Givargizov and coworkers[151, 152]. Both of the electrodeposition and the VLS techniques can be combined with the template' approach where a precise control of wire dimensions (height and width) can be achieved [153-157]. In addition, nanostructures with extraordinary small diameters can be prepared and the wires can moreover be freed from the template membrane and collected for further use. Moreover, segmented multi-metal rods of controllable length can be easily prepared by sequential electroplating. More interestingly, growth within templates provides a controlled method for contact formation. The objective of this part is to address the application of the template-assisted electrodeposition approach in the synthesis of Cobalt silicide ($CoSi_x$) nanowire contacts to SiNWs for nanoelectronics applications [153, 156]. A combination of the electrodeposition, the template and the vapor-liquid-solid growth approaches are all combined to synthesize this type of contacts in which, the contact interface is formed along the cross-section of the wire. Let us first overview the main features of both of the metal-catalyzed VLS and the template approaches before talking about their merge to assemble the nanocontacts.

7.4.1. The Vapor-Liquid-Solid (VLS) Growth Mechanism [147-152]

The VLS mechanism serves as a basis for controlled growth of whiskers (crystals or wires) from vapors. In this mechanism, an impurity, usually called a catalyst, is used to form a liquid alloy droplet of relatively low freezing temperature. The crystal material should be soluble in the liquid droplet that is situated between the vapor and the growing crystal. The surface of the liquid has a large accommodation coefficient, and therefore, becomes a preferred site for deposition from the vapor. The deposition continues until the liquid droplet becomes supersaturated with the crystal material. The whisker grows by precipitation of the crystal material from the liquid droplet. Therefore, one simply can assume that the VLS mechanism occurs in two steps. The first step is a deposition from the vapor directly on a liquid solution at a vapor-liquid interface. The second and subsequent step, occurring in a liquid-solid system, is a precipitation from the supersaturated liquid solution at the liquid-solid interface.

An important feature of the VLS growth is that the rate of supply of material from the vapor should be kept low. Too rapid vapor transport can locally raise the supersaturation in the liquid and terminate the controlled VLS growth. The liquid droplet can be completely covered by silicon crystals or the droplet can break up into numerous small droplets giving rise to a "brushlike" growth of whiskers. Another key feature of the VLS growth method is that equilibrium phase diagrams can be used to predict catalysts and growth conditions, thereby enabling rational synthesis of new nanowire materials. The presence of a nanoparticle catalyst with a diameter 1.5-2 times that of the connected nanowire at one end of the nanowire, is a third essential feature of VLS growth [158]. Finally, the VLS growth can only occur within a specific range of temperatures, where liquid metal-containing droplets can form. Above this range, evaporation will occur, and the liquid droplet cannot form. Below this range, the gas cluster is solidified directly and again no liquid droplet is formed [158].

7.4.2. Advantages of VLS Technique [149-151]

1. The methods can be applied to crystals of many substances.
2. Using this technique, it is possible to control the locations and dimensions of the growing crystals.
3. Because VLS uses a lower temperature than is used in the case of direct deposition from vapor onto a solid, diffusion is also slower in the case of VLS.
4. A clean interface can be obtained since the initial action of the liquid droplet or layer is to dissolve surface layers.

7.4.3. Properties of the Alloying Agent (Catalyst) [147, 148]

A certain alloying agent must meet a number of requirements to be suitable for the VLS growth.

1. It must form a liquid alloy with the crystal material to be grown at the deposition temperature.
2. The distribution coefficient of the agent, $k = C_s/C_l$, must be less than 1, where C_s and C_l are the solubilites of the agent in the solid and liquid, respectively, at the deposition temperature. In the absence of a flux of agent from the vapor, the k value determines the length of a crystal grown by VLS. For whisker growth k should be about 10^{-4} or smaller. For large area growth, values as large as 0.1 can be tolerated for k.
3. The equilibrium vapor pressure of the agent over the liquid alloy should be small. Evaporation of the agent from the liquid solution does not change the composition, but changes the volume and the cross section of the crystal.
4. The agent must be inert towards the chemical reaction products.
5. The vapor-solid, vapor-liquid, and liquid-solid interfacial energies influence strongly the shape of the growing crystal. A small contact angle is desirable for large area growth; a large contact angle leads to whisker formation.

6. In case of using VLS in the growth of compound semiconductors, such as GaAs or SiC, the agent can be one of the components, Ga or Si.
7. In some cases, the agent and the deposition temperature must be selected to avoid the formation of a solid intermediate phase of the agent with one of the constituents of the vapor.
8. For controlled unidirectional growth by VLS, the solid-liquid interface should be well-defined crystallographically.

7.5. Growth of Si NWs via the VLS Mechanism

In case of using gold as a catalyst for the growth of Si whiskers, a small particle of Au is placed on the (111) surface of a Si wafer and heated to a temperature high enough to form a liquid droplet of Au-Si alloy. A Si containing gas is then introduced, and the liquid alloy acts as a preferred sink for arriving Si atoms. The Si enters the liquid and freezes out, with a very small concentration of Au in the solid solution, at the interface between solid Si and the liquid alloy. The process is continued, and the alloy droplet becomes displaced from the substrate crystal and "rides" (lifts off the horizontal surface) atop the growing whisker. The whisker grows on the (111) direction in length by this mechanism until the Au is consumed or until the growth conditions are changed. A Au-Si alloy of hemispherical shape is clearly visible at the termination of the wire. Most of the wires grown via VLS are found free of dislocations and stacking faults.

7.5.1. Template Approach

Synthesis of metallic and semiconductors nanowires has long been attracting the attention of several interdisciplinary and international research groups as they have been regarded as the most promising building block for nanoscale electronic and optoelectronic devices [159]. Indeed, nanosystems can easily be built from semiconductor nanowires using metallic nanowires as interconnects. One of the most convenient procedures for the controlled synthesizing of nanowires in a large quantity is the *in*-template deposition approach [153, 156, 160, 161]. This process involves synthesizing a certain material using the direct catalytic growth methods within the pores of a porous membrane [162]. Because the membranes that are used have cylindrical pores of uniform diameter, a nanocylinder of the desired material is obtained in each pore.

There are several interesting and useful features associated with template synthesis. The most useful one originate from its being extremely general in terms of the types of materials that can be prepared [162]. Therefore, a large set of nanostructures with small diameters can be prepared. Moreover, the grown wires obtain monodispersed diameters similar to that of the pores of the membrane and can moreover be freed from the template membrane and collected for further use [163, 164]. Alternatively, if the nanostructure-containing membrane is attached to a surface and the membrane is removed, an ensemble of micro- or nano-structures that protrude from the surface like the bristles of a brush can be obtained [161]. Finally, segmented multi-metal rods of controllable length can be easily prepared by sequential electrodeposition [165, 166]. Two different types of membranes have largely been used for this sake; the "track-etch" polymeric [167, 168] and the porous alumina membrane [153, 156]. In the "track-etch" method, a non-porous thin sheet of the desired membrane material is bombarded with nuclear fission fragments to create damage tracks in the material, and then chemically etching these tracks into pores. A drawback of this process originates from the random nature of the pore-production, which often results in pore intersection [161]. The porous alumina membranes, which are produced using anodization of aluminium metal in acidic media, on the other hand, have an isolating-non-connecting pore structure [161]. Recently, the technology to create porous alumina membranes of long pores and well-defined diameter has matured to the point of commercialization (available at Whatman)[160]. Figure 7.22.a, displaying the scanning electron microscopy (SEM) top view of the commercially available porous alumina membrane (pore diameter ~ 200 nm and pore length ~ 60 μm), indicates the homogeneity of the pore distribution. The pores are not always straight but sometimes kinked which results in structurally defective wires (see Figure 7.22.b).

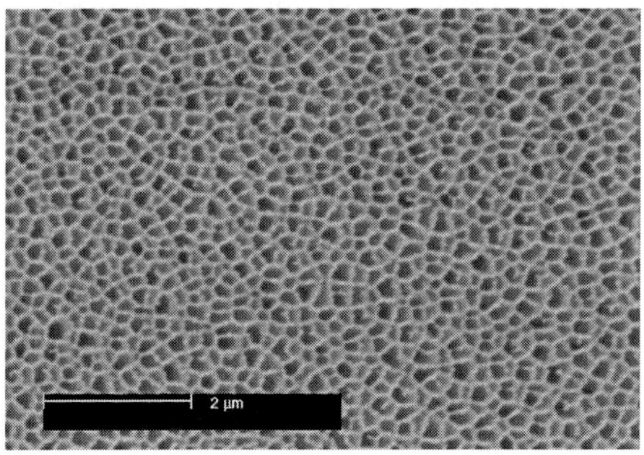

Figure 7.22.a. The SEM surface top view of a commercially available porous alumina membrane purchased from whatman (pore diameter ~ 200 nm and pore length ~ 60 μm).

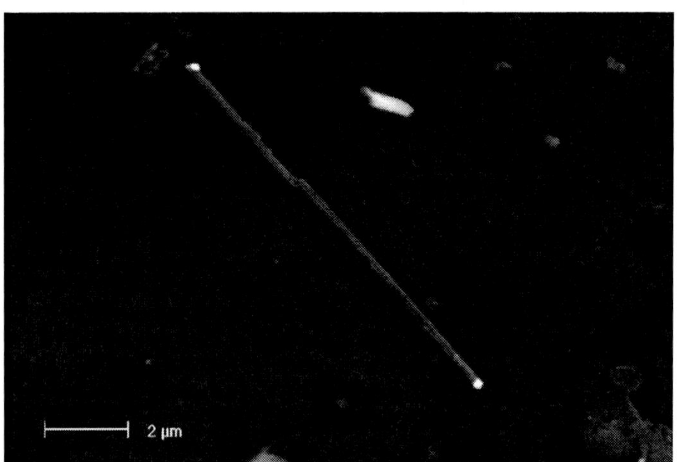

Figure 7.22.b. The SEM for a single isolated SiNW gold nanoparticles tips at both sides. This wire was grown within the commercially available porous alumina templates.

7.5.2. Fabrication of CoSix Nanowire Contacts to SiNWs[153, 156]

Figure 7.23. shows the fabrication scheme of the cobalt silicide nanowire contacts to the SiNWs using the porous alumina membranes. The metal deposition in the membrane, which is an insulator, was achieved electrochemically. Hence, a thin film of approximately 150 nm Ag was sputtered on one side of the membrane to provide a conductive layer for electrodeposition within the pores. To fully cover this side with a compact layer of Ag, the membrane was subsequently placed in an electrochemical cell, and Ag was electrodeposited at 22.27 mA/cm^2 for 1 min on the conductive side of the membrane. To electrodeposit within the pores, the membrane was flipped upside down and the plating solutions were added from the open side of the membrane. A graphite rod served as a counter electrode with no separate compartment. The temperature of the plating solutions was kept constant at about 20°C. Ag, Co and Au were then sequentially electroplated within the pores. The Ag (1.125 tr. oz/qt) and Au (0.25 tr. oz/qt) plating solutions were purchased ready from Technic, Inc., while the Co bath was made by mixing 500 g/l hydrated cobalt sulfate, 17 g/l sodium chloride and 45 g/l boric acid at 25 ^{o}C. The pH of Co bath was then adjusted to 5[169]. Ag was electroplated inside the membrane at 7.32 mA/cm^2 for 2 h to control the position of the Co and Au layers. Nevertheless, Co was electroplated at 89.09 mA/cm^2 for 5 min. The Au was electroplated at 2.22 mA/cm^2 for various lengths of time to participate as a catalyst in the VLS growth of the Si nanowires. The template was next placed in a chemical vapor deposition (CVD) chamber, which was purged with hydrogen at room temperature. The temperature of the CVD furnace was then raised to 500°C and left at this temperature for about 20 min to reach thermal equilibrium. After that, a mixture of 5% silane in hydrogen was allowed to flow over the template at 500°C, and Si nanowires were grown by the VLS mechanism. At this temperature, the Si portion also reacted with the Co portion of the nanowire to form one or more of the cobalt silicide phases. After growth, the nanowires were released from the membranes using 8.0 M nitric acid to dissolve the Ag and the remaining metallic Co, and further treatment with 1.0 M sodium hydroxide was used to remove the alumina membrane. After several centrifugation and rinsing steps, the nanowires were then stored in de-ionized water.

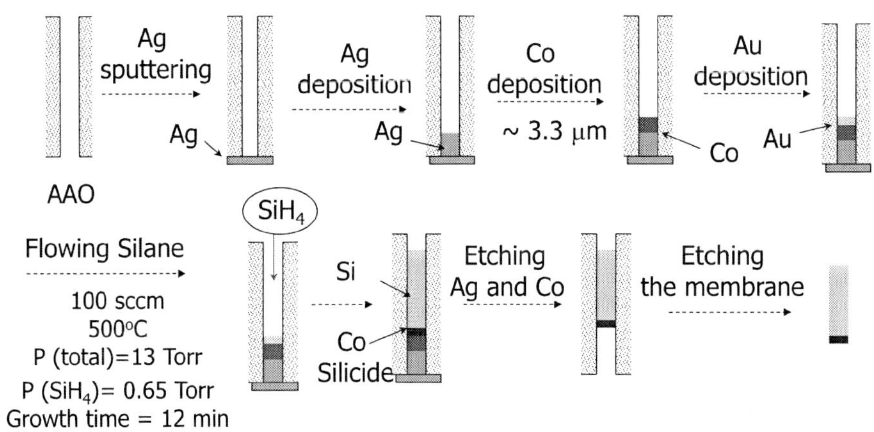

Figure 7.23. Fabrication scheme of the cobalt silicide contacts to the SiNWs.

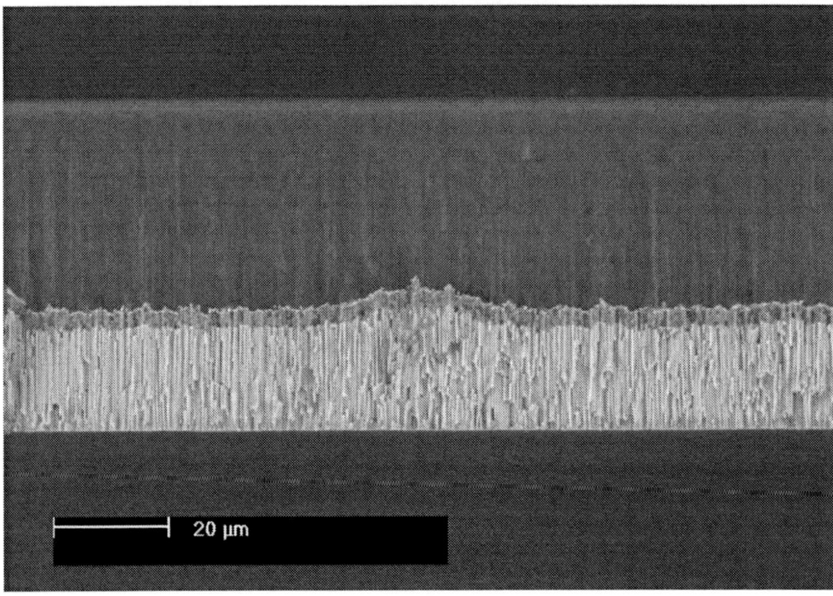

Figure 7.24.a. A cross-sectional SEM view for a piece of the alumina membrane after plating Ag, Co and Au sequentially inside the pores of the membrane but before the Si growth.

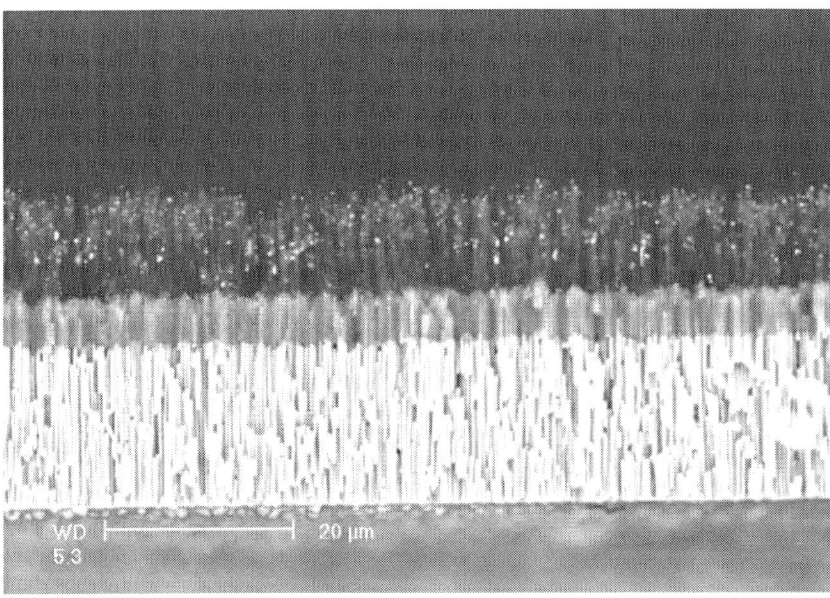

Figure 7.24.b. A cross-sectional SEM image for a piece of the alumina membrane after the growth of SiNWs.

Figure 7.24.a shows a cross-sectional SEM view of a piece of the alumina membrane after plating Ag, Co and Au sequentially inside the pores of the membrane but before the Si growth. Silver, cobalt and gold were electroplated for 2 h, 5 min, and 7 min, respectively, yielding layers approximately 20 μm, 3.5 μm and 0.23 μm, respectively. Since the pores of the membrane are not all the exact same diameter, the plated segments in some area were a little bit longer than others, as seen in Figure 7.24.a.

Following the electrodeposition steps, the rest of the membrane was transferred to a chemical vapor deposition (CVD) chamber fitted with a mechanical pump. The chamber was then purged with H_2 at room temperature to ensure a reductive atmosphere. While H_2 is purging, the furnace temperature of the chamber was raised gradually to 500 °C and remained at this temperature until equilibrium. Silane gas at a partial pressure of 0.65 Torr along with the carrier gas H_2 was allowed to flow over the membrane at a total pressure of 12.8 Torr at 500 °C. The binary phase diagram of Au and Si indicates that when the Si concentration with respect to Au exceeds 18.6 % and at temperatures above 363 °C, Si and Au should form a liquid alloy [170].

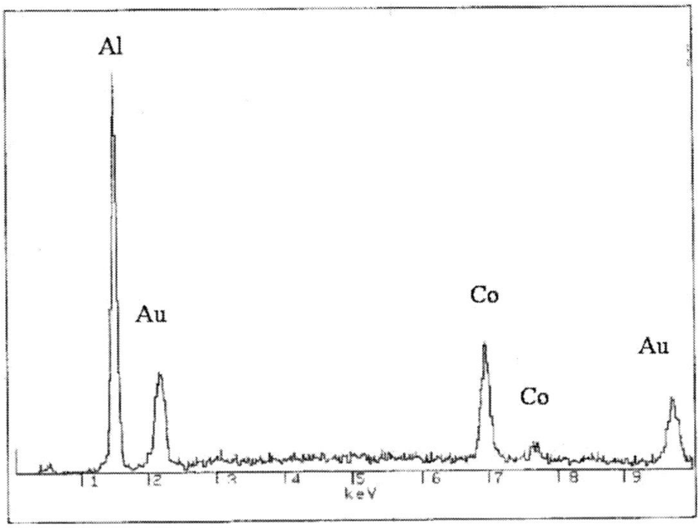

Figure 7. 25.a. The EDS output for the cobalt layer (the lower layer). The spectrum was obtained from the spectrometer attached to the scanning electron microscope.

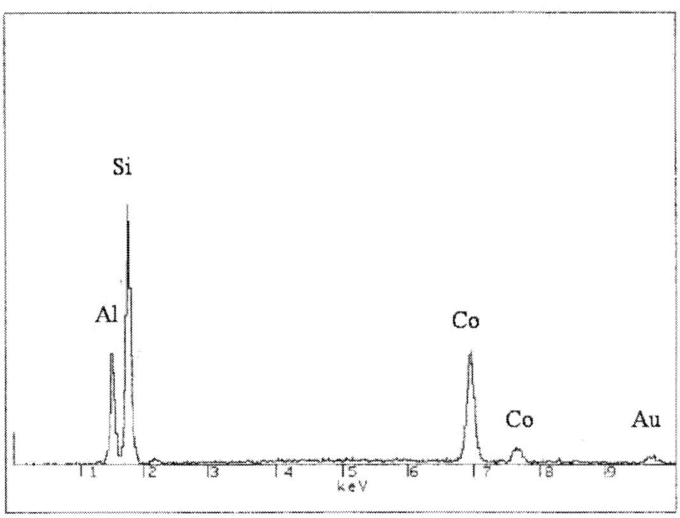

Figure 7. 25.b. The EDS output for the cobalt silicide layer (the upper layer). The spectrum was obtained from the spectrometer attached to the scanning electron microscope.

Therefore, when the decomposition of silane into silicon and hydrogen at the surface of gold took place, a ball of Au-Si liquid alloy was formed. In this study, we intended to keep the rate of silane supply as low as 100 sccm to avoid raising the supersaturation locally in the liquid and terminating the controlled VLS growth. After the liquid alloy became supersaturated with Si, Si nanowire growth occurred by precipitation at the cobalt-liquid alloy interface. Si atoms continued entering the liquid droplet and freezing out at the cobalt-liquid alloy interface, and the alloy droplet became displaced from the cobalt layer and rode atop the growing SiNW. After 12 min, the SiH_4 gas tank was turned off, and H_2 flowed over the membrane. The chamber was then cooled gradually. Figure 7.24.b. shows the cross-sectional SEM image for another piece of the membrane after growing the SiNWs. The image reflects the existence of three distinct segments with a bright nanoparticle in each individual nanowire. It was clear that Ag did not participate in the VLS reaction as the eutectic temperature in the Ag-Si binary phase diagram is 835 °C[170], which is much higher than the growth temperature. Therefore, the same length for the bottom Ag segment was measured after growth. By EDS, the intermediate segment was found to consist of two separate layers. One of them (the bottom one) reflected only the Co peaks and the other one (the one above) contained two equivalent set of peaks for Co and Si. Figure 7.25. shows the EDS outputs for both layers. We used to mount the SEM samples on aluminum substrates and coat them with gold to be conducting. Therefore, peaks for the aluminum and gold appear as background in the EDS characterization. We propose that when the grown silicon crystal came in contact with cobalt at 500 °C, a silicidation reaction took place between the silicon and cobalt to produce a cobalt silicide, as reported earlier [171]. Since Co and Si have comparable atomic weights, the cobalt and cobalt silicide layers had a similar contrast and appeared as one layer in Figure 7.24.b. The top segment and the bright particles were also identified by EDS and found to be pure silicon and gold, respectively. Figure 7.26. shows the EDS outputs for the silicon layer and the gold particles.

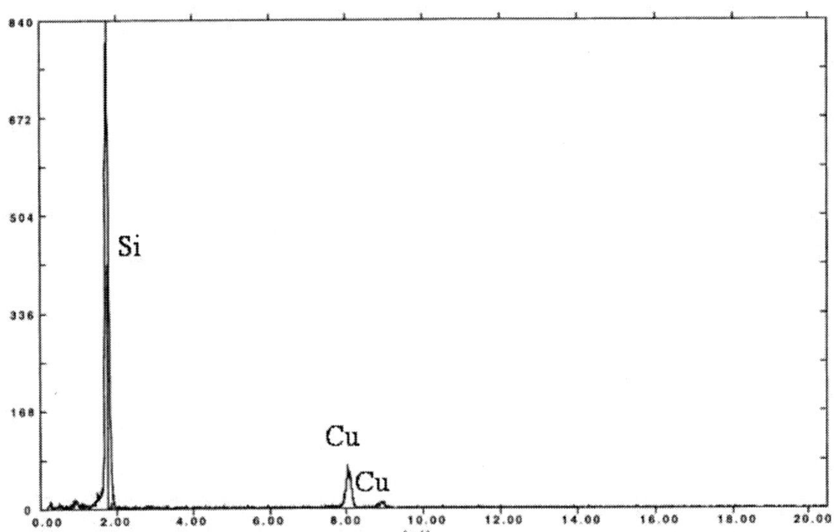

Figure 7. 26.a. The EDS output of the silicon layer. Cu peaks appeared since the sample was mounted in a Cu grid. The spectrum was obtained from the spectrometer attached to the transmission electron micrscope.

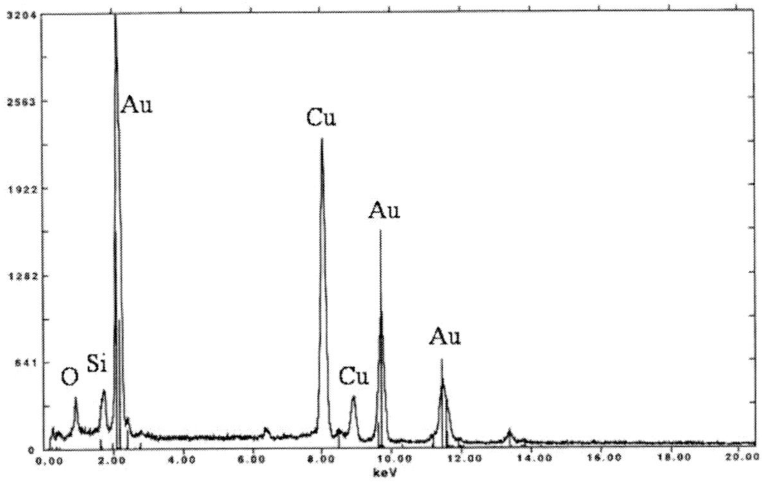

Figure 7 26.b. The EDS output of the Au particles. Cu peaks appeared since the sample was mounted in a Cu grid. The spectrum was obtained from the spectrometer attached to the transmission electron microscope.

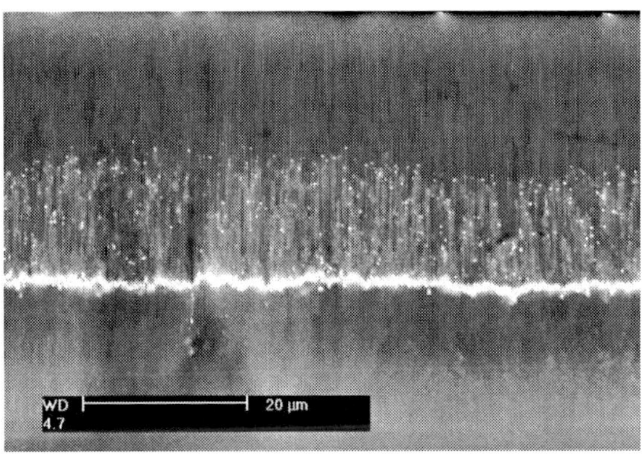

Figure 7.27.a. A cross-sectional SEM view for a piece of the alumina membranc after removing Ag and the remaining elemental Co.

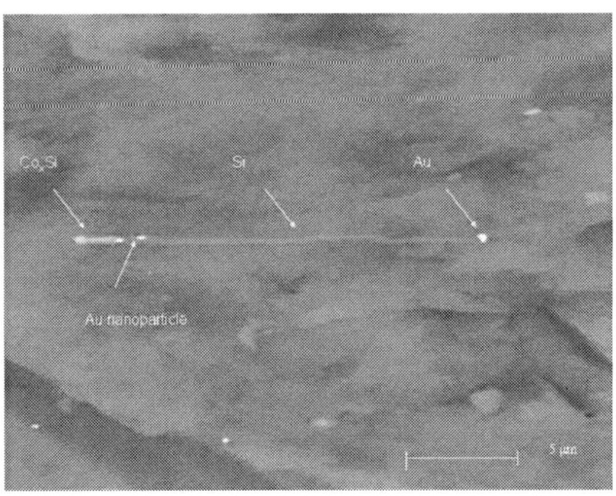

Figure 7.27.b. SEM micrograph of one of the released nanowires. (A Au thickness of 230 nm was initially used to grow this nanowire).

Table 7.7. The measured gold thicknesses, z, as a function of deposition

Deposition time, t, min	Au thickness, z, nm
2.00	80.00
3.00	128.00
3.50	160.00
5.00	200.00
7.00	230.00
25.00	780.00

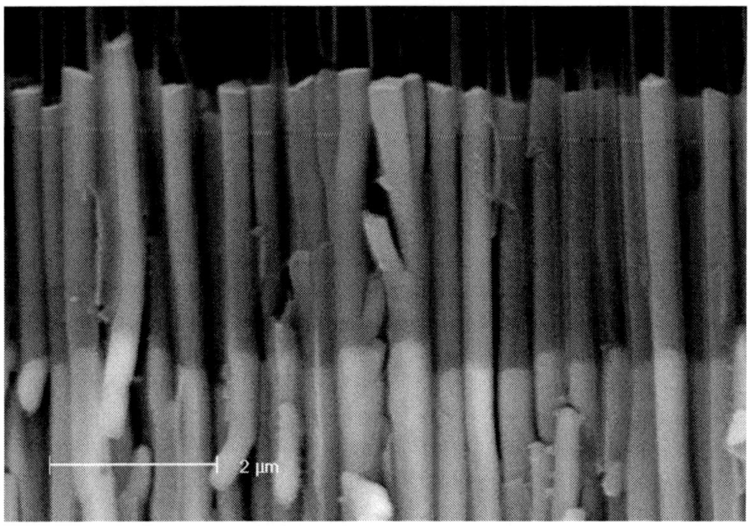

Figure 7.28. A cross-sectional SEM view for a piece of the membrane after plating gold inside for 2 min.

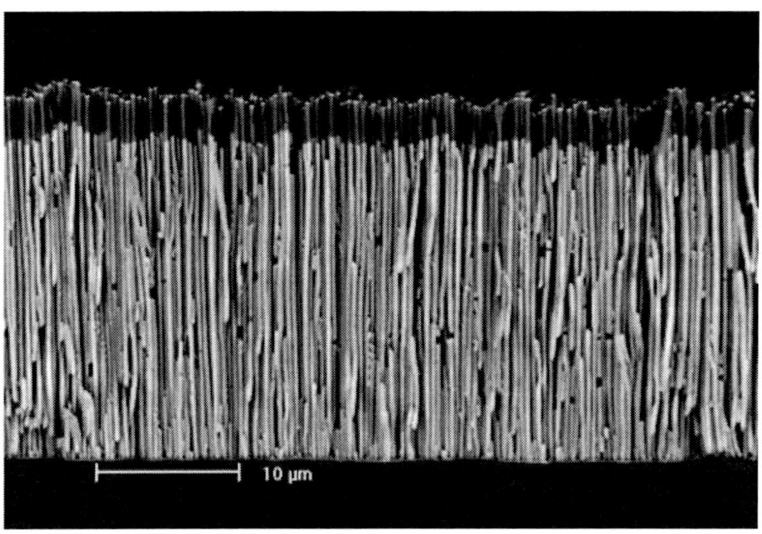

Figure 7.29. A cross-sectional SEM view for a piece of the membrane after plating gold inside for 3.5 min.

Figure 7.30. A cross-sectional SEM view for a piece of the membrane after plating gold inside for 5 min.

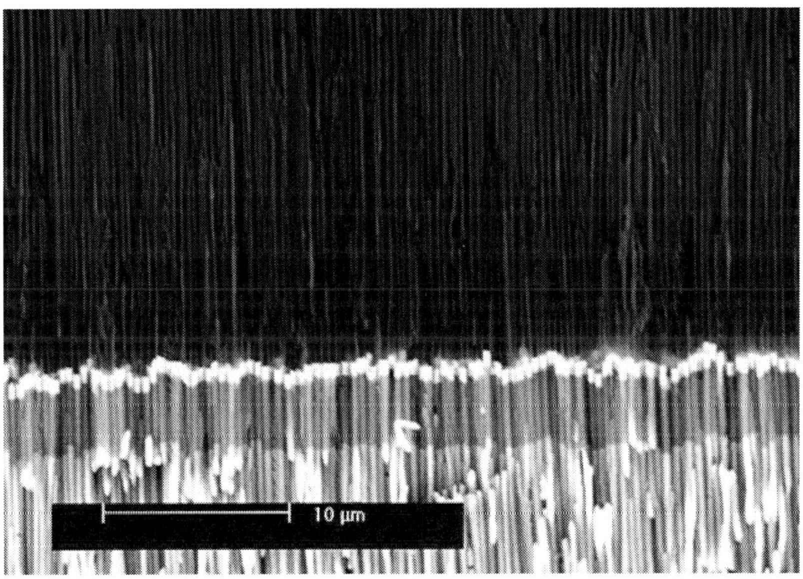

Figure 7.31. A cross-sectional SEM view for a piece of the membrane after plating gold inside for 25 min.

Figure 7.27.a. shows the cross-sectional SEM view for a piece from the template after dissolving silver and the remaining elemental cobalt in 8.0 M nitric acid. Silicon, Au and the cobalt silicide phase are resistant to HNO_3 and were not attacked. We noticed in this study, consistent with a report on the rate of reaction in Co/Si diffusion couples by Jan *et al.* [172] which stated that the kinetics of the silicidation reaction is relatively slow, so all the cobalt could not be consumed within the short time for growth of the SiNWs. Hence, only a short cobalt silicide segment (less than a micron) was obtained. The images in Figures 7.24.b. and 7.27.a. also revealed the existence of retained nanoparticles near the cobalt silicide segment.

Gold was the only element that the EDS could detect for these nanoparticles. These bright nanoparticles appeared obviously in the released nanowires. Figure 27.b. shows a typical nanowire from our early attempts to grow multilayered nanowires, with the segments of the wires labeled according to the EDS analysis. We have used 1.0 M NaOH to dissolve the alumina membrane and release the nanowires. Because these retained Au particles can obviously affect the electrical performance of the wire, we were interested in avoiding the retained Au particles to improve the structural quality of the nanowires [153], and consequently, effort was focused on this problem.

The approach we took to address this problem was to search for an optimal Au thickness for the nanowires to be grown without leaving any Au particles along the wires. Membranes with various Au lengths were prepared. Figures 7.28-7.31 show the cross-sectional SEM views of several alumina templates after electrodepositing gold inside for 2, 3.5, 5, and 25 min, respectively. Table 7.7. summarizes the measured gold thicknesses, z, as a function of deposition time, t, and Figure 32 shows a calibration curve for this data. The following straight-line equation fitted the calibration curve with a regression coefficient of 0.997

$$z = 38.07 + 29.66\ t \tag{5}$$

To determine the effect of Au thickness on the structural quality of the obtained nanowires, membranes with Au thicknesses of 80, 160, 200, and 780 nm above the Co segment were investigated after growing the nanowires for the same period. Inspection of the cross-sectional SEM images (images are not included) for the templates with various Au lengths after the growth of SiNWs and after removing Ag indicated that Au was not retained in the membranes with Au thicknesses of 80 and 160 nm and retained for the other samples. It was also obvious that the density of retained Au for samples with Au thicknesses of 200 nm and higher increased as the initial Au thickness increased. To confirm the above observations, we released the nanowires from the membranes and studied each individual nanowire separately.

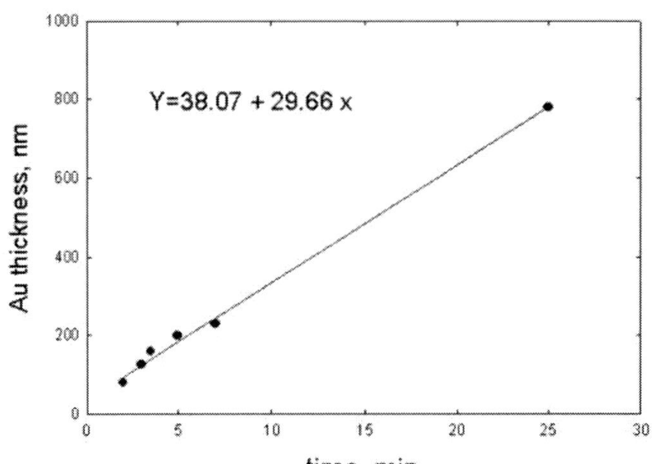

Figure 7.32. A calibration curve for the Au thickness within the alumina membrane as a function of the plating time.

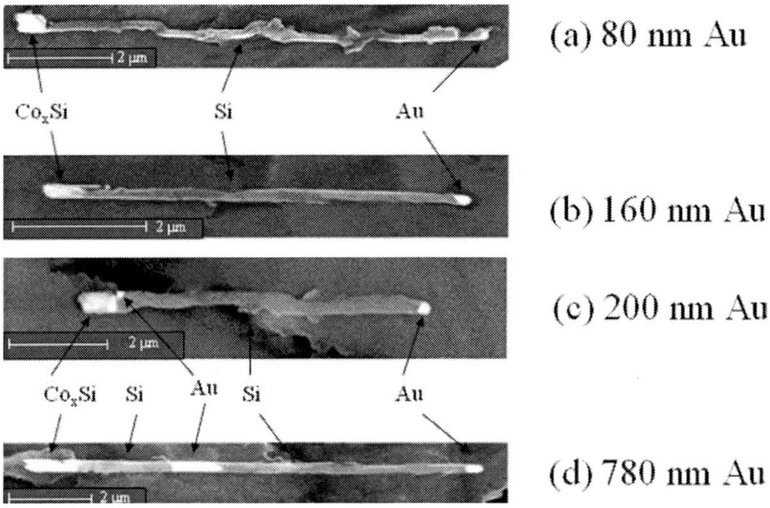

Figure 7.33. The SEM micrographs of a single nanowire grown with different Au thicknesses of (a) 80 nm Au, (b) 160 nm Au, (c) 200 nm Au, and (d) 780 nm Au.

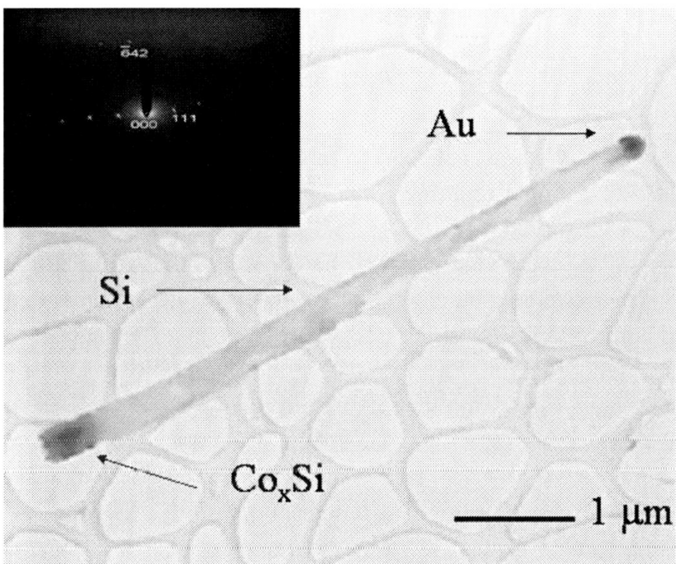

Figure 7.34. A TEM image for one of the nanowires, and the inset is a diffraction pattern from the Si segment.

With a Au thickness of 80 nm, the retained Au particles disappeared, as shown in Figure 7.33.a. for a nanowire with a diameter of 152 nm. The silicon nanowire diameter was smaller than the diameters of both the cobalt plug and the alumina membrane pore, and accordingly the wires did not take the exact shape of the pores during growth. They were curved and bent rather than straight. Thus, the shape of the SiNWs was not desirable. In fact, and at a given amount of Au, the liquid alloy ball the Au forms with Si is expected to span the whole diameter of the pore, and the growth of SiNWs occurs after that. However, when an insufficient amount of Au participates in the VLS reaction, there is not enough liquid available to span the entire diameter of the pore of the membrane. Therefore, the diameter of the nanowire will be smaller than that of the membrane's pore, and structural defects are expected to occur.

A gold thickness of 160 nm was found to be optimal in this study. No Au particles appeared within the wires, as shown in Figure 7.33b. Furthermore, the structure of the wires was found to be good. With Au thicknesses of 200 and 780 nm, some Au was retained within the SiNWs. Not surprisingly, the amount of retained Au became greater as the thickness of the electrodeposited Au increased, as shown in Figure 7.33c and 7.33d. On the other hand, the structure of the wires was good, and the silicon nanowires, the cobalt plugs and the pore diameters are nearly the same.

Using TEM, we found that the nanowires that were released from the alumina membranes containing the optimal Au thickness had silicon segments that were single crystals or crystalline with a small number of defects. Figure 7.34 shows a TEM image for one of the nanowires, and the inset is a diffraction pattern from the Si segment. From examination in the TEM of 13 nanowires, an average diameter of 249 nm with a standard deviation of 32 nm was measured. The wires have a greater diameter than the nominal pore diameter, along with a spread in diameters, because the pores are not always the exact same size as the nominal value stated by the manufacturer, nor are they all the same size as each other. The SEM and TEM studies also show a little roughness along some of the wires that might be due to the use of NaOH to release the wires from the alumina membranes. In fact, the silicon segment of the nanowires is expected to be attacked by NaOH but with a much slower rate than alumina.

Several approaches were next taken to provide a formula to describe the optimal thickness (or length) of Au plug required for VLS growth. We assumed that the liquid Au-Si alloy present during VLS growth forms either a hemispherical or a spherical bead of the same diameter as the growing Si nanowire, and that the volume of the Au-Si alloy is close to that of the solid Au plug. We therefore calculated that we would need Au plugs that are either 1/3 D or 2/3 D thick for hemispherical or spherical beads, respectively, where D is the diameter of the nanowire. In fact, the thickness of the Au plug that worked well in our experiments, 160 nm, is very near the value of 166 nm that we would predict using the formula for a spherical bead for our nanowires of average diameter 249 nm. On the other hand, the prediction we would make assuming a hemispherical cap (83 nm) provides a Au thickness very close to the 80 nm thickness that was shown to be insufficient in our study. As we see from Figure 7.34, however, the Au bead at the end of the Si nanowire is neither a hemisphere nor a sphere. More careful examination in the TEM of 13 nanowires was next performed. The diameter of each nanowire was measured, and the volume of the Au bead at the end of each nanowire was approximated from the TEM images. From these experiments, the starting thickness of the Au plug L was estimated to vary from 0.5–0.9 D. Note that the value 2/3 D falls within this range. It is interesting that nothing has been reported in the literature concerning Au retention in SiNWs when they are grown on Si wafers using the VLS technique. In this case, the shape of the liquid bead is not constrained by a pore, so the bead can form a portion of a sphere of any diameter depending on the amount of Au present. Hence, no excess Au is left behind during VLS growth. Furthermore, the silicon-containing gas has easier access to the surface of the Au catalyst than it does for SiNWs growing within pores, so Au is again less likely to be retained along the length of a SiNW.

It is also worth noting that laser ablation and physical evaporation methods usually produce SiNWs that are sheathed by amorphous SiO_2 layer. That happens due to the reaction of SiNWs with the residual oxygen in the surrounding atmosphere. In our experiment, similar to the observation by Zhang *et al.*[173] which suggested that the reducing atmosphere and the low growth temperature prevented a thick amorphous coating.

CONCLUSION

This chapter introduces a comprehensive review on our recent experimental approaches for the fabrication of metallic nanostructures in various geometries using the electrodeposition technique. That is, the fabrication of catalytically active metallic Ni or Cu thin films on reticulated vitreous carbon (RVC), and Pd or black Ni onto copper screens has been achieved for use as cathodes for the hydrogen evolution reaction (HER). Also, the electrodeposition of metallic (e.g., Au) and metal oxide (e.g., manganese oxide) nanostructures has been reported for applications in the oxygen reduction and evolution reactions as well as for formic acid oxidation. Additionally, the fabrication of one-dimensional nanostructures, i.e., nanowires, has been reviewed and some methodologies are outlined for this sake, including the electrodeposition in templates.

ACKNOWLEDGMENT

Several parts of this work have been carried out while one of the authors (M. S. El-Deab) was holding a JSPS fellowship at Tokyo Institute of Technology (Japan) and other parts were done during his Alexander von Humboldt fellowship at Ulm University (Germany). Similarly, a considerable part of this work was carried out while the second author (A. M. Mohammad) was holding a fellowship at the Pennsylvania State University (USA). The authors are very grateful for the sincere supervision of the whole research activities outlined herein by the late Professor Dr. Badr G. Ateya, Professor of Physical Chemistry at Cairo University.

REFERENCES

[1] H. Huang and X. Yang, Colloids Surf. A: Physicochem. *Eng. Aspects* 255, 11 (2005).
[2] M. O. Finot, G. D. Braybrook, and M. T. McDermott, *J. Electranal. Chem.* 466, 234 (1999).
[3] M. S. El-Deab, T. Sotomura, and T. Ohsaka, *J. Electrochem. Soc.* 152, C1 (2005).
[4] V. Srinivasan and J. W. Weidner, *J. Electrochem. Soc.* 144, L210 (1997).
[5] H. Y. Lee, S. W. Kim, and H. Y. Lee, Electrochem. *Solid-State Lett.* 4, A19 (2001).
[6] M. S. Hong, S. H. Lee, and S. W. Kim, Electrochem. *Solid-State Lett.* 5, A227 (2002).
[7] J. Spatz, S. Mossmer, and M. Moller, *Chem. Eur. J.* 2, 1552 (1996).
[8] R. Glass, M. Moller, and J. P. Spatz, *Nanotechnology* 14, 1153 (2003).
[9] M. Magnusson, K. Deppet, J.-O. Malm, J.-O. Bovin, and L. Samuelson, *Nanostruct. Mater.* 12, 45 (1999).
[10] P. J. Collier, J. A. Iggo, and R. Whyman, *J. Mol. Catal. A: Chem.* 146, 149 (1999).

[11] A. P. Alivisatos, *Science* 271, 933 (1996).

[12] A. P. Alivisatos, K. P. Johnsson, X. Peng, T. E. Wilson, C. J. Loweth, M. P. Bruchez, and P. G. Schultz, *Nature* 382, 609 (1996).

[13] G. Schmid and B. Corain, *European J. Inorg. Chem.*, 3081 (2003).

[14] D. G. Schultz, X.-M. Lin, D. Li, J. Gebhardt, M. Meron, J. Viccaro, and B. Lin, *J. Phys. Chem. B* 110, 24522 (2006).

[15] M.-C. Daniel and D. Astruc, *Chem. Rev.* 104, 293 (2004).

[16] M. Chen and D. W. Goodman, *Acc. Chem. Res.* 39, 739 (2006).

[17] R. Zanella, A. Sandoval, P. Santiago, V. A. Basiuk, and J. M. Saniger, *J. Phys. Chem. B* 110, 8559 (2006).

[18] J. Zhang and M. Oyama, *Anal. Chim. Acta* 540, 299 (2005).

[19] M. Haruta, *Catal. Today* 36, 153 (1997).

[20] J.-D. Grunwaldt, C. Kiener, C. Wogerbauer, and A. Baiker, *J. Catal.* 181, 223 (1999).

[21] F. Boccuzzi, A. Chiorino, and M. Manzola, *Mater. Sci. Eng.*, C 15, 215 (2001).

[22] M. S. El- Deab and T. Ohsaka, El*ectrochemistry Communications* 4, 288 (2002).

[23] R. R. Adzic, A. V. Tripkovic, and W. E. O'Grady, *Nature* 296, 137 (1982).

[24] A. M. El-Aziz and L. A. Kibler, *J. Electroanal. Chem.* 534, 107 (2002).

[25] I. A. Raj, *Appl. Surf. Sci.* 59, 245 (1992).

[26] D. L. Stojic´, M. P. Marc?eta, S. P. Sovilj, and S? . S. Miljanic´, *J. Power Sources* 118, 315 (2003).

[27] M. Treuba, S. Trasatti, and S. Trasatti, *Mater. Chem. Phys.*, in press (2005).

[28] O. Bo¨hme, F. U. Leidich, H. J. Salge, and H. Wendt, *Int. J. Hydrogen Energy* 19, 349 (1994).

[29] S. Rausch and H. Wendt, *J. Electrochem. Soc.* 143, 2852 (1996).

[30] I. J. Brown and S. Sotiropoulos, *J. Appl. Electrochem.* 30, 107 (2000).

[31] M. S. El-Deab and M. M. Saleh, *International Journal of Hydrogen Energy* 28, 1199 (2003).

[32] F. Coeuret, V. E. Oliveira, and C. E. Bezerra, *J. Appl. Electrochem.* 32, 1175 (2002).

[33] M. M. Saleh, *Journal of Applied Electrochemistry* 30, 939 (2000).

[34] M. M. Saleh, M. H. El-Ankily, M. S. El-Deab, and B. E. El-Anadouli, *Bulletin of the Chemical Society of Japan* 79, 1711 (2006).

[35] A. J. Appleby, M. Chelma, H. Kita, and G. Bronoel, in "Encyclopedia of the electrochemistry of elements" (A. J. Bard, ed.). Marcel Dekker, New York, (1982).

[36] Vector L. Streeter, "*Fluid Mechanics*", p. 186. McGraw-Hill, New York, (1962).

[37] R. B. Bird, W. E. Stewart, and E. N. Lightfoot, "Transport Phenomena", p. 411. John Wiley and Sons, NY, (1960).

[38] J. Newman, "*Electrochemical Systems*", p. 454. Prentice-Hall, Englewood Cliffs, NJ, (1991).

[39] V. G. Levich, "*Physicochemical Hydrodynamics*". Prentice-Hall, Englewood Cliffs, NJ, (1991).

[40] R. C. Weast, (ed.), CRC Press Inc., Boca Raton, Florida, (1988).

[41] J. O'M. Bockris and A. Reddy, "*Modern Electrochemistry*", p. 382, Plenum/Rosetta, NY, (1973).

[42] A. J. Bard and L. R. Faulkner, "*Electrochemical Methods: Fundamentals and Applications*". John Wiley and Sons, Inc., NY, (1980).

[43] L. Mammar and G. Wranglen, *Electrochim. Acta* 9, 1 (1964).

[44] M. S. El-Deab, M. M. Saleh, B. E. El-Anadouli, and B. G. Ateya, *Egypt. J. Chem.* 9, 67 (2009).
[45] M. S. El-Deab, T. Sotomura, and T. Ohsaka, *Electrochemistry Communications* 7, 29 (2005).
[46] M. S. El-Deab and T. Ohsaka, *Electrochimica Acta* 47, 4255 (2002).
[47] M. S. El-Deab, T. Okajima, and T. Ohsaka, *Journal of the Electrochemical Society* 150, A851 (2003).
[48] M. S. El-Deab and T. Ohsaka, *Journal of Electroanalytical Chemistry* 553, 107 (2003).
[49] M. S. El-Deab, T. Sotomura, and T. Ohsaka, *Electrochimica Acta* 52, 1792 (2006).
[50] Y. Zhang, S. Asahina, S. Yoshihara, and T. Shirakashi, *Electrochim. Acta* 48, 741 (2003).
[51] I. Yagi, T. Ishida, and K. Uosaki, *Electrochem. Commun.* 6, 773 (2004).
[52] N. Pradhan, A. Pal, and T. Pal, *Colloids Surf. A: Physicochem. Eng. Aspects* 196, 247 (2002).
[53] S. Schimpf, M. Lucas, C. Mohr, U. Rodemerck, A. Bruckner, J. Radnik, H. Hofmeister, and P. *Calus, Catal. Today* 72, 63 (2002).
[54] A. D. Sarkar and B. C. Khanra, *J. Mol. Catal. A: Chem.* 299, 25 (2005).
[55] M. Haruta and M. Date, *Appl. Catal. A: General* 222, 427 (2001).
[56] A. Shiga and M. Haruta, *Appl. Catal. A: General* 291, 6 (2005).
[57] A. Doron, E. Katz, and I. Willner, Langmuir 11, 1313 (1995).
[58] A. A. Michri, A. G. Pshchenichikov, and R. K. Burshtein, *Sov. Electrochem.* 8, 351 (1972).
[59] S. Trasatti and O. A. Petrii, *Pure Appl. Chem.* 63, 711 (1991).
[60] H. A. Kozlowska, B. E. Conway, A. Hamelin, and L. Stoicoviciu, *J. Electroanal. Chem.* 228, 429 (1987).
[61] Y. Tian, L. Mao, T. Okajima, and T. Ohsaka, *Biosensors Bioelectronics* 21, 557 (2005).
[62] W. F. McClune, (ed.), "Powder Diffraction File". International Centre for Diffraction Data (ICDD), USA, (2003).
[63] T. Kinoshita, S. Seino, K. Okitsu, T. Nakayama, T. Nakagawa, and T. A. Yamamoto, *J. Alloys Comp.* 359, 46 (2003).
[64] K. Nogita and A. K. Dahle, *Mat. Character.* 46, 305 (2001).
[65] M. S. El-Deab, *Electrochimica Acta* 54, 3720 (2009).
[66] E. Budevski, G. Staikov, and W. J. Lorenz, "*Electrochemical Phase Formation and Growth*". VCH, Weinheim, (1996).
[67] A. K.-Veneva and M. R. Koblischka, *Mat. Sci. Eng. B* 151, 60 (2008).
[68] J. M. Thomas and W. J. Thomas, "*Principle and Practice of Heterogeneous Catalysis*". VCH, Weinheim, (1997).
[69] J. Klikovits, M. Schmid, J. Gustafson, J. Mikkelsen, A. Resta, E. Lundgren, J. N. Andersen, and P. Varga, *J. Phys. Chem. B* 110, 9966 (2006).
[70] A. Mills, N. Elliott, G. Hill, D. Fallis, J. R. Durrant, and R. L. Willis, *Photochem. Photobiol. Sci.* 2, 591 (2003).
[71] C. J. Barbe, F. Arendse, P. Comte, M. Jirousek, F. Lenzmann, and J. Shklover, *J. Am. Ceram. Soc.* 80, 3157 (1997).
[72] M. Gratzel, *J. Sol-Gel Sci. Technol.* 2, 7 (2001).
[73] L. R. Skubal, N. K. Meshkov, and M. C. Vogt, *J. Photochem. Photobiol. A: Chem. Rev.* 148, 103 (2002).

[74] T. Watanabe, H. Hayashi, and H. Imai, *Solar Energy Mater. Solar Cells* 90, 640 (2006).
[75] B. O'Regan and M. Gratzel, *Nature* 353, 737 (1991).
[76] B. Hunsche, M. Vergohl, and A. Ritz, *Thin Solid Films* 502, 188 (2006).
[77] J. Livage, M. Henry, and C. Sanchez, *Prog. Solid State Chem.* 18, 259 (1998).
[78] L. Zhao, Y. Yu, L. Song, H. X., and A. Larbot, *Appl. Surf. Sci.* 239, 285 (2005).
[79] M. K. Lee, H. C. Lee, and C. M. Hsu, *Semicond. Sci. Technol.* 21, 604 (2006).
[80] G. Zhao, Q. Tian, Q. Liu, and G. Han, *Surf. Coat. Technol.* 198, 55 (2005).
[81] I. Truijen, M. K. Van Bael, H. Van den Rul, J. D'Haen, and J. Mullens, *J. Sol-Gel Sci.* Technol. 41, 43 (2007).
[82] N. Setter and R. Waser, *Acta Mater.* 48, 151 (2000).
[83] S. Nishiyama, T. Kubota, and K. Kimura, *J. Mol. Catal. A: Chem.* 120, 17 (1997).
[84] M.M. Yusuf, Y. Chimoto, H. Imai, and H. Hirashima, *J. Sol-Gel Sci. Technol.* 26, 635 (2003).
[85] B. H. Kim, J. Y. Lee, Y. H. Choa, M. Higuchi, and N. Mizutani, Mater. Sci. Eng. B. Solid State Mater. *Adv. Technol.* 107, 289 (2004).
[86] F.D. Torre, H. V. Swygenkoven, and M. Victoria, *Acta Mater.* 30, 3957 (2002).
[87] M.R. Hoffmann, S.T. Martin, W. Choi, and D.W. Bahnemann, *Chem. Rev.* 95, 69 (1995).
[88] P.V. Kamat, *Chem. Rev.* 93, 267 (1993).
[89] A. M. Mohammad, A. I. Abdelrahman, M. S. El-Deab, T. Okajima, and T. Ohsaka, *Colloids and Surfaces A: Physicochemical and Engineering Aspects* 318, 78 (2008).
[90] M. S. El-Deab, T. Sotomura, and T. Ohsaka, *Journal of the Electrochemical Society* 152, C730 (2005).
[91] E.G. Derouane, *Stud. Surf. Sci. Catal.* 5, 5 (1980).
[92] J.A. Martens, M. Tielen, P. A. Jacobs, and J. Weitkamp, *Zeolites* 4, 98 (1984).
[93] P. B. Weisz and V.J. Frilette, *J. Phys. Chem.* B 64, 382 (1960).
[94] L. Mao, T. Sotomura, K. Nakatsu, N. Koshiba, D. Zhang, and T. Ohsaka, *J. Electrochem. Soc.* 149, A504 (2002).
[95] B. Klapste, J. Vondrak, and J. Velicka, *Electrochim.Acta* 47, 2365 (2002).
[96] J. Vondrak, B. Klapste, J. Velicka, M. Sedlarikova, V. Novak, and J. Reiter, *J. New Mater. Electrochem. Syst.* 8, 1 (2005).
[97] R. Yang, Z. Wang, L. Dai, and L. Chen, *Mater. Chem. Phys.* 93, 149 (2005).
[98] M.-S.Wu and P.-C. J. Chiang, Electrochem. *Solid State Lett.* 7, A123 (2004).
[99] M. J. Garrido, *Seances Acad. Sci.* 200, 69 (1935).
[100] M. S. El-Deab and T. Ohsaka, Angewandte Chemie - International Edition 45, 5963 (2006).
[101] A. M. Fundo and L. M. Abrantes, *Russ. J. Electrochem.* 42, 1291 (2006).
[102] M. S. El-Deab, L. A. Kibler, and D. M. Kolb, *Electrochemistry Communications* 11, 776 (2009).
[103] L. J. Zhang, Z. Y. Wang, and D. G. Xia, *J. Alloys Compd.* 426, 268 (2006).
[104] S. Wang, N. Kristian, S. Jiang, and X. Wang, *Electrochem. Commun.* 10, 961 (2008).
[105] J. M. Feliu and E. Herrero, in "*Handbook of Fuel Cells*" (A. Lamm, ed.), Vol. 2, p. 625. Wiley, NY, (2003).
[106] T. J. Schmidt, R. J. Behm, B. N. Grgur, N. M. Markovic, and P. N. Ross, Langmuir 16, 8159 (2000).

[107] M. Arenz, V. Stamenkovic, T. J. Schmidt, K. Wandelt, P. N. Ross, and N. M. Markovic, Phys. Chem. Chem. Phys. 5, 4242 (2003).
[108] T. J. Schmidt, B. N. Grgur, N. M. Markovic, and P. N. Ross, *J. Electroanal. Chem.* 500, 36 (2001).
[109] Y. Wu, H. Yan, M. Huang, B. Messer, J. H. Song, and P. Yang, *Chem. Eur. J.* 8, 1261 (2002).
[110] X. Lu, W. Z. Shen, and J. H. Chu, *J. Appl. Phys.* 91, 1542 (2002).
[111] G. D. J. Smit, S. Rogge, and T. M. Klapwijk, *Appl. Phys. Lett.* 80, 2568 (2002).
[112] C. A. J. Foss, G. L. Hornyak, J. A. Stockert, and C. R. Martin, *J. Phys. Chem.* 96, 7497 (1992).
[113] C. A. J. Foss, G. L. Hornyak, J. A. Stockert, and C. R. Martin, *J. Phys. Chem.* 98, 2963 (1994).
[114] Y. Cui, X. Duan, J. Hu, and C. M. Lieber, *J. Phys. Chem. B* 104, 5213 (2000).
[115] T. Toriyama, Y. Tanimoto, and S. Sugiyama, in "*IEEE Micro Electro Mechanical Systems (MEMS)*", p. 305, Interlaken, 2001.
[116] Y. Cui and C. M. Lieber, *Science* 291, 851 (2001).
[117] H. Ahmed, in "*Technical Digest - International Electron Devices Meeting*", p. 363. IEEE, Washington DC Piscataway NJ, 1999.
[118] Y. Cui, Q. Wei, H. Park, and C. M. Lieber, *Science* 293, 1289 (2001).
[119] . Zhong, D. Wang, Y. Cui, M. W. Bockrath, and C. H. Lieber, *Science* 302, 1377 (2003).
[120] T. Toriyama, Y. Tanimoto, and S. Sugiyama, *J. Microelectromech.* S 11, 605 (2002).
[121] Y. Huang, X. Duan, Q. Wei, and C. M. Lieber, *Science* 291, 630 (2001).
[122] J. Schmelzer Jr., S. A. Brown, A. Wurl, M. Hyslop, and R. J. Blaikie, *Phys. Rev. Lett.* 88, 2268021 (2002).
[123] Y. F. Zhang, Y. H. Tang, N. Wang, D. P. Yu, C. S. Lee, I. Bello, and S. T. Lee, *Appl. Phys. Lett.* 72, 1835 (1998).
[124] U. Landman, R. N. Barnett, A. G. Scherbakov, and P. Avouris, *Phys. Rev. Lett.* 85, 1958 (2000).
[125] J.-Y. Yu, S.-W. Chung, and J. R. Heath, *J. Phys. Chem. B* 104, 11864 (2000).
[126] S.-W. Chung, J.-Y. Yu, and J. R. Heath, *Appl. Phys. Lett.* 76, 2068 (2000).
[127] F. C. K. Au, K. W. Wong, Y. H. Tang, Y. F. Zhang, I. Bello, and S. T. Lee, *Appl. Phys. Lett.* 75, 1700 (1999).
[128] E. I. Givargizov, *J. Vac. Sci. Technol. B* 11, 449 (1993).
[129] J. D. Holmes, K. P. Johnston, R. C. Doty, and B. A. Korgel, *Science* 287, 1471 (2000).
[130] Z. G. Bai, D. P. Yu, J. J. Wang, Y. H. Zou, W. Qian, J. S. Fu, S. Q. Feng, J. Xu, and L. P. You, *Mater. Sci. Eng.,* B 72, 117 (2000).
[131] K. Kurihara, K. Iwadate, H. Namatsu, M. Nagase, H. Takenaka, and K. Murase, *Jpn. J. Appl. Phys.* 1 34, 6940 (1995).
[132] L. Pescini, A. Tilke, R. H. Blick, H. Lorenz, J. P. Kotthaus, W. Eberhardt, and D. Kern, *Nanotechnology* 10, 418 (1999).
[133] E. Giovine, A. Notargiacomo, L. Di Gaspare, E. Palange, F. Evangelisti, R. Leoni, G. Castellano, G. Torrioli, and V. Foglietti, *Nanotechnology* 12, 132 (2001).
[134] T. Tsutsumi, E. Suzuki, K. Ishii, H. Hiroshima, M. Yamanaka, I. Sakata, S. Hazra, and K. Tomizawa, *Superlattices Microstruct.* 28, 453 (2000).

[135] H. Namatsu, S. Horiguchi, M. Nagase, and K. Kurihara, *J. Vac. Sci. Technol.*, B 15, 1688 (1997).
[136] Y. Yin, B. Gates, and Y. Xia, in "*Materials Research Society (MRS) Symposium*", Vol. 636, p. D421, Boston, MA, 2001.
[137] J. L. Liu, Y. Lu, Y. Shi, S. L. Gu, R. L. Jiang, F. Wang, H. M. Bu, and Y. D. Zheng, Phys. Status Solidi A 168, 441 (1998).
[138] N. Wang, Y. H. Tang, Y. F. Zhang, C. S. Lee, and S. T. Lee, *Phys. Rev. B* 58, R16024–R16026 (1998).
[139] W.-S. Shi, H.-Y. Peng, Y.-F. Zheng, N. Wang, N.-G. Shang, Z.-W. Pan, C.-S. Lee, and S.-T. Lee, *Adv. Mater.* 12, 1343 (2000).
[140] Y. F. Zhang, Y. H. Tang, C. Lam, N. Wang, C. S. Lee, I. Bello, and S. T. Lee, *J. Cryst. Growth* 212, 115 (2000).
[141] Y. H. Tang, Y. F. Zhang, N. Wang, W. S. Shi, C. S. Lee, I. Bello, and S. T. Lee, *J. Vac. Sci Technol. B* 19, 317 (2001).
[142] S. T. Lee, Y. F. Zhang, N. Wang, Y. H. Tang, I. Bello, C. S. Lee, and Y. W. Chung, *J. Mater. Res.* 14, 4503 (1999).
[143] Y. F. Zhang, Y. H. Tang, N. Wang, C. S. Lee, I. Bello, and S. T. Lee, *J. Cryst. Growth* 197, 136 (1999).
[144] S. T. Lee, N. Wang, and C. S. Lee, *Mater. Sci. Eng.,* A 286, 16 (2000).
[145] G. Zhou, Z. Zhang, and D. Yu, *J. Cryst. Growth* 197, 129 (1999).
[146] A. M. Morales and C. M. Lieber, *Science* 279, 208 (1998).
[147] R. S. Wagner and W. C. Ellis, *Appl. Phys. Lett.* 4, 89 (1964).
[148] R. S. Wagner, in "*Whisker Technology*" (A. P. Levitt, ed.), p. 47. Wiley, New York, (1970).
[149] R. S. Wagner and W. C. Ellis, Trans. *Metall. Soc. AIME* 233, 1053 (1965).
[150] R. S. Wagner and C. J. Doherty, *J. Electrochem. Soc.* 115, 93 (1968).
[151] E. I. Givargizov, *J. Cryst. Growth* 31, 20 (1975).
[152] E. I. Givargizov and A. A. Chernov, *Sov. Phys. Crystallogr.* 18, 89 (1973).
[153] A. M. Mohammad, S. Dey, K.-K. Lew, J. M. Redwing, and S. E. Mohney, *J. Electrochem.* Soc. 150, 577 (2003).
[154] C. Y. Kuo and C. Gau, *Thin Solid Films* 519, 3603 (2011).
[155] A. M. Mohammad, A. M. Abdullah, B. E. El-Anadouli, and S. E. Mohney, *Int. J. Nanomanufacturing* 4 (2009).
[156] S. M. Dilts, A. Mohmmad, K.-K. Lew, J. M. Redwing, and S. E. Mohney, in "*Mater. Res. Soc. Symp. Proc.*" (I. M., ed.), Vol. 832, p. 287, Boston, MA, 2005.
[157] D. Shir, B. Z. Liu, A. M. Mohammad, K. K. Lew, and S. E. Mohney, *J. Vac. Sci. Technol.* B 24, 1333 (2006).
[158] H. Y. Peng, Z. W. Pan, L. Xu, X. H. Fan, N. Wang, C.-S. Lee, and S.-T. Lee, *Adv. Mater.* 13, 317 (2001).
[159] M. Z. Wang, (ed.), "*One-Dimensional Nanostructures*". Springer Science+Business Media, LLC, New York, (2008).
[160] S. Roth and D. Carroll, "*One-Dimensional Metals*". WILEY-VCH Verlag GmbH and Co. KGaA,, Weinheim, (2004).
[161] J. C. Hulteen and C. R. Martin, *J. Mater. Chem.* 7, 1075–1087 (1997).
[162] G. Cao and D. Liu, *Advances in Colloid and Interface Science* 136, 45 (2008).
[163] C. R. Martin, *Science* 266, 1961 (1994).

[164] C. R. Martin, *Chem. Mater.* 8, 1739 (1996).
[165] V. Scarani, B. Doudin, and J.-P. Ansermet, *J. Magn. Magn. Mater.* 205, 241 (1999).
[166] B. R. Martin, D. J. Dermody, B. D. Reiss, F. Mingming, L. A. Lyon, M. J. Natan, and T. E. Mallouk, *Adv. Mater.* 11, 1021 (1999).
[167] S. Zein El Abedin and F. Endres, *Electrochimica Acta* 54, 5673 (2009).
[168] L. Velleman, J. G. Shapter, and D. Losic, *Journal of Membrane Science* 328, 121 (2009).
[169] W. Blum and G. B. Hogaboom, "*Principles of Electroplating and Electroforming (Electroplating)*", p. 361. McGraw-Hill Book Company, Inc., New York, (1949).
[170] T. B. Massalski, "*Binary Alloy Phase Diagrams*". ASM International, Metals Park, OH, (1990).
[171] S. P. Murarka, "*Silicides for VLSI Applications*". Academic Press, Inc., Orlando, Florida, (1983).
[172] C.-H. Jan, C.-P. Chen, and Y. A. Chang, *J. Appl. Phys.* 73, 1168 (1993).
[173] X. Y. Zhang, L. D. Zhang, G. W. Meng, G. H. Li, N. Y. Jin-Phillipp, and F. Phillipp, *Adv. Mater.* 13, 1238 (2001).

In: Electrodeposition: Properties, Processes and Applications ISBN: 978-1-61470-826-1
Editor: Udit Surya Mohanty

Chapter 8

ELECTRODEPOSITION OF MOLYBDENUM CARBIDE FROM MOLTEN SALTS

Alain Robin[1]* and Antonio Fernando Sartori[1, 2]
[1]Departamento de Engenharia de Materiais - Escola de Engenharia de Lorena
Universidade de São Paulo, SP, Brazil
[2]Departamento de Informática, Matemática e Física –
Universidade de Taubaté, SP, Brazil

ABSTRACT

The refractory carbides are characterized by their high melting points, high values of hardness, high tensile strength and chemical stability. Due to these remarkable properties they have a wide range of applications such as cutting and grinding tools, bearings, textile-machinery parts and oxidation-resistant gas burners. Metal carbide coatings are mainly used to improve the mechanical (wear resistance, strength and hardness) and chemical (corrosion and oxidation resistance) properties of metallic parts. The principal coating processes for refractory carbides are chemical-vapor deposition, physical-vapor deposition and thermal spray, but such coatings can also be obtained by electroplating. In the present work molybdenum carbide layers were electrodeposited from the LiF-NaF-KF eutectic melt containing K_2CO_3 and Na_2MoO_4 salts. The coatings were obtained by both direct and pulse electroplating processes. The influence of bath composition, temperature and electrical parameters of electrodeposition on the morphology of the deposits was studied using scanning electron microscopy. The composition of the coatings was determined by X-ray diffractometry. The nucleation mechanism of the molybdenum carbides was analyzed by chronoamperometry.

INTRODUCTION

The refractory metal carbides exhibit unique properties (high melting point, high value of hardness, high tensile strength, chemical stability) [1] which makes them suitable for

*E-mail address: alain@demar.eel.usp.br.

applications demanding both high temperature wear and corrosion resistances. The coatings of refractory carbides have a wide range of applications in electronic components, cutting tools, gas-turbine vanes and blades, precision bearings, extruders, for example [1]. The major processes for obtaining refractory carbide coatings are chemical-vapor deposition (CVD), physical-vapor deposition (PVD) and thermal spray. These methods have some disadvantages, such as high temperature involved in the process, difficulty for plating parts of complex shape and high cost.

The electrodeposition process has none of these disadvantages. Nevertheless, the electrodeposition of the refractory metal carbides is only possible if the potential for the formation of its constituents is more positive than that required for the solvent reduction. The electrodeposition of tantalum [2-6], tungsten [3,7-10], chromium [11], vanadium [12], niobium, [4,12], titanium [12], zirconium [12] and molybdenum [3,7-8,13-17] carbides was carried out in molten alkali halides that present high decomposition voltages due to very negative Gibbs free energies of formation. Among all the refractory metal carbides, the molybdenum carbide coatings are promising materials for coating the container and positive-electrode current collector of high-temperature power sources such as sodium/sulfur and lithium/FeS_2 batteries [18]. It has been also reported that the molybdenum carbides exhibit high activity in removal of sulfur-containing compounds by hydrodesulfurization in petroleum refineries and have the potential to replace the commercial sulfided molybdenum catalysts [19] .The first work about molybdenum carbide electrodeposition was published by Andrieux [8] in the 40´s decade of last century. This author used a mixture of LiF and $NaBO_2$ as solvent and Na_2CO_3 and MoO_3 as sources of C and Mo, respectively. Afterwards, Heinen [13] and Suri [14] electrodeposited molybdenum carbides from NaF-KF-NaB_2O_7 melts containing Na_2CO_3 and MoS_2 or Na_2MoO_4 salts. None of these studies led to the formation of compact and coherent carbide layers.

Nevertheless, in the 80´s and 90´s Stern [3], Shapoval [15] and Topor [16] claimed the production of molybdenum carbide electroplates. Both Stern [3] and Topor [16] used the LiF-NaF-KF eutectic mixture (known as FLINAK) as solvent. K_2CO_3 and MoF_6 (or K_3MoCl_6) were employed as solutes by Stern [3] and K_2CO_3 and Na_2MoO_4 by Topor [16]. Shapoval [15] used a NaCl-LiF melt containing K_2CO_3 and Na_2MoO_4 and obtained adherent and non-porous Mo_2C coating of nearly 50 μm thickness with current efficiency ranging from 40 to 50%. More recently, Malyshev [17] deposited uniform and continuous Mo_2C coatings with low porosity from a Na_2WO_4-Li_2MoO_4-Li_2CO_3 electrolyte. The rate of deposition of the coatings in the investigated range of current densities was 5-25 μm/h and the current yield, calculated with respect to Mo_2C, was up to 50%. In another work Malyshev [7] also studied the initial stages of nucleation of Mo_2C in Na_2WO_4-MoO_3-Li_2CO_3 melt.

Uptodate, the reactions that led to the formation of the molybdenum carbides by electrolysis of FLINAK- K_2CO_3-Na_2MoO_4 melts are not well understood.

For C formation, the direct reduction of carbonate ions according to [16]:

$$CO_3^{2-} + 4e \rightarrow C + 3O^{2-} \quad (1),$$

the reduction of carbonate ions by an alkaline metal (such as Li) formed during cathodic polarization [20]:

$$Li^{+} + e \rightarrow Li \quad (2)$$

$$CO_3^{2-} + 4Li \rightarrow C + 3O^{2-} + 4Li^{+} \quad (3),$$

and a two-step mechanism involving an intermediate product CO_2^{2-} [16]:

$$CO_3^{2-} + 2e \rightarrow CO_2^{2-} + O^{2-} \quad (4)$$

$$CO_2^{2-} + 2e \rightarrow C + 2O^{2-} \quad (5)$$

have been proposed.

For Mo deposition, a two-step mechanism according to:

$$Mo\ (VI) + 3e \rightarrow Mo(III) \quad (6)$$

$$Mo\ (III) + 3e \rightarrow Mo\ (0) \quad (7)$$

was reported [12].

The subsequent reaction would be:

$$2Mo + C \rightarrow Mo_2C \quad (8)$$

In electrodeposition of metals and compounds, the cathodic current density is a measure of the combined rate of nucleation and growth of the mature nuclei [21]. Under potential controlled conditions, the current-time relationships are indicative of the nucleation and growth mechanisms. Some relationships between current density and time were established for one- (1-D), two- (2-D) and three- (3-D) dimensional growth of nuclei. The 1-D growth is associated for example to needles formation, 2-D growth to the formation of cylindrical forms and 3-D growth to the case of hemi-spherical or spherical growth [22]. Sharifer and Hills [23] derived dimensionless expressions of the current transient for multiple 3-D growth for two limiting cases:

(i) nuclei form simultaneously

$$\left(\frac{I}{I_m}\right)^2 = \frac{1.9542}{(t/t_m)}\left[1-\exp\left(-1.2564.(t/t_m)\right)\right]^2 \quad (9)$$

(ii) nuclei form progressively during electrodeposition

$$\left(\frac{I}{I_m}\right)^2 = \frac{1.2254}{(t/t_m)}\left[1-\exp\left(-2.3367.(t/t_m)^2\right)\right]^2 \quad (10)$$

where I_m is the maximum of the current transient which occurs at time t_m.

This chapter reports the electrodeposition of molybdenum carbide layers from the LiF-NaF-KF eutectic melt containing K_2CO_3 and Na_2MoO_4 salts. The coatings were obtained by both direct and pulse electroplating processes. The influence of bath composition, temperature and electrical parameters of electrodeposition on the morphology of the deposits was studied using scanning electron microscopy. The composition of the coatings was determined by X-ray diffractometry. The nucleation mechanism of the molybdenum carbides was analyzed by chronoamperometry.

8.1. Electrodeposition Set up

8.1.1. Electrolytic Cell

The electrolytic cell was made of a stainless steel cylindrical vessel and a Pyrex cap with connections for vacuum system and argon line (figure 8.1). A nickel crucible containing the electrolytic solution was placed at the lower stainless steel part whose wall was protected by a nickel line. The bottom of nickel crucible was isolated from the cell by an alumina disc. Heating was supplied by a vertical electric furnace whose temperature could reach up to 1000°C. The Pyrex cap was designed to allow the introduction of the electrodes.

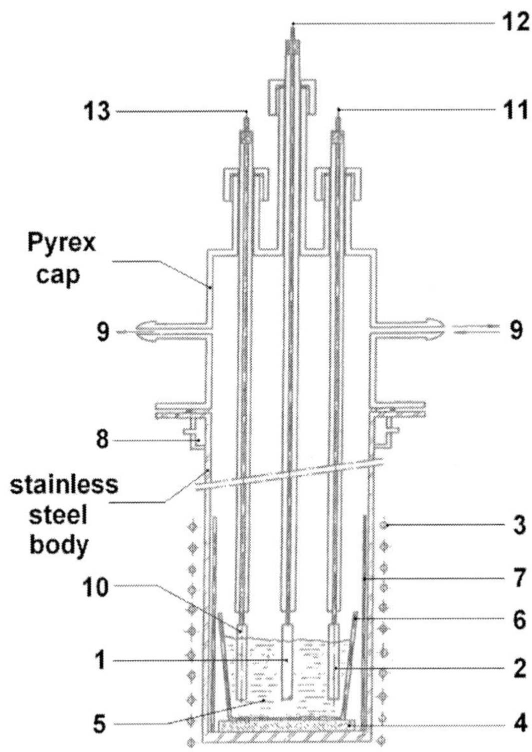

Figure 8.1. Electrolytic cell used for molybdenum carbide electrodeposition: 1. cathode; 2. anode; 3. heating elements; 4. alumina insulation; 5. electrolyte; 6. nickel crucible; 7. nickel liner; 8. water inlet and outlet; 9. argon and vacuum lines; 10. pseudo-reference electrode; 11. anode lead; 12. cathode lead; 13. pseudo-reference lead.

8.1.2. Electrolytic Solution and Electrodeposition Parameters

Electrodeposition of molybdenum carbide was achieved by electrolysis of LiF-NaF-KF eutectic melt containing K_2CO_3 and Na_2MoO_4 salts. All individual fluorides were of analytical grade.The eutectic melt that contained 29.3 wt % LiF, 11.7 wt % NaF and 59.0 wt % KF (FLINAK) was prepared by mixing pure and finely crushed fluorides, placing them in the nickel crucible and introducing it in the electrolysis cell. The mixture was maintained at 400°C for 8 hours under vacuum for deoxygenation and dehydration. Then, the fluoride mixture whose melting point is 454°C was heated to 550°C under argon atmosphere and maintained at this temperature for 12 hours. The bath was then cooled until room temperature.

The K_2CO_3 and Na_2MoO_4 salts were then added to the solid eutectic melt and dried under vacuum at 400 °C for 8 hours. Afterwards, argon was introduced into the cell and the temperature was elevated to the working temperature. The electrodeposition runs were performed using 5 mm-diameter molybdenum rods as anodes, and nickel sheets of dimensions 5x 50 x 0.25 mm, as cathodes. The electrodes were previously degreased, rinsed with distilled water and dried by rinsing in acetone and exposure to hot air. Before electrodeposition, the electrodes were placed just above the bath level for 10 min in order to reach the working temperature, and dipped in the electrolyte. After electrolysis the electrodes were withdrawn from the bath, left just over the melt for 30 min in order to drain the entrapped salt, lifted to the upper part of the cell and after cooling removed from the cell. Afterwards, the cathodes were successively washed in current tap hot water and in water under ultrasonication, rinsed with acetone and dried in hot air.The morphology of the deposits was analysed by scanning electron microscopy using a JEOL TECHNICS T200 microscope. The phase composition of the coatings was determined by X-ray diffraction in a SHIMADZU XRD-6000 diffractometer using Ni filtered Cu-Kα radiation. The coatings were obtained by direct and pulse electroplating processes using both potentiostatic and galvanostatic modes. For the potentiostatic direct and pulse electrodeposition, a 5 mm-diameter molybdenum rod was used as comparison electrode (pseudo-reference electrode).

Table 8.1. Electrolysis parameters used for direct electrodeposition of molybdenum carbide in FLINAK

Potentiostatic eletrodeposition	
Bath composition	1 wt % K_2CO_3 +1.5 wt % Na_2MoO_4 4 wt % K_2CO_3 + 6 wt % Na_2MoO_4 6 wt % K_2CO_3 + 7.5 wt % Na_2MoO_4
Temperature	750, 800, 850 °C
Potential vs pseudo-reference	-0.25 to -1.2 V
Electrodeposition time	60 min
Galvanostatic eletrodeposition	
Bath composition	7.5 wt % K_2CO_3 + 7 wt % Na_2MoO_4
Temperature	800 °C
Cathodic current density	25 to 250 mA cm^{-2}
Electrodeposition time	variable

A THOMSON ELECTROCHEM model 259 potentiostat with a PAR model 175 programmer, and an ADVANCE ELECTRONICS model PP3 DC supply were used. The investigated experimental conditions were summarized in tables 8.1 and 8.2 for direct and pulse plating, respectively. The various pulse parameters are defined in Figure 8.2. The influence of bath composition, temperature and electrical parameters of direct and pulse electrodeposition on the characteristics of the coatings was evaluated.

For the study of molybdenum carbide nucleation, the chronoamperometric technique was used. A nickel sheet of 0.5 cm^2 exposed area was used as working electrode, a 5 mm-diameter molybdenum rod as counter-electrode and another 5 mm-diameter molybdenum rod as pseudo-reference electrode. The experiments were carried out using a THOMSON ELECTROCHEM model 259 potentiostat and a PAR model 175 programmer.

Table 8.2. Electrolysis parameters used for pulse electrodeposition of molybdenum carbide in FLINAK

Potentiostatic eletrodeposition	
Bath composition	5 wt% K_2CO_3 + 7.5 wt% Na_2MoO_4
Temperature	800°C
E_{OFF} vs pseudo-reference	-0.4V
E_{ON} vs pseudo-reference	-0.6 to -1.0 V
T_{ON} and T_{OFF}	0.1 to 100 ms
Electrodeposition time	20 and 45 min
Galvanostatic eletrodeposition	
Bath composition	5 wt% K_2CO_3 + 7.5 wt% Na_2MoO_4
Temperature	800 °C
i_{OFF}	0 mA cm^{-2}
i_{ON}	160 to 300 mA cm^{-2}
T_{ON} and T_{OFF}	0.6 to 36 ms
Electrodeposition time	20 and 80 min

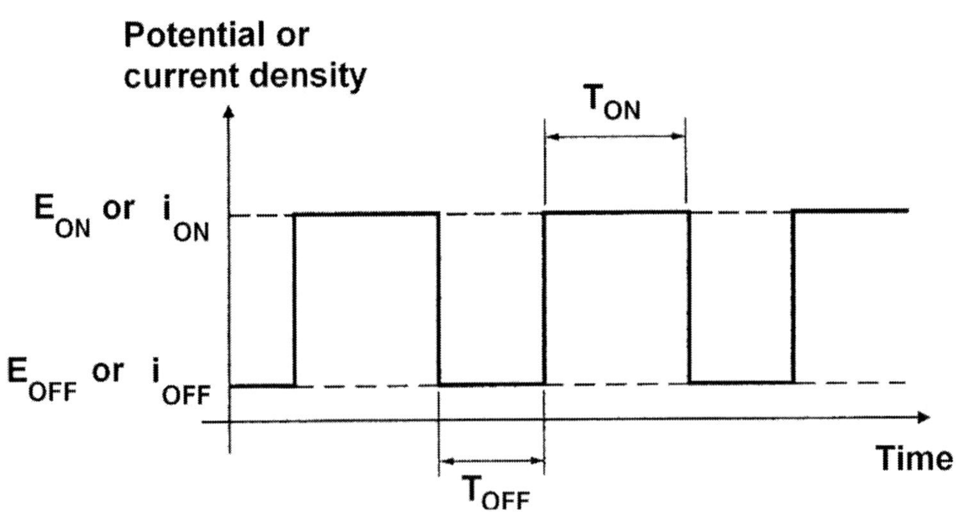

Figure 8.2. Parameters for pulse plating.

8.2. Surface Morphology of Coatings

8.2. 1. Potentiostatic Deposition

Figures 8.3. to 8.5. show MEV photographs of the surface of some electrocoatings obtained in FLINAK-K_2CO_3-Na_2MoO_4 mixtures under different experimental conditions. All the investigated conditions led to the production of a deposit on the cathode but the morphology of the electrocoating was highly dependent on the temperature (figure 8.3), bath composition (figure 8.4) and applied potential (figure 8.5). The increase of the temperature from 750 to 850°C led to the increase of the deposit grain size but at 850°C grains of irregular shape were produced (figure 8.3.a, 8.3.b and 8.3.c). The deposits with the best superficial characteristics (more regular grain size and shape, lower porosity and lower roughness) were obtained at 800°C (fig 8.3b).Poor deposits with high porosity and dendritic growth were obtained in FLINAK + 1 wt % K_2CO_3 + 1.5 wt % Na_2MoO_4 at 800°C (figure 4a). For the same temperature and applied potential, homogeneous coatings with low porosity were deposited in FLINAK containing higher concentrations of solutes, 4 wt % K_2CO_3 + 6 wt % Na_2MoO_4 and 6 wt % K_2CO_3 + 7.5 wt % Na_2MoO_4 (figure 8.4b and 8.4c, respectively). No significant difference in the surface morphology of the coatings obtained in both these baths was observed.

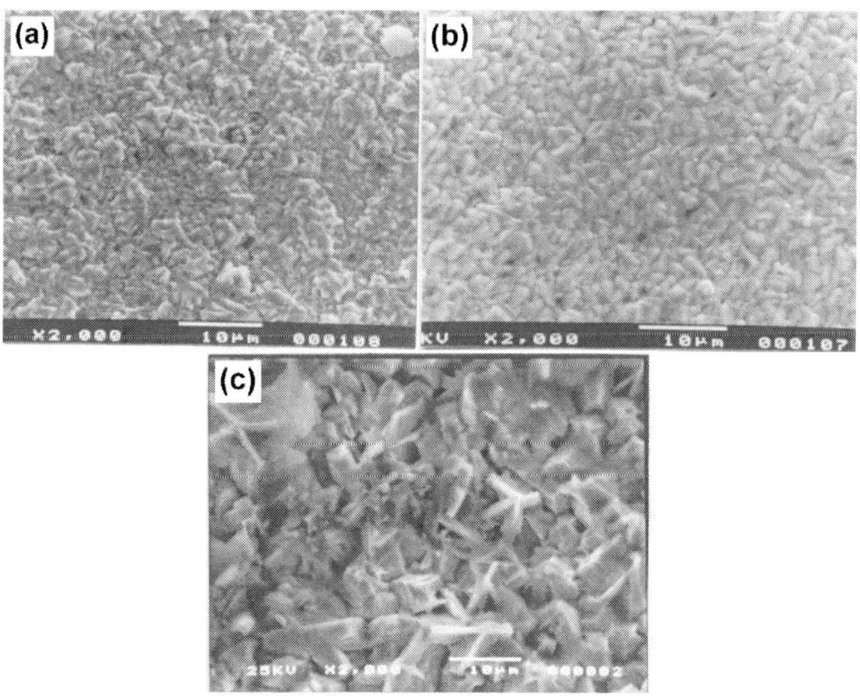

Figure 8.3. Surface morphology of electrocoatings obtained at -0.6 V applied potential during 1 h in FLINAK + 6 wt % K_2CO_3 + 7.5 wt % Na_2MoO_4 at (a) 750°C, (b) 800°C and (c) 850°C.

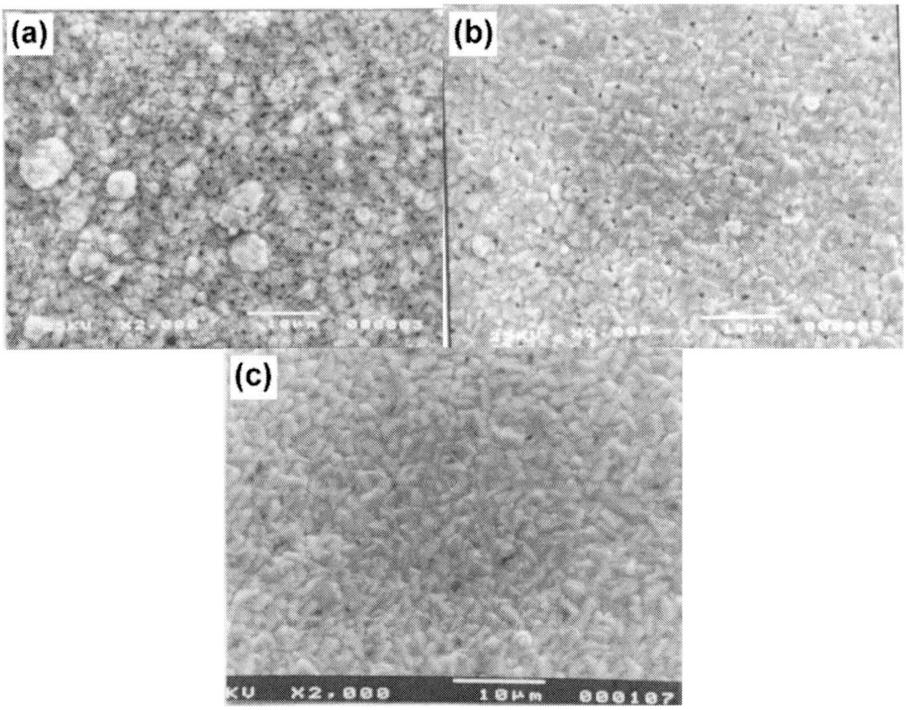

Figure 8.4. Surface morphology of electrocoatings obtained at -0.6 V applied potential during 1 h in (a) FLINAK + 1 wt % K_2CO_3 +1.5 wt % Na_2MoO_4 , (b) FLINAK + 4 wt % K_2CO_3 + 6 wt % Na_2MoO_4 and (c) FLINAK + 6 wt % K_2CO_3 + 7.5 wt % Na_2MoO_4 at 800°C.

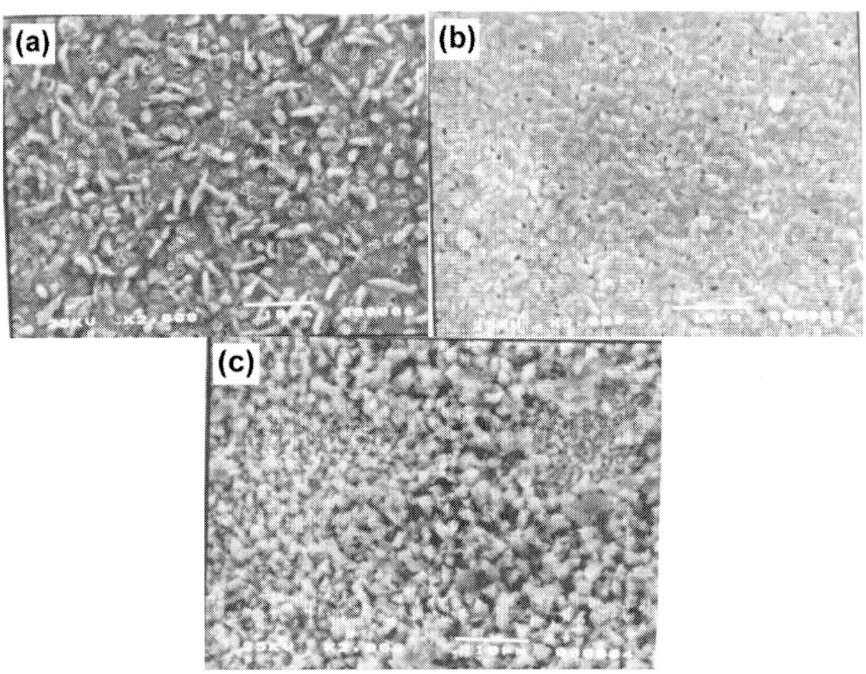

Figure 8.5. Surface morphology of electrocoatings obtained during 1 h in FLINAK + 4 wt % K_2CO_3 + 6 wt % Na_2MoO_4 at 800°C at (a) -0.5 V, (b) -0.6 V and (c) -0.8 V applied potential.

Coatings were performed in FLINAK + 4 wt % K_2CO_3 + 6 wt % Na_2MoO_4 at 800°C varying the applied potential between -0.25 and -1.2 V. For potentials nobler than -0.5 V, needle-like coatings were obtained, whereas for potentials lower than -0.8 V the coatings presented a disordered growth and a dendritic morphology. Non-homogeneous coatings were also obtained at -0.5 V and -0.8 V. For -0.5 V a more or less compact layer covered by numerous needles was obtained (figure 8.5.a), whereas for -0.8 V an inhomogeneous and porous coating was achieved (figure 8.5.c). On the contrary, a low porosity and well distributed deposit constituted of well-joined grains was obtained at -0.6 V (figure 8.5.b).

8.2.2. Galvanostatic Deposition

Since electrodeposition at 800°C using the potentiostatic mode led to coatings of good quality, the same temperature was chosen for studying the influence of current density on the morphology of deposits obtained by galvanostatic plating. Figures 8.6. to 8.8 demonstrate MEV photographs of the surface of some electrocoatings obtained in FLINAK+7.5 wt% K_2CO_3+7 wt% Na_2MoO_4 mixture under different experimental conditions.

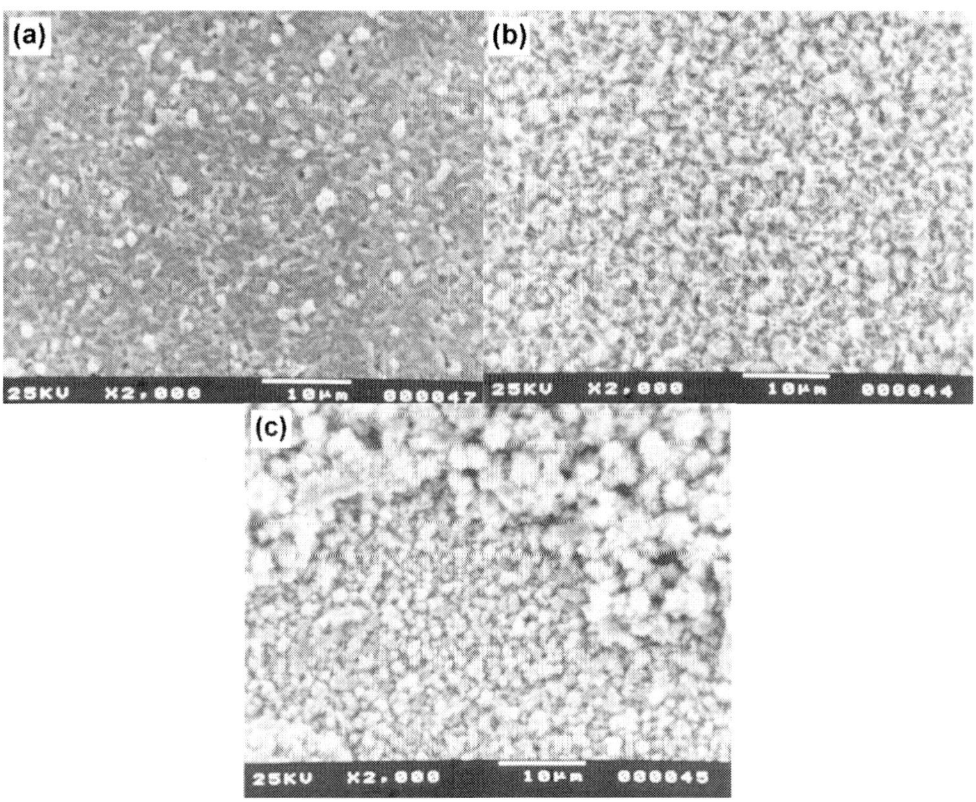

Figure 8.6. Surface morphology of electrocoatings obtained during 30 min in FLINAK + 7.5 wt % K_2CO_3 + 7 wt % Na_2MoO_4 at 800°C using (a) 36 mA cm^{-2}, (b) 77 mA cm^{-2} and (c) 200 mA cm^{-2} cathodic current density.

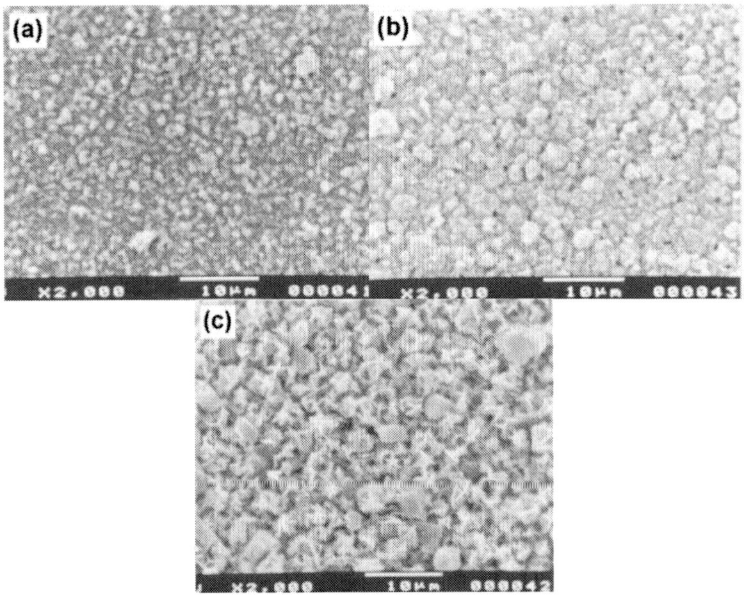

Figure 8.7. Surface morphology of electrocoatings obtained during (a) 20 min, (b) 40 min and (c) 60 min in FLINAK + 7.5 wt % K_2CO_3 + 7 wt % Na_2MoO_4 at 800°C using 77 mA cm^{-2} cathodic current density.

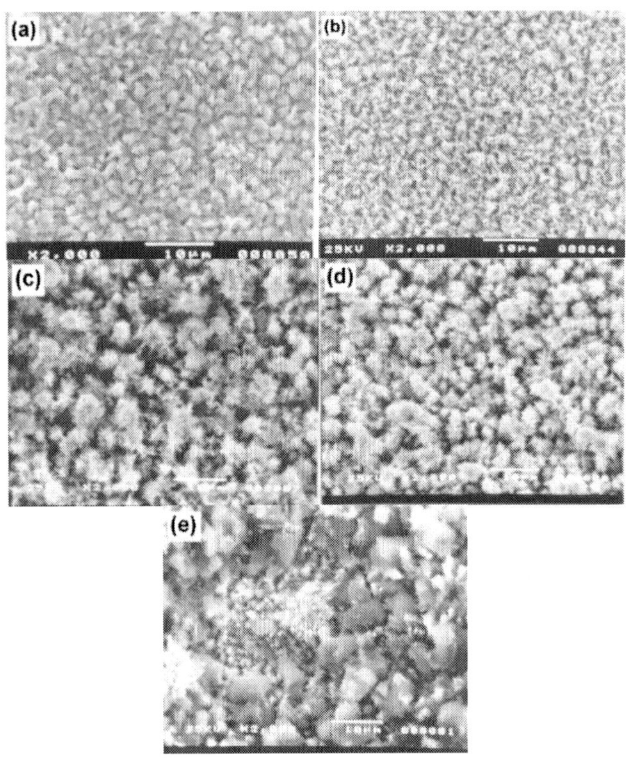

Figure 8.8. Surface morphology of electrocoatings obtained in FLINAK + 7.5 wt % K_2CO_3 + 7 wt % Na_2MoO_4 at 800°C using (a) 36 mA cm^{-2} for 1 h, (b) 80 mA cm^{-2} for 30 min, (c) 115 mA cm^{-2} for 20 min, (d) 150 mA cm^{-2} for 15 min and (e) 192 mA cm^{-2} for 12 min (total electrical charge density 138 A s cm^{-2}) .

Using the same electrodeposition time, more compact and more homogeneous coatings were obtained at low current densities. The porosity of the deposits increased with increasing current density and the coatings turned less compact (figure 8.6.a. and 8.6.b). For very high current densities (200 mA cm^{-2} for example), coarse and non-homogeneous coatings were obtained (figure 8. 6.c).Using the same cathodic current density, the electrodeposition time also affects the morphology of the deposited layers. Indeed, increasing electrolysis time generally led to the increase of the grain size of the coatings (figure 8.7) and eventually to dendritic growth (figure 8.7.c). Electrocoatings were obtained at different cathodic current densities but using the same electrical charge density in order to obtain deposits of very close thicknesses. As depicted in Figure 8.8, a great degradation of the deposits in terms of both porosity and grain size occurred with the increase of current density.

8.2.3. Pulse Plating

The pulse plating experiments were performed in FLINAK + 5 wt% K_2CO_3 + 7.5 wt% Na_2MoO_4 at 800°C under potential or current density control. The pulse parameters were presented in Figure 8.2. In this study, E_{OFF} and i_{OFF} were established to -0.4V and 0 mA cm^{-2}, respectively. The effects of the pulsed potential E_{ON} (or pulsed current density i_{ON}) and pulse duration T_{ON} and off time T_{OFF} on the morphology of the deposits were evaluated. Only the main results of the study are shown in this chapter.

8.2.3.1. Controlled E Pulse Plating

(a) Effect of $E_{ON.}$ Figure 8.9. shows the variation of the coating surface morphology for E_{ON} pulsed potential ranging from -0.6 V to -1.0 V. Smooth deposits with regular grain size and shape were obtained for the less negative applied potentials, -0.6 and -0.7 V (figure 8.9.a and 8.9.b). For -0.8 V the deposit showed non-uniform grain size distribution (figure 8.9.c), whereas for much more negative potential (-1.0 V), coarse morphology of the coating was observed (figure 8.9.d).

(b) Effect of T_{ON} and $T_{OFF.}$ The effect of T_{ON} pulse duration and T_{OFF} on the surface morphology of the coatings was analyzed using a E_{ON} pulsed potential that led to a good quality coating (-0.6 V).Low values (≤ 1 ms) of both T_{ON} and T_{OFF} led to needle-like deposits (figure 8.10.a, 8.10.b and 8.11.a), but smooth and compact layers were obtained for higher values of T_{ON} and T_{OFF} (figure 8.10.c, 8.10.d, 8.10.e, 8.11b and 8.11c). For T_{OFF} of 3 ms, the grain size of the deposits was smaller than that observed for the coatings obtained under potentiostatic conditions (figure 8.4.c). Increasing T_{OFF} led to an increase of the coating grain size (figures 8.11.b and 8.11.c).

8.2.3.2. Controlled i pulse plating

(a) Effect of $i_{ON.}$ Little reproducibility of the coating morphology was observed for the experiments performed using i pulse plating mode. Thus, only few results will be presented here.Figure 8.12. shows the variation of the coating surface morphology for i_{ON} pulsed current density ranging from 160 to 300 mA cm^{-2}. As T_{ON} and T_{OFF} had the same value, 3 ms, the mean current density ranged from 80 to 150 mA cm^{-2}.

Smooth and small grain-sized deposit was obtained for 80 mA cm^{-2} mean current density (figure 8.12.a). This morphology is finer than that observed for the coating performed under galvanostatic mode using a constant 80 mA cm^{-2} current density (figure 8.8.b). With increasing the pulsed current density to 200 and 300 mA cm^{-2}, coarse coatings and coatings with irregular grain shape were obtained (figure 8.12.b and 8.12c).Though the quality of the coating obtained using a 150 mA cm^{-2} mean current density (figure 8.12.c) was not so good, it was better that the deposit obtained under galvanostatic mode using a constant 150 mA cm^{-2} current density (figure 8.8.d).

(b) Effect of T_{ON} and $T_{OFF.}$ Maintaining the same value for T_{ON} and T_{OFF}, no significant influence of T_{ON} (or T_{OFF}) on the coating morphology was observed up to 9 ms value (figure 8.13.a, 8.13.b and 8.13.c). Higher values of T_{ON} (or T_{OFF}) led to highly porous layers (figure 8.13.d).

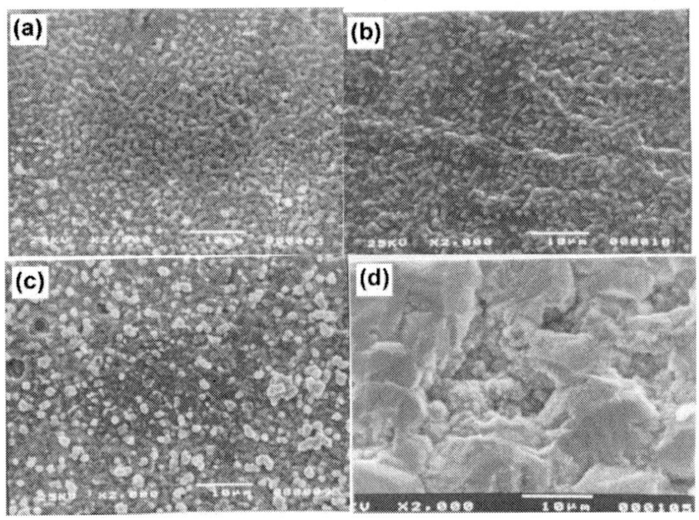

Figure 8.9. Surface morphology of electrocoatings obtained in FLINAK + 5 wt % K_2CO_3 + 7.5 wt % Na_2MoO_4 at 800°C by E controlled pulse plating using the following pulse parameters : E_{OFF} = -0.4 V ; $T_{ON} = T_{OFF}$ = 3 ms; deposition time = 45 min;E_{ON} = (a) -0.6 V; (b) -0.7 V; (c) -0.8 V; (d) -1.0 V.

8.3. Cross-Sectional Observation of the Coatings

Information about the coating quality can also be obtained from observations of the cross-sections. Figure 8.14 shows typical SEM photographs of the cross-sections of compact and porous layers obtained by electrolysis of FLINAK - K_2CO_3 -Na_2MoO_4 melts. The compact coatings demonstrated almost uniform thickness over the cathode surface and was pore-free (figure 8.14a). No porosity was also noted at the coating/substrate interface which is indicative of the good adhesion of the deposit. Differently, the porous deposits usually showed irregular thickness, porosity and sometimes dendritic growth (figure 8.14b and 8.14c).The thickness of the electrocoatings was dependent on the electrical parameters of deposition (i.e current density and applied potential) and on the cathodic efficiency. The current efficiencies calculated considering necessary ten electrons to form one Mo_2C were in the 21 to 48% range.

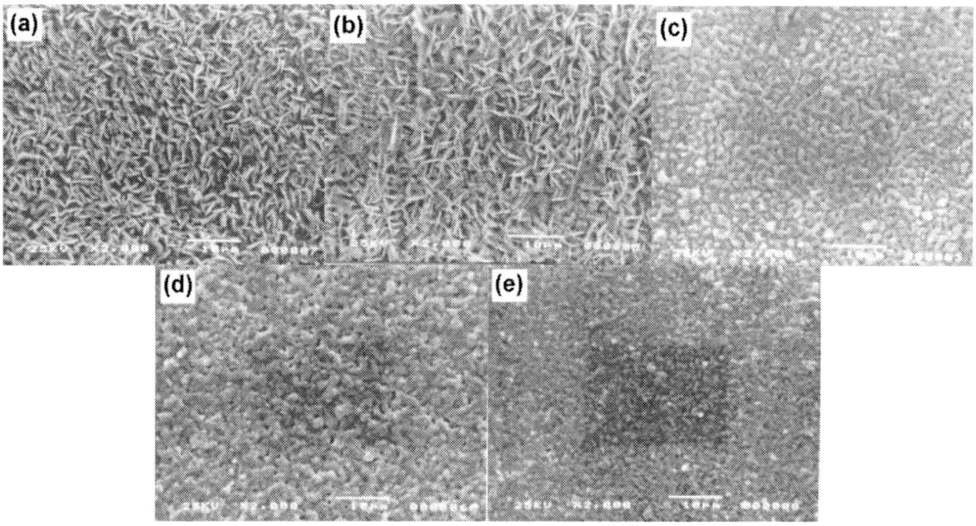

Figure 8.10. Surface morphology of electrocoatings obtained in FLINAK + 5 wt % K_2CO_3 + 7.5 wt % Na_2MoO_4 at 800°C by E controlled pulse plating using the following pulse parameters : E_{OFF} = -0.4 V ; E_{ON} = -0.6 V ; T_{OFF} = 3 ms; deposition time = 45 min;T_{ON} = (a) 0.5 ms; (b) 1 ms; (c) 3 ms; (d) 6 ms; (e) 9 ms.

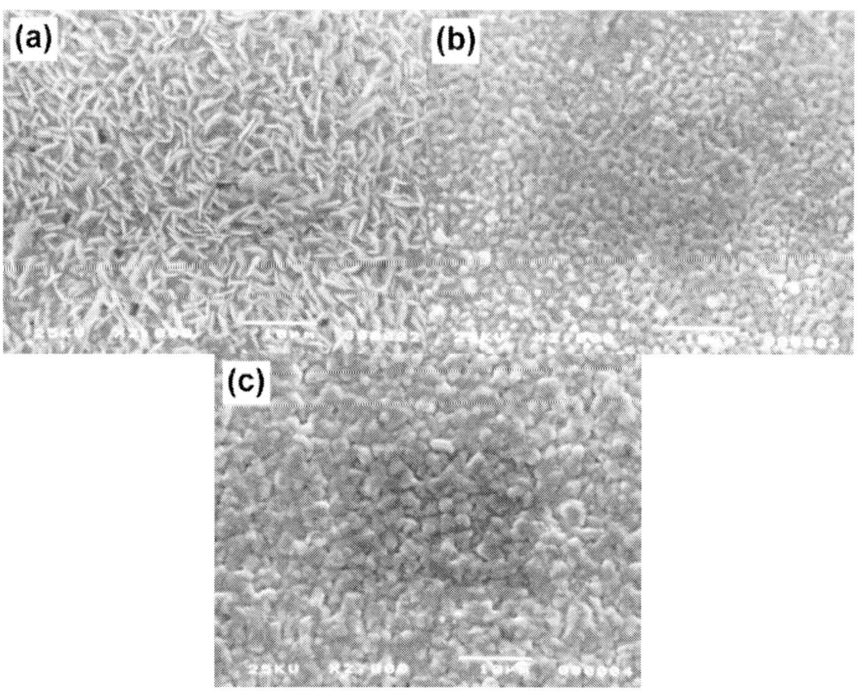

Figure 8.11. Surface morphology of electrocoatings obtained in FLINAK + 5 wt % K_2CO_3 + 7.5 wt % Na_2MoO_4 at 800°C by E controlled pulse plating using the following pulse parameters : E_{OFF} = -0.4 V ; E_{ON} = -0.6 V ; T_{ON} = 3 ms; deposition time = 45 min;T_{OFF} = (a) 1 ms; (b) 3 ms; (c) 9 ms.

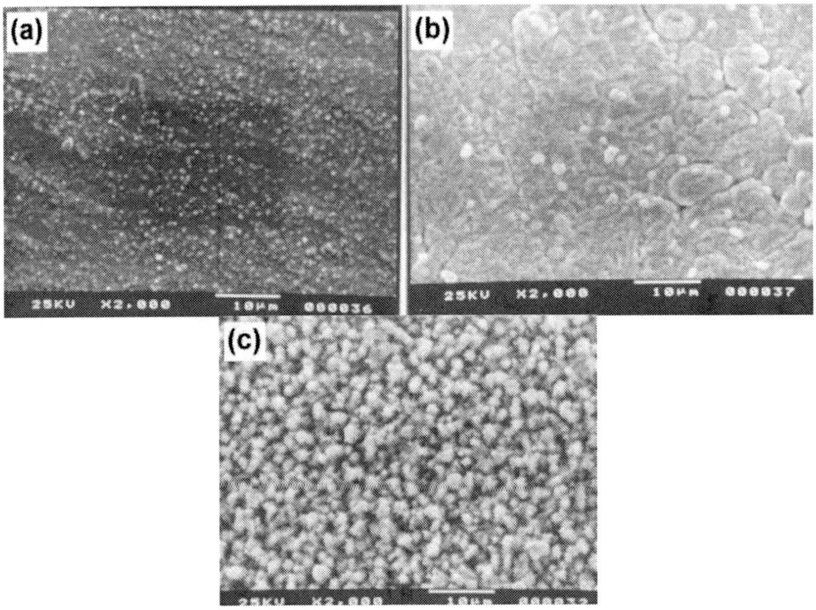

Figure 8.12. Surface morphology of electrocoatings obtained in FLINAK + 5 wt % K_2CO_3 + 7.5 wt % Na_2MoO_4 at 800°C by i controlled pulse plating using the following pulse parameters : i_{OFF} = 0 mA cm^{-2} ; $T_{ON} = T_{OFF}$ = 3 ms; deposition time = 20 min; i_{ON} = (a) 160 mA cm^{-2}; (b) 200 mA cm^{-2}; (c) 300 mA cm^{-2}.

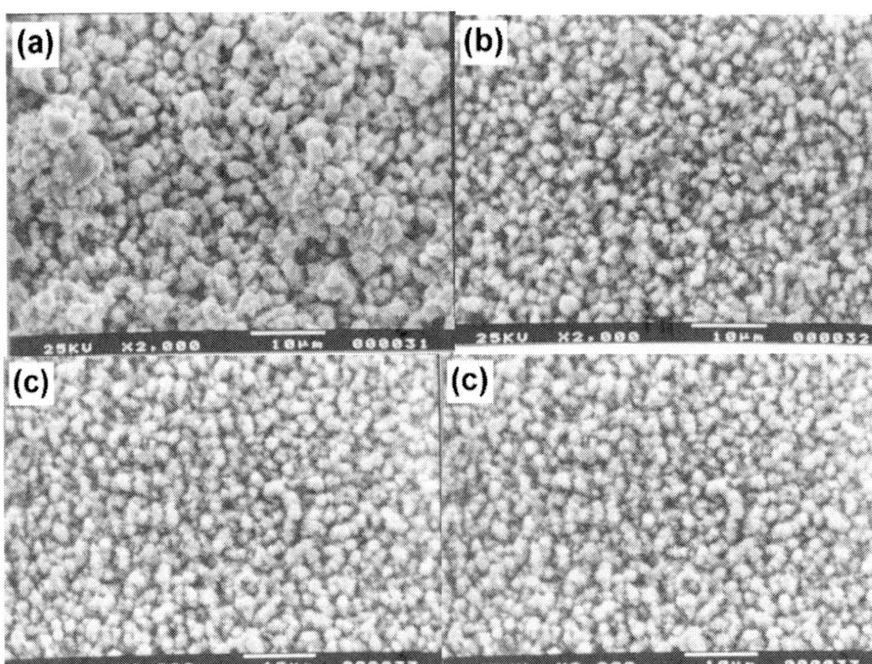

Figure 8.13. Surface morphology of electrocoatings obtained in FLINAK + 5 wt % K_2CO_3 + 7.5 wt % Na_2MoO_4 at 800°C by i controlled pulse plating using the following pulse parameters : i_{OFF} = 0 mA cm^{-2}; i_{ON} = 300 mA cm^{-2}; deposition time = 80 min;$T_{ON} = T_{OFF}$ = (a) 0.6 ms; (b) 3 ms; (c) 9 ms; (d) 36 ms.

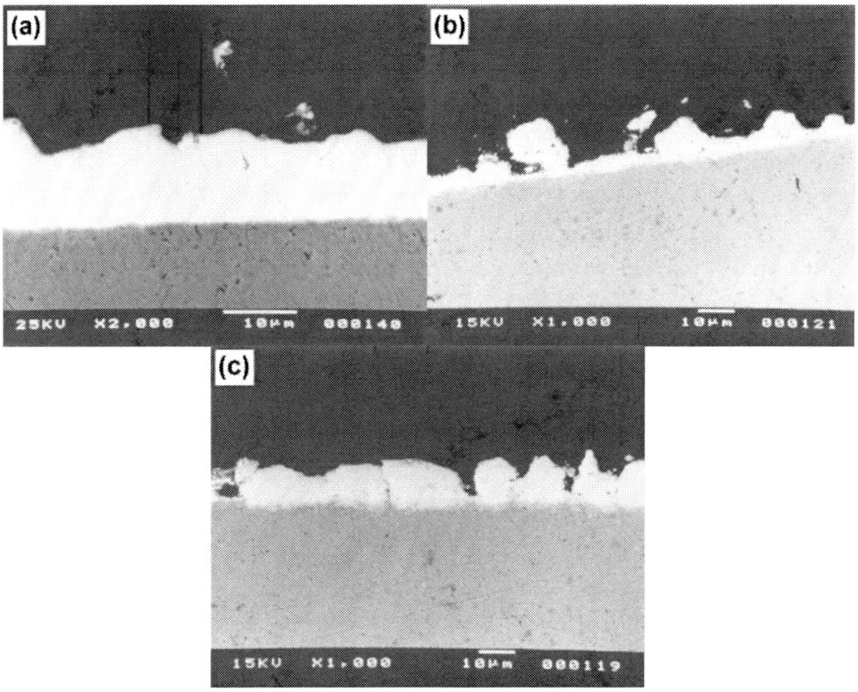

Figure 8.14. Typical cross-section of (a) compact and (b,c) porous molybdenum carbide electrocoatings obtained in FLINAK-K_2CO_3-Na_2MoO_4.

8.4. Phase Composition

Figure 8.15. represents a typical XRD pattern of electrocoatings obtained by electrolysis of FLINAK-K_2CO_3-Na_2MoO_4 melts. Most coatings showed similar XRD spectra. The Mo_2C molybdenum carbide phase whose crystallographic planes were indexed in figure 8.15. is the principal constituent of the coatings.

Sometimes, some peaks of the Ni phase corresponding to the substrate were present. Some XRD patterns also revealed the presence of some peaks of the Mo phase (figure 8.15), which evidenced that some amount of molybdenum has not reacted according to the reaction (8).

8.5. Nucleation Process

Typical current-time transients obtained for different applied potentials in FLINAK - K_2CO_3 -Na_2MoO_4 melts at 800°C are presented in figure 8.16. The current increases rapidly to a maximum (I_m,t_m) due to the nucleation and growth of nuclei and then decreases gradually with time due to linear diffusion.

The shape of the transients is indicative of a 3D-nucleation behavior with diffusion controlled growth of molybdenum carbide. Similar behaviour was observed by Malyshev [7]. Kinetics information about electrocrystallization process can be obtained analyzing the rising portion and the maximum of the transients. Figure 8.17. compares the dimensionless experimental data and the data obtained from the theoretical models for instantaneous and progressive nucleation (Equations 9 and 10). It is clear that the nucleation of molybdenum carbide is predominantly progressive. Another criterion for nucleation process given by Sharifer and Hills [23] is based on the slope of the curve Log I versus Log t on the rising portion of the transient. The slopes are 3/2 and 1/2 for progressive and instantaneous nucleation respectively.

The plot of Log I against Log t (figure 8.18) obtained from the experimental data of Figure 8.16. shows two linear segments with slopes close to 1.5 and 0.5, respectively. This indicates that the nucleation is not progressive over the whole transient but at a certain time it changes from progressive to instantaneous. This behaviour is frequently observed for metal electrodeposition under diffusion control [21,24].

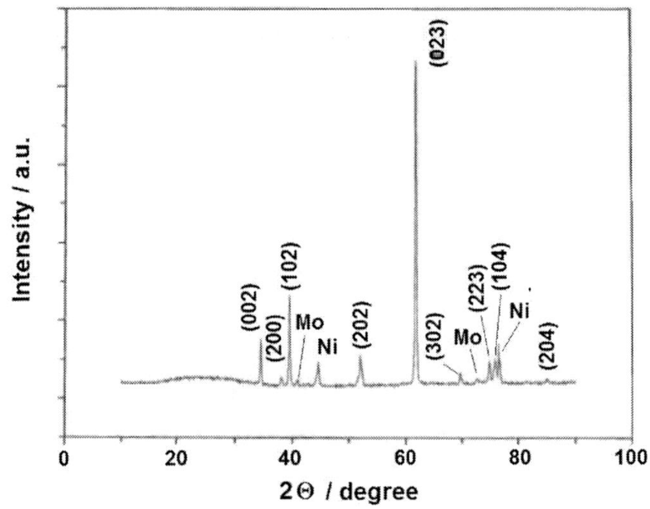

Figure 8.15. Typical XRD pattern of an electrocoating obtained in FLINAK-K_2CO_3-Na_2MoO_4 melt at 800°C .

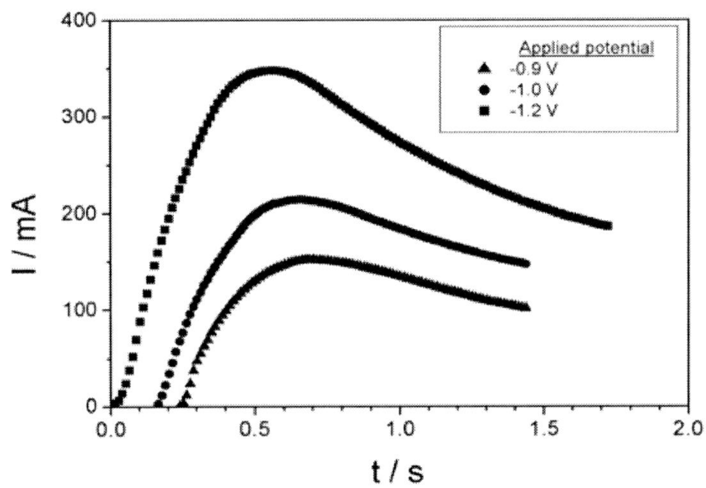

Figure 8.16. Current-time transients obtained at different potentials in FLINAK + 0.1 wt % K_2CO_3 + 0.15 wt % Na_2MoO_4 at 800°C; área = 0.5 cm^2.

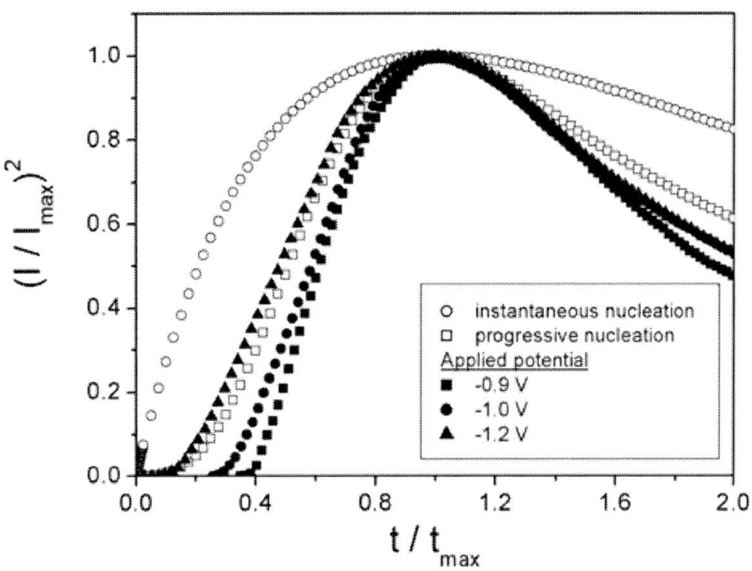

Figure 8.17. Dimensionless experimental data deduced from figure 8.16. and theoretical data calculated using equations (9) and (10).

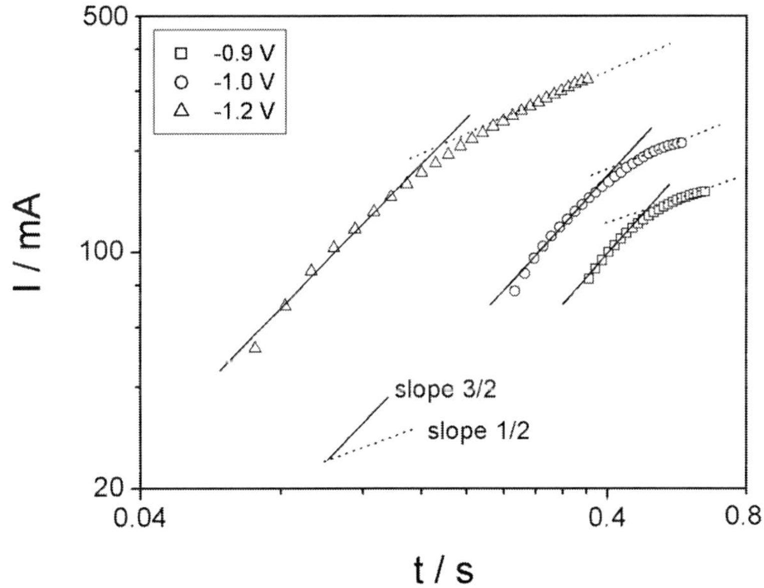

Figure 8.18. Log I against log t plots deduced from the experimental data of figure 8.16.

CONCLUSION

Mo_2C molybdenum carbide coatings with low porosity and good adhesion can be electrodeposited from FLINAK containing K_2CO_3 and Na_2MoO_4 at 800°C.The morphology of the coatings was shown to be highly dependent on temperature, electrolyte composition and electrical parameters (current density and applied potential).Pulse plating generally produced coatings of finer grain size than those obtained under direct plating conditions. The current efficiencies ranged from 21 to 48%.

Acknowledgments

This work was supported by the RHAE/CNPq (Brazil) Program. A.F. Sartori wants to express his gratitude to Professor D. Inman, from Imperial College, for his helpful advice.

References

[1] H.O. Pierson, *Handbook of Refractory Carbides and Nitrides*, Noyes Publications, Westwood (1996).
[2] K.H. Stern and S.T. Gadomski, *J. Electrochem. Soc.* 130, 300 (1983).
[3] Kurt H. Stern, *United States Patent* No. 4,430,170, Feb. 7, 1984.
[4] A.J. Hackman and R.S. Feigelson, *J. Electrochem. Soc.* 130, 221 (1983).
[5] K.H. Stern and D.R. Rolison, *J. Electrochem. Soc.* 136, 3760 (1989).
[6] L. Massot, P. Chamelot and P. Taxil, *J. Alloys Compds* 424, 199 (2006).
[7] V. Malyshev, A. Gab and M. Gaune-Escard, *J. Appl. Electrochem.* 38, 315 (2008).
[8] L. Andrieux and G. Weiss, *Bull. Soc. Chim.* 15, 598 (1948).
[9] K.H. Stern and M.L. Deanhardt, *J. Electrochem. Soc.* 132, 1891 (1985).
[10] H. Yabe, Y. Ito, K. Ema and J. Oishi, "Electrodeposition of tungsten and tungsten carbide from molten halides", Proceedings of the 6th International Symposium on Molten Salts, 1987, p804.
[11] K.H. Stern and D.R. Rolison, *J. Electrochem. Soc.* 137, 178 (1990).
[12] K.H. Stern, *J. Appl. Electrochem..* 22, 717 (1992).
[13] H.J. Heinen, C.L. Barber and D.H. Baker, "Conversion to metal of dimolybdenum carbide electrosynthesized from molybdenite", Bureau of Mines, Department of the Interior, USA, 6590 Report of investigations, 1965.
[14] A.K. Suri, T.K. Mukherjee and C.K. Gupta, *J. Electrochem. Soc.* 120, 622 (1973).
[15] V.I. Shapoval, Kh.B. Khushkov, V.V. Malishev, V.T. Vesna and V.P. Maslov, *Poroshkovaya Metallurgiya* 7, 43 (1987).
[16] D.C. Topor and J.R. Selman, *J. Electrochem. Soc.* 135, 384 (1988).
[17] V.V. Malyshev, *Mater. Sc.* 34, 211 (1998).
[18] J.R. Selman, "Corrosion-resistant coatings for high-temperature high-sulfur-activity applications". Final report. Illinois Institute of Technology, LBL-35339, 1-33, 1994.
[19] P. Liu and J.A. Rodriguez, *Catalysis Lett.* 91, 247 (2003).
[20] M.L. Deanhardt, K.H. Stern and A. Kende, *J. Electrochem. Soc.* 133, 1148 (1986).
[21] G. Gunawardena, G. Hills, I. Montenegro and B. Scharifker, *J. Electroanal. Chem.* 138, 225 (1982).
[22] G.J. Hills, D.J. Schiffrin and J. Thompson, *Electrochim. Acta* 19, 657 (1974).
[23] B. Scharifker and G. Hills, *Electrochim. Acta* 28, 879 (1983).
[24] M. Sanchez Cruz, F. Alonso and J.M. Palacios, *J. Appl. Electrochem.* 23, 364 (1993).

In: Electrodeposition: Properties, Processes and Applications ISBN: 978-1-61470-826-1
Editor: Udit Surya Mohanty

Chapter 9

AQUEOUS ELECTRODEPOSITION OF NON-FERROUS METALS

Tondepu Subbaiah and Kali Sanjay*
Institute of Minerals and Materials Technology, Bhubaneswar, Orissa, India

ABSTRACT

This chapter outlines the electrodeposition of non-ferrous metals such as copper, zinc, nickel, cobalt, chromium and lead from their aqueous solutions. Effects of organic and inorganic impurities generally present in the aqueous solutions during electrodeposition are discussed. Energy reduction techniques during electrodeposition and augmentation of mass transfer in electrochemical cells are also included.

INTRODUCTION

In recent years there is immense attention in adopting hydrometallurgical routes for producing non-ferrous metals from multi-metal sulfides, concentrates, secondaries and lean oxide ores. Several hydrometallugical processes have been developed for extracting metal values of copper, zinc and lead from complex sulfide ores, copper, nickel and cobalt from copper converter slag and deep sea polymetallic nodules, nickel and cobalt from lateritic nickel ores and secondaries, chromium from chromic acid and lead from Jarosite. In all the above processes, the unit operations involved are leaching, solvent extraction and electrodeposition. Obtaining optimum conditions for electrodeposition (ED) of various metals using the synthetic solutions, fixing the tolerance limits of organic and inorganic impurities, testing with actual solution, coupling of electrodeposition conditions with solvent extraction and finding out the hydrodynamic conditions prevailing in the electrolytic cell are some of the major areas that are required for electrodeposition of these metals. The experience gained over three decades of processing wide variety of raw materials for producing high pure metals

* E-mail addresses: tsubbaiah@yahoo.com / kalisanjay@gmail.com.

as end products through aqueous electrodeposition is presented in this chapter. As it is difficult to detail all the work carried out, some of the salient features are discussed.

Table 9.1. Major uses of non-ferrous metals

Metals	Major uses
Copper	electrical and electronic industry, heat and corrosion resistant alloys
Zinc	galvanized steel, painting, buildings, alloys
Nickel	super alloys, wear resistant coatings, corrosion resistant alloys, low expansion alloys in metallurgical industries
Cobalt	soft magnetic materials, catalysts, pigments, dies, electronics, ceramics, enamels and agriculture
Chromium	ballistics and aviation, electric resistance heating elements, hard facing alloys, cutting tools, ferrous alloys, non-ferrous alloys, chromium carbide, cermets
Lead	storage vessels of corrosive liquids, batteries, solders, type metal, bearings, pipe, cable covering, plumbing and ammunition

The ever increase in demand for metals due to rapid industrial growth resulted in development of flow sheets for recovery of metal values by recycling or exploiting low grade resources. Secondaries and low grade ores / concentrates can be successfully processed by employing hydro and electrometallurgical techniques. Major uses of Cu, Zn, Ni, Co, Cr and Pb are shown in Table 9.1.

The major steps of hydrometallurgical flow sheet consist of (a) leaching (atmospheric / pressure / bacterial) (b) solution purification (precipitation / cementation / solvent extraction) and (c) production of high pure metals through electrodeposition. A typical flow sheet for production of non-ferrous metals is shown in Figure 9.1. Process development for a commercially viable electrodeposition circuit requires laboratory and bench scale studies followed by pilot plant demonstration.

9.1. Laboratory Scale Studies

In the laboratory studies, conditions for electrodeposition are optimized with respect to leach liquors from either precipitation / cementation circuit or from solvent extraction circuit with a criterion of maximizing current efficiency (CE), minimizing energy consumption (EC) and obtaining quality deposits.

Usually, these experiments are carried out in batch scale and metals are produced in gram scale. The aqueous electrolysis studies are carried out in glass containers / cells made of Perspex or plastics using different anodes and cathodes. Details of the experimental setups are available in the literature [1-2]. Electrodeposition experiments are carried out using synthetic solutions prepared with pure salts dissolved in deionized water. Reagent grade chemicals are used for conducting these experiments. Effect of various parameters such as bath composition, current density, temperature, pH are studied to optimize the electrodeposition conditions for each metal independently.

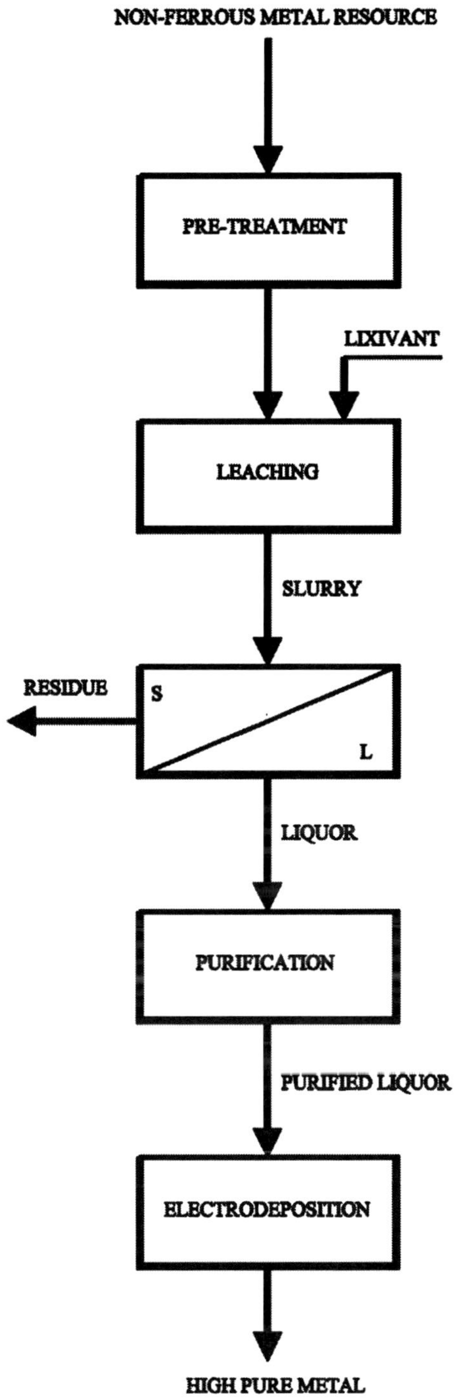

Figure 9.1. Flow sheet for production of non-ferrous metals.

The experimental setups for mass transfer studies using limiting current density technique [3], tracer technique [4] and the physicochemical properties of the electrolytes were reported elsewhere [5].The optimum conditions for electrodeposition of metals copper, zinc, nickel, cobalt, chromium and lead are shown in Table 9.2.

Table 9.2. Optimum conditions for electrodeposition of metals

Metals	Metal ion in the electrolyte (g/L)	Acidity	Current density/ A/m^2	Current efficiency (CE)/%	Cell voltage (V)	Purity />%	Temp. /°C
Copper	50	150g/L	200	95	2.2	99.9	30
Zinc	65	110g/L	400	85	3.4	99.9	30
Nickel	50	~2.0pH	200	70	4.5	99.9	60
Cobalt	50	~2.0pH	220	75	4	99.9	60
Chromium	60	1 g/L	800	10	5 – 6	99.9	90
Lead	30	0.7–0.9 pH	50	90	2	99.9	70

Bench and Pilot Scale Studies

Three primary objectives of bench scale study are scaling up of the coupling circuit, testing the coupling circuit in continuous mode and generating data for the preparation of material balance. These studies are carried out at feed rates of 5 – 25 L/h for 10 - 100 h to assess feasibility of the coupling circuit, control of different streams, building of impurities in the circuit, effect of duration on deposit quality and purity. Actual solutions obtained after leaching and purification operations are tested in continuous bench scale runs at optimal conditions. Testing of the process in pilot scale helps the engineers to understand the performance of unit operations and equipment in large scale and generate adequate data necessary for scaling up and designing of commercial plants. Pilot scale testing also helps in estimating the requirements of utilities and power for commercial plants.

9.2. Electrodeposition of Copper

The electrolyte for copper deposition consists of 50 g/L copper and 150g/L sulfuric acid. The energy requirement for electrodeposition is estimated to be 1800 - 2000 kWh/tone of copper. Impurities such as Fe, Al, M, Ca, suspended solids and gelatinous materials not only affect the quality of the copper deposit, but reduce the current efficiency. It is well-known [6] that solvents such as LIX64N cause organic burn on the cathode during copper electrodeposition. Presence of LIX64N beyond 50mg/L not only causes a drastic fall in the CE but also deteriorates the quality of the deposit. Even 10 mg/L of LIX64N in presence of Fe^{3+} changes the surface quality of copper cathode. When the LIX64N concentration crosses 50 mg/L, then a netted deposit is formed. Presence of either LIX64N or Fe^{3+} favors growth of pyramidal structure whereas their combination favors a ridge type of growth. However, addition of thiourea to the above combination causes the structure to change from ridge to pyramidal thus improving both CE and deposit quality. Electrolysis parameters have significant influence on both current efficiency and cathode deposit due to combined presence of LIX64N and Fe^{3+} in the copper electrolyte. The interaction of LIX65N with Cl^- during electrodeposition of copper has been reported [7].Studies have been carried out on electrodeposition of copper at higher current densities in the presence of Fe^{3+} [8].

Table 9.3. Tolerance limits of various inorganic impurities in the copper bath

[Cu] = 50 g/L; [H_2SO_4] = 150 g/L; Temperature = ambient

Impurity	Tolerance Limit (ppm)
Mn^{+2}	4000
Ni^{+2}	7000
Zn^{+2}	600
Fe^{+2}	10
Fe^{+3}	200
Sb	20

At all flow rates, CE dropped when concentration of Fe^{3+} exceeds beyond 1 g/L. However, it is possible to achieve nearly 90% CE by maintaining Fe^{3+} ratio below 1. Higher CE (>90%) with acceptable quality of cathode deposit can be achieved at higher Fe^{3+} concentration (2 g/L) only if copper is electrowon at higher current density (900A/m^2) and elevated temperature (60°C). Examination of cathode deposit shows that pyramidal type growth is favored under almost all the experimental conditions except at lower flow rate where a ridge type structure is promoted with the deposit. The deposit is more crystalline and compact at higher temperatures.

During copper deposition, the anions also affect the electrodeposition process by blocking the deposition sites by their adsorption. The extent of the effect depends on the anion size. A detailed study on the effect of Cl^- and interaction of Cl^- and Fe^{3+} has been undertaken. Presence of Cl^- affects current efficiency and surface quality of deposited copper [9]. At lower Cl^- concentration, a lacy type of deposit is formed and (220) plane is preferred. It also significantly changes the microscopic morphology of copper deposit. When the electrolyte is contaminated with Fe^{3+}, the adverse effects due to Fe^{3+} are neutralized when Cl^- concentration exceeds 500 mg/L [10]. In order to find out the effect of various impurities present in the electrolyte, studies were carried out to fix the tolerance limits of organic and inorganic impurities (Table 9.3). As the solutions are contaminated with organic, solvent was removed by using 5% activated carbon powder prior to studies with inorganic contaminates.

9.3. Augmentation of Ionic Mass Transfer Rates by Forced Convection

In industrial operations, reduction in equipment size, increasing the productivity along with deposit morphology and purity forms the major criteria. Mass transfer caused by forced convection has been given more importance because of its application in industrial units. In the conventional electrodeposition cell the current density affects a number of factors like current efficiency, steam consumption, energy consumption, building area, electrode spacing as well as quality and purity. An excessive current density causes the cathode to be rough, impure and porous. It is, therefore, required to judiciously select the operating current density for economic operation of an electrolytic cell and production of quality cathode.

In order to achieve these objectives, augmentation of ionic mass transfer coefficient is essential. This is achieved by (i) forced convection flow of the electrolyte, (ii) use of inert promoters and cross flow elements in the flow streams, and (iii) presence of solids either in packed or fluidized beds etc. Some methods for achieving the objectives are discussed. Continuous electrodeposition of copper in conjunction with solvent extraction was carried out in an open channel electrochemical cell. To enhance the mass transfer conditions in the electrodeposition cell turbulence promoters similar to those adopted by Andersen [11] were used and the results showed that smooth and compact deposits could be obtained when the current density was as high as 300 A/m^2. Without promoters good quality of the copper can be deposited at 180 A/m^2.

9.3.1. Air Sparging

Limiting current density technique was used to find out the mass transfer conditions inside the electrolytic cell. The variables that affect the mass transfer such as concentration of the electrolyte, temperature, inter electrode distance, electrode height etc. were studied. Effect of air sparing on mass transfer conditions are reported in Table 9.4.

In order to assess the behavior of the electrolyte, the physicochemical properties i.e. density, viscosity and conductivity were measured [5] in presence of Mn^{2+}, Co^{2+}, Ni^{2+}, Fe^{2+} and Fe^{3+}. The results showed that in the presence of the above impurities, conductivity of the electrolyte decreases whereas the viscosity and density increases. The decrease in the conductivity would adversely affect the power consumption in copper cell house and increase the density and viscosity values. On the other hand, a fall in the limiting current density 'i_L' values would occur, by inhibiting diffusion of copper ions.

Table 9.4. Effect of air sparing on limiting current density

[Cu] = 50 g/L; [H_2SO_4] = 150 g/L; Temperature = ambient;
Flow rate = 0.5 L/min; Inter-electrode distance: 4 cm

Compressed air flow rate (L/min)	Limiting current density, i_L (mA/cm^2)
-	25.86
1.0	64.65
2.0	71.12
4.0	73.28
6.0	77.58

9.3.2. Open Channel

Copper electrodeposition has been studied using a rectangular open channel cell under forced circulation of electrolyte [3]. Based on the data generated by varying electrolysis parameters, a general correlation among dimensionless number i.e. Sh (Sherwood number) Re (Reynolds number) and Sc (Schmidt number) has been established which is as follows.

$$Sh = K\,Re^a\,Sc^b \; [3] \qquad (1)$$

where, K (mass transfer coefficient) = 19.95, a= 0.5 and b= 0.33 (a and b are constants)

The reliability of this correlation has been verified by testing the process at a higher scale using a model cell. In addition to this, the following empirical equation in terms of mass transfer coefficient and Reynolds number has also been proposed.

$$K_L / K_{LO} = 0.5077\ Re^{0.2956} \quad (2)$$

where, K_L is the mass transfer coefficient under forced circulation and K_{LO} is under natural convection. These studies have been carried out at low Reynolds number. [9]

9.3.3. Submerged jet

The impinging jet system has many practical applications since they yield high rates of heat and mass transfer. Submerged jets play an important role in electrodeposition cells as they improve mass transfer.

A study on the local mass transfer on the electrode surface, using submerged jet parallel to the electrode was carried out. An important finding of this study was that up to about 3 cm height from the bottom edge of the cathode plate there is negligible increase in i_L value and this zone acts almost as a dead zone. The cell with a jet system may also be used for Cu electrorefining (ER) at a higher current density as in case of air-sparged cell. The local mass transfer data can be correlated using the following equation [12].

$$[Sh / Sc^{0.33}] = 0.0004\ (Re)^{1.5}\ [10] \quad (3)$$

The relationship is valid for $y/d < 6$ where y = height of the electrode plate from the orifice of the jet and d = orifice diameter.

9.3.4. Turbulence Promoters

The presence of turbulence promoters increase the intensity of local turbulence which enhances the limiting current density and, hence, mass transfer coefficient at the transfer surface. The similarity of their geometrical configurations induces pseudo-uniformity of spatial distribution of the local mass transfer rates. A detailed study on Cu electrodeposition was carried out using rectangular turbulence promoter and the results were compared with those of cylindrical and triangular promoters. Based on the data generated following generalized equation with an average deviation of 6.2% and standard deviation of 7.96% has been established [13].

$$J_D = 390\ Re^{-0.87}\ (S/H)^{-0.15}\ [11] \quad (4)$$

where J_D is the mass transfer factor, and S and H are spacing and height of the promoters respectively. The increase in the value of the coefficients due to rectangular promoters is nearly four-fold over the coefficient obtained with natural convection. Mass transfer coefficient increases with promoter height and decreases with promoter spacing.

9.3.5. Energy Considerations

Copper electrodeposition requires nearly ten times more energy than copper electrorefining. This is mainly due to the wide difference in the cell voltages [2.0 - 2.5V for ED *vs* 0.2 - 0.25V for ER]. In the copper electrodeposition cell, nearly 70% of the cell voltage is due to the anodic oxygen evolution reaction which requires high voltages (~2.0V). For a reduction in the cell voltage one can consider use of an alternative anode reaction which occurs at lower potential. Anodic oxidation of SO_2 has been found to be one of the most suitable anodic reactions which can be used in copper electrodeposition. A detailed study on the effect of SO_2 oxidation on energy consumption and deposit characteristics during copper electrodeposition was carried out. The results indicate that SO_2 oxidation requires catalytic anodes. The catalytic activity of the anodes investigated are graphite>Pb-Ir-O_2>Ti-Ir-O_2>Pb. It is possible to reduce energy consumption by >50% using a graphite anode. Higher SO_2 concentration reduces energy consumption but at the same time affects the cathode deposit.

9.4. ELECTRODEPOSITION OF ZINC

More than 90% of zinc is produced by electrodeposition from its sulfate solution. As the cathodic reduction potential of zinc is negative (-0.76 V), it is essential to polarize the hydrogen formation through aluminium cathodes during electrodeposition of zinc. Zinc electrodeposition is complicated by the presence of more electropositive ions than zinc, such as Ni, Co, Sb, Ge, etc, if present. Most of the impurities are removed during electrolyte purification, but a certain amount that is sufficient to initiate a process known as ''zinc redissolution'' remains. By addition of organic compounds to the electrolyte, it is possible to inhibit this process and to produce zinc with higher current efficiency and good quality [14].

Zinc electrodeposition consumes ~80% of total energy required for zinc production. In order to reduce energy consumption, studies were conducted either by controlling and optimizing additive concentration or replacing usual additives employed in zinc electrodeposition process by newer ones. It has been observed that by controlling gum arabic concentration in the zinc electrolyte, energy can be saved [15]. Studies with a series of surfactants indicate that Maganafloc-455, Superfloc A-30 and Superfloc S-8 are superior to gum arabic [15]. Similarly results on tetraalkylammonium compounds [16, 17], pyridine and its derivatives [18, 19], perfluoro carboxylic acids [20], sodium lauryl sulphate [21] etc. indicate that cetyltrimethyl ammonium bromide, triethylbenzylammonium chloride, methyl substituted pyridines and sodium lauryl sulphate are superior additives than glue for saving energy.

9.4.1. Augmentation of Ionic Mass Transfer Rates by Forced Convection

Effect of electrolyte circulation on mass transfer coefficient ($K_{Zn^{2+}}$) and limiting current density (i_L) are given in Table 9.5. It is not usually possible to get the limiting current density plateau in zinc electrodeposition. Therefore, a tracer technique developed by Ettel et al. [22] has been used for calculating the limiting current density ($i_{L \cdot Zn^{2+}}$) and mass transfer coefficient (KZn2$^{+)}$ for zinc. Copper has been used as the tracer. For calculating $K_{Zn}{}^{2+}$ the following equations have been used [4].

$$W_{Cu} = i_{L.Cu}{}^{2+}.e.t/F \quad (5)$$

where 'e' is the electrochemical equivalent of copper and 't' is the time period, and

$$K_{Cu}{}^{2+} = i_{L.Cu}{}^{2+} / ZF.C_{Cu}{}^{2+} = D_{Cu}{}^{2+} / \delta_{Cu}{}^{2+} \quad (6)$$

$$K_{Zn}{}^{2+} = K_{Cu}{}^{2+} [D_{Zn}{}^{2+} / D_{cu}{}^{2+}]^{2/3} \quad (7)$$

Table 9.5. Effect of electrolyte circulation during zinc electrodeposition

[Zn] = 60 g/L; [H_2SO_4] = 100 g/L; Temperature = ambient

Electrolyte circulation rate (L/min)	$K_{Zn}{}^{2+}$ (cm/s x 10^4)	i_L Zn^{2+} (mA/cm^2)
0.275	7.872	139.4
0.550	7.946	140.7
0.825	7.971	141.1
1.060	9.952	176.2
1.320	12.808	226.8

The effects of parameters which affect the industrial zinc electrolytic cell have been studied and based on the results, the following correlation among the dimensionless numbers i.e. Sh, Re and Sc has been proposed.

$$Sh = 6.32\ Re^{0.39} .\ Sc^{0.33} \quad (8)$$

9.5. Electrodeposition of Nickel

Nickel is extracted from various raw materials such as lateritic nickel ores, manganese nodules, secondaries etc. Electrodeposition of nickel from synthetic bath containing boric acid and sodium sulfate was carried out. Different variables such as bath pH, concentration of nickel content, temperature, current density, additives boric acid and sodium sulfate were studied to find out the effect on CE and nature of deposit. The tolerance limits of various impurities are listed in Table 9.6.

The phenomenon of electrodeposition of nickel is very sensitive to presence of foreign ions. The effects of cations such as Mn^{2+}, Zn^{2+}, Al^{3+}, Co^{2+}, Mo^{6+} Fe^{2+} and Fe^{3+} on the nickel electrodeposition have been studies [23 - 25]. These impurities reduce current efficiency, affect the deposit quality and contaminate the cathode deposit. At the tolerance limits there is no deviation of nickel structure from FCC in the presence of these impurities and the crystallographic orientations are found to be (111), (200), and (220). The surface morphology of cathode nickel changes with the change in the crystallographic orientations. The effects of anions like Cl^- and NO_3^- have also been studied during nickel electrodeposition. Presence of NO_3^- in the nickel electrolyte has more deleterious effect than Cl^-. A concentration of >50gm/l NO_3^- affects both current efficiency and deposit quality. CE decreased significantly by almost by 20% at 500 mg/L of Mo^{6+} in the electrolyte. The tolerance limit of Mo^{6+} with regard to surface quality of the nickel deposit is 40 mg/L. Addition of thiourea (40 mg/L) to bath resulted in decrease of CE by 4%[26].

Systematic studies have been carried out to understand the effects of organic impurities such as LIX64 [27] and D2EPHA during electrodeposition of nickel. The results show that presence of LIX64N above 3 mg/L causes fall in the CE and affect the deposit quality adversely. D2EPHA, on the other hand, does not affect CE up to 100 mg/L but the deposit quality is adversely affected beyond 20 mg/L of D2EPHA.

Synergistic effect of inorganic impurities and LIX64N on nickel electrodeposition has been studied using a nickel electrolyte containing Mg^{2+} 500, Mn^{2+} 220, Zn^{2+} 30, Co^{2+} 180, Fe^{2+} 5, Fe^{3+} 10 and Cu^{2+} 30 mg/L. Increase in LIX64N concentration causes fall in the CE, deteriorates the cathode deposit quality and contaminates the cathode with impurities. A remarkable change in the microscopic structure and crystallographic orientations was also observed. The behavior of D2EPHA is similar to that of LIX64N in the presence of the above mentioned metallic impurities as regards the quality of the cathode deposit. Though there is a fall in CE in the presence of D2EPHA, it is not as significant as that caused by LIX64N.

9.6. Electrodeposition of Cobalt

It is difficult to electrodeposit cobalt from an aqueous cobalt sulfate-sulfuric acid medium since the process is controlled by competition between cobalt and hydrogen production. Electrodeposition of cobalt was studied using sulfate bath in the presence and absence of additives such as H_3BO_3 and NaF. A combination of additives NaF and H_3BO_3 in the bath was found to give bright deposits with about 86% CE (Table 9.7). But a pure cobalt sulfate solution seems to be preferable for coupling with solvent extraction. Fixing the pregnant electrolyte pH at 2, spent electrolyte pH at 1.5 and with cobalt concentration more than 50 g/L, continuous solvent extraction and electrodeposition was carried out using actual leach solutions. Smooth, uniform, comparatively bright and malleable deposits were obtained at a CE of 85% and the purity more than 99%.

Table 9.6. Tolerance limits of various cations and anions in the nickel bath

[Ni] = 60 g/L; Temperature = 60 °C

Impurity	Tolerance limit (ppm)
Mg^{+2}	500
Al^{+3}	5
Mn^{+2}	250
Fe^{+2}	5
Fe^{+3}	100
Zn^{+2}	100
Co^{+2}	500
Cu^{+2}	250
Cl^-	250
NO_3^-	50

Table 9.7. Effect of various additives in the bath during electrodeposition of cobalt [28]

[Co] = 40 g/L; pH = 2; Temperature = ambient

Bath composition g/l	CE %	Nature of deposit
Co 40	75	dull
Co 40 H_3BO_3 20	77	brighter than case 1
Co 40 NaF 1	79	bright
Co 40 Na_2SO_4 8	74	dull
C0 40 H_3BO_3 20, Naf 1	86	bright
Co 40, H_3BO_3 20 Na_2SO_4 8	75	bright same as case 2

A systematic study on the effects of Zn^{2+}, Cd^{2+}, Cr^{6+}, Fe^{2+}, and Fe^{3+} during electrodeposition of cobalt has been carried out. The effects due to these impurities are, in general, similar to those found during nickel electrodeposition.

9.7. ELECTRODEPOSITION OF CHROMIUM

Chromium ferro-alloys and silicon reduced chromium are of an adequate degree of purity for the production of commercial alloys (those of Fe, Ni and Co or mixtures of these). They are no means of high purity. Apart from their content of small quantities of Fe, Si and Al, they often contain slag inclusions and substantial quantities of oxygen. Various methods have been suggested for the production of a high degree of purity. The easiest of these methods to operate is the electrolytic process, thought the limiting factor being very low CE.

Two types of electrolytic baths are in use to produce electrolytic chromium metal -- chromium alum bath and chromic acid bath.

Chromium alum bath: Here the chrome alum $NH_4Cr(SO_4)_2.12H_2O$ is used as the source for chromium electrodeposition. The chromium is deposited in a diaphragm cell and the catholyte and anolyte are renewed to keep the bath conditions constant at a composition of 15 g/L Cr^{2+}, 15 g/L Cr^{3+}, 42 g/L NH_4^+ and 27 g/L Na^+. The pH is maintained in the range of 1.8 – 2.2 and the CE of 45 – 60% could be obtained.

Chromic acid bath: In the chrome alum bath, the starting raw material is either chromite or high carbon ferro-chromium. Since they contain appreciable amount of impurities that contaminate the electrolyte, the chromium produced in this process has a purity of about 99.5%. The chromium electrodeposited from a chromic acid bath is more pure (99.9%). In such a bath, the starting raw material is chromic acid which is produced by chemical process. Thus, it is commercially available in pure form. Chromium deposited from this bath is intrinsically pure because few foreign metals co-deposit even when present in the electrolyte. Further, the foreign metals are generally present as cations and can be removed by ion exchange techniques. Electrodeposition of chromium from chromic acid bath is not so easy. A number of acids can be added in very small amounts to chromic acid to permit chromium deposition. These acids act as catalysts in the bath. After surveying a number of possible catalysts [29], it was found that SO_4^{2-}, F^- and SiF_6^{2-} ions were the most suitable.

Current efficiency during electrodeposition in chromic acid bath is too low in comparison with other electrodeposition processes due to evolution of hydrogen at the cathode. This occurs because the hydrogen overvoltage on chromium is significantly more positive than that at which chromium ion deposition would be expected to commence. In case of SO_4^{2-} ion catalyst, either H_2SO_4 or soluble sulphate form may be added. Since the chromium acid is a strong acid, it probably liberates sulphuric acid from any soluble sulphates. But in actual practice H_2SO_4 is used in all commercial processes. In sulphate bath, the chromium deposition takes place at current efficiency of about 8 – 10%. Although some improvement in current efficiency was noted by varying the deposition conditions in the conventional sulphate electrolyte, a review of the literature disclosed that higher efficiencies could be obtained with fluoride or silico fluoride baths [30]. Ryan [31] has reported investigation of chromic acid electrolytes containing fluoride ion operated at boiling point. In this bath current efficiencies as high as 40 – 45% was achieved. Although the fluoride baths have better throwing power and a higher CE than baths containing sulphate additive, the metal deposited from sulphate electrolytes appears to be of better quality because it contains slightly less oxygen and impurities from anodes. Besides, plant construction, maintenance and control with fluoride baths are problematic.

The electrodeposition of chromium is one of the most complicated and at the same time one of the most interesting subject. The mechanism of chromium electrodeposition is especially of great interest, since the deposition of chromium from chromic acid baths involve the cathodic discharge of a complex multivalent ion (CrO_7^{2-}) whose reduction may lead to the formation of compounds of intermediate valence. Moreover, in the cathodic deposition of chromium, several reactions take place simultaneously. The electrodeposition of chromium is further complicated by the strong tendency of chromium to become passive, and also because chromate ions can be reduced only in the presence of certain foreign anions. In the presence of sulphate ion, the following reactions can take place simultaneously at cathode.

$$Cr_2O_7^{-2} + 14\,H^+ + 12e \rightarrow 2Cr + 7\,H_2O \qquad (9)$$

$$Cr_2O_7^{-2} + 14\,H^+ + 6e \rightarrow 2Cr^{3+} + 7\,H_2O \qquad (10)$$

$$2H_3O^- + 2e \rightarrow H_2 + 2H_2O \qquad (11)$$

9.8. Electrodeposition of Lead

Lead is produced by pyrometallurgical method since age old and the method is still in practice. However, there have been many attempts to produce lead metal by electrolysis route, either by fused salt electrolysis or by aqueous electrolysis. In the fused salt electrolysis, lead compound used is $PbCl_2$. $PbCl_2$ is obtained either by leaching PbS with $FeCl_3$ or by $Cl_2 + O_2$ and crystallized on cooling. Different eutectics like $PbCl_2$ – KCl – NaCl, $PbCl_2$ – KCl – LiCl, $PbCl_2$ – NaCl etc. have been used in the fused salt bath in the temperature range of 400 – 600°C.

Lead is electrowon from aqueous solution usually from two type of baths i.e alkaline or chloride. In the alkaline bath, the lead compound used is either $PbSO_4$ or Pb-oxide and the alkali used is usually NaOH. Although several research papers are available on the electrodeposition of lead from the chloride bath, the issues of lead deposit either as sponge, powder or flakes and not a compact sheet resulted in the process far from perfection. Electrolysis of chloride solution necessarily involves the use of diaphragm to separate chlorine evolved at the anode from the cathode product except where soluble anodes such as mild steel are employed. Porcelain diaphragm is more effective than terelyne cloth or porous plastics. Additives such as β-napthol, glue, Magnafloc 455, rosin (dissolved in ethyl alcohol) have been tried for improving the deposit quality. Use of β-natthol and animal glue jointly showed some improvement in nature of deposit than each one individually and other combinations. Electrodeposition of brine lead liquor obtained from Jarosite decomposition was carried out. The only impurity present was Fe to an extent of 0.17 g/L which had no effect on the deposition of lead. The parameters of the study [32] were: [Pb]: 30-20 g/L; [Glue] = 2 g/L; [β-naphthol] – 0.1 g/L; Current density: $50A/m^2$; Temperature: 70^oC; pH = 0.7 – 0.9; Cell voltage = 2 V. The electrolysis of such liquor produced a bright sheet deposit without leaflet at a current efficiency more than 90%.

CONCLUSION

Electrodeposition processes offer clean technology for the production of high pure metals. The criteria of this unit operation are to produce metal with smaller equipment size at lower energy consumption with good morphology and purity. As electrodeposition is operated in conjunction with purification circuit, impurity levels in the bath affects the deposition process. Engineering hydrodynamic issues in cell can considerably help reducing the mass transfer problems that can result in improvement in performance of the cell. The chapter discusses the issues involved in aqueous electrodeposition processes for copper, zinc, nickel, cobalt, chromium and lead.

ACKNOWLEDGMENTS

The authors thank the colleagues of Hydro and Electrometallurgy Department of Institute of Minerals and Materials Technology (IMMT), Bhubaneswar, India for providing the data. Authors are also thankful to the Director, IMMT for his kind permission to publish this work.

REFERENCES

[1] S. C. Das and T. Subbaiah, *Hydrometallurgy*, 12, 317 (1984).
[2] S. C. Das and T. Subbaiah, *Journal of applied electrochemistry*, 17, 675 (1987).
[3] T. Subbaiah, S.C. Das and R. P. Das, Erzmetall, 39, 501 (1986).
[4] T. Subbaiah and S.C. Das, *Erzmetall*, 41 (10), 518 (1988).
[5] T. Subbaiah and S. C. Das, *Met. Trans.*, 20B, 375 (1989).
[6] W. R. Hopkins, G. Eggett and J. B. Scuffham, *Int. Symp. On HM*, AIME, N. York, 127 (1973).
[7] D.J. Mackinon, V. I. Lakshmanan and J. M. Brannen, *Trans. Inst. Min. Metall. (Sec.C)* 85, C 184 (1976).
[8] S.C. Das and P. Gopalkrishna, *Int. Miner. Processes*, 46, 91 (1996).
[9] N. Pradhan , P. Gopalakrishna, and S.C. Das, *Symp. Plating and Surf. Fin.*, 56, (1996).

[10] N. Pradhan , P. Gopalakrishna, T. Subbaiah and S.C. Das, *Symp. Proc. Hydrotech. – 93*, IMMT- Bhubaneswar, 247 (1993).
[11] T. Balberyszski and A.K. Andersen. *Proc. Aust. Inst. Min.*, 244, 12 (1972).
[12] P. Gopala Krishna, S. C. Das, T. Subbaiah and R.P.Das, *J. Chem. Tech. Biotech.*, 55, 65 (1992).
[13] T. Subbaiah, P. Venkateswarlu, R.P.Das and G.J.V.J. Raju, *Hydrometallurgy*, 42, 93 (1996).
[14] Ivan Ivanov, *Hydrometallurgy*, 72, 73 (2004).
[15] B. C. Tripathy, I. N. Bhattachraya, P. Gopal Krishna and S. C. Das, *Trans. Ind. Inst. Met.* 51, 303 (1998).
[16] B.C. Tripathy, S. C. Das, G. T. Hefter and P. Singh, *J. Apple. Electrochem.*, 28, 915 (1998).
[17] B.C. Tripathy, S. C. Das, G. T. Hefter and P. Singh, *J. Apple. Electrochem.*, 29, 1229 (1999).
[18] S. C. Das, P. Singh and G. T. Hefter, *J. Apple. Electrochem.*, 26, 1245 (1996).
[19] S. C. Das, P. Singh and G. T. Hefter, *J. Apple. Electrochem*, 27, 738 (1997).
[20] B.C. Tripathy, S. C. Das, P. Singh and G. T. Hefter, *Bull. Electrochem.*, 15, 340 (1999).
[21] B.C. Tripathy, S. C. Das, G. T. Hefter and P. Singh, *Bull. Electrochem.*, 27, 673 (1997).
[22] V.A. Ettel, A.S. Gendron and V.B. Tilak, *Metallurgical and Materials Transactions B,* 6(1), 31 (1975).
[23] S.K. Gogia and S.C. Das, Met. *Trans.,* 19B, 823 (1988).
[24] S.K. Gogia and S.C. Das, *J. Appl. Electrochem.*, 21, 64 (1991).
[25] U. S. Mohanty, B. C. Tripathy, P. Singh, S. C. Das and V. N. Misra, *J Appl Electrochem* 38, 239 (2008).
[26] U.S. Mohanty, B.C. Tripathy, S.C. Das, and V.N. Misra, *Metallurgical and Materials Transactions B,* 36B, 737 (2005).
[27] S.K. Gogia, E. Kuzeci and R. Kammel, *Min. Proc. and Extr. Met. Rev.*, 10, 57 (1992).
[28] S.C. Das and T. Subbaiah., *J. Applied Electrochemistry*, 17, 675 (1987).
[29] J.L. Griffin, Plating, 53, 196 (1966).
[30] T.A. Hood, Metal Finishing, 50, 103 (1950).
[31] N.E. Ryan, *Report ARL/Met.26* (1956)
[32] S.C. Das, T. Subbaiah, P. K. Shoo, R. P. Das and P. K. Jena., *Hydrometallurgy*, 21, 373 (1989).

In: Electrodeposition: Properties, Processes and Applications ISBN: 978-1-61470-826-1
Editor: Udit Surya Mohanty

Chapter 10

ELECTRODEPOSITED BIOMIMETIC HYDROXYAPATITE FOR OSTEO-INTEGRATION AND DRUG DELIVERY

Norberto Roveri** and Marco Lelli
University of Bologna, Department of Chemistry "G Ciamician", Italy

ABSTRACT

In medical devices such as orthopaedic metallic bone implants or coronary stents, mechanical strength can only be achieved by metals, yet these lack the required biocompatibility and bio-integration. Surface treatments to improve the metal prosthesis interface with biological tissues have been extensively studied. The electrochemically-assisted surface deposition on the metallic prosthesis of a biomimetic coating appears as the most interesting resolution. This method allows both to overcome the difficulty of depositing protein components by plasma spray or physical vapour deposition, and to control the coating process easily. In this chapter we review the obtained results and the future perspective about the electrodeposited biomimetic hydroxyapatite.

Coatings of hydroxyapatite nanocrystals mimicking bone nanocrystals in composition, structure, morphology, nano size and bioactivity have been obtained on a titanium surface, by an electrochemically assisted deposition in order to improve the surface bioactivity. With a view to reduce the thrombogenic potential of artificial blood-contact devices and natural tissues, a new anti-thrombogenic coating, consisting of an hydroxyapatite nanocrystals-heparin conjugate, has been electrodeposited on metallic coronary stents. It is also possible to obtain calcium-phosphate/collagen coating on a titanium surface with an electrochemical cell containing a slightly acidic collagen molecule suspension in a Ca^{2+} and PO_4^{3-} ions aqueous solution. In such a process, the collagen/calcium phosphate composite formation involves the self-assembly of collagen molecules into reconstituted fibrils during the contemporary crystallization of calcium phosphate mineral on the electrode surface.

Following the biomimetic approach, it is possible to realize a nano-structured hydroxyapatite/collagen biomimetic coating. This allows the formation of a coating of

* E-mail address: norberto.roveri@unibo.it.

reconstituted collagen microfibers inside calcified by hydroxyapatite nanocrystals, closely resembling bone calcified collagen fibres.

Natural wood templates have been selected as a starting material to obtain open-pore geometries with wide surface area and microstructures allowing cell in-growth and reorganization and providing the necessary space for vascularisation. In fact, the alternation of fibre bundles and channel-like porous areas makes wood an elective material to be used as template in starting the development of new bone substitute biomaterial. Particularly, Biomorphic Silicon Carbide (BioSiC) wood-derived structures have been optimized to be employed as bio inert bone scaffolds representing a new generation of light, tough and strong material for biomedical applications.The good biocompatibility and biological response of BioSiC coated by electrochemically-assisted surface deposition of nano-structured biomimetic hydroxyapatite allow to consider this porous material both as a bone substitute in orthopaedic, odontoiatric, dental and maxillo-facial implantation.

Here we report a detailed description of the electrochemical assisted deposition of a biomimetic apatite coating on prosthetic surface in order to show its potentiality in improving tissue integration and in realising bioactive molecules with kinetic determination studies.

INTRODUCTION

Chemists, biologists, physicists and engineers interested in material science and in the design of innovative technological materials are amazed by the high degree of sophistication, miniaturization, hierarchical organization, hybridization, reliability, efficiency, resistance and adaptability that characterize natural biogenic materials.

Present synthetic processes allow us to obtain materials possessing only part of the peculiar properties, which biogenic materials have achieved through specific building principles selected by evolution. For this reason nature is the ultimate architect and represents a school for material science. Mimicking nature and designing bio-inspired materials represent a promising way for the design and the synthesis of innovative materials and devices. [1-4] Nature produces soft and hard materials exhibiting remarkable functional properties by controlling the hierarchical assembly of simple molecular building blocks from the nano- to the macro-scale. [5]

Biogenic mineral morphogenesis is related to specific strategies for the long-range chemical construction of well-organized architectures from preformed nano or micro-crystalline inorganic building blocks. In fact, many biological complex structures are obtained by promoting specific links induced by the conformation variability at the nanometric scale of biological macromolecules. The concept of 'interfacial molecular recognition' observed in the biogenic materials led many scientists to perform synthesis by functionalized organic matrices for the template directed control of inorganic compound nucleation and crystal growth. Bio systems reveal a high level of integration of three fundamental aspects: the nano–micro 'spatial confinement' of biochemical reactions, the inorganic and organic 'hybridisation' compounds and the 'hierarchy' from nano- to macro-scale, in order to produce a biomaterial able to exhibit the right chemical–physical properties at any different scale level. [6-9]

Biogenic materials are nucleated in defined nano– micro-dimensioned sites inside the biological environments in which chemistry can be spatially controlled. The spatial delimitation is essential to biological mechanisms employed to control the size, shape and structural organization of biomaterials.

Biomimetism of synthetic materials for biomedical applications can be carried out at different levels according to the composition, structure, morphology, bulk, surface and chemical-physical properties. Furthermore the synthetic biomimetic processes have to be carried out without organic solvents, but in water, in mild conditions and using light and soft reagent concentrations at low temperature and pressure. Among biomimetic inorganic nano-structured biomaterials synthetic hydroxyapatites probably have received the highest attention and interest as matrices for drug delivery. Also silica xerogels represent a very interesting inorganic matrix for drug delivery, but the bioactive molecules undergo physical interaction with the inorganic surface and the drug release kinetic appears to be controlled with more difficulty. On the contrary, nanostructured hydroxyapatites link the bioactive molecules chemically so that these can be released through a kinetic study strictly related to the surface charge disorders distributed on the inorganic surface.

Hydroxyapatite matrices easily bind bioactive molecules on their surface and help in balancing incomplete neutralization of the cations and anions which constitute the hydroxyapatite crystalline ionic structure. The Ca/P molar ratio of 1.7 can be obtained only in the hydroxyapatite crystal bulk, while the Ca/P ratio decrease to 1.3 on the hydroxyapatite crystal surface, not only for Ca deficiencies, but also for phosphate and hydroxyl groups replaced by carbonate anions. The lack of neutralized surface charges induces the binding of biomolecules and drugs on the hydroxyapatite surface. These can be released spontaneously in the physiological liquids for this reason, but also as a consequence of pH changes, ionic strength or temperature. Changes in pH, ionic strength and temperature can be observed in biological environment under physiological and pathological stimuli. The chemical physical characteristics of biomimetic nano-structured hydroxyapatite allow us to consider this inorganic matrix particularly suitable for utilization in drug delivery stimuli [10,11] induced on the hydroxyapatite surface.

Nano-structured hydroxyapatite matrices can be synthesized as nano sized crystals with morphology, structure and surface area which mimic closely bone apatite crystals, but also as porous scaffolds mimicking bone spongiosity. Biomimetic nano-structured hydroxyapatite are not only biocompatible and bio-resorbable, but can be synthesized by biomimetic chemical methods.

In fact, nano-structured hydroxyapatite can be synthesized in water without organic solvents, in mild and soft conditions similar to biological mechanistic studies in the template mediated synthesis. In order to obtain a high biomimetic matrix, biomimetic hydroxyapatite nanocrystals can be associated to proteins and biocompatible macromolecules. The combination of hydroxyapatite and collagen fibrils probably represents the hybrid compound at the highest biomimetic level closely resembling the bone composition.

10.1. Electrochemically Assisted Deposition

The process for electro depositing an adherent biomimetic coating of nano-structured hydroxyapatite on a metallic surface can be described as follows:

10.1.1. Electrolytic Cell

The cell was composed of three electrodes (Working electrode, Reference electrode and Counter Electrode) or two electrodes (Working and Reference Electrode – scheme 1). Applying a DC potential between Working and Reference electrode, a pH value rise was observed from the initial value of 3.5 – 4.0 to the final value of about 8.0 – 11.0). This basic environment was favourable for precipitation of an apatitic phase formed on the working electrode surface.

The Reference electrode used was a platinum plate; the Working electrode was made of a metal or a conductive material, such as Ti, Ta, Nb, V, Al, Pt. Different geometries were proposed for this work 1) Reference electrode in platinum (composed of platinum plate or string) was positioned in front of Work electrode; 2) Reference electrode used was a platinum string modelled like a ring all around the Working electrode which was positioned in the centre of the same platinum ring. It is possible to realize the biomimetic hydroxyapatitic nano-structured coating by a rotation of the Working electrode during the cathodic electrodeposition of hydroxyapatite, which can be applied both at higher (10-20 laps/minute) or slower speed (0.5-1 laps/minute). By this method, migration of calcium ions took place at the beginning towards the cathode followed by the formation of a phosphate group (in the second phase) on the same Working electrode, resulting in the formation of an apatite coating.

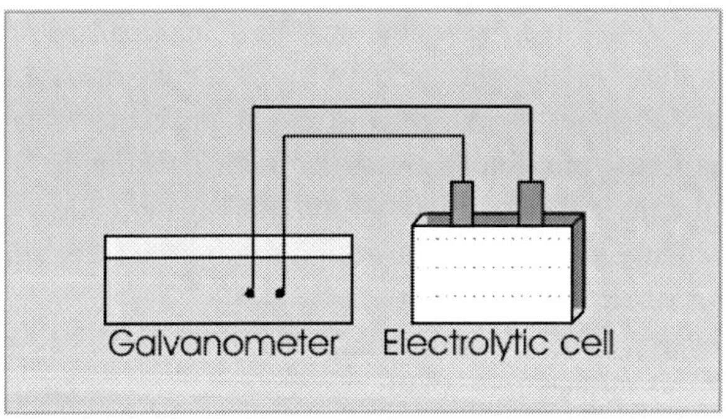

Scheme 1. Electrochemical system with galvanometer (on the left), and electrolytic cell with two electrodes (on the right).

10.1.2. Electrolytic Solutions

The electrolytic salt solution commonly used were :

Calcium ions: $Ca(NO_3)_2*4H_2O$, $Ca(NO_3)_2*2H_2O$, $Ca(CH_3COO)_2$, or a mixture of these salts;

Phosphate ions: $NH_4H_2PO_4$, K_2HPO_4, KH_2PO_4, H_3PO_4, or a mix of these salts;

Carbonate ions: $CaCO_3$, CO_2 (gas), Na_2CO_3, K_2CO_3, or a mix of these salts;

The concentrations of electrolytic solutions varied between 0.5 M and 0.001 M.

The commonly used experimental conditions were:

DC current ranging in a range between 10 nA and 300 mA;

Voltage applied ranged between 1 V and 100 V;

Temperature was between 2°C and 70°C;

Total electrodeposition time ranging between 5 seconds and 3 hours.

The application of fixed experimental conditions (Voltage, intensity of current, temperature of electrochemistry cell, time of electrodeposition) can be realized only in a single step of electrodeposition. However, it is also possible to perform progressive single short steps of electrodeposition under different or same experimental conditions thus realizing an overall electrodeposition time that it is the sum of the single progressive steps; the length of a single step can be of few seconds, but even fraction of a second. Electrodeposition process started with mixing of salt solution into a electrochemistry cell. Then the two or three electrode system were immersed into the mixing solution and the instrument was set for the experimental conditions with DC potential ranging between 1 V to100 V and more preferably between 10 V and 60 V; current was adjusted between 10 nA and 100 mA, and more preferably between 4 mA and 40 mA. The geometry of the system and the temperature of reaction were taken into consideration to perform the electrochemical process for the HA substituted biomimetic nano-structured coating.

During electrodeposition two important reactions at the cathode and anode are depicted below:

Cathodic reaction (H_2 evolution)

$$2H_2PO_4^- + 2e^- \rightarrow 2HPO_4^{2-} + H_2 \quad (1)$$

$$HPO_4^{2-} + 2e^- \rightarrow 2PO_4^{3-} + H_2 \quad (2)$$

$$2H^+ + 2e^- \rightarrow H_2 \quad (3)$$

$$2H_2O + 2e^- \rightarrow H_2 + 2OH^- \quad (4)$$

Anodic reaction (O_2 evolution reaction)

$$4OH^- \rightarrow O_2 + 2H_2O + 4e^- \quad (5)$$

The cathodic reaction obtained involved the reduction of H^+ ions to molecular hydrogen like gas, and consequently the elimination and water reduction and therefore the production of OH^- ions. These two events determine a local raise of pH (that increases from initial value of pH solution (about 3.5 – 4.0) to final pH (about 8.0 – 11.0) only in a thin μm of solution thickness around the Work electrode), the optimal condition for obtaining the precipitation of an apatitic phase on the cathode.The formation of an inorganic phase on the metallic cathode is not due to the precipitation, but related to the migration of ions on the surface of the same electrode. At first we have the migration of calcium ions towards the cathode. After this, around the electrode surface there is a thin μm of solution thickness charged positively which induces a migration of a phosphate group towards the Working electrode. Moreover, the basic conditions that can be found close to Working electrode is determined by the presence of OH^- and these hydroxylic ions with phosphate ions which neutralize the Ca^{2+} ions forming hydroxyapatite nano-crystals on the Working electrode. If we have CO_3^{2-} ions in the solution, these can also be included in the hydroxyapatite crystals replacing phosphate ions like it was observed in natural biogenic hydroxyapatite nano-crystals.

Modulating these experimental conditions and following different electrodeposition systems it is possible to realize different coating thickness as a function of the final applications of the coated implant. Moreover, it is possible to choose the morphology of the same biomimetic HA-substituted coating, the degree and the kind of ions substitutions, the degree of crystallinity as a function of the temperature of solution during the electrochemistry process and the possibility to have the functionalization of biomimetic nano-crystals with bioactive molecules.

10.2. Biomimetic Hydroxyapatite Coating to Improve Osteo-Integration

In medical implanted devices, such as metallic ones or coronary stents, the necessary mechanical properties can be achieved only with metals, which surely lack the required bioinduction and biointegration. Surface treatments to improve metal proteases biointegration have been extensively studied in order to modify their surface characteristics. One of the most promising treatment is the production of a biomimetic coating by the electrochemically-assisted deposition method on the prosthetic metallic surface. The electrochemically-assisted deposition method not only allows to overcome the difficulty of depositing protein components by plasma spray or physical vapour deposition, but also allows to control the coating process easily.

10.2.1. Biomimetic Nano-Structured Hydroxyapatite Coating

The electrochemically-assisted deposition process (ELD) to form a biomimetic nano-structured hydroxyapatitic coating on dental titanium screws has been patented in 2010 [12]. The process can be carried out in a two-electrode electrochemical cell where the Working electrode and cathode is made of the titanium prosthesis, while a platinum sheet is used as a counter electrode (anode). The ELD process can be carried out using a typical three-electrode electrochemical cell in which the Working electrode is the metallic prosthesis, the Reference electrode is a standard silver, sliver-chloride reference electrode (Ag/AgCl in saturated KCl) and the Counter electrode is a platinum flag. Prior to deposition, titanium prosthesis needs to be cleaned by a standard protocol: it is ultrasonically cleaned in cold acetone and then in distilled water to remove impurities. Figure 10.1. reports the X-Ray diffraction patterns of the hydroxyapatite coating electrodeposited on titanium plate using a different temperature of the electrochemical cell. It an increase in the intensity of various crystal planes of the apatite is observed with increase in temperature from 10 to 700 °C.

Increasing the degree of crystallinity of the apatite phase, the morphology of the crystals also changes. In fact, the hydroxyapatite crystal morphology of the electrodeposited coating can be determined by changing both the temperature and the electrolytic solution. Figure 10.2 shows the morphology of the hydroxyapatite crystals obtained by electrodeposition at 50°C using an electrolytic solution of $NH_4H_2PO_4$ and $Ca(NO_3)_2$*$2H_2O$, while Figure 10.3 reports the TEM image of hydroxyapatite nanocrystals obtained at room temperature using an electrolytic solution of $NH_4H_2PO_4$ and $Ca(CH_3COO)_2$.

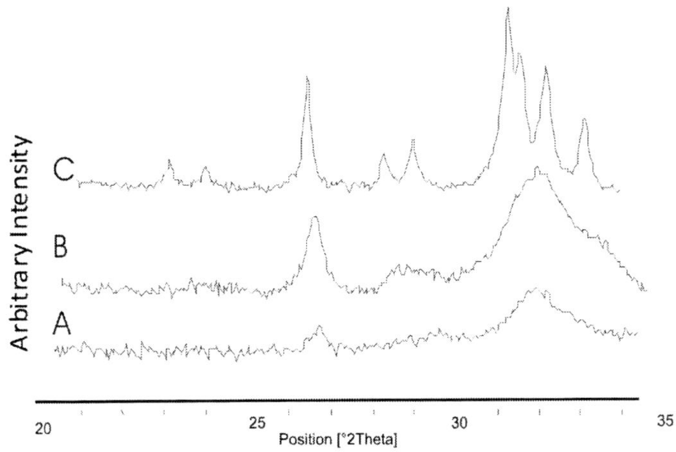

Figure 10.1. X-Ray diffraction of hydroxyapatitic coating on metallic surface obtained at 10 °C (a), at room temperature of 25 °C (b) and at the high temperature of 70 °C (c).

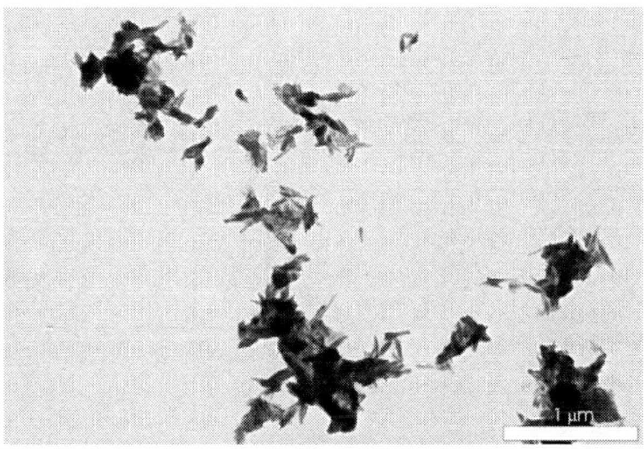

Figure 10.2. TEM Image of hydroxyapatite nano-crystals obtained at high Temperature (50°C) 34 mA 1 step of 10 minutes of ELD. Solution used are $NH_4H_2PO_4$ and $Ca(NO_3)_2{*}2H_2O$.

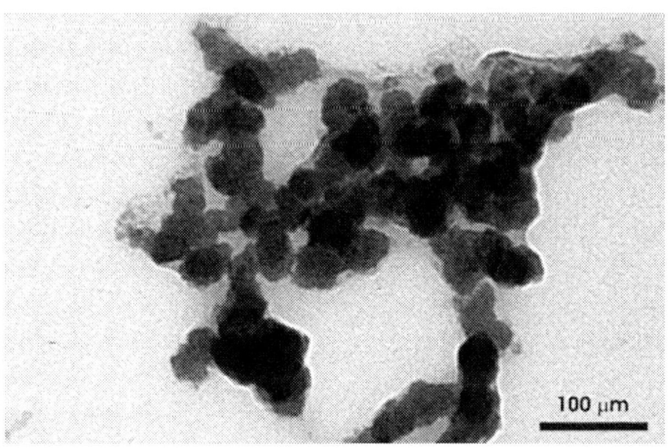

Figure 10.3. TEM Image of hydroxyapatite nanocrystals obtained at Room Temperature 34 mA 1 step of 10 minutes of ELD. Solution used are $NH_4H_2PO_4$ and $Ca(CH_3COO)_2$.

The high bio-reactivity of the electrodeposited nano-structured hydroxyapatite coatings induces proliferation of quick cells on the coated titanium surface. However, proliferation was enormously reduced on uncoated titanium plate.

10.2.2. Hydroxyapatite/Collagen Nano-Structured Biomimetic Hybrid Coating

Biomimetic apatite or collagen/apatite composite coating formed on the Working electrode at shorter time intervals using an appropriate electrolytic solution. [13]. During the preparation of the electrolyte, a proper amount of Ca(NO3)2*2H2O and NH4H2PO4 were dissolved separately in distilled water to make two different solutions: the first one with 42 mM of [Ca2+] and the second with 25 mM of [PO43-]. These two solutions were mixed in the same ratio.

In order to obtain a biomimetic hybrid coating containing apatite and collagen, a molecular suspension of type I collagen (10% wt) had to be added to the amount of about 5-15 % in respect to the global volume of the solution. During electrolytic deposition, the electrochemical cell with the electrolyte mixture was kept at room temperature and a constant current of about 30- 60 mA was applied. During the electrochemical process the ion migration was directed towards the cathode favouring both the formation of hydroxyapatite nanocrystals and the nano fibrils reconstituted from collagen molecules on the same electrode (Figure 10.4). Furthermore, to obtain a uniform and homogeneous coating deposition on the whole cathode surface, a slow rotation to the cathode was applied by an engine application.

In order to well visualize the collagen fibrils electrochemically deposited on the cathode, the collagen fibrils–hydroxyapatite nano-structured hybrid coating was decalcified in a 10 wt% EDTA solution for 24 h before being examined by SEM (Figure 10.5).

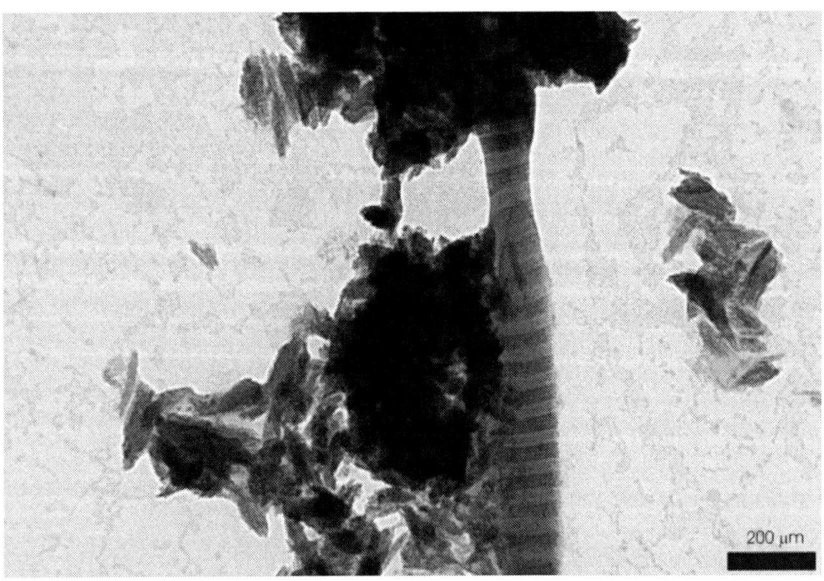

Figure 10.4. TEM image of nanocrystal hydroxyapatite/collagen after electrodeposition.

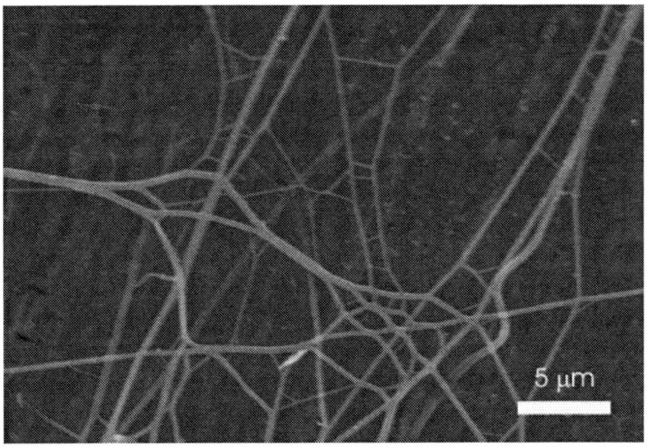

Figure 10.5. SEM image of collagen fibrils–hydroxyapatite nanocrystals hybrid coating after decalcification in a 10-wt% EDTA solution for 24 h.

10.3. Biomimetic Nano-Structured Hydroxyapatite Coating for Drug Delivery

10.3.1. Hydroxyapatite Nanocrystals-Heparin Coatings

With the aim of reducing the thrombogenic potential of artificial blood-contact devices and natural tissues, a new anti-thrombogenic coating material, consisting of an hydroxyapatite nanocrystals-heparin conjugate has been developed. A biomimetic surface activated coating, made of carbonate hydroxyapatite nanocrystals, functionalized with heparin, has been performed by an electrochemically-assisted deposition on titanium plate (Figure 10.6). [14,15]

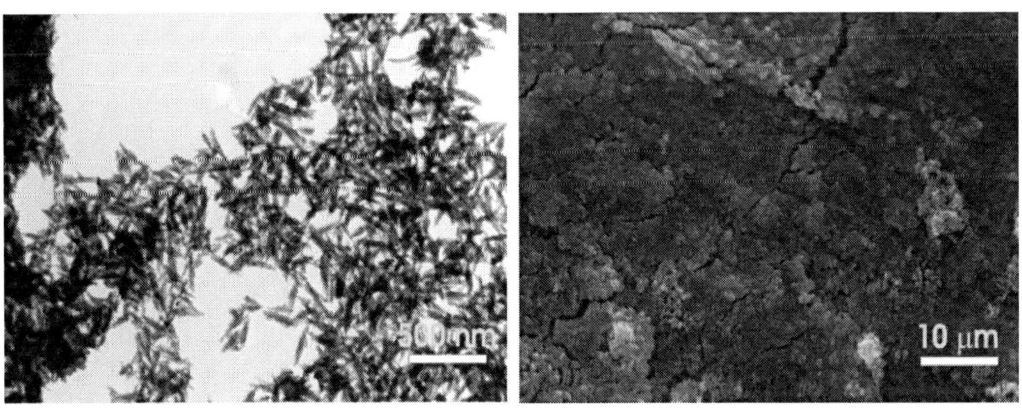

Figure 10.6. TEM image of electrodeposited biomimetic hydroxyapatite nanocrystal surface functionalized by heparin (left) and SEM image of the hydroxyapatite-heparin functionalized coating on titanium plate (right).

10.3.2. Hydroxyapatite Nanocrystals-Lactoferrin Coating

A biomimetic surface activated coating, made of hydroxyapatite nanocrystals, functionalized with lactoferrin has been performed by an electrochemically-assisted deposition on titanium plate in order to induce to prosthesis surface antibacterial, anti radicals, anti antioxidant, cariostatic, anti-carcinogenic, anti-inflammatory and promoting bone growth and healing properties. [16] In fact, lactoferrin is a non-heme iron-binding protein belonging to the transferrin protein family, whose role is to transport iron in the blood plasma. Lactoferrin is found in mucosal secretions, including tears, saliva, vaginal fluids, semen, nasal and bronchial secretions, bile, gastrointestinal fluids, urine, amniotic fluid and at higher concentrations in milk and colostrums (7 g/L), which makes it the second most abundant protein in milk, after caseins. It is involved in several physiological functions: regulation of iron absorption in the bowel, immune response, antioxidant, cariostatic, anti-carcinogenic and anti-inflammatory properties; and protection against microbial infections. Recently it has been shown that at physiological concentrations of lactoferrin potently stimulates the proliferation and differentiation of primary osteoblasts and acts as a survival factor. Lactoferrin also affects osteoclasts, potently inhibiting their formation. In vivo, local injection of lactoferrin results in substantial increases in bone formation and bone area. In a critical bone-defect model in vivo, lactoferrin was shown to promote bone growth. These data suggest that lactoferrin could have a physiological role in bone growth and healing and a potential therapeutic role as an anabolic factor in osteoporosis. The study of the interaction of lactoferrin purified from cow's milk with biomimetic hydroxyapatite surface at two different pH values (7.4 and 9.0) has been recently reported. [17]

An electrodeposited coating of biomimetic hydroxyapatite surface functionalized by lactoferrin favours the development of bone-implantable biomaterials for hard-tissue engineering and regeneration technologies with a bioactive surface coating. In this context the implanted hydroxyapatite nano-structured coating can act as a surface for migrating cells aimed at the formation of functional tissue. In the meantime the lactoferrin loaded on hydroxyapatites could promote bone growth and act as anti-inflammatory agent promoting enhancement of their osteo-integration and the osteo-induction properties.

10.4. Biomimetic Nano-Structured Hydroxyapatite Coating on Biomorphic Bone Scaffolds

Following the biomimetic approach, inspiring to Nature, it is possible to realize [18] a nano-structured hydroxyapatite/collagen biomimetic coating. This is an innovative hybrid material to prepare innovative bone substitute and bone tissue engineering scaffold, with the possibility of synergistically joining the porous bio inspired morphology and mechanical property of Biomorphic Silicon Carbide.

Natural wood templates have been selected as a starting point to obtain open-pore geometries with wide surface area and microstructure allowing cell in-growth and reorganization and providing the necessary space for vascularisation. [19] In fact, the alternation of fibre bundles and channel-like porous areas makes the wood an elective material to be used as template in starting the development of new bone substitute biomaterials by an ideal biomimetic hierarchical structure. Particularly, Si/SiC wood-derived structures have been optimized to be employed as bio-inert bone scaffolds. In fact, during the last decade, Biomorphic Silicon Carbide (BioSiC) prepared by Si vapour or by Si melted reactive infiltration of carbon templates, previously obtained by pyrolysis of different kinds of wood, have received great attention. [20-23]. The reason is not only due to the good thermo mechanical properties of wood, but also to its large availability, renewability, low cost and very low environmental impact. [24,25]

The structural and morphological characteristics of BioSiC can change according to the different kind of wood utilized for its preparation.

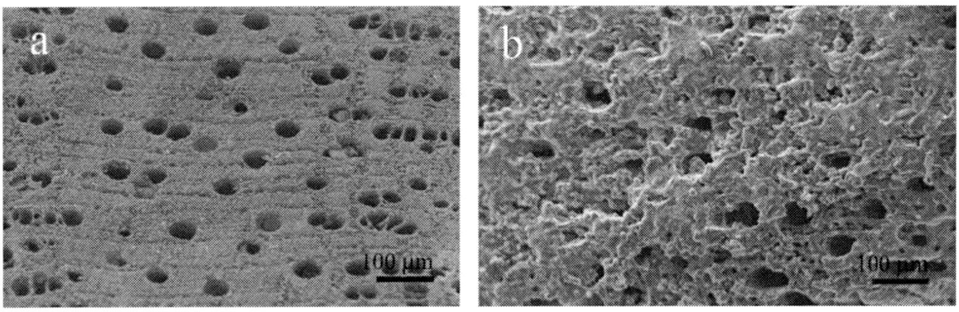

Figure 10.7. SEM image of wood silicon carbide. Transversal cross-sectioned sample (a) before and (b) after electrochemically assisted biomimetic hydroxyapatite/collagen surface deposition.

In fact, the macroscopic organization and structural hierarchical arrangement of cells and channels in coniferous, oak, bamboo and many other woods appear greatly different than the nano-micro size level. [19] Biomorphic Silicon Carbide is a siliconized carbon material produced by reactive infiltration of molten Si into a carbon template. It is obtained from wood pyrolysis and represents a novel kind of porous ceramic constituted of elongated tubular cells with diameter of some hundred micrometers, preferentially aligned with the axis of tree trunk. BioSiC cannot be prepared by traditional technologies in porous ceramics manufacturing, but it represents a new generation of light, tough and strong material for biomedical applications. BioSiC's good biocompatibility and biological response allow us to consider this porous material as a bone filler and substitute in orthopaedic, odontology, dental and maxillo-facial implantations.

An innovative hybrid biomaterial, like bone substitute and bone tissue engineering scaffold, with the possibility of synergistically joining the porous bio-inspired morphology and mechanical property of BioSiC has been recently synthesized. BioSiC exhibiting the surface bioactivity of a nano-structured hydroxyapatite/collagen biomimetic coating has been recently obtained by the electrochemically assisted deposition (Figure 10.7). [26,27]

CONCLUSION

The biomimetism of biogenic materials based hydroxyapatite could be carried out at different levels of composition, structure, morphology and surface bio-reactivity. The aim of a biomimetic research is to realise a biomaterial which is biomimetic in all these characteristics, in order not only to optimise the interaction of synthetic prosthetic materials with biological tissues, but also, and more ambitiously, to mimic the biogenic materials in its functionality. Electrochemically assisted biomimetic hydroxyapatite surface deposition represents an easy controlled process to transform a synthetic prosthetic surface into a bioactive interface with biological tissues. Furthermore, prosthetic materials for implantation and scaffold porous materials for tissue engineering can be strongly improved by an electrochemically assisted biomimetic hydroxyapatite surface deposition process which can deliver bioactive molecules with a controlled kinetics. By changing the operative conditions of the electrochemically assisted biomimetic hydroxyapatite surface deposition we can appreciably modify the thickness, structure, degree of crystallinity, morphology, nano-micro structuring, surface area, bio-reactivity and kinetics of bioactive molecules. The possibility to adapt the chemical-physical and functional properties of apatite coating to different parameters provides the electrochemical process with many challenges for biomedical applications in the future.

ACKNOWLEDGMENTS

We thank the University of Bologna, (funds for selected research topics), the Inter University Consortium for Research on Chemistry of Metals in Biological Systems (C.I.R.C.M.S.B), the Chemical Center S.r.l for financial and experimental supports and MiPAAF "Finale-Qualifu" Project D.M.2087/7303/09.

REFERENCES

[1] Mann S. *Biomimetic materials chemistry*. Weinheim. Wiley-VCH; (1997).

[2] Sarikaya, M. and Aksay. *Biomimetics: Design and Processing of Materials*, I. A., Am. Inst. Phys., New York (1995).

[3] Bensatude-Vincent B, Arribart H, Boulligand Y and Sanchez C, *New J. Chem,* 26, 1, (2002).

[4] Sanchez C, Hervé Arribart H, and Guille MMG, *Nature Materials,* 4, 277, (2005).

[5] Tamerler C and Sarikaya, *Acta Biomater,* 3, 289, (2007).

[6] Weiner S and Addadi L,. *Journal of Materials Chemistry,* 7, 689, (1997).

[7] Sarikaya M, Tamerler C. and Jen Aky, *Nature Materials,* 2, 585, (2003).

[8] Vriezema DM, Aragonès MC and Elemans JAAW, *Chem. Rev,* 105, 1445, (2005).

[9] Mann S, and Ozin GA, *Nature,* 382, (1996).

[10] Roveri N, Palazzo B, and Iafisco M, *Expert Opinion Drug Delivery,* 5(8), 861, (2008).

[11] Iafisco M, Palazzo B, Marchetti M, Margiotta, N, Ostuni R, Natile G, Morpurgo M, Gandin V, Marzano C, Roveri N, *Journal of Materials Chemistry,* 19(44), 8385, (2009).

[12] M. Lelli, G. Balducci, E. Foresti, I.G. Lesci, M. Marchetti, F. Pierini and N. Roveri *Italian Patent No. 21.C0460.12.IT.2,* Nov, 9, 2010.

[13] Manara S, Paolucci F, Palazzo B, Marcaccio M, Foresti E, Tosi G, Sabbatini S, Sabatino P, Altankov G and Roveri N, *Inorganic Chimica Acta,* 361(6), 1634, (2008).

[14] M. Lelli, B. Palazzo, F. Paolucci, B. Parma, S. Vismara, and N. Roveri, *Journal of Materials Chemistry*, in press, (2011).

[15] M. Lelli, B. Palazzo, S. Manara, M. Iafisco, C. Bruno, F. Paolucci, S. Vismara, B. Parma, and N. Roveri, "Electrochemically assisted deposition of nanostructured hydroxyapatite coatings as a novel method to prepare drug-activated surface" *shortlisted for Journal of Material Chemistry,* 2-4. July. 2007 (London UK) p. 21.

[16] Lelli M, Foresti E, Marchetti M and Roveri N, *Chemistry of Materiels,* in press, (2011).

[17] M. Iafisco, M. Di Foggia, S. Bonora, M. Prat, and N. Roveri, *Dalton Trans,* 39, 1, (2010).

[18] Foltran, E. Foresti, M. Lelli, and N. Roveri. "Electrodeposition and electrospinning of Collagen and Hydroxyapatite in composite coatings" First International Conference of Multifunctional, Hybrid and Nanomaterials. 15–19, March.2009 (Tours, France) p. A27.

[19] X. Li, T. Fan, Z. Liu, J. Ding, Q. Guo, and D. Zhang, *J. Eur.Ceram.,* 26, 3657, (2006).

[20] M. Singh, J. Martinez-Fernandez and A. R. de Arellano-Lopez, *Curr. Opin. Solid State Mater. Sci.,* 7, 247 (2003).

[21] R. de Arellano-Lopez, J. Martinez-Fernandez,P. Gonzalez, C. Dominguez, V. Fernandez-Quero and M. Singh, *Int. J. Appl. Ceram. Technol.,* 1, 56, (2004).

[22] Martinez-Fernandez, F. M. Varela-Feria, S. Lopez-Pombero, A. R. De Arellano-Lopez and M. Singh, *Ceramics Engineering and Science,* 22, 135, (2001).

[23] R. Sepulveda, M. J. Lopez Robledo, A. R. De Arellano-Lopez, J. Martinez-Fernandez and C. Dominguez, *Bol. Soc. Esp. Ceram. Vidrio,*44, 357, (2005).

[24] F. M. Varela-Feria, M. J. Lopez-Robledo, J. Martinez- Fernandez, A. R. De Arellano-Lopez and M. Singh, *Ceramics Engineering and Science Proceedings,* 23, 854, (2002).

[25] F. M. Varela-Feria, J. Martinez-Fernandez, A. R. De Arellano-Lopez and M. Singh, *J. Eur. Ceram. Soc,* 22, 2719, (2002).

[26] M. Lelli, I. Foltran, E. Foresti, J. Martinez-Fernandez, C. Torres-Raya, F. M. Varela-Feria and N. Roveri, *Adv Eng Mater.,*12(8), B348, (2010).

[27] M. Carano, M. Lelli, I. G. Lesci, M. Marcaccio, F. Paolucci and N. Roveri, *RSC Advanced,* in press, (2011).

In: Electrodeposition: Properties, Processes and Applications ISBN: 978-1-61470-826-1
Editor: Udit Surya Mohanty

Chapter 11

ELECTRODEPOSITION OF HYDROXYAPATITE-NANODIAMOND COMPOSITE COATING ON METALS, INTERACTION WITH PROTEINS AND OSTEOBLAST-LIKE CELLS

Emilia Pecheva,[1,*] Lilyana Pramatarova,[1] Todor Hikov,[1] Kamelia Hristova,[2] George Altankov,[3,4,5] Paul Montgomery[6] and Takao Hanawa[7]

[1]Laboratory of Biocompatible Materials, Institute of Solid State Physics, Bulgarian Academy of Sciences, Sofia, Bulgaria
[2]Institute of Biophysics and Biomedical Engineerings, Bulgarian Academy of Sciences, Acad. G. Bonchev str., Sofia, Bulgaria
[3]Associate Member of the Institute of Biophysics and Biomedical Engineerings, Bulgarian Academy of Sciences, Sofia, Bulgaria
[4]ICREA (Institucio Catalana de Recercia i Estudias Avançats), Barcelona, Spain
[5]Institute for Bioengineering of Catalonia (IBEC), Barcelona, Spain
[6]Institut d'Electronique du Solide et des Systèmes, UDS-CNRS, Strasbourg, France
[7]Institute of Biomaterials and Bioengineering, Tokyo Medical and Dental University, Tokyo, Japan

ABSTRACT

Hydroxyapatite (HA) is the main component of human bones, a highly bioactive and biocompatible material; however, it has poor mechanical properties. Carbon-based coatings are found to significantly improve the mechanical properties of apatite, increase its adhesion, prevent metal ion release from metal implants and inhibit the formation of fibrous tissue and blood clotting upon implantation. In this chapter, homogeneous nanodiamond-reinforced hydroxyapatite (HA-ND) composite coating with improved mechanical strength and ductility was developed to enhance the biological properties of

*E-mail: emily@issp.bas.bg.

metal surfaces (stainless steel and titanium). The CO^{3}- and HPO^{4}-containing HA was deposited by electrodeposition from simulated body fluid with dispersed nanodiamond particles. Study of the initial interaction of osteoblast-like MG-63 cells revealed that cells attached well on all plain samples (HA-ND, pure HA and stainless steel). However, pre-coating with fibronectin (FN) even at low adsorption concentrations (1mg/ml) strongly improved cell adhesion and preferentially spreading on the HA-ND samples as indicated by the flattened cell morphology and pronounced vinculin positive focal adhesions. This effect correlates with the observed higher affinity for FN. Moreover, osteoblasts tended to rearrange both adsorbed and secreted FN in a fibril-like pattern, suggesting improved FN matrix organization on HA-ND samples.

INTRODUCTION

Hydroxyapatite (HA), found in bones and teeth in mammals, is a highly bioactive and biocompatible inorganic material. However, the benefits of HA coatings are constrained by their poor mechanical properties and weak adhesion to implant surfaces [1,2]. Metal materials, on the other hand, have excellent mechanical strength and corrosion resistance in aggressive media; therefore, various composite materials of HA coated on metals, such as titanium (Ti), Ti alloys, austenitic stainless steel (SS), Co-Cr alloys, etc. have been developed [3-5]. A frequently used method for coating of metals with HA is the electrodeposition (ED) technique, which has been widely applied to produce desirable composite coatings and gives the possibility of combining the beneficial properties of each individual material used in the composite [6-8].

Carbon-based coatings such as diamond-like carbon, carbon nanotubes and amorphous carbon have shown favorable properties: they improve the mechanical properties of the coatings, increase their adhesion to metal implants and prevent metal ion release from the metal surface [9-11]. On the other hand, nanodiamond (ND) particles are attracting increasing interest because of their unique properties. Also, due to their very small particle size (2-10 nm); they have high hardness, inertness to chemical attack and biological compatibility, and low friction coefficient [12-17]. ND particles can be easily synthesized by detonation and they are a promising material for obtaining mechanically strong composites [16-19]. They are truly nanoscale materials and also present the advantage of extremely high surface area, surface charge and the existence of surface functional groups (OH^-, $COOH^-$, NH^{2+} or SO_3H^-) as a result of the detonation process [14,15]. NDs are considered to be potential medical agents due to their high adsorption capacity, high specific surface area and chemical inertness, and their use for the design of biosensors has been suggested [20]. The interest in using nanocrystalline diamonds for medical and biological applications is also prompted by their spherical surface morphology, which is stable with respect to cage opening under in-vivo environments. For these applications, surface-functionalization will aid these carbon nanomaterials in becoming biocompatible, improving their solubility in physiological solutions and selective binding to bio-targets [20].

In this chapter, the ED technique was utilized to modify the surface of metal substrates (SS and Ti) with an HA layer from a supersaturated precursor solution (simulated body fluid, SBF), containing ND particles (this prepared solution is named 'ND-SBF' throughout the text). The goal of this chapter was to obtain a stress-free HA-ND coating with ductility and improved hardness in comparison to pure HA by incorporating ND particles as a minor phase into the HA. Two groups of samples were prepared:

(i) by ED from ND-SBF solution on SS or Ti (samples named 'HA-ND');
(ii) by ED from pure SBF solution on SS or Ti (samples named 'HA' and used as controls of the first group).

As mentioned above, biomimetic HA is a highly biocompatible inorganic material from the apatite (AP) family and it osseointegrates well due to its chemical resemblance to mammalian bone and teeth [21-23]. HA adsorbs many proteins and interacts well with osteoblast precursor cells, therefore metal implants are frequently coated with HA in order to facilitate bone adaptation, firmer implant-bone attachment, reduced healing time and enhanced bone apposition in comparison to uncoated implants [21,22]. In order to examine the biological compatibility of the HA-ND coating deposited on SS, we studied *in vitro* the initial interaction of osteoblast-like MG 63 cells with plain substrata, including HA-ND, pure HA and control SS surfaces, as well as after their pre-adsorption with fibronectin (FN) and vitronectin (VN) – the main adhesive proteins in the human blood plasma. Osteoblasts are the principal cells in the bone matrix and their successful interaction with a material provides insights on its osseointegration [23]. We investigated the overall cell morphology and quantified the initial cell adhesion and spreading of MG-63 cells. The particular effects of main serum proteins, fibronectin (FN) and vitronectin (VN) were also examined.

11.1. Coating Deposition and Characterization

Substrates were cut from rods of austenitic SS (AISI 316L, ø15 mm) or commercially pure Ti (ASTM grade 2, ø14 mm). They were subjected to further standard metallographic polishing to obtain a mirror-finished surface (SiC papers with grit sizes #150, #320, #600, and #800, followed by alumina suspension with grain size of 0.05 μm). Finally, the samples were ultrasonically rinsed in acetone, dried under a stream of pure nitrogen (99.9 %), and stored in a dessicator.

ND particles were synthesized by the shock-wave propagation method through the detonation of trinitrotoluene and hexogen at high pressure and high temperature produced in the detonation [18,19]. Subsequent purification from graphite by applying oxidation with potassium dichromate in sulphuric acid was carried out, and after several washings with hydrochloric acid and water, the as-obtained ND powder was dried [18,19]. The purification method led to an oxidation of the ND surface, which was found to be covered with carboxyl, and to a lesser extent with carbonyl and hydroxyl groups as revealed by infrared spectra [19]. Analyses have shown that the particle size was 4-6 nm, density was 3.2 g/cm^3, growth surface was 350-400 m^2/g and the content of diamond was 97-99% [19].

A potentiostat (Hokuto Denko HA150G) was used for the cathodic deposition of the coatings. The deposition was performed in a three-electrode electrolytic cell, consisting of the studied sample (SS or Ti) as the cathode, a Pt foil as the anode and a saturated calomel electrode (SCE) as the reference electrode. The SBF electrolyte resembled the ion composition, concentrations and pH of human blood plasma. It was prepared by mixing the reagent grade chemicals NaCl, $NaHCO_3$, KCl, K_2HPO_4-$3H_2O$, $MgCl_2$-$6H_2O$, Na_2SO_4-$10H_2O$, and $CaCl_2$-$2H_2O$ with distilled water [24,25]. As-prepared SBF was adjusted to pH 7.4 with tris-hydroxymethyl-aminomethane or hydrochloric acid. The ND particles were added to the SBF electrolyte to a concentration of 0.5 g/l and ultrasonically agitated for 20 min before the ED process ('ND-SBF' solution). Cathodic ED was performed with the electrolyte at 37^{0}C, applying a potential of -1.5V (SCE) for 15, 30 or 60 min on the SS substrates or -1.8V for the same times on the Ti substrates. After the deposition, the samples were washed under a flow of distilled water, dried in air and stored in a desiccator.

The morphology of the electrodeposited layers was studied by scanning electron microscopy (SEM; Hitachi S-3400 NX, 12 kV), coupled with energy dispersive X-ray spectroscopy (EDX; performed over five areas with size of 20x25 μm, as well as from five points taken on white aggregates). Topography, roughness and thickness parameters were investigated by atomic force microscopy (AFM; SOLVER SPM, NT-MDT, non-contact mode) and coherence probe microscopy (CPM, Z scan mode, Leitz Linnik microscope, 50x objective, white light illumination in spectral range of 350-1100 nm, lateral resolution of 0.45 μm, vertical resolution of 10 nm over a dynamic range of 10 μm, [26]). The layer structure was examined by Fourier transform infrared spectroscopy (FTIR; Bomem IR spectrometer FTLA2000, adsorption mode, 100 scans, resolution of 4 cm^{-1}) and micro-Raman spectroscopy (Jobin Yvon Horiba, 632 nm, resolution 2 cm^{-1}). A micro hardness tester (HMV-1, Shimadzu) was utilized to estimate the Vickers hardness (HV) by applying a load of 245 mN for 10 s and performing five indents on each sample (for the selected load and time, the effect of the substrate on the hardness can be ignored). X-ray photoelectron spectroscopy measurements (XPS; Surface Science Inc. electron spectrometer SSX100, USA; Al Kα excitation source, E= 1486.6 eV) were performed to investigate the surface chemistry of the composites (calibration of the binding energy was done by setting the C 1s line to 285.0 eV).The mineralization ability of the composite coatings (i.e. their ability to induce new AP formation) was tested in SBF at the body temperature (37^{0}C) for up to 7 days. For this purpose, a batch of samples was immersed in SBF (the solution was renewed every day) and each day several samples were taken out, washed with Milli-Q pure water, dried in air and analysed by SEM, EDX and FTIR.

11.2. Biological Studies with Osteoblast-Like Cells

11.2.1. Preparation of FITC-Labeled Fibronectin

Human plasma FN (Sigma, F2006) was dissolved in 0.1M sodium bicarbonate buffer (pH = 9.0) at 1 mg/ml, then 10 μl of fluorescein isothiocyanate-labeled FN (FITC-FN, Sigma, F7378) in dimethyl sulfoxide from a stock of 10 mg/ml was added. The mixture was incubated for 2 h at 37ºC. Labeled FN was separated from unreacted dye on a Sephadex G-25 column equilibrated with phosphate buffered saline (PBS) solution. The final protein concentration was estimated by measuring the absorbance at 280 nm, while the degree of FITC-labeling was calculated against the absorbance at 494 nm. The aliquots were then stored at -20 ºC.

For cell adhesion studies, samples from the three groups (HA-ND, HA and SS) were coated with non-labeled FN at low (1 μg/ml) and high (20 μg/ml) concentrations, as well as with VN (1 μg/ml, Sigma) dissolved in PBS for 30 min at 37ºC before washing twice with PBS. To obtain serum-coated samples, the slides were covered with pure fetal bovine serum (FBS; Gibco) and further processed under identical conditions.

11.2.2. Cell Culture and Cell Adhesion Assay

A human osteoblast-like MG-63 cell line (ATCC, USA) was used as model system to examine the effects of surface coatings on osteoblast adhesion, spreading and overall morphology, as well as on FN matrix organization. Cells were maintained in Dulbecco´s modified Eagle medium (Gibco, 11960-044) supplemented with 10% fetal bovine serum (FBS), 1% penicillin/streptomycin, 2 mM L-glutamine and 1 mM sodium pyruvate (Gibco, 11360-039), in a humidified atmosphere of 5% CO_2 in air. The culture medium was exchanged every second day. Upon reaching confluence the cells were detached with 0.05% trypsin-EDTA (Gibco, 25200-072) inactivated with FBS after detachment (approx. 5 min), then cells were re-cultured or used for the experiments.

To investigate the overall morphology, initial cell adhesion and spreading, $2x10^4$ cells/well were seeded onto each surface in 2.0 ml serum free medium. One set of samples was pre-coated with proteins as described above and another one served as plain, uncoated controls. After 2 h of incubation, the living cells were stained with fluorescein diacetate (FDA) by adding of 10 μl/ml from a stock of 5 mg/ml FDA in acetone. FDA can be transported across cell membranes and deacetylated by nonspecific esterases of the living cells. Resultant fluorescein accumulates within cells and allows direct visualization by fluorescent microscopy. Representative pictures of the adherent cells were taken with a fluorescent microscope (Zeiss, Axiovert 40, Germany, 20x), equipped with a digital camera. At least three representative pictures of each sample were made and morphological parameters such as the adhesion and mean spreading area of the cells were evaluated using automated image analysis software (analySIS v. 3.0, Soft Imaging System GmbH). Focal adhesions formed after 2 h of incubation of MG-63 cells with the samples were visualized using specific mouse monoclonal antibody (Sigma) against vinculin (1:800 dilution) in 1% bovine serum albumin (BSA). It was followed by goat anti-mouse Cy3 conjugated secondary antibody (Jackson ImmunoResearch). After incubation with the antibodies, the samples were mounted and viewed on inverted fluorescent microscope and at least three representative images were obtained.

11.2.3. Reorganization of Adsorbed Fibronectin (Early FN Matrix)

The ability of MG-63 cells to reorganize adsorbed FN (i.e. early matrix formation) was monitored on samples coated with FITC-labeled FN before seeding with $2x10^4$ cells/well in serum containing medium. After 5 h of incubation the samples were fixed using 3% paraformaldehyde, mounted in Mowiol and viewed and photographed with a fluorescent microscope. Regular round-shaped glass coverslips (Menzel GmbH and Co KG, D 15 mm) coated with FITC-FN were used as a positive control.

11.2.4. Fibronectin Matrix Formation (Late FN Matrix)

The ability of MG-63 cells to secrete and deposit FN into the ECM fibrils (i.e. late FN matrix formation) was examined via immunofluorescence for FN. For that purpose $3x10^4$ cells /well were cultured on the different samples for 3 days in 10% serum containing medium. At the end of the incubation, the samples were rinsed with PBS and fixed with 3% paraformaldehyde for 5 min. The samples were then washed and saturated with 1% BSA for 15 min. Subsequently they were stained with a polyclonal rabbit anti-FN antibody (Sigma) dissolved in 1% albumin in PBS for 30 min, followed by goat anti-rabbit Alexa Fluor 555-conjugated secondary antibody (Invitrogen) for 30 minutes before washed and mounted with Mowiol.

11.3. Coating Deposition and Characterization

During the ED of calcium phosphates (CaPs) from an aqueous electrolyte on a metal cathode, reduction of water and dissolved oxygen occurs at the surface of the cathode [27]. As a consequence, hydroxide ions are generated and hydrogen gas evolution arises simultaneously, according to the reactions:

$$2H_2O + 2e^- \rightarrow H_2 + 2OH \qquad [1]$$

$$2H_2O + O_2 + 4e^- \rightarrow 4OH^- \qquad [2]$$

The cathodic current in the experiments was recorded and used to calculate the current density over the specimen area (3 cm^2) exposed to the electrolyte. As shown in Figure 11.1. for ED carried out for 15 min in ND-SBF and in pure SBF solutions on steel substrates, the current density increased with the initial increase of the potential in the cathodic direction and became almost constant after reaching the target potential of -1.5 V (the same behavior was observed for 30 and 60 min of ED, as well as for 15, 30 and 60 min of ED on Ti and therefore these graphs were not presented). The increase of the current density just after the cathodic charge was due to the ED of a CaP layer. During this increase water was decomposed to H_2 and OH^- groups and gas evolution was observed. A decrease of the current density during the increase and a peak were observed for both curves in Figure 11.1. The current density decrease (i.e. the minimum in the current density) was generated by the reduction of nickel in the SS. After reaching the steady-state current, H_2 evolution continued through the electric reduction of water. The region of steady current density (J_{const}) in both curves was attributed to the passivation of the metal surface due to the formation of an electrodeposited layer after 15, 30 or 60 min. The current density recorded for the samples with ND particles in the SBF was much smaller than for the samples without ND. The reason was most probably a higher electric resistance of the deposited composite layer, which worked as a protective film for the current, and consequently as a barrier for the mass transportation. After ED for 60 min on steel and Ti, J_{const} for both curves showed a slight decrease, thus indicating that the deposited layer increased in thickness to such a degree that it started to isolate the metal electrode. The SEM image of the composite coating electrodeposited for 15 min from ND-SBF solution on SS or Ti substrates revealed the formation of a dense and uniform coating with platelet-like morphology (Figure 11.2.a). White aggregates of platelets were also found, randomly distributed on the underlying homogeneous coating. Increasing the duration of the ED to 30 or 60 min did not change the coating morphology and the same platelet-like coating with attached aggregates was observed on both materials. The coating on the HA control samples obtained from pure SBF after 15, 30 or 60 min on steel or Ti had the same morphology, i.e. the incorporation of the ND particles from the SBF did not influence the morphology. AFM imaging was used for complementary information and it showed the same picture of the coatings (Figure 11.2.b), i.e. homogeneously distributed platelet crystals forming a dense layer and random aggregates with lateral sizes of a few μm. Peak-to-peak and rms roughness of the coatings were about 1 μm and 99 nm, respectively. CPM technique was used as a complement to the other surface measurement techniques such as AFM and SEM to estimate the platelet height (for more details on the CPM application for study of HA layers [23]). Thus, knowing that the coating thickness was about 1 μm according to the AFM data, the height of the platelets was estimated by a Z scan along the optical axis to be about 0.66 μm as seen in Figure 11.2.c. where peak 1 comes from the air-platelet interface and peak 2 from the coating-substrate interface. Comparison of the EDX spectroscopy measurements performed over areas or from point measurements taken on white aggregates on the coated SS and Ti substrates (Table 11.1) revealed that the platelet aggregates in the composite coating had more carbon than the coating itself (see area measurements) and than the control HA coating. Probably during the process of ED, the ND particles agglomerated in the white aggregates despite the preliminary sonication of the precursor solution. As reported in [15], ND particles produced by a detonation method easily form conglomerates with a characteristic size of the order of 1 μm because of their high surface energy. It was observed that ND particles present in a water solution form conglomerates ranging from a few hundred to even a few thousand nm, depending on the solution pH. Thus, the aggregates in the composite coatings can be attributed to ND conglomerates deposited from the ND-SBF precursor solution, which corresponds well to the sizes of the observed aggregates, i.e. a few μm (Figure 11.2). Apart from the high C content, Ca and P were detected in the platelet crystals forming the aggregates by EDX. Control electrodeposited coatings grown in pure SBF on steel (i.e. HA samples) had the highest Ca:P ratio (Table 11.1). It decreased with increasing time of deposition as follows: 1.61, 1.46 and 1.40 after 15, 30 and 60 min of ED, respectively. Further, after deposition from ND-SBF on steel, the Ca:P values of HA-ND coatings were lower than those obtained for the HA control coatings, namely: 0.83, 1.09 and 1.29 for 15, 30 and 60 min, respectively. The decrease in the Ca:P ratio in these samples could be attributed to some degree of amorphization, which the ND particles introduced in the already non-stoichiometric coatings. In addition, the Ca:P value increased with the time of ED, showing that the HA coatings matured with time [24,25]. The Ca:P ratio of the composite coatings on Ti was similar for 15, 30 and 60 min of ED from the ND-SBF solution as seen in Table 11.1. (about 1.15). The coating on the HA control samples, obtained from the pure SBF after 15, 30 or 60 min also had a low Ca:P ratio. As the value for synthetic stoichiometric HA is 1.67, the lower values calculated for the control coatings could be attributed either to low crystallinity apatite, or to a combination of HA and other CaP phases, which also contribute to a decrease of the coating crystallinity [28,29]. The lower values match well with the observed morphology of coatings consisting of platelet crystals (stoichiometric HA has a needle-like morphology [28]).

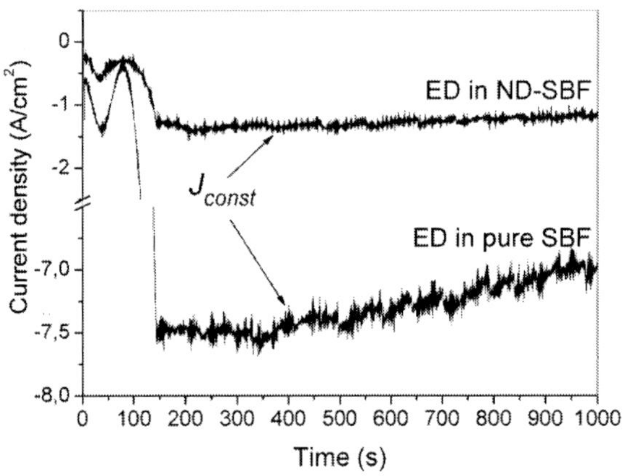

Figure 11.1. Cathodic current density during ED of CaP for 15 min with/without the addition of ND particles in the SBF electrolyte.

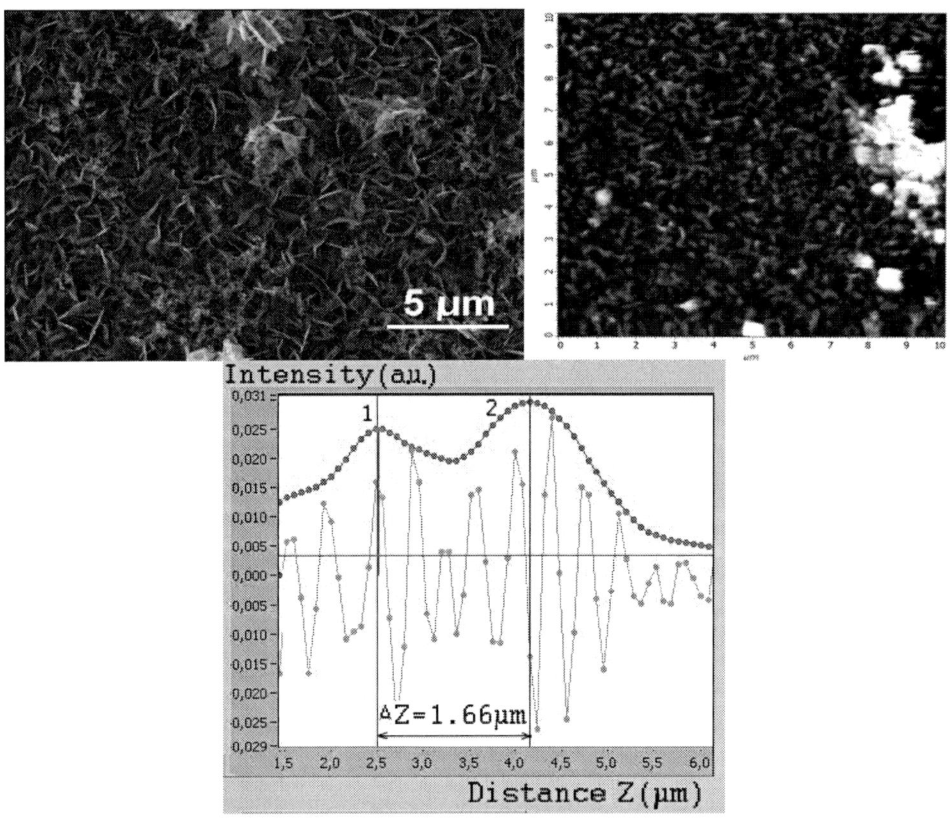

Figure 11. 2. (a) SEM and (b) AFM images of the coating obtained after ED from ND-SBF for 15 min on stainless steel: a dense and uniform coating with platelet-like morphology and randomly distributed white aggregates of platelets was grown; (c) CPM Z scan on a platelet in direction of the optical axis.

Table 11.1. EDX results for the Ca:P ratio and C content (at.%), measured either over areas (20 x 25 μm in size) or from point measurements taken on white aggregates on the HA-ND composite and HA control coatings deposited on stainless steel and Ti substrates

Samples	Steel substrate: C concentration [at%]		Ti substrate: C concentration [at%]		Steel: Ca:P ratio	Ti: Ca:P ratio
	Area	Aggregate	Area	Aggregate		
HA-ND15	2.50	35.45	4.57	47.74	0.83	1.15
HA-ND30	4.18	22.00	1.84	27.74	1.09	1.13
HA-ND60	5.66	8.81	11.23	49.65	1.29	1.15
HA15	3.96	12.33	1.49	1.94	1.61	1.14
HA30	5.43	9.91	0.46	0.99	1.46	1.21
HA60	4.88	11.96	1.98	2.21	1.40	1.10

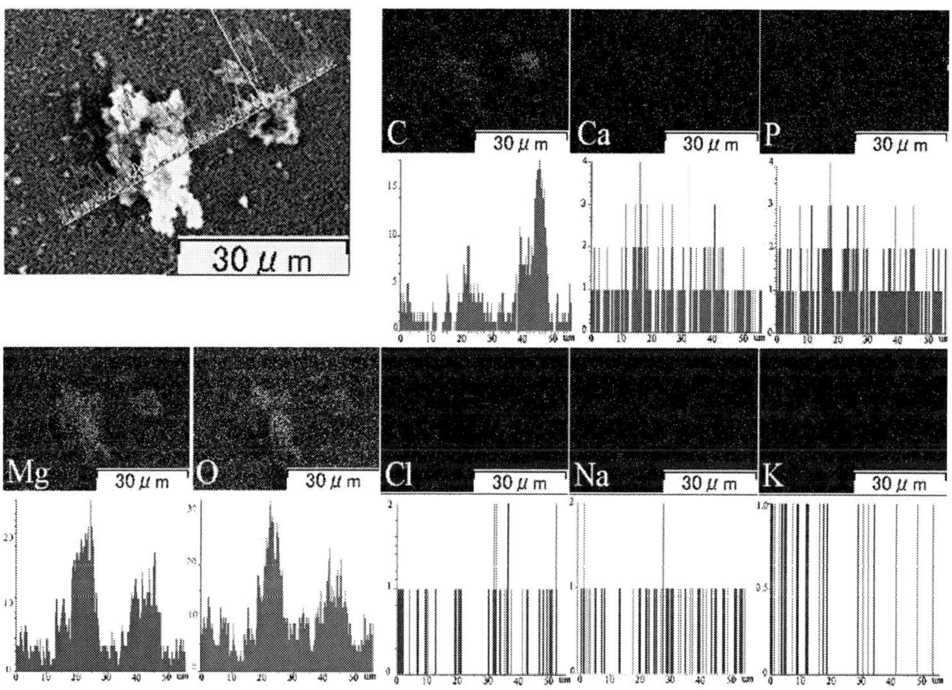

Figure 11.3. EDX elemental mapping over area (20x25 μm) on the HA-ND15 coating of stainless steel substrate (images) and over a line crossing the white aggregates (spectra below the images).

Apart from the Ca and P signals, K, Na and Cl were detected in the composite and control coatings in minor amounts (Na = 0.02-0.27 at%, Cl = 0.02-0.19 at% and K = 0.04-0.10 at%), independently on the substrate used for the ED and similarly to the natural bone composition [24,28,29]. Comparison to our previous results on HA deposition by a precipitation process [24] revealed a strong increase in the Mg content in the two types of electrodeposited coatings (composite and control) studied in this chapter. According to a literature survey, increased Mg content in the case of electrodeposited AP coatings was usually observed. Thus, the non-stoichiometry of the HA-ND composite coating could be attributed to its high Mg content as it is known that Mg is an element which introduces imperfection in the HA lattice [28,29]. EDX elemental mapping over area (20x25 μm) on the

HA-ND15 samples (Figure 11.3; SS substrates) confirmed the higher concentrations of C, Mg and O with respect to the HA samples, especially in the white aggregates, as well as uniform distribution of Ca, P, Na, Cl and K. The same results were obtained for the Ti substrate coated with HA-ND composite.

The composite structure was investigated by FTIR and micro-Raman spectroscopy. Figure 11.4.a. shows the FTIR spectrum of the HA-ND coating obtained on steel for 15 min of ED and it is representative also for 30 and 60 min of ED of the composite. A highly intense envelope of peaks due to $\nu_{2,4}$ PO_4^{3-} stretching modes was detected in the frequency region of 400-700 cm^{-1}. A broad peak envelope with lower intensity was observed at wavenumbers of 950-1180 cm^{-1}. Peak fitting revealed that it was an envelope of underlying peaks due to $\nu_{1,3}$ PO_4^{3-} stretching, ν_2 CO_3^{2-} stretching and HPO_4^{2-} vibrational modes. In addition, two absorption peaks due to ν_3 CO_3^{2-} stretching at 1430 and 1620 cm^{-1}, characteristic of partial CO_3 substitution of PO_4 ions in the HA structure were detected [28,29]. Thus, it was concluded that the coating was non-stoichiometric, CO_3- and HPO_4-containing HA. As known, incorporation of carbonate and acid phosphate is typical for the HA in bones [28]. Both ions are found to induce disorder in the stoichiometric HA thus leading to the formation of Ca-deficient and low crystalline HA, which explains the broad peaks in the FTIR spectra. The spectrum intensity of the HA-ND coatings was lower than those of the corresponding HA coating (spectra not shown since they were identical), thus revealing lower thickness (as known, peak intensity is proportional to the coating thickness according to the Beer's Law [30]). Probably the HA growth was retarded with the incorporation of the ND particles. The same peaks characteristic of CO_3- and HPO_4-containing HA were observed for the composite coatings on Ti substrate.

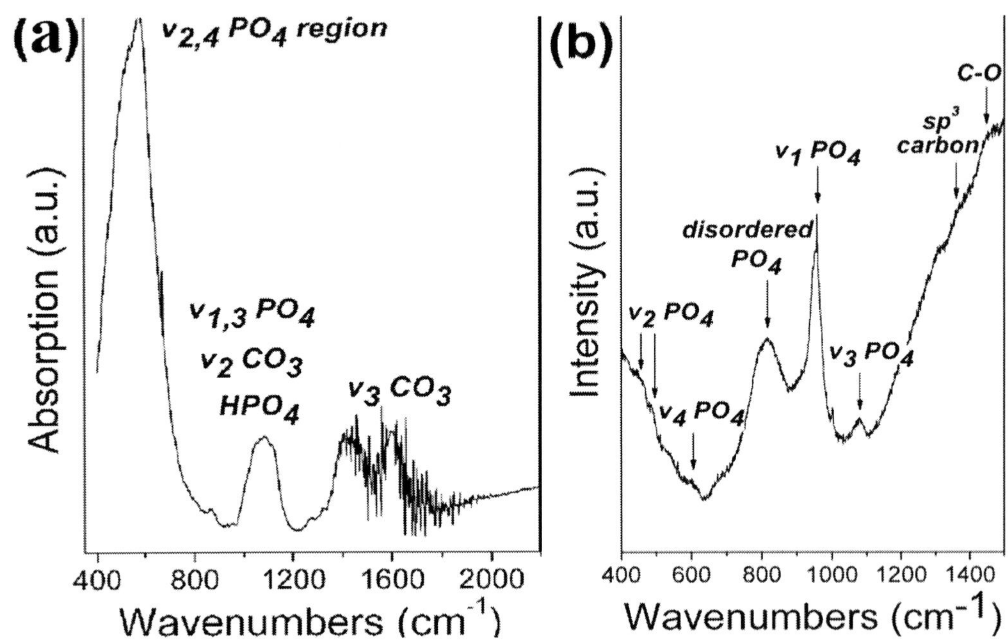

Figure 11.4. Representative FTIR (a) and Raman (b) spectra of the HA-ND composite coating observed in Figure 11.2 (stainless steel substrate).

Representative coated samples were subjected to annealing in vacuum ($2x10^{-2}$ Pa) at 750^0C for 60 min. This treatment was found to improve the crystallinity of the layers as observed by FTIR spectroscopy through the increase of peak intensity and the decrease of peak width [31]. Thus, the peaks lying under the envelope in the $PO_4/CO_3/HPO_4$ region were better resolved. Decrease of the coating thickness was detected by the decreased intensity of the spectra as discussed in [31]. Due to the fact that Raman and FTIR spectroscopies complement each other, they were both used in this chapter. The characteristic spectrum of the HA-ND composite layer deposited on SS for 15 min of ED is depicted in Figure 11.4.b. and is representative for 30 and 60 min of ED of the composite. Vibrational modes due to P-O in PO_4, typical for HA were detected: strong ν_1 symmetric stretching at 960 cm^{-1}, ν_2 asymmetric bending at 440 cm^{-1} with a shoulder at 480 cm^{-1}, shoulders due to ν_4 symmetric bending at 540 and 600 cm^{-1}, and ν_3 asymmetric stretching at 1078 cm^{-1}. Vibrations due to P-O in a disordered continuous phosphate network at 808 cm^{-1} with a shoulder at 670 cm^{-1} [32] were found. A shoulder at 1360 cm^{-1} could be due to sp^3 hybridized carbon coming from the presence of the ND in the coatings. C-O vibrations due to CO_3 in the HA or carboxyl (COOH) groups on the ND surface were observed as a shoulder at 1460 cm^{-1} [33]. Raman and FTIR spectra of the corresponding HA-ND and HA coatings on Ti substrates were identical and thus were not shown here.

More information on the chemical state of the composite coating was obtained by XPS. The deconvoluted XPS spectrum of Ca 2p on the surface layer obtained by ED from the ND-SBF electrolyte on Ti substrates (Figure 11.5) confirmed the formation of CaP, carbonate, as well as calcium oxide (under the common envelope of Ca $2p_{3/2}$, centered at 348.5 eV [34]). Typically, Ca 2p has a doublet structure and the second peak situated at 350.5 eV (not shown) was unambiguously attributed to $CaCO_3$ compound. CaP was also observed in the P 2p spectrum through the peak at 133.5 eV [34]. In addition, the deconvolution of the O 1s spectrum revealed peaks originating from O^{2-} at 530.2 eV, adsorbed water from H_2O at 532.6 eV and the predominant existence of OH^- groups (peak at 531.5 eV), and hence a hydrated surface layer, which was also an indication of CaP formation. Metal oxides were also detected from the O 1s spectrum by a peak at 528.0 eV [34].

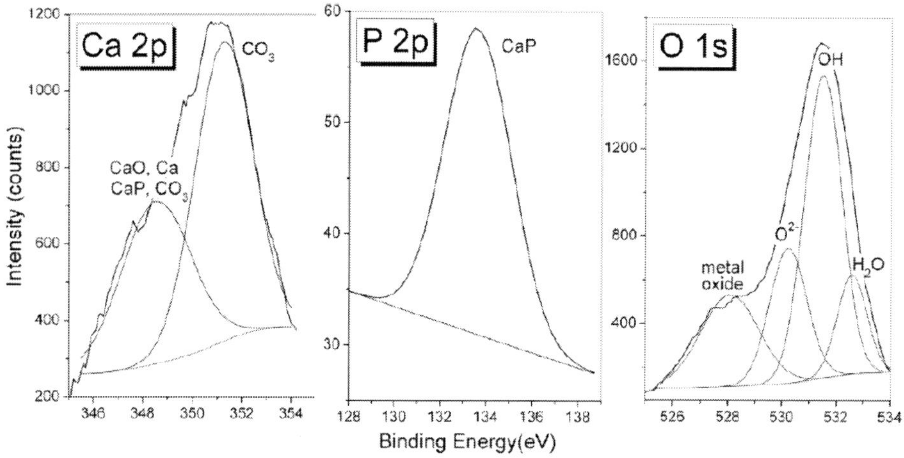

Figure 11.5. Deconvoluted XPS spectra of Ca 2p, P 2p and O 1s for the layers obtained by ED from the ND-SBF electrolyte.

Table 11.2. Vickers hardness (HV, mean ± SD) of the coatings obtained by ED from ND-SBF and pure SBF solutions on stainless steel and Ti. The results showed higher hardness for the HA-ND coatings

	Vickers hardness					
Sample	HA-ND15	HA-ND30	HA-ND60	HA15	HA30	HA60
Steel	310.0±7.8	394.0±42.6	453.8±66.7	335.0±15.3	355.0±29.2	373.7±32.3
Ti	218.2±9.9	246.0±15.9	205.0±13.4	186.6±14.6	193.6±11.1	197.4±17.1

We also examined the coatings' **hardness** since this parameter is a guarantee of good wear resistance, which is crucial for HA implant coatings. In this chapter, according to the HV results presented in Table 11.2, deposition of the composite coating for 15 min from ND-SBF on SS did not increase the HA hardness (HV310 versus HV335 for the HA coating). However, the composite obtained after 30 or 60 min (HV394 and HV454, respectively) showed higher HV values than the HA coating deposited for the same time from pure SBF (HV355 and HV374, respectively). Thus, despite the low initial concentration of the ND particles in the ND-SBF electrolyte and the sedimentation of the bigger particles with the ED time, the HA hardness was improved with the higher content of ND particles. HV results for the layers obtained on Ti substrates (Table 11.2) revealed that the deposition from ND-SBF yielded an increase of the HA-ND coating hardness in comparison to pure HA. The highest value for these samples was measured after ED for 30 min and was HV246. For comparison, the control samples deposited after 15, 30 or 60 min of ED from pure SBF, all had hardness values less than HV200. The substrate used for the ED influenced the HV values, which were lower on the Ti substrates. SAICAS measurements confirmed the AFM and CPM data on the coating thickness to be about 1 μm.

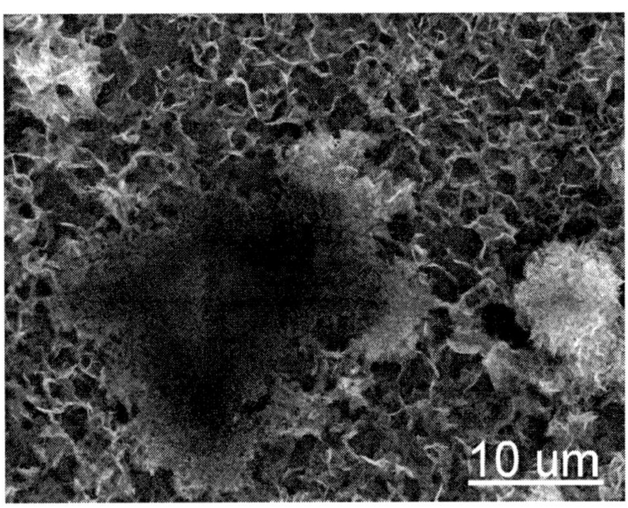

Figure 11.6. Qualitative data obtained by SEM image of the imprint left by the diamond indenter revealed no cracks in the HA-ND layers.

ED in this chapter was performed at low temperatures in order to simulate the body environment and to avoid thermal stress at the metal-coating interface, which could lead to cracks in the coating and phase transformations. Low temperatures during ED are known to yield enhanced binding strength between substrate and coating. The growth of thin homogeneous layers as in this experiment (thickness of 1 μm) is advantageous since there is less possibility of layer delamination and crack formation, usually observed with thick HA layers (tens of micrometers). SEM images of the electrodeposited coatings obtained in ND-SBF and pure SBF solutions, independently on the substrate showed that cracks were actually absent. A representative image of the imprint left by the diamond indenter (Figure 11.6) revealed no cracks extending from the imprint corners for the HA-ND coating, which testified to the good ductility and the lack of residual stress in the composite. No delamination of the coating was observed even at testing with the highest available load (980 mN).

To reveal the mineralization activity of the new composite coatings, i.e. their ability to form new HA layer, they were immersed in SBF supersaturated solution at the body temperature of 37^0C for up to 7 days. Representative SEM images shown in Figure 11.7 showed the overgrowth of a layer with a thickness and Ca:P ratio increasing with the time of precipitation in the SBF. The Ca:P ratios calculated for the layers on the 1st, 3rd, 5th and 7th day of the mineralization were 1.21, 1.45, 1.59 and 1.65, respectively. This increase revealed an improvement in the coating stoichiometry, as the Ca:P ratio steadily approached the value of stoichiometric HA (i.e. 1.67). FTIR spectra of the newly formed layers after the mineralization experiment (Figure 11.8) revealed characteristic bands of CO_3- and HPO_4-containing HA with very well defined narrow peaks in the $PO_4/CO_3/HPO_4$ region, i.e. the coating crystallinity was improved. The strong peaks of CO_3 (1520 cm^{-1}) and adsorbed water (1690 cm^{-1}) observed in Figure 11.4.a. disappeared at the 5th day of the experiment. The same SEM and FTIR results observed for Ti suggested that the composite coating is potentially bioactive independent on the underlying substrate and thus it is attractive as a surface modification for metal medical device materials.

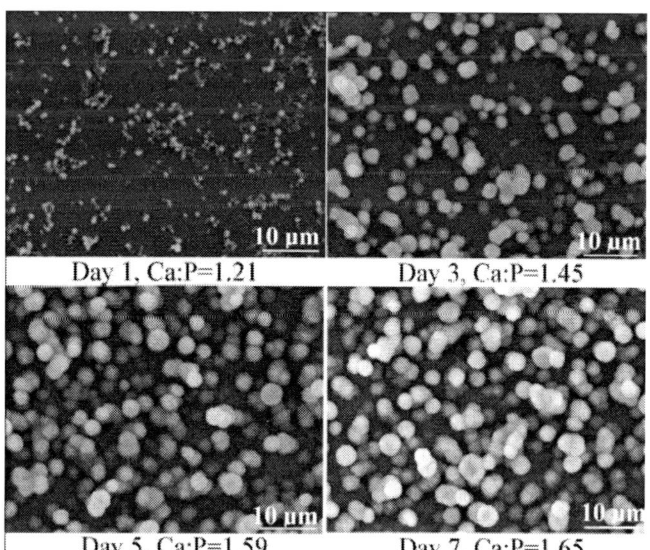

Figure 11.7. SEM results for the mineralization activity of HA-ND60 coatings on stainless steel (up to 7 days) revealed the overgrowth of a layer with a thickness and Ca:P ratio increasing with the time of precipitation in SBF.

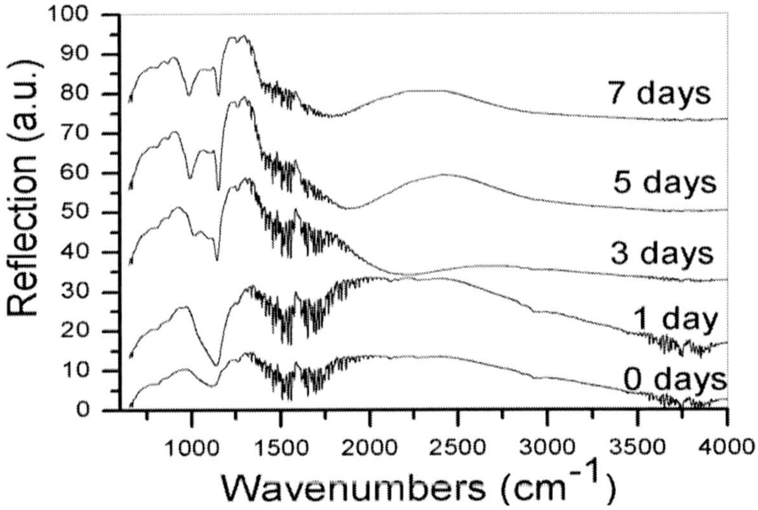

Figure 11.8. FTIR spectra of the HA-ND60 coatings during the mineralization experiment on steel showed that the overgrown layer was HA with improved crystallinity.

11.4. Biological Studies with Osteoblast-Like Cells

For a comprehensive characterization of a novel material with potential biomedical applications, it is necessary to test its biological compatibility with living cells. This was the reason to study the initial adhesion of MG-63 cells on plain and serum-coated samples after 2 h of incubation *in-vitro.* The overall cell morphology on plain HA-ND, HA and SS is shown in Figure 11.9. (a-c). Cells attached slightly better to the HA-ND coated SS but they did not spread on non-coated samples, exhibiting predominantly round morphology. Quantitative data for cell adhesion (Table 11.3) showed a non-significant increase (about 20%) of the cell number on HA-ND ($p<0.05$) (Figure 11.9.a). The adhesion also did not differ with plain HA and SS samples (Figure 11.9.b-c, see also Table 11.3). Serum coating substantially improved cell interaction to all samples (Figure 9.d-f). Quantitative data presented in Table 11.3 showed that both cell adhesion and cell spreading area increased significantly in comparison with the plain surfaces ($p<0.05$). However, adhesion of MG-63 cells was the highest on the HA-ND coating (Figure 11.9d, Table 11.3) in comparison with the HA and SS samples (Figure 11.9 e,f, respectively). The increase of cell spreading observed on the HA-ND samples was not statistically significant ($p<0.05$).

The main adhesive proteins in serum are FN and VN. Assuming that the amount of FN is relatively low as it binds to fibrin during blood clotting, a probable reason for the improved cell adhesion to HA-ND pre-coated with serum could be the different adsorption of VN. To test this assumption, we studied the adhesion of MG-63 cells to VN in a single protein system. The samples were coated with pure VN and then the cellular interaction was studied in serum free medium. Indeed, VN promoted cellular interaction (Figure 11.10 a-c), however no significant difference between samples was found, which was confirmed by the quantitative data for cell adhesion (see Table 11. 3), a result suggesting that VN is involved in the adhesion process but this effect cannot explain the observed difference between serum coated samples. Thus, as next step we studied the effect of FN adsorption. The results presented in Figure 11.10. (d-i) demonstrated that FN in a saturating coating concentration of 20 μg/ml also improved the cellular interaction equally to all studied surfaces. However, the effect appeared to be dependent on the coating concentration. Surprisingly, a significant increase in the adhesion (Table 11.3) of MG-63 cells to HA-ND coating compared to pure HA and SS was found when lower concentration of 1 μg/ml was used (Figure 11.10. g-i), a result suggesting either higher affinity for the FN or higher avidity of this protein on the HA-ND coatings.

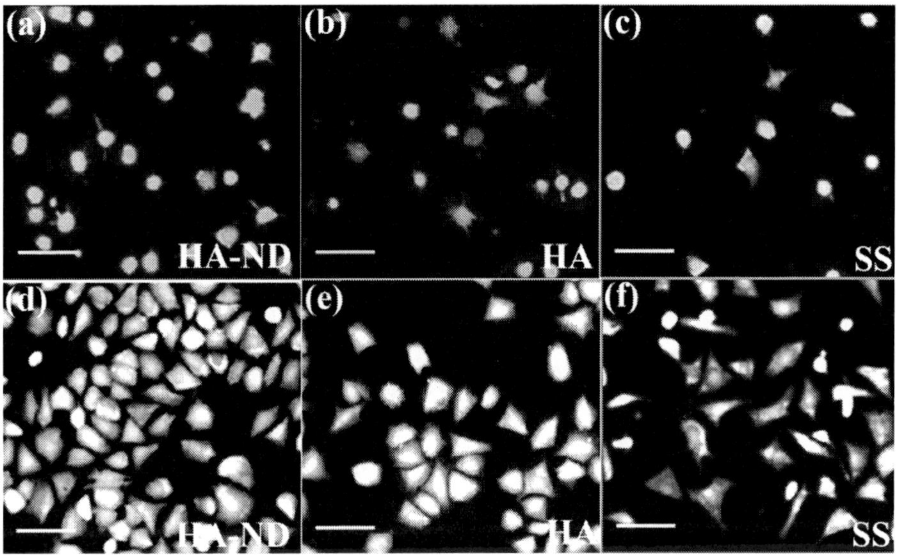

Figure 11.9. Overall morphology of MG-63 cells adhering for 2 h on plain (a-c) and serum-coated HA-ND, HA or SS (d-f) samples (bar 100 μm).

Table 11. 3. Cell adhesion and spreading to various samples after 2h of incubation in serum-free medium (significantly different values are marked with an asterisk)

Sample		HA-ND	HA	SS
VN	Cell number (mm^2)	163 ± 10	121 ± 19	78 ± 13
1μg/ml	Spreading area (μm^{-2})	1277 ± 635	1078 ± 484	1123 ± 24
FN	Cell number (mm^2)	171 ± 6*	134 ± 10	86 ± 19
1μg/ml	Spreading area (μm^{-2})	1533 ± 697	1431 ± 603	1276 ± 582
FN	Cell number (mm^2)	200 ± 14	183 ± 17	75 ± 22
20μg/ml	Spreading area (μm^{-2})	1963 ± 759	1836 ± 718	1682 ± 458
	Cell number (mm^2)	257 ± 7*	20 ± 23	133 ± 28
Serum	Spreading area (μm^{-2})	1414 ± 521	1284 ± 393	1142 ± 602
	Cell number (mm^2)	69 ± 11	57 ± 10	70 ± 17
Plain	Spreading area (μm^{-2})	800 ± 315	715 ± 312	673 ± 375

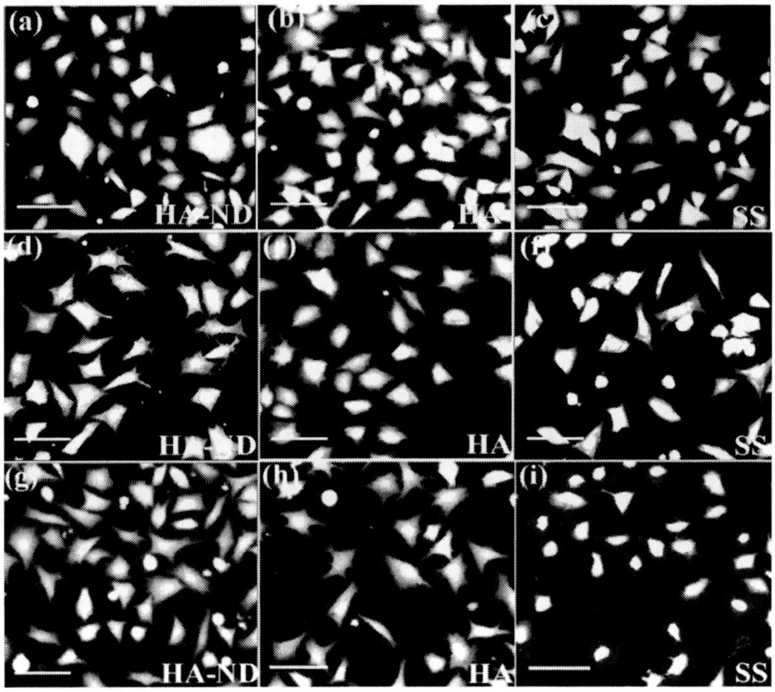

Figure 11.10. Overall morphology of MG-63 cells adhering for 2 h on VN-coated (a-c), FN-coated with high (20μg/ml; d-f) or low concentration (1μg/ml; g-i) HA-ND, HA or SS samples (bar 100 μm).

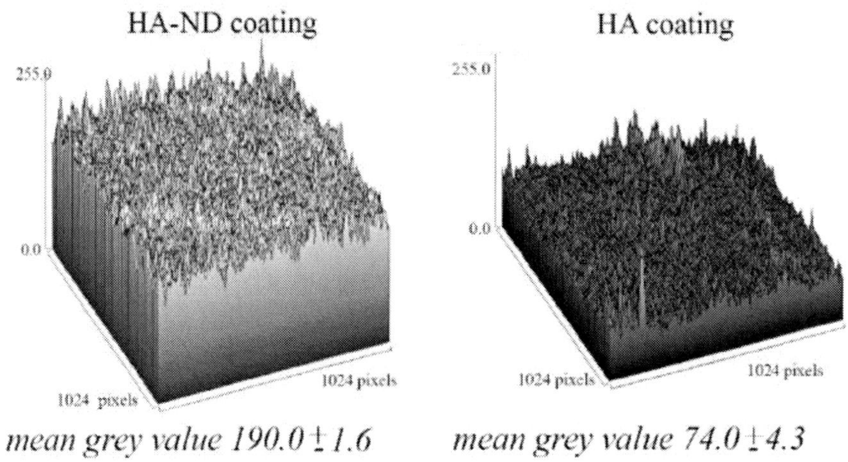

Figure 11.11. 3D confocal images of adsorbed FITC-FN on HA-ND (a) and pure HA (b) coatings. The fluorescent intensity in grayscale pixels is analyzed to gain information for the FITC-FN adsorption.

Quantitative analysis of the fluorescent intensity in grayscale pixels based on the confocal images of adsorbed FITC-FN (Figure 11.11) show a substantially higher adsorption of FN on HA-ND in comparison to pure HA, suggesting a higher affinity of these samples for FN. The three dimensional reconstruction of the images also showed that the adsorption of FN on both HA and HA-ND coatings was rather heterogeneous, presumably following the organization of the HA crystals. As these linear features were pronouncedly expressed on the HA-ND samples, we anticipated that such topography could promote the interaction with cells [35-

37]. Although no FN fibrils were observed it does not necessarily mean an absence of small micro/nanofibrils, as the characteristic scale could lie below the optical resolution. Based on these data we concluded that both FN and VN are involved in the adhesion but the improved cellular interaction with HA-ND coating should be attributed to its higher affinity for FN. Immunofluorescence studies on SS samples were not performed because of the higher refractivity of their surface.

Formation of focal adhesion complexes is another approach to learn more about the effectiveness of cellular interaction with materials. Focal adhesions are the places where real physical contact with the material surface occurs [38]. To visualize these structures we used immunofluorescence for vinculin, a main cytoskeletal protein constituent of the focal adhesion complex. As can be seen in Figure 11.12. (e,f) no significant formation of focal contacts was found on plain surfaces, consistent with the low cell attachment and spreading found (Figure 11.9). Nevertheless, there were more cells on the HA-ND samples, also in agreement with the improved adhesion to plain samples although they looked smaller. However, on both serum- and FN-coated samples cells acquired flattened morphology as in Figure 11.9. (d-f) correlating with the well developed focal adhesion complexes (Figure 11.12. a-d). Focal adhesions were better expressed on HA-ND samples, suggesting stronger cellular interaction on both serum- and FN-coated surfaces, particularly when compared to non-coated samples.

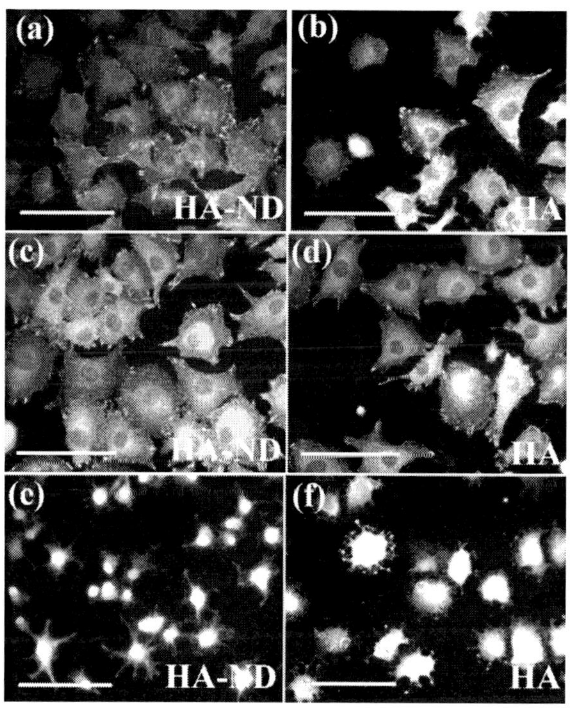

Figure 11.12. Focal adhesion formation of MG-63 cells adhering for 2 h on HA-ND (a,c,e) and pure HA (b,d,f). Samples are coated with serum (a,b), with FN (c,d) or studied as plain (e,f) (bar 100 μm).

It is well documented that many type of cells such as fibroblasts and endothelial cells not only attach to adsorbed FN but also tend to rearrange it in a fibril-like pattern, presumably as an attempt to organize their own matrix [39-41]. As some surfaces can promote FN reorganization which relates positively to their biocompatibility [40], while others cannot, we decided to learn more about the adsorbed FN on the studied samples. For that purpose we cultured the cells for 5 h on FITC-FN coated SS, HA and HA-ND, as well as on control glass samples as a positive control. As seen in Figure 11.13 (a-c), cells on HA-ND rearranged FN particularly well in an almost indistinguishable pattern from the glass control. FN removed by osteoblasts appeared as dark streaks on the otherwise bright fluorescent background of adsorbed FN, and this FN was further organized in a linear matrix-like structure located above or beneath the cells. When fluorescent and phase contrast images were merged, more than 90% of the reorganized FN was associated with osteoblasts, which was in accordance with previous studies [39-41] and confirm the cellular dependence of the process. This result also suggested that FN is relatively loosely adsorbed on the HA-ND coating, at least in such a way that the cells can easily remove and reorganize it in a matrix-like structure, which we consider as beneficial for its biocompatibility [39-41]. In contrast, both HA coating and SS surfaces (Figure 11.13 b,d) represented only dark streaks of removed FN and only small patches of rearrangement, indicating reduced matrix-forming activity. It has to be pointed out that this effect might be relative, e.g. caused by lowered background of adsorbed FN.

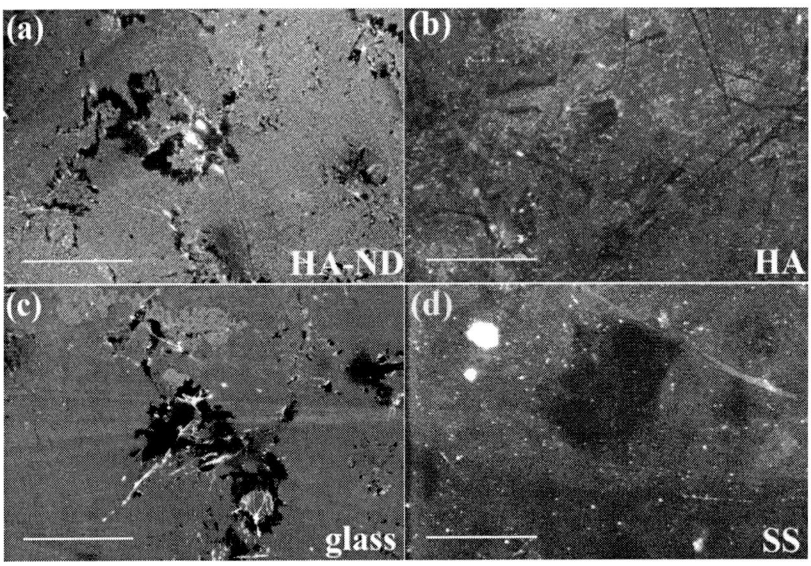

Figure 11.13. Early FN matrix formation on HA-ND (a) and HA (b) coatings compared with control glass (c) and SS (d) surfaces (bar 100 μm).

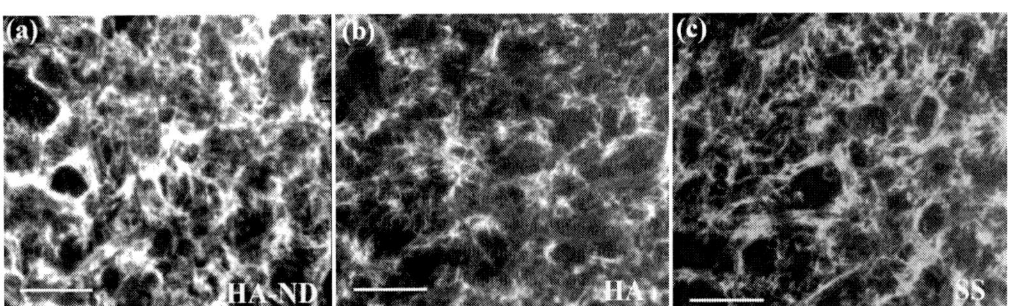

Figure 11.14. Late FN matrix deposition on HA-ND (a), HA (b) and SS (c) samples (bar 100 μm).

At longer incubation, osteoblasts produce their own osseogenic matrix [42] containing collagens, proteoglycans and FN [43]. To learn whether the initial difference in FN adsorption will affect the subsequent development of FN matrix we cultured MG-63 cells for 72 h and visualized FN via immunofluorescence. Multiple bright fibrils forming networks were observed on all samples (Figure 11.14), indicating good cell functionality and a rather materials-independent mechanism for matrix organization. However, a trend for stronger deposition of FN matrix on HA-ND samples (Figure 11.14.a) might be assumed, when comparing to plane HA (Figure 11.14.b) or SS surfaces (Figure 11.14.c).

CONCLUSION

A dense and homogeneous, non-stoichiometric hydroxyapatite-nanodiamond composite coating, containing CO_3 and HPO_4 groups is prepared by electrodeposition on austenitic stainless steel and titanium. The non-stoichiometry of the composite is attributed to the presence of CO_3 and HPO_4 ions in the coating structure, as well as to the high Mg content resulting from the electrodeposition process. Additionally, the nanodiamond particles retard the transformation of the coating to more crystalline apatite which can be changed by applying thermal treatment in vacuum. The hydroxyapatite-nanodiamond composite is characterized by ductility and better Vickers hardness in comparison to pure hydroxyapatite coating. No residual stress and cracks are observed in the composite. A mineralization experiment in a simulated body fluid revealed that the composite coating is potentially biologically active since it can induce the formation of a hydroxyapatite layer with improved crystallinity. Thus, the nanodiamond-reinforced hydroxyapatite coating can be considered as an attractive surface modification for metal materials with biomedical applications.

The overall cell morphology of osteoblast like MG63 cells and focal adhesions formation reveal improved cellular interaction on the composite hydroxyapatite-nanodiamond coating deposited on stainless steel resulting presumably from the increased fibronectin adsorption. This fact combined with the sufficient organization of fibronectin matrix by osteoblasts is another indicator of the good biocompatibility of hydroxyapatite-nanodiamond coating in an osseogenic environment.

ACKNOWLEDGMENTS

This chapter was supported by a national project with the Agency of Innovation of Bulgaria (NIF 02-54/2007), two projects with the Bulgarian National Scientific Research Fund (TK-X-1708/2007 and DO 02-167/2008) and a project with the Spanish Ministry for Education and Science (MAT2009-14440-2). NATO grant CBP.EAP.RIG.982693, JSPS grant 18-06728, ANNA Transnational Activities project (026134(RI3)) and E!3033 Bionanocomposit are also acknowledged. Gratitude to Assoc. Prof. E. Kobayashi (Tokyo Institute of Technology, Dept. of Metallurgy and Ceramics Science) for the useful discussions.

REFERENCES

[1] K. Burg, S. Porter, J. Kellam, *Biomaterials* 21, 2347 (2000).

[2] L.L. Hench, *J. Am. Ceram. Soc.* 81, 1705 (1998).

[3] D. Mears, *Materials and Orthopaedic Surgery*, Williams and Wilkins Company, Baltimore (1972).

[4] M. Sivakumar, U. Kamachi Mudali, S. Rajeswari, *J. Mater. Sci.* 28, 6081 (1993).

[5] G. Rondeli, B. Vicentini, A. Cigada, *Br. Corros. J.* 32, 193 (1997).

[6] S. Rossler, A. Sewing, M. Stolzel, R. Born, D. Scharnweber, M. Dard, H. Worch, *J. Biomed. Mater. Res.* 64A, 655 (2003).

[7] Zhitomirsky, *Bull. Am. Ceram. Soc.* 79, 57 (2000).

[8] Zhitomirsky, *Adv. Colloid Interface Sci.* 97, 277 (2002).

[9] R. Narayan, *Mater. Sci. Engin. C* 25, 405 (2005).

[10] R. Lappalainen, M. Selenius, A. Anttila, Y. Konttinen, S. Santavirta, *J. Biomed. Mater. Res.* 66B, 410 (2003).

[11] T. Akasaka, F. Watari, Y. Sato, K. Tohji, *Mater. Sci. Engin. C* 26, 675 (2006).

[12] J. Wilson, W. Kulisch (Eds.), *Diamond thin films*, Akademie Verlag, Berlin (1996).

[13] O. Goudouri, S. Stavrev, X. Chatzistavrou, N. Kantiranis, T. Zorba, P. Koidis, K. Paraskevopoulos, Proc. XXII Panhellenic Confer. Solid State Phys. Mater. Sci., 2006, 24-27 Sept. 2006, Patra, Greece, p.1-4.

[14] O. Mykhaylyk, Y. Solonin, D. Batchelder, R. Brydson, *J. Appl. Phys.* 97, 074302 (2005).

[15] M. Baidakova, A. Vul, *J. Phys. D: Appl. Phys.* 40, 6300 (2007).

[16] E. Ekimov, E. Gromnitskaya, S. Gierlotka, W. Lojkowski, B. Palosz, B, A. Swiderska-Sroda, J. Kozubowski, A. Naletov, *J. Mater. Sci. Lett.* 21, 1699 (2002).

[17] S. Catledge, M. Fries, Y. Vohra, Nanostructured surface modifications for biomedical implants, in: H.S. Nalwa (Ed.), *Encyclopedia of Nanoscience and Nanotechnology*, Volume 7, American Scientific Publishers (2003).

[18] S. Stavrev, S. Lazarov, Kh. Stoev, L. Markov L, V. Ivanov, *US patent* No 5353708, 1994; *BG patent* No 49267A, 1991.

[19] S. Stavrev, J. Karadjov, L. Markov, Shock-wave methods for synthesis and modification of extrahard and new carbon materials with a nanometer particle size, in: A. Proykova (Ed.), *Meetings in Physics at University of Sofia*, Volume 2, Heron Press, Sofia (2001).

[20] V. Khabashesku, J. Margrave, E. Barrera, *Diamond Relat. Mater.* 14, 859 (2005).

[21] K. Kilpadi, P-L. Chang, S. Bellis, *J. Biomed. Mater. Res. A* 57, 258 (2001).

[22] J. Xie, M. Baumann, L. McCabe, *J. Biomed. Mater. Res. A* 71, 108 (2004).

[23] L. Baxter, V. Frauchiger, M. Textor, I ap. Gwynn, R. Richards, *Eur. Cells Mater.* 4, 1 (2002).

[24] L. Pramatarova, E. Pecheva, Modified Inorganic Surfaces as a Model for Hydroxyapatite Growth, in: *Materials Science Foundations*, Volume 26, Trans Tech Publications, Switzerland (2006).

[25] T. Kokubo, *Mater. Sci. Engin. C* 25, 97 (2005).

[26] E. Pecheva, P. Montgomery, D. Montaner, L. Pramatarova, *Langmuir* 23, 3912 (2007).

[27] M. Pourbaix, *Atlas of Electrochemical Equilibria in Aqueous Solution*, National Association of Chemical Engineers, Houston, TX (1974).

[28] J. Elliott, *Structure and chemistry of the apatites and other calcium orthophosphates*, Elsevier Science, Amsterdam (1994).

[29] R.Z. LeGeros, *Clin. Orthop. Relat. Res.* 395, 81 (2002).

[30] L. Socrates, *Infrared Characteristic Group Frequencies*, Wiley, New York (1980).

[31] E. Pecheva, L. Pramatarova, T. Hikov, D. Fingarova, Y. Tanaka, H. Sakamoto, H. Doi, Y. Tsutsumi, T. Hanawa, *Surf. Interf. Anal.* 42, 475 (2010).

[32] S. Leeuwenburgh, J. Wolke, J. Schoonman, J. Jansen, *Biomaterials* 25, 641 (2004).

[33] X. Wang, Y. Li, J. Wei, K. de Groot, *Biomaterials* 23, 4787 (2002).

[34] C. Wagner, W. Riggs, L. Davis, J. Moulder, G. Muilenberg, *Handbook of X-ray photoelectron spectroscopy*, Perkin-Elmer Corporation, Physical Electronics Division, Eden Prairie, Minn. (1979).

[35] K. Anselme, *Biomaterials* 21, 667 (2000).

[36] R.G. Lebaron, K.A. Athanasiou, *Tissue Eng.* 6, 85 (2000).

[37] C. Reyes, T. Petrie, A. García, *J. Cell Physiol.* 217, 450 (2008).

[38] I.D. Campbell, *Biochem. Soc. Trans.* 36, 263 (2008).

[39] G. Altankov, Th. Groth, *J. Biomater. Sci. Polym. Edn.* 8, 299 (1997).

[40] Th. Groth., G. Altankov, In: Frontiers in Biomedical Polymer Applications, R.M. Ottenbrite, I. Sunamoto (Eds.), Technomics Publisher Inc., Lancaster-Basel (1998).

[41] Th. Groth, G. Altankov, In: New Biomedical Materials P. Harris, D. Chapman (Eds.), IOS Press (1998).

[42] C. Barrias, M. Cristina, L. Martins, G. Almeida-Porada, M. Barbosa, P. Granja, *Biomaterials* 30, 307 (2009).

[43] T. Velling, J. Risteli, K. Wennerberg, D. Mosher, S. Johansson, *J. Biol. Chem.* 277, 37377 (2002).

In: Electrodeposition: Properties, Processes and Applications ISBN: 978-1-61470-826-1
Editor: Udit Surya Mohanty

Chapter 12

ELECTRODEPOSITION OF $CuInSe_2$ THIN FILMS

M. H. Valdés* and M. Vázquez

División Corrosión, INTEMA, Facultad de Ingeniería, Universidad Nacional de Mar del Plata, Juan B. Justo FDQ Mar del Plata, Argentina

ABSTRACT

Copper indium chalcogenides are materials under permanent investigation due to their potential use for photovoltaic applications, given that they show a high absorption coefficient, coupled to a band gap energy value that matches that of the solar radiation. Many techniques have been reported for preparation of $CuInSe_2$ (CISe) as a thin film.

Among them, electrodeposition presents many advantages: it is cost-effective, mainly because it does not require high vacuum, it may be used to copy intricate geometries while being applied even on top of flexible substrates and it can be scale-up into industrial production.

$CuInSe_2$ (CISe) thin films have been prepared by electrodeposition from a single bath on top of conductive glass as well as on conductive glass coated with thin layers of TiO_2 and In_2S_3. Potentiostatic and pulsed electrodeposition produced films with different properties. The electrodeposition conditions were decided after carrying out cyclic voltammetry in acidic electrolytes. CISe precursors films need to be subjected to different thermal treatments to enhance the crystallinity of the deposit. The presence of sulfur vapor during annealing leads to substantial changes in the composition of the chalcogenide.

The crystallinity, morphology and stoichiometry of the annealed films are characterized by XRD, microRaman spectroscopy and SEM, coupled with Energy dispersive scanning (EDS). Etching the films in KCN solution is a key step, enabling a final adjustment in the stoichiometry. Photoelectrochemical measurements are employed to obtain semiconductor properties of the CISe films and to elucidate the influence of the post-deposition treatments.

* E-mail address: mvaldes@fi.mdp.edu.ar, Tel +54 223 481 6600, Fax +54 223 481 0046.

INTRODUCTION

For the last few decades chalcopyrites such as $CuInSe_2$ (CISe) and $CuInS_2$ (CIS), have attracted permanent attention due their excellent properties for optoelectronic applications. For instance, their high absorbing coefficients make them promising materials for terrestrial solar cells, and for this kind of application, low cost and large area processing are crucial [1-3]. Among the many techniques used to prepare CISe and CIS as a thin film, electrodeposition presents many advantages: it is cost-effective, it does not require high vacuum, it can copy intricate geometries and it can be applied even on top of diverse types of substrates [4].

However, there are still some problems associated to electrodeposition. The stability of the electrodeposition bath over long periods of time can be improved using complexing agents [3, 5]. The presence of secondary phases, such a $CuSe_x$, is another well-known disadvantage which can be solved by immersion in KCN solutions [6]. Finally, the crystallinity of the films needs to be improved by thermally treating the samples. During the annealing stage, an oxygen-free atmosphere avoids the formation of oxides within the film [7]. But due to the low vapor pressure of Se, films annealed in vacuum or inert atmospheres may present high levels of Se vacancies [8]. For this reason, various Se sources (many of them toxic compounds) have to be present in the annealing chamber to adjust the final stoichiometry of the CISe [3, 9]. Although these are efficient solutions, they impose serious environmental issues and increase costs [10].

In the present chapter results will be shown for $CuInSe_2$ (CISe) thin films preparation and characterization. The samples were prepared by electrodeposition from a single bath on top of conductive glass as well as on conductive glass coated with TiO_2 and In_2S_3. Films produced using potentiostatic and pulsed electrodeposition will be shown to present different properties. In this context, the variables associated with the electrodeposition process and thermal annealing will be explored with emphasis to the composition and optical properties of the films.

The experimental conditions for the present study are described below:

$CuInSe_2$ (CISe) was electrodeposited from a single bath. The primary substrate was glass coated with fluorine-doped tin oxide (TCO). The substrate was degreased by immersion in an ultrasonic bath with ethanol, 3 times during 15 minutes.

For some experiments, a dense TiO_2 layer (d-TiO_2) was deposited by spray pyrolysis deposition (SPD) on top of the TCO [11-12]. The precursor solution contained 2.4 mL titanium IV isopropoxide (TTIP) as a titanium source, 3.6 mL acetylacetone (AcAc) as a stabilizer and 54 mL ethanol as solvent. The solution was atomized by a pneumatic spray system with oxygen as carrier. The films were deposited by pulsing solution while the substrate temperature was held at 350°C. Each pulse consisted of 1 minute of continuous spray and 1 minute of delay. On top of this compact TiO_2 layer, a nanocrystalline TiO_2 coating (nc-TiO_2) was applied by doctor blading when indicated. The particles have 25 nm as average diameter. For the doctor blade deposition, a viscous paste of TiO_2 was prepared using commercial TiO_2 (Degussa AG, Germany). P25 and P9 were used, which present different particle size. In order to break the aggregates into separate particles, the powder (6 g) was grinded in a porcelain mortar with a small amount of water (2 mL) containing acetylacetone

(0.2 mL) to prevent re-aggregation of the particles. After the powder had been dispersed, it was diluted by slow addition of water (8 mL). Finally, 0.1 mL Triton X-100 was added to facilitate the spreading of the colloid on the substrate. The substrate was covered on two parallel edges with adhesive tape to control the thickness of the nanoporous TiO_2 film. The colloid was distributed by means of a glass rod sliding over the tape-covered edges. Once dry, the sample was treated for 6 hours at 450 ºC in air in order to obtain a highly crystalline electron-conducting matrix. The resulting film thickness was 2 μm, which could be varied by changing the colloid concentration or the adhesive tape thickness.A three-electrode cell, with a Pt mesh as counter electrode and a saturated calomel electrode (SCE) as reference, was used. The geometrical area of the working electrode was 1.13 cm^2. The cell has a quartz window conveniently placed to perform optical characterizations. The electrodeposition was undertaken employing a PGP 201 Voltalab® potentiostat/galvanostat. When CISe was electrodeposited on bare TCO the precursor electrolyte consisted of an aqueous solution containing 2.5 mmol L^{-1} $CuCl_2$, 10 mmol L^{-1} $InCl_3$ and 5 mmol L^{-1} SeO_2 with 0.2 mol L^{-1} KCl as supporting electrolyte. The pH was adjusted between 2-2.5 with a stock HCl solution. The solution was purged with nitrogen to remove the oxygen before starting the electrodeposition.

Potentiostatic electrodeposition (PSED) was carried out holding the potential at a given value between -0.7 and -0.9 V_{SCE} during a certain time. Times ranged between 400 and 3600 s. After completing the electrodeposition the samples were rinsed with distilled water and dried in air.When carrying out cyclic pulsed electrodeposition (CPED), each pulse was defined by a highly negative potential (E_1) and a slightly negative potential (E_2). E_1 varied between -0.7 and -0.9 V_{SCE} and E_2 was fixed at -0.1 V_{SCE}. One cycle consisted of E_1 and then E_2, each applied during 10 s each. 20 to 180 cycles were used in order to produce CISe films with different thicknesses.When CISe was electrodeposited on top of TCO/TiO_2 substrates, the electrolytic bath consisted of an aqueous solution containing 2.5 10^{-3} mol L^{-1} $CuCl_2$, 1 10^{-2} mol L^{-1} $InCl_3$ and 5 10^{-3} mol L^{-1} SeO_2. Also, 0.2 mol L^{-1} KCl was used as a supporting electrolyte and the pH was adjusted between 2-2.5 using HCl. Potentials ranging from -0.8 to -1.0 V_{SCE} were explored. The deposition time was 5 to 60 min and again, the solution was purged with nitrogen and the samples were rinsed with distilled water and dried in air.

The thickness of the electrodeposited films (T) can be calculated using Faraday's law: T= (Q M) / (n F A ρ), where M is the formula weight and the other symbols have their usual meanings. Q results from integrating the current profile over the total deposition time. According to reaction (1), 13 electrons per mol are transferred (n):

$$Cu^{+2} + In^{+3} + 2\ SeO_3^{-2} + 12\ H^+ + 13e^- \rightarrow CuInSe_2 + 6\ H_2O \quad (1)$$

CISe molar mass and density are taken as M = 336.28 g/mol and ρ = 5.77 g cm^{-3} [13].

To improve the crystallinity of the CISe films, the films were annealed in a three-temperature zone thermal reactor, which is vacuum pumped and flushed with argon to remove the oxygen inside the quartz tube. Most of the $CuInSe_2$ samples were annealed in argon atmosphere at 500ºC during 30 minutes [14]. These samples will be referred as "annealed CISe". A group of samples underwent a sulfurization process. This step changes the composition of the film from $CuInSe_2$ to $CuInSe_{2-x}(S)_x$ (with x close to 2), modifying also the optoelectronic properties of the films. First, the samples were placed in the cold zone (T_{amb}) of the tube. Then the second zone was heated to 500ºC and in the third zone the sulfur powder was allowed to reach 300ºC (melting starts at around 200-250ºC). At this point, the samples were moved to the second zone where the sulfurization process started as soon as the vaporized sulfur carried by the argon flux got in contact with the samples. After 30 minutes the samples were let to cool down by placing them back in the "cold zone". The samples receiving this treatment will be labeled as CISeS.

When indicated, CISe films were chemically etched by a 2 or 5 min immersion in0.5 mol/L KCN. Etching can be done before or after the annealing the CISe films. When as-deposited films were etched in KCN, partial detachment of the film occurred already after 1 minute. On the other hand, when annealed CISe films were etched, detachment was no longer observed. Therefore, all etching experiments described in this chapter were performed on annealed CIS films.The crystalline structure of the films was analyzed by X-ray diffraction using a Philips 3020 goniometer with PW 3710 controller, using Cu-α radiation and Ni filter at 40 KV and 30mA. The samples were scanned between 10º and 70º with a step size of 0.01º. Some of the films were analyzed in grazing incidence configuration (GXRD) using a PANalytical X'Pert PRO diffraction system employing Cu-α radiation at 40 KV and 40 mA. The crystallographic data for each phase were taken from the literature [15].Raman spectroscopy measurements were performed using an Invia Reflex confocal Raman microprobe using a 50x objective. Excitation was provided with the 514 nm emission line of an Ar^+ laser. The power of the laser was reduced to 10% to prevent damage by heating. In this configuration the laser power on the sample was 0.2 mW, as measured with a silicon photodiode (Coherent Inc.). Raman micro-mapping was performed by recording 29 spectra along the diagonal of 20 x 20 μm squares.

To evaluate the morphology of the samples, a scanning electron microscope JEOL JSM-6460LV coupled with an EDAX Genesis XM4 - Sys 60 Multichannel Analyzer for X-ray Energy Dispersive Spectroscopy (EDS) was used. A Shimadzu UV-160A spectrophotometer was employed to register absorption spectra in the wavelength range 350-1100 nm, so as to evaluate the band gap energy (E_g) by extrapolation.Photocurrent and photopotential were evaluated forming a semiconductor/electrolyte junction by immersion of the FTO/CISe film in 0.5 mol L^{-1} KCl solution. The light source is a 150 W Xe lamp, chopped using an electronic shutter (Uniblitz model T132). A filter removed the ultraviolet radiation.

12.1. Cyclic Voltammetric Studies

Cyclic voltammograms were recorded in order to investigate the redox reactions of each of the various components of the electrolytic bath. The voltammograms were recorded at a scan rate of 0.01 V s^{-1}. The potential was scanned first in the negative direction. 0.2 mol L^{-1} KCl is added as supporting electrolyte in order to avoid the effect of having different ionic strengths. Bare TCO and TCO coated with a sprayed film of TiO_2 were used as substrate and working electrode.

Figure 12.1. shows the voltammograms recorded in three different solutions containing $1.0x10^{-2}$ mol L^{-1} $InCl_3$, $2.5x10^{-3}$ mol L^{-1} $CuCl_2$ and $5.0x10^{-3}$ mol L^{-1} SeO_2, using TCO and TCO/TiO_2 as working electrodes. The voltammogram produced by bare TCO is also shown, for the sake of comparison. As can be seen, the currents are lower and some peaks are displaced in potential when the layer of TiO_2 is present, probably due to the more resistive nature of this substrate, as compared to bare TCO. Oxidation and reduction currents are higher in the case of the In^{3+} solution because the concentration of indium ions used was higher. In the case of In^{3+}, one reduction and one oxidation peak can be observed (reaction 1), as has been described before by other authors using SCN^- as complexing agent [16]. Chlorides and thiocianates are well-known complexing agents for indium ions. The potential of the peaks is very similar with both substrates, confirming that conductive glass coated with TiO_2 is a suitable substrate.

$$In^{3+} (aq) + 3\ e^- \leftrightarrows In\ (s) \qquad (1)$$

In the case of the Cu^{2+} solution, one of the reduction peaks in the voltammogram probably corresponds to the formation of Cu^+ (reaction 2a) and the other to metallic copper formation (reaction 2b).

$$Cu^{2+} (aq) + e^- \leftrightarrows Cu^+ (aq) \qquad (2a)$$

$$Cu^+ (aq) + e^- \leftrightarrows Cu\ (s) \qquad (2b)$$

In the case of $HSeO_3^-$, the first cathodic peak probably corresponds to the reduction to elemental Se (reaction 3a) and the second to further reduction to H_2Se (reaction 3b).

$$HSeO_3^-(aq) + 5H^+ (aq) + 4\ e^- \leftrightarrows Se(s) + 3`H_2O \qquad (3a)$$

$$Se(s) + 2\ H+ (aq) + 2\ e^- \leftrightarrows H_2Se\ (aq) \qquad (3b)$$

Comparing the three voltammograms shown in Figure 12.1.b, 12.1.c. and 12.1.d. the following reduction order can be established: $HSeO_3^-$ reduces first, then Cu^{2+} and finally In^{3+}. All three reductions take place between -0.6 and -1.2 V on both substrates. Figure 12.2 illustrates the first, third and fifth cycles of a cyclic voltammogram carried out on TCO/TiO_2 and using a solution containing a combination of all three ions together: $1.0x10^{-2}$ mol L^{-1} $InCl_3$, $2.5x10^{-3}$ mol L^{-1} $CuCl_2$ and $5.0x10^{-3}$ mol L^{-1} SeO_2. The first negative scan is very similar than that for $HSeO_3^-$. Apparently Se forms first, then inducing $Cu_{2-x}Se$ formation and finally incorporating In^{3+} into $CuInSe_2$, as suggested before [17-18]. As the number of cycles increases, the maximum negative current is registered between -0.8 and -1.2 V, which ends up being the potential range chosen to perform the electrodeposition of CISe. The presence of a thin layer of TiO_2 placed on top of the TCO does not present a substantial influence in the chosen interval.

12.2. CISE ELECTRODEPOSITION STUDIES

12.2.1. CISe Potentiostatic Electrodeposition on TCO and Characterisation Studies

Figure 12.3. presents GXRD diffractograms of as-deposited CISe films deposited at fixed potential on TCO substrates. The TCO diffraction pattern is included as blank. Unfortunately, the main diffraction plane of CISe (112) appears superimposed with that of SnO_2, which is the main component of the conductive glass surface film. This seriously complicates the analysis of the material cristallinity.The CISe films in Figure 12.3. do not seem to show any characteristic diffraction peak corresponding to the CISe chalcopyrite phase. This is an indication of the low crystallinity, typical of as-deposited samples of this material. Other peaks could be related to indium compounds revealing an excess of indium that was confirmed by EDS. The presence of In_2Se_3 [19-20] and CuIn alloys [21-22] in XRD measurements has been reported before.

Figure 12.4. displays scanning electron microscopy (SEM) images of as-deposited samples, where the effect of deposition time is compared. The films grown at constant potential were very porous, presenting a cauliflower–like structure frequently found in electrodeposited CISe films [23-25]. This type of morphology became more characteristic as the time increased, which is most likely associated to a deposition processes controlled by mass transport. Using less negative potentials more compact structures can be obtained [26].

The thickness of the films can be calculated using Faraday's law. SEM cross-sections resulted in good agreement with the thickness values given in Table 12.1. However, the values calculated using Faraday's law should be taken as an approximation, since the formula weight and density vary with the composition and the efficiency of electrodeposition rarely reaches 100%.

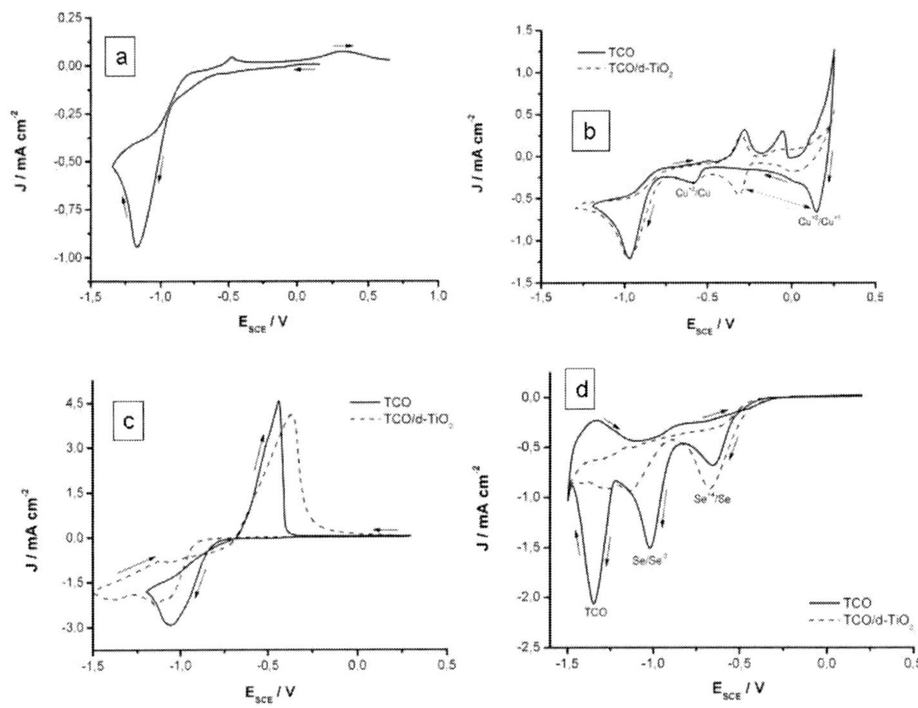

Figure 12.1. Cyclic voltammogram of (a) bare TCO electrode in 0.2 mol L^{-1} KCl supporting electrolyte; (b) TCO and TCO/d-TiO_2 electrodes in 2.5×10^{-3} mol L^{-1} $CuCl_2$; (c) TCO and TCO/d-TiO_2 electrodes in 10^{-2} mol $L^{-1}InCl_3$; (d) TCO and TCO/d-TiO_2 electrodes in 5×10^{-3} mol L^{-1} SeO_2. The sweep rate is $10mVs^{-1}$.

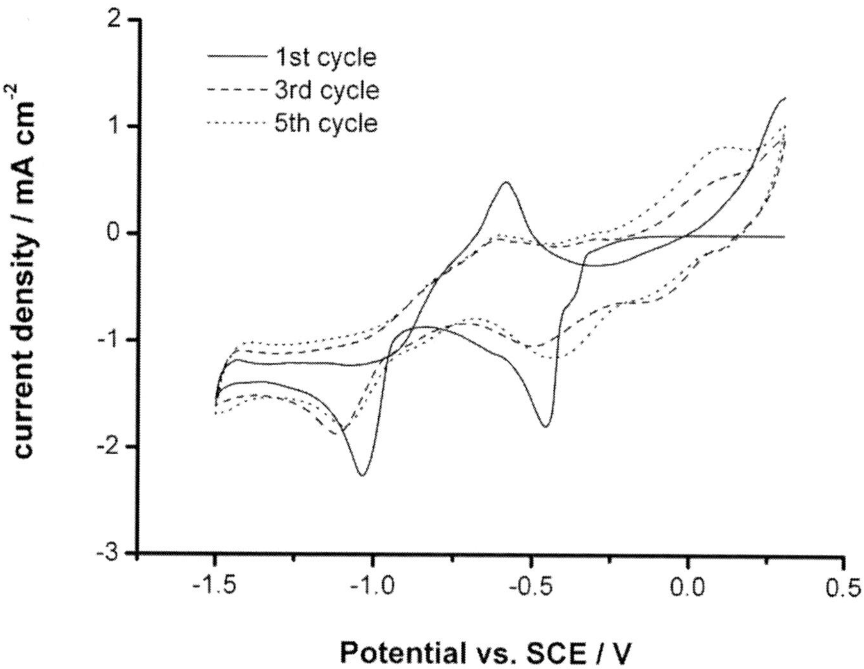

Figure 12.2. Cyclic voltammogram of a TCO/TiO_2 coated electrode in the precursor electrolyte containing 2.5×10^{-3} mol L^{-1} $CuCl_2$, $10-^{2}$ mol L^{-1} $InCl_3$, 5×10^{-3} mol L^{-1} SeO_2 . KCl 0.2 mol L^{-1} was used as supporting electrolyte (pH = 2.5). The sweep rate is $10mVs^{-1}$. Taken with permission from [14].

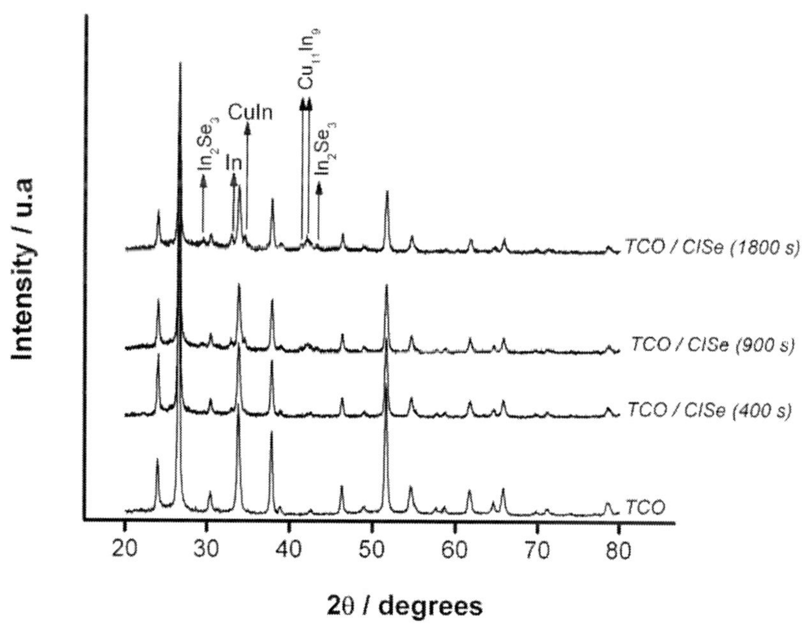

Figure 12.3. Grazing incidence X-rays diffractrograms (GXRD) of "as deposited" CISe films prepared using different deposition times. The diffraction pattern of the TCO substrate is included as a reference. CISe films were deposited using at -0.9 V_{SCE}.

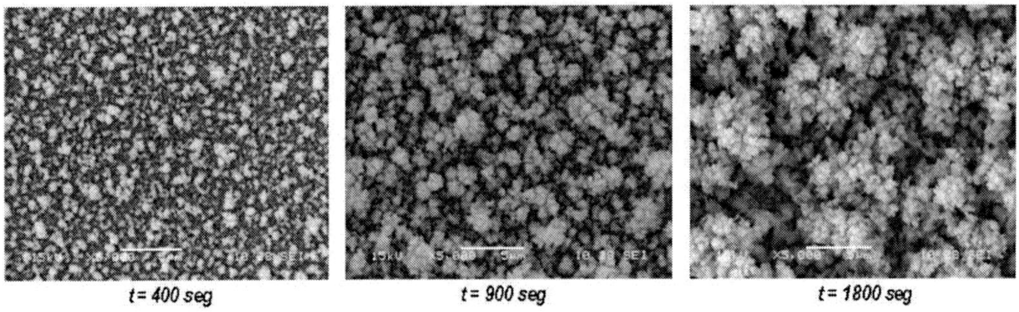

Figure 12.4. SEM images of CISe films deposited using at -0.9 V_{SCE} on TCO substrate at different growing times. The scale bar represents 5 μm and the magnification is 5000x.

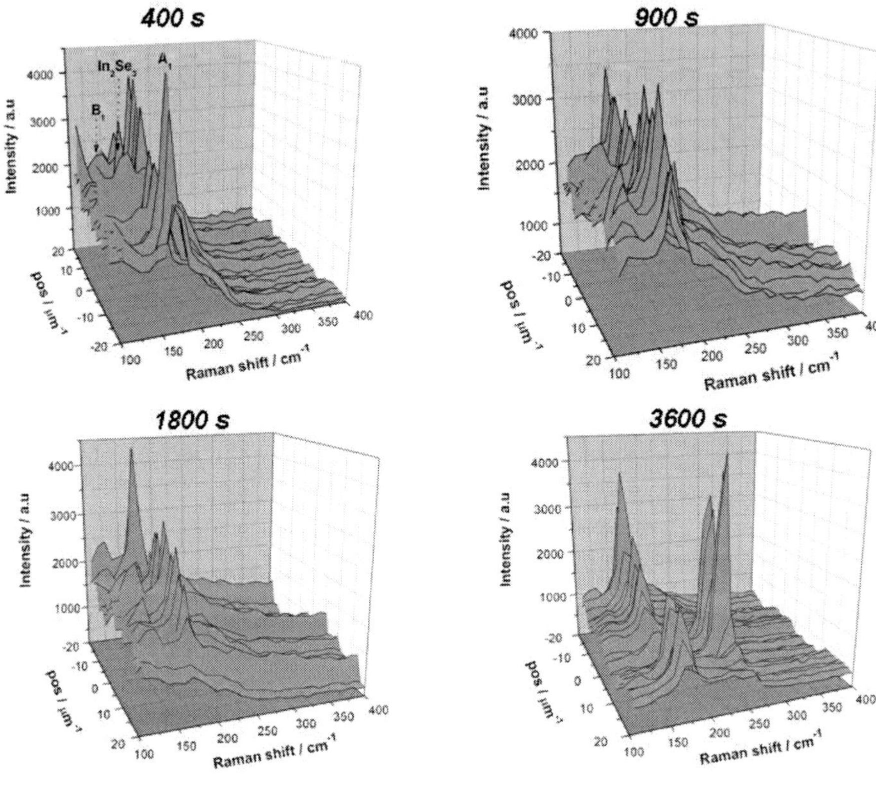

Figure 12.5. Raman micro-mapping of "as deposited" CISe films prepared by applying E= -0.9 V_{SCE} for different times.

Table 12.2. presents the composition of as-deposited samples, as determined by EDS. As can be seen, the as-deposited sample prepared at -0.9 V_{SCE} is highly rich in indium. This could be associated to the composition of the electrolytic bath, to the excessively negative potential, or both. It should be taken into account that when using conducting substrates, such a Mo-coated glass, the potential used for potentiostatic deposition is rarely more negative than -0.8 V_{SCE} [2, 27]. XRD diffractograms confirm this result. In the case of the sample electrodeposited at -0.7 V_{SCE}, the global composition seems to be well-adjusted to that of CISe. Nevertheless, it should be kept in mind that this analysis does not identify the different phases and for this reason, it has to be complemented with other surface techniques.

To evaluate the compositional homogeneity of the samples, Raman micro-mapping was performed by recording 29 spectra along the diagonal of 20 x 20 μm regions. Figure 12.5. displays Raman mapping for CISe films prepared at -0.9 V_{SCE} at different times. In every spectrum CISe is the main phase present although the signal at 173 cm^{-1} is not evenly distributed. At lower deposition times (400 s) the indium excess, previously seen, is also evident in the Raman maps with signals at 144 cm^{-1} which can be assigned to γ-In_2Se_3 [28-29]. Also present in these maps is the B_1 vibration mode produced by Cu-In bonds. This might also be related to an In-rich composition of the film. At longer times (900 and 1800 s), the signals from γ-In_2Se_3 are absent in the spectra. As it was shown before in relation to EDS analysis, the composition of the film deposited during 3600 s is close to stoichiometric CISe. However, the Raman map shows that binary Cu_xSe has a strong presence in the film, as indicated by the intense signals around 260 cm^{-1}.The band gap values (Eg) of semiconductor films were determined from transmittance spectra by plotting and extrapolating $(-\ln(T).h\nu)^2$ versus ($h\nu$), and are shown in Figure 12.6. The calculated Eg presents average values close to 1.0 eV, in good agreement with values reported for CISe films obtained and by fixed potential deposition in different conditions [11, 30]. It is worth pointing out that no significant difference in the band gap energy value could be attributed to the use of various potentials or to the annealing stage.

Table 12.1. Electrodeposition of CISe at constant potential on TCO. Average thicknesses calculated using Faraday's law

time/s	Potential / V_{SCE}	Thickness / μm
400	-0.9	0.230
900	-0.9	0.532
1800	-0.7	0.711
1800	-0.8	0.814
1800	-0.9	1.017

Table 12.2. Chemical composition of different as-deposited CISe films

Potential	time / s	Cu	In	Se	Cu/In	Se/(Cu+In)
- 0,7 V_{SCE}	1800	23,53	23,48	52,99	1	1,13
- 0,9 V_{SCE}	400	19,94	29,95	50,11	0,66	1
- 0,9 V_{SCE}	900	21,19	42,96	35,85	0,49	0,56
- 0,9 V_{SCE}	1800	21,39	48,95	29,66	0,44	0,42
- 0,9 V_{SCE}	3600	26,06	24,56	49,38	1,06	0,98

12.2.2. Cise Pulsed Electrodeposition on TCO and Characterization Studies

The potential profile and the response in current for both potentiostatic (PSED) and cyclic pulsed electrodeposition CPED are compared in Figure 12. 7a and 12.7b respectively. Figure 12.8. presents GXRD diffractograms of as-deposited CISe films deposited by applying a varying amount of potential cycles on TCO substrates. The TCO diffraction pattern is included as blank. Using 20 cycles produced an extremely thin CISe film, so that only diffraction peaks from the TCO can be identified. As commented above, the main diffraction plane of CISe (112) appears superimposed with peaks from the conductive glass surface film. As the number of cycles increases, particularly with 90 cycles, new diffraction peaks are registered.

There is a wide peak around 44° which can either be attributed to $Cu_{1.8}Se$ or also to a small diffraction from the characteristic trigonal plane splitting (204/220) of the $CuInSe_2$.

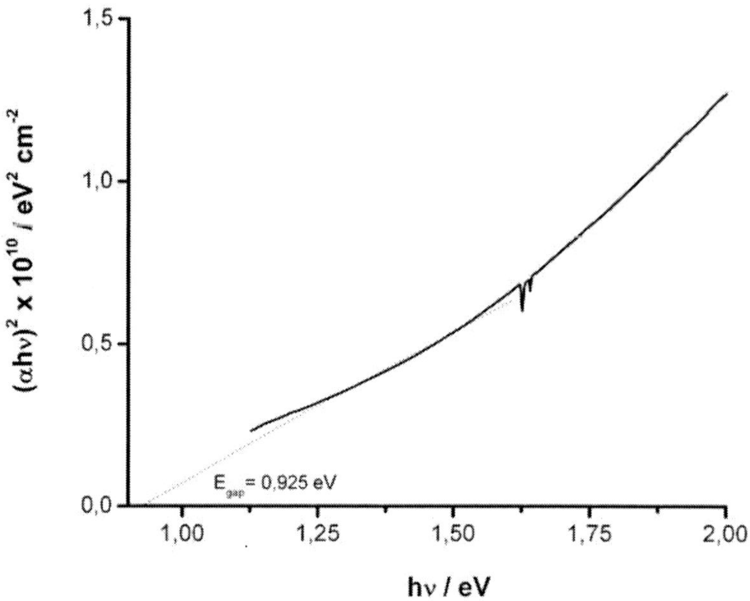

Figure 12.6. Plot of $(\alpha h\nu)^2$ vs. $(h\nu)$, normally employed for band gap energy (Eg) calculation. The CISe film was deposited at -0.9 V_{SCE} for 1800 s.

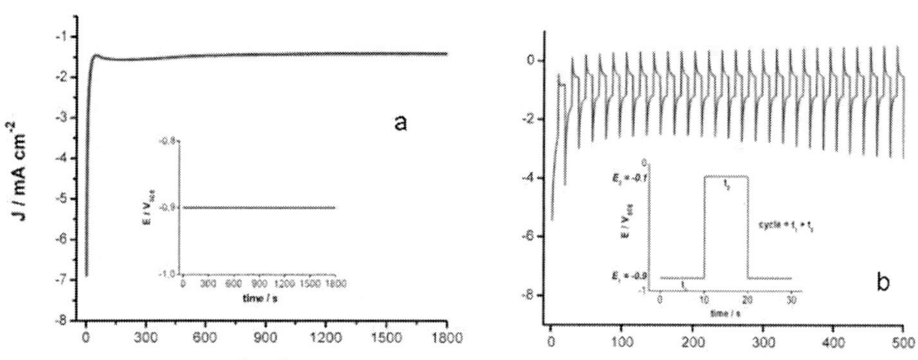

Figure 12.7. Comparison between the potential – time imposed and the current – time response obtained during the deposition of CISe films for (a) potentiostatic electrodeposition (PSED) and (b) cyclic pulse electrodeposition (CPED).

Figure 12.9 displays scanning electron microscopy (SEM) images of as-deposited samples, where the effect of varying amount of cycles is compared. The films prepared by pulsed electrodeposition exhibited a more compact and homogenous coverage with submicron-size grains. Table 12.3 presents the thickness of the films, calculated using

Faraday's law. SEM cross-sections resulted in good agreement with the thickness values given in Table 12.3.

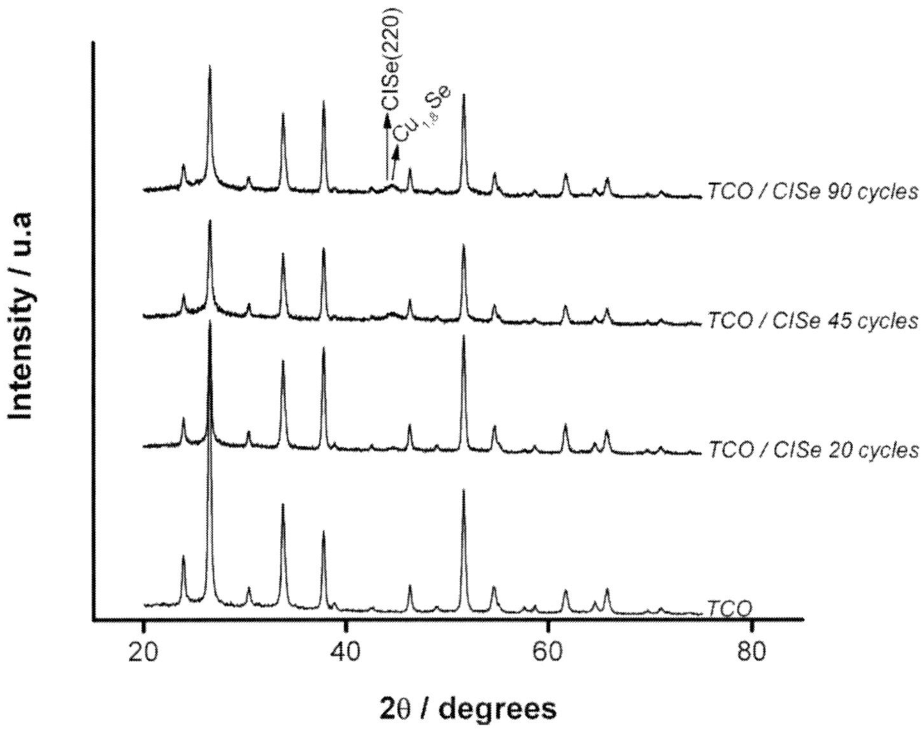

Figure 12.8. GXRD diffractograms of "as-deposited" CISe films deposited on TCO substrates varying the number of cycles. The TCO diffraction pattern is included as blank.

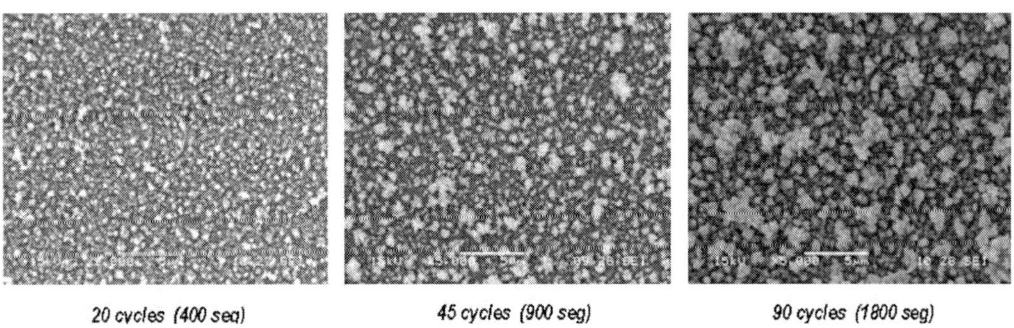

Figure 12.9. SEM images of CISe films deposited on TCO substrate using different number of cycles. **The scale bar represents 5 μm and the magnification is 5000x.**

Table 12.4 presents the composition of as-deposited samples, as determined by EDS. The effect of annealing in Ar and etching in KCN solution is also shown. When using 90 cycles of pulsed electrodeposition, as-deposited samples are highly rich in selenium, in agreement with the XRD results. Others authors have found that an excess in selenium promotes the formation of Cu_XSe compunds [23-24]. However, as selenium is readily vaporized when the temperature increases, the annealing stage seems to improve the stoichiometry of the CISe layer, particularly in the film deposited at -0.9 V_{SCE}. In this case, etching seems to produce no further improvement in the average composition of the film. The evolution of the indium and selenium contents seemed to be regular with increase in the number of cycles. The amount of indium increases and that of selenium decreases, while the content in copper remains mostly constant. Low indium contents for 20 and 45 cycles could be related to slow deposition kinetics. It has been demonstrated that the formation of $CuInSe_2$ is preceded by the deposition of Cu_XSe, which then reacts with the indium ions present in the solution [31-32]. After 180 cycles the composition corresponds closely to the stoichiometric formula. The chemical composition of both potentiostatic and cycle pulsed potential electrodeposition films indicates that longer times favor the formation of nearly stoichiometric $CuInSe_2$. These results are complementary with those from Raman spectroscopy.

Table 12.3. Electrodeposition of CISe cycling the potential on TCO.Average thicknesses calculated using Faraday's law

Number of cycles	Potential/V_{SCE}	Thickness, μm
20	-0.9	0,139
40	-0.9	0,311
90	-0.9	0,705
180	-0.9	1,004
360	-0.9	2,14
90	-0.7	0.398
90	-0.8	0.570

Table 12. 4. Chemical composition of different CISe films

As deposited	cycles	Cu	In	Se	Cu/In	Se/(Cu+In)
- 0,7 V_{SCE}	90	21,32	17,41	61,27	1,22	1,58
- 0,9 V_{SCE}	20	22,09	15,86	62,05	1,39	1,635
- 0,9 V_{SCE}	45	24,21	17,65	58,14	1,37	1,39
- 0,9 V_{SCE}	90	20,74	21,04	58,22	0,99	1,39
- 0,9 V_{SCE}	180	27,06	25,16	47,7	1,07	0,91
Annealed Ar (500ºC – 30 min)						
-0,7 V_{SCE}	90	22,33	16,25	61,42	1,37	1,59
-0,8 V_{SCE}	90	23,2	20,1	56,8	1,15	1,31
-0,9 V_{SCE}	90	22,58	21,88	55,54	1,03	1,25
Annealed + KCN etching						
-0,9 V_{SCE} / 2 min	90	23,31	21,78	54,90	1,07	1,22
-0,9 V_{SCE} / 5 min	90	23,36	21,97	54,67	1,06	1,21

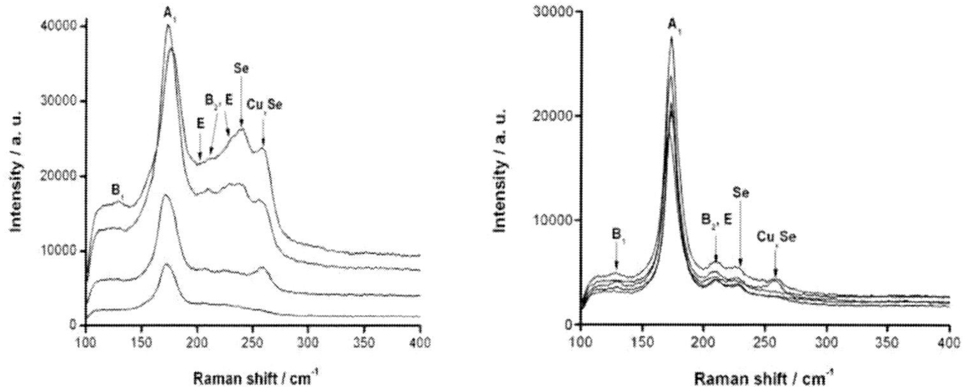

Figure 12.10. Comparison of Raman spectra of (a) as deposited and (b) annealed in argon (500 °C – 30 min) CISe films prepared using CPED. The Raman spectra were taken randomly in each sample using the same Raman set-up experimental conditions.

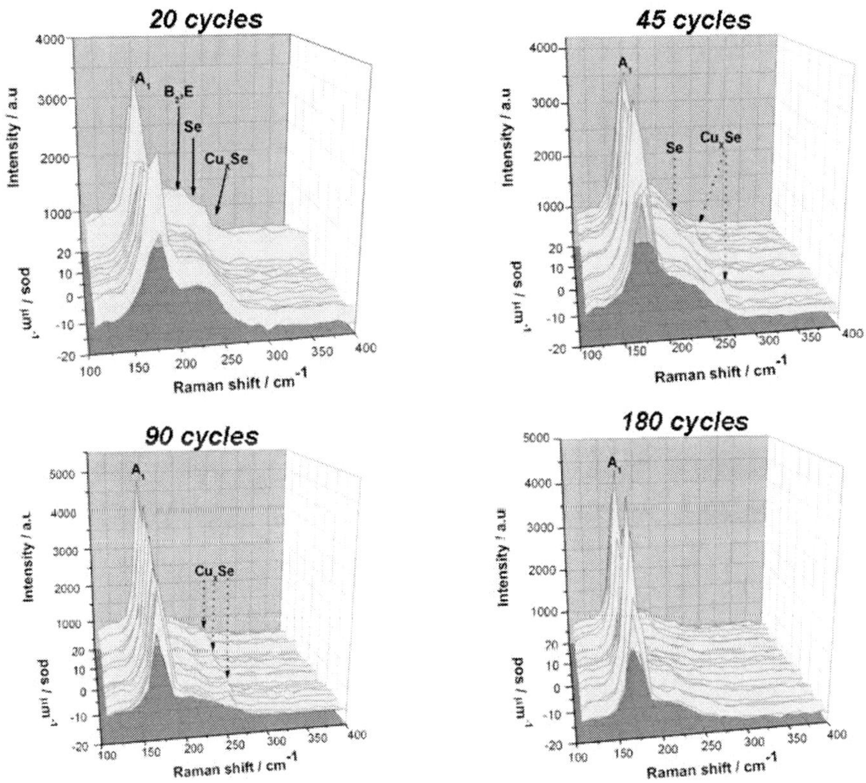

Figure 12.11. Raman micro-mapping of "as deposited" CISe films prepared using E_1= -0.9 V_{SCE}/ E_2= -0.1 V_{SCE} varying the number of cycles.

The influence of annealing is analyzed in Figure 12.10, where in Figure 12.10.a spectra recorded at random spots on as-deposited samples prepared using E_1 = -0.9 V_{SCE} and 90 cycles are compared with other group of spectra in Figure 12.10.b, recorded on samples annealed at 500ºC in argon. As-deposited CISe films (Figure 12.10.a) show broad and low intensity signals of the characteristic CISe vibration modes. Additional Raman signals, at 240 cm^{-1} and 260 cm^{-1}, can be attributed to trigonal selenium [33] and Cu_XSe compounds, respectively. The presence of Se in excess (also evidenced by EDS analysis) could be related to the higher current densities reached when cycling the potential, which favors the deposition of the element with the most noble electrochemical reduction potential. The fact that the presence of Se in excess may enhance the formation of Cu_XSe in the film has been discussed before [23-24]. After annealing, the intensity of the signal increases and the CISe A_1 Raman mode becomes sharper. Moreover, the relative intensities of the various signals in each spectrum are in agreement with Raman analysis performed on $CuInSe_2$ obtained using high vacuum and high temperatures techniques [34-35]. The remnant Se is partially vaporized during the annealing stage as evidenced by the lower intensity of the signal attributed to trigonal Se. Also, during the thermal treatment more CISe may be formed, with the consequent decrease in the amount of Se. Figure 12.11 displays Raman mapping for CISe films prepared using different numbers of cycles. It can be established that the composition is spatially homogeneous. All the spectra show that the A_1 mode is the most intense signal. Besides this mode, the film obtained using 20 cycles shows a significant contribution of Cu_xSe secondary phase, extended all over the sample. When using more cycles, the bands due to binary phases tend to show lower intensities and are randomly distributed throughout the mapped zone. This can be explained as due to the incorporation of indium by reaction with Cu_XSe, which is the mechanism leading to the formation of $CuInSe_2$. With 90 cycles the homogeneity of the film is clear, with the mode A_1 evenly distributed along the whole mapped area. Remnant Cu_XSe small signals are still present. When 180 cycles are used these signals seem to be absent and CISe is the only phase in the film. This is an important result taking into account the benefits of avoiding the use of toxic KCN as etchant to eliminate residual Cu_XSe.

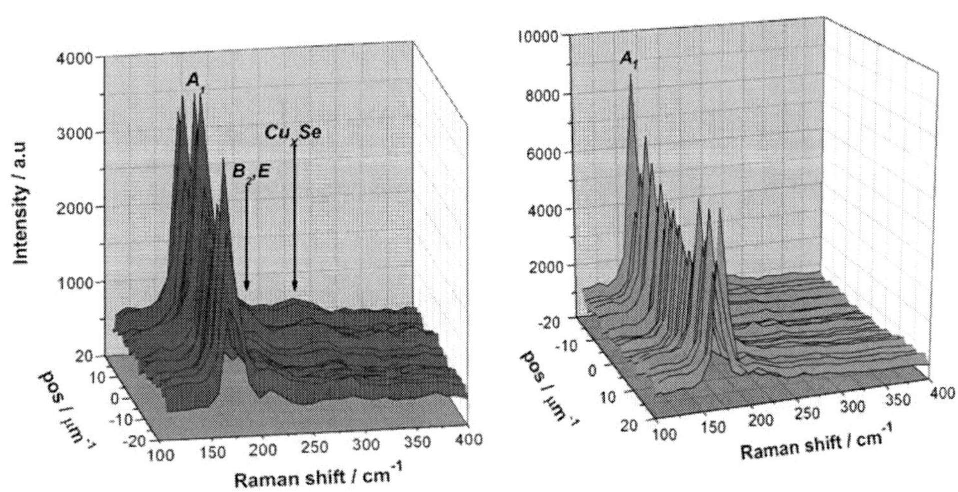

Figure 12.12. Raman maps of annealed CISe films (500°C – 30 min) comparing samples with (c) and without (d) 5 min etching in KCN. The CISe films were prepared by CPED in the same condition of figure 12.11.

Figures 12.12.a and 12.12b represent micro-mapping of CISe films, showing the effect of annealing in Ar and etching in KCN solution. Both films were annealed in argon flux (500 ºC – 30 minutes) but only one (Figure 12.12.b) was etched in KCN solution. As before, the composition of both films is clearly homogeneous. Both types of films present sharp signals of the A1 mode indicating that the film is mainly composed by $CuInSe_2$. However, small lumps appear in the annealed samples in the region around 260 cm^{-1} and these can be attributed to disperse Cu_XSe compounds that are still present in the film.

In the case of the etched film, the region close to 260 cm^{-1} is flatter which confirms the partial dissolution of the Cu_XSe secondary phase. The band gap energy values (Eg) of semiconductor films were determined from transmittance spectra and are shown in Figure 12.13 for samples prepared using 20 to 180 cycles. The Eg values improve with annealing, and vary between 0.9 and 1 eV, in close agreement with values reported for CISe films obtained by three-potential pulsed electrodeposition [36]. Again, no significant difference in the band gap energy value could be attributed to the use of various E_1 values or to the annealing stage.

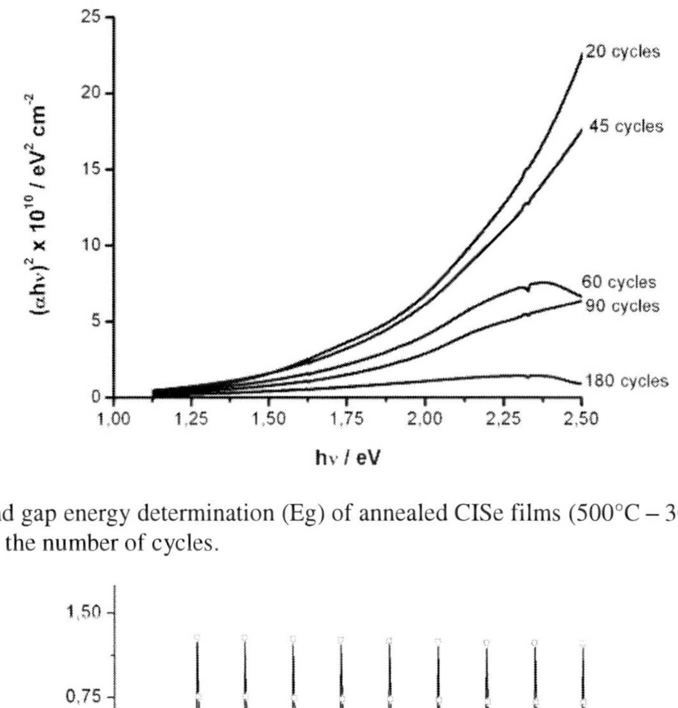

Figure 12.13. Band gap energy determination (Eg) of annealed CISe films (500°C – 30 min) prepared by CPED varying the number of cycles.

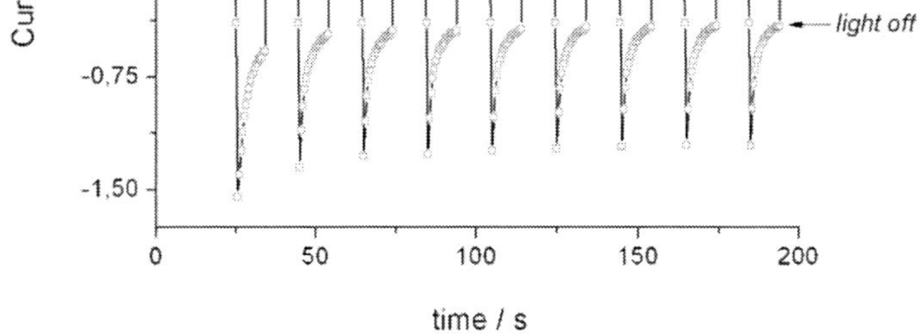

Figure 12.14. Photocurrent – time response of CISe film using chopped xenon light (10 s on / 10 s off). CISe films were prepared using 90 cycles, with E_1= -0.9 V_{SCE} and E_2= -0.1 V_{SCE}, then annealed in argon (500°C – 30 min) and finally etched in KCN for 5 minutes.

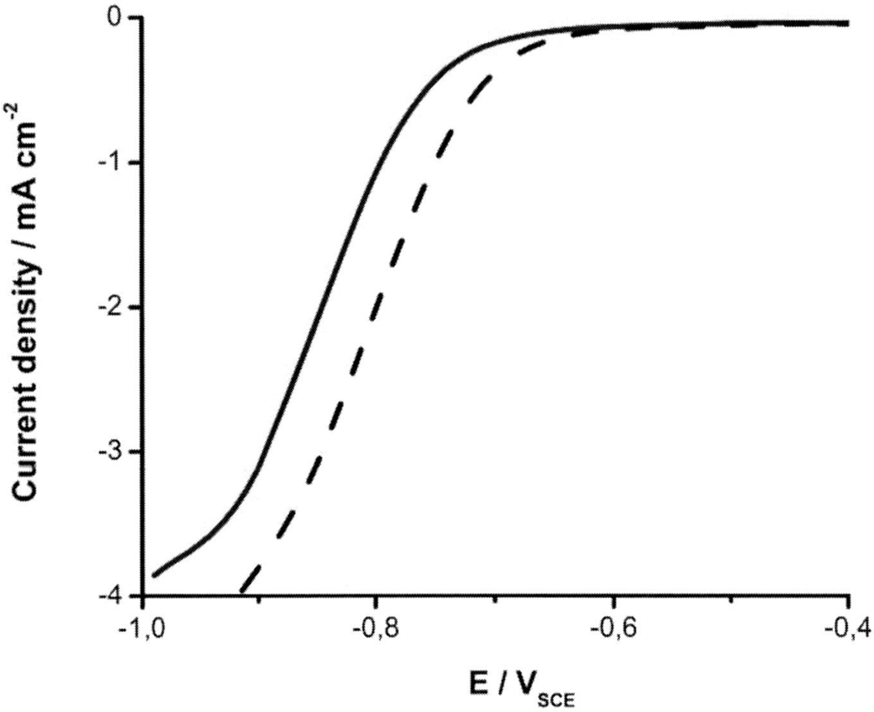

Figure 12.15. Potentiodynamic polarization curves of $CuInSe_2$ film in contact with KCl solution (pH~2) in dark (solid line) and under simulated solar light (dashed line). The CISe film was prepared using 90 cycles, with E_1= -0.9 V_{SCE} and E_2= -0.1 V_{SCE}, then annealed in argon (500°C – 30 min) and finally etched in KCN for 5 minutes.

It is well known that the semiconducting properties of chalcopyrites depend on the non-stoichiometry and are governed by the presence of intrinsic defects such as vacancies and interstitials [37-38]. Furthermore, the predominant type of conduction of CISe films can be either n-or p-type depending on preparation methods. Figure 12.14 shows the photocurrent registered at open circuit potential. The cathodic nature of the photocurrent confirms the p-type character of the film. The potentiodynamic polarization curves, in the dark and under white light irradiation, are shown in Figure 12.15. The dark current density (J_d) is relatively small indicating that CISe forms a blocking contact with the electrolyte. Upon illumination, the curve exhibits a cathodic photocurrent J_{ph}, in agreement with the p-type character expected for the Cu-rich side in the ternary Cu–In–Se phase diagram [39-40]. The onset of the photocurrent can be taken as a good estimation for the flat band potential (E_{FV}=-0.58 V_{SCE}) according to the relation that $J_{ph} \propto (V_{FB} - V)$ [41]. Other authors have found similar results in electrodeposited p-type CISe films [40].

12.2.3. CISe Potentiostatic Electrodeposition on TCO/TiO_2

In this section, the characterization of an inorganic heterojunction based on TiO_2 (n-type semiconductor) and electrodeposited $CuInSe_2$ (p-type semiconductor) will be presented. Some experiments involve a compact TiO_2 layer and others combine this with a nanostructured ~~of~~ TiO_2. This last case intends to explore a pn junction distributed in space, aiming a preparing a solar cell where the minority carrier diffusion length is reduced down to a few tens of nanometers [42]. The TiO_2 layers were prepared as described in section 2 and the CISe layer was electrodeposited at constant potential, using the potential (-0.8 to -1.0 V) and time (5 to 60 min) indicated in each case. Figure 12.16 shows typical GXRD patterns of CISe films electrodeposited on compact TiO_2, illustrating the effect of annealing in Ar. The films were deposited at -0.9 V during 30 min. Figure 12.16 presents the diffractogram of an as-deposited film, together with annealed CISe (Ar flush, 30 minutes at 500 ºC) and the TCO substrate. Without annealing, the main peaks belong to the substrate (SnO_2:F) and anatase from the sprayed TiO_2 films. Annealing significantly improves the crystallinity of the CISe films and the characteristic peaks of the chalcogenide structure, as (312), (204) and (400) can be clearly distinguished in the diffractograms. As discussed above, the mean peak (112) is superimposed with a signal from the substrate. Also, some other peaks appear that can be attributed to the formation of secondary phases such as $Cu_{2-x}Se$ (JCPD 06-0427) and CuO (JCPD 03-0884). Other authors have reported the presence of these secondary phases, detected by XRD analysis [8, 11, 43]. Figure 12.17 displays scanning electron microscopy (SEM) images of as-deposited and annealed samples. The films were porous, presenting a cauliflower–like structure frequently found in electrodeposited CISe films [23, 25].The cracks were most likely associated to the thermal treatment.

Table 12.5 displays the thickness of the films, **calculated using Faraday's law.** Random Raman spectra of CISe films as-deposited on compact TiO_2 at different deposition times are presented in Figure 12.18. All the spectra contain the A1 mode between 175-178 cm^{-1}, and an additional peak at 258 cm^{-1} [44]. The latter peak is related to the presence of a Cu_xSe secondary phase. At short deposition times (5 and 15 minutes) Cu_xSe seems to be dominant with respect to the CIS phase in the film but at longer deposition times (45 and mainly 60 minutes) the fraction of Cu_xSe phase becomes smaller.

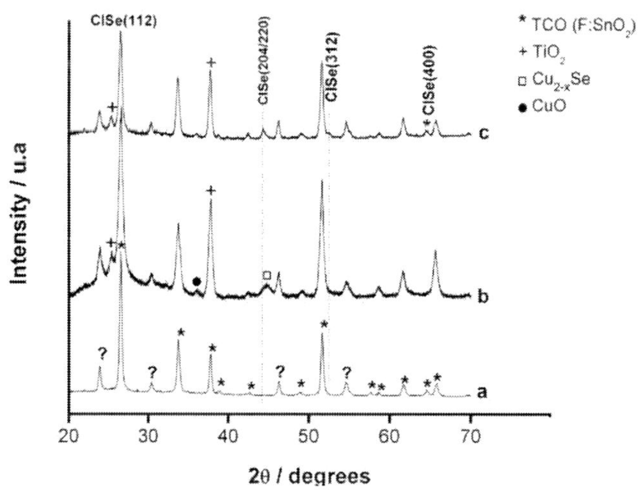

Figure 12.16. GXRD diffractograms of (a) TCO/d-TiO_2 substrate; (b) as-deposited CISe film; (c) CISe film annealed in argon (500°C – 30 min). CISe films were deposited imposing E= -0.9 V_{SCE} –during 30 min on top of TCO/d-TiO_2 coated substrates.

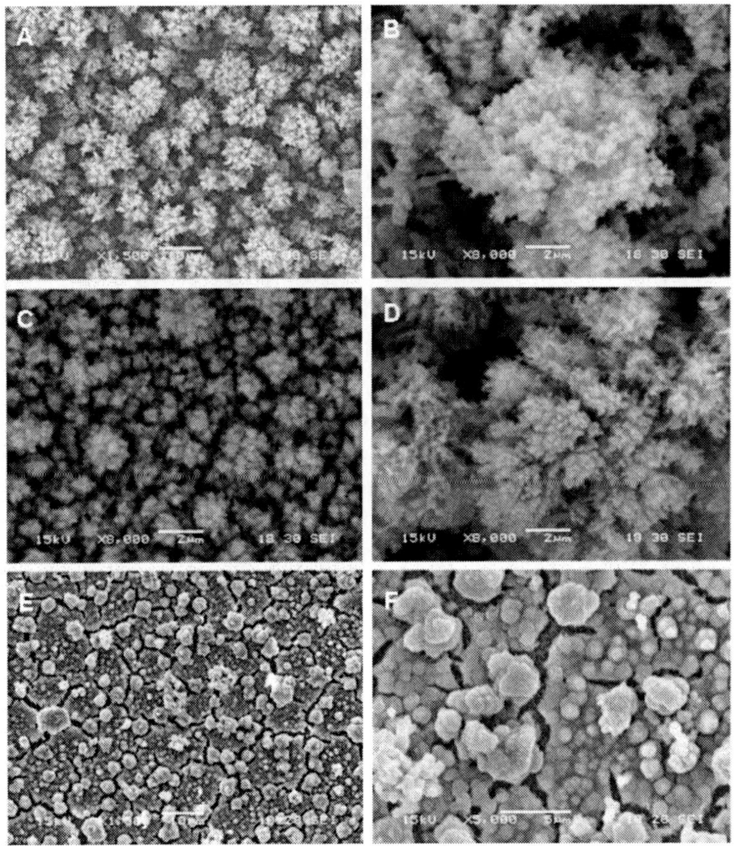

Figure 12.7. SEM photographs of (A- B) TCO/d-TiO_2/n-TiO_2(P9)/CISe (as deposited); (C-D) TCO/d-TiO_2/n- TiO_2(P25)/CISe (as deposited); (E-F) TCO/d-TiO_2/n-TiO_2(P25)/CISe (annealed in Ar, 400º - 30 min). CISe films were deposited applying a potential of -0.8 V_{SCE} during 1 h.

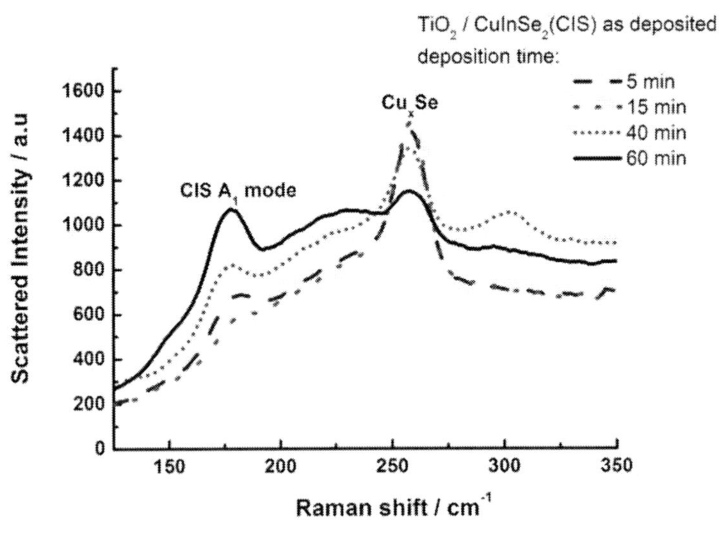

Figure 12.18. Raman spectra of as-deposited CISe films at different deposition times. CISe films were deposited on top of TCO/d-TiO_2 substrates, using a potential of -0.9 V_{SCE}. Taken with permission from [11].

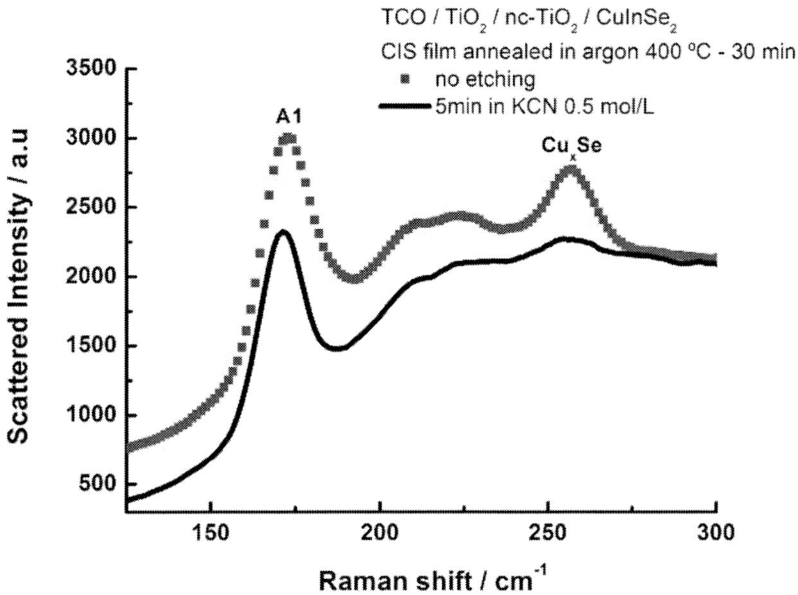

Figure 12.19. Raman spectra of annealed CISe films, before and after etching in KCN. Taken with permission from [11].

Table 12.5. Electrodeposition of CISe at constant potential on TCO/d-TiO_2. Average thicknesses calculated using Faraday's law

time/min	Potential / V_{SCE}	Thickness / μm
30	-0.7	0,83±0,08
30	-0.8	0,88±0,07
30	-0.9	0,99±0,1
30	-1.0	1,15±0,13
60	-0.8	1,98±0,15
60	-0.9	2,11±0,14
60	-1.0	2,55±0,2

This is in agreement with the findings described above confirming that the electrodeposition of CISe is induced by an initial deposition of $CuSe_x$ which later transforms into $CuInSe_2$. As we showed before, A well-known method to remove the Cu_xSe phase is etching the film in KCN.

Table 12.6. Band-gap energy values of CISe electrodeposited at constant potential on TCO/d-TiO_2, calculated from absorption spectra

Potential	as-deposited	Ar 300 °C, 30 min	Ar 400 °C, 30 min	Ar 500 °C, 30 min
-0,7 V_{SCE}	0,97	0,97	X	X
-0,8 V_{SCE}	0,99	0,99	1,03	0,93
-0,9 V_{SCE}	1,04	1,04	0,95	0,86

Table 12.7. Chemical composition of CISe electrodeposited at constant potential on TCO/d-TiO_2, calculated from energy dispersive spectroscopy

Deposition conditions	Treatment	KCN etching	EDS (%) Cu - In - Se - S
-1 VSCE / 1 hour	x	x	32,34 - 22,01 - 45,65
-1 VSCE / 1 hour	Argon	x	36,83 - 20,62 - 42,55
-1 VSCE / 1 hour	sulfurized	x	31,22 - 16,17 - 8.02 - 44,59
-1 VSCE / 1 hour	sulfurized	1 min	26,15 - 22,85 - 8,39 - 42,61
-1 VSCE / 1 hour	sulfurized	5 min	18,55 - 26,78 - 8,45 - 46,22

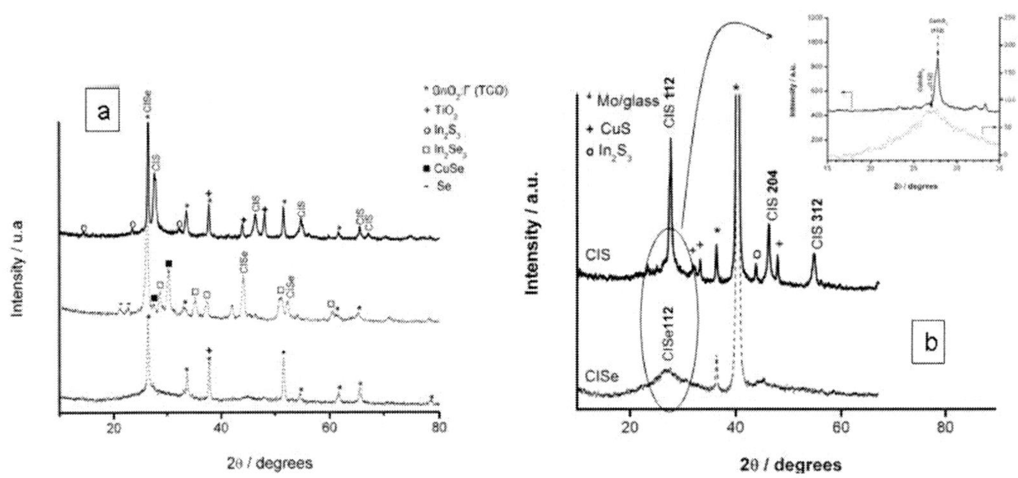

Figure 12.20. Figure 3. X-ray diffractogram of CuInSe electrodeposited on a TCO/TiO_2 /In_2S_3 electrode. In_2S_3 was prepared by spray pyrolysis as described elsewhere [14]. From bottom to top the diffractograms represent an as-deposited sample, a sample annealed in Ar (CISe) and a sample annealed in S (CISeS) and b) X-ray diffractogram of CuInSe electrodeposited on Mo/glass electrode showing the change of the crystalline structure before (bottom) and after (top) the sulfurization process. The inset in figure (b) shows a magnification of the (112) diffraction plane. Taken with permission from [14].

Figure 12.19 shows Raman spectra of an annealed CIS film before and after a KCN etch. The signal at 258 cm^{-1}, related to a Cu_XSe phase, is absent after etching and only CISe ~~and CIS~~ vibration modes were still present. Absorption spectra were recorded for as-deposited and annealed samples prepared using various potentials and deposition times. These data were employed to determine the band gap energy values (Eg) of the films. The calculated Eg values of as-deposited CIS or for samples annealed under various conditions vary between 0.86 and 1.04 eV and are shown in Table 12.6. Eg values of films produced at different potentials or for different deposition times were found to be always close to 1.0 eV, in agreement with the reported values for the electrodeposited CIS thin films [45]. The Eg values decrease with the increment in the temperature of annealing, probably due to some degree of selenium loss.

Figure 12.21. SEM images of CISe films deposited on a $TCO/TiO_2/In_2S_3$ substrate using different thermal treatments. A. As-deposited; B. Annealed in Ar 30 min a 500ºC; C. sulfurized at 500ºC for 30 min; D. sulfurized at 500ºC for 30 min and etched in 0.5 mol L^{-1} KCN solution for 1 min; E. cross section showing $TCO/d\text{-}TiO_2/In_2S_3/CuInSe_2$.

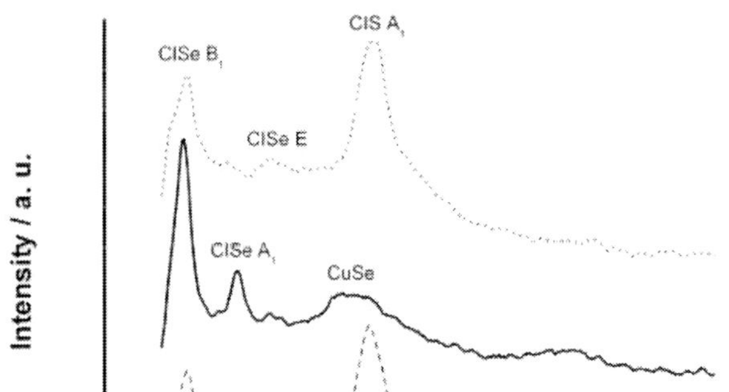

Figure 12.22. Raman spectra of CISe and CIS films. (Full line) CISe film annealed in argon at 500 ºC during 30 minutes, (dashed line) CISeS sulfurized at 500 ºC during 30 minutes and (dotted line) CISeS sulfurized at 500 ºC during 30 minutes and etched in 0.5 mol L^{-1} KCN for 5 minutes. Taken with permission from [14].

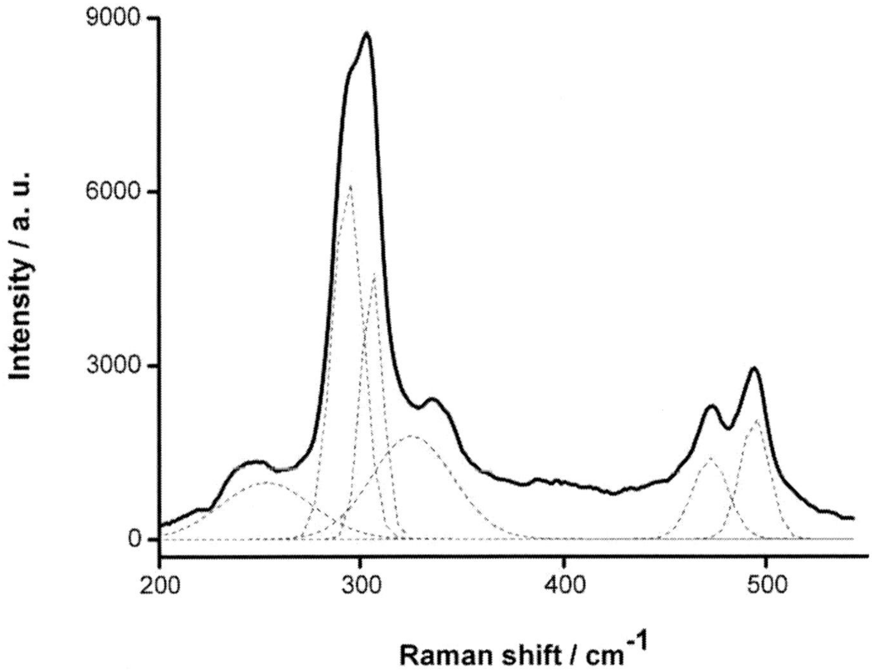

Figure 12.23. Raman spectrum of CISeS deposited at –1 V and sulfurized at 500 °C during 30 minutes. The doted lines represent the deconvolution of the spectrum. Taken with permission from [14].

12.3. Sulfurization of CISe Electrodeposited on TCO/TiO_2

As selenium can be vaporized during the thermal treatment, it is a common practice to carry out the annealing stage incorporating various Se sources to the gaseous phase. However, the high toxicity of gaseous Se imposes risks for mass production of CISe devices. To overcome this drawback, $CuInSe_2$ (CISe) can be transformed into $CuInSe_{2-x}(S)_x$ (CISeS) using S sources during the annealing step. An important side effect of this transformation is that the band gap energy of the film increases from 1.1 eV to 1.4 eV, better matching the band gap energy to the solar spectrum. This section comprises results describing the influence of a post sulfurization treatment with molten S on the transformation of electrodeposited $CuInSe_2$ into $CuInSe_{2-x}(S)_x$. Izquierdo-Roca et al. [43, 46] have thoroughly studied the annealing of electrodeposited CISe under sulfurizing conditions by means of Raman spectroscopy. They found that if the annealing time is longer than 20 minutes the chemical composition of the layer changes, resulting in a higher incorporation of sulfur into the chalcopyrite lattice. Figure 12.20 shows typical XRD patterns of electrodeposited CISe films, illustrating the effect of annealing in S atmosphere. The films were deposited at -1.0 V and using TCO/TiO_2/In_2S_3 substrates. As a result from the sulfurization treatment, $CuInS_2$ characteristic peaks appear in the diffractogram, while the peaks corresponding to $CuInSe_2$ can no longer be seen. This can be taken as proof of the replacement of Se atoms by S atoms. Furthermore the sulfurization seems to reduce the amount of secondary phases in the film in comparison to the annealing in argon. However, it is difficult to evaluate by XRD how much residual CISe remains after sulfurization. The main reflection of the (112) of the $CuInSe_2$ is located at 26.6° (JCPD N° 40-

1487), too close to that of the SnO_2 main peak at 26.48° (JCPD N° 77-0452). To avoid this superposition, Mo glass has been used as substrate. In this case, the main peaks of CISe and CIS are well separated from those of the substrate, so figure 12.20.B can be taken as evidence for the complete replacement of Se atoms by S atoms during the sulfurization step. No peaks from $CuInSe_2$ were identified after sulfurizing film. The inset in figure 12.20.B shows a magnification of the change in the (112) diffraction plane from the CISe structure to the CIS structure, after sulfurization.

SEM photographs of as-deposited samples, together with CISe, CISeS and KCN etched CISeS can be seen in Figure 12.21 to illustrate the processes of sulfurizing and etching electrodeposited CISe samples. Following the sequence from Figure 12.21.a to Figure 12.21.d it can be seen that after 5 minutes of etching the relative amount of secondary phases is greatly reduced. The cross-section image (Figure 12.21e) is included to illustrate average thicknesses of the TiO_2 film (0.3 μm), buffer layer (0.5 μm) and the CISeS absorbing layer (2.6 μm). The elemental composition obtained by EDX is presented in Table 12.7. The composition shows Cu in excess, probably associated to the presence of binary phases such as Cu_XSe o Cu_XS. Also, by comparing the S and Se amounts, it can be seen that Se is replaced by S almost completely. In parallel, the etching time has a strong influence on the global composition of the film. With 1 min, the film is rich in copper (Cu/In=1.14) while a 5 min immersion ends up with an indium rich film (Cu/In=0,69). The optimum time is most likely close to the shortest evaluated time.

Figure 12.22 shows the effect of the sulfurizing process, together with the effect of annealing, investigated by Raman spectroscopy. Although the intensity of the Raman signal depends on the surface morphology and the phases present within the CISe films, all the spectra contain the A_1 mode between 172-178 cm^{-1} related to the vibration of Se-Se bonds (anions) and a peak at 120 cm^{-1} of the Cu-In bonds (cations).

In the case of annealed CISe an additional peak at 263 cm^{-1} can be seen, related to the presence of Cu_XSe. After sulfurizing, the A_1 mode shifts to the A_1 frequency of the $CuInS_2$ at 292 cm^{-1} thereby leading to the transformation of CISe into CISeS. Moreover, a new peak appears in the region at 470 cm^{-1}, related to Cu_xS compounds. This can be taken as a proof that the Cu_xSe secondary phase is also transformed during the annealing in sulfur, changing its composition. After immersion in KCN 0.5 mol L^{-1} for 5 minutes, the spectrum (Figure 12. 22) shows the dissolution of the secondary phases, with only CISe and CIS vibration modes still present.

The Raman spectrum of CISeS is presented in more detail in Figure 12.23. Performing an in-depth analysis, characteristic modes of the $CuInS_2$ structure can be identified, such as the previously mentioned A_1 mode (292 cm^{-1}) and a signal at 470 cm^{-1} related to the presence of covallite (CuS) structure [47].

Additional modes are found: a low frequency signal at 240 cm^{-1} can be assigned to the E vibration mode, a signal at 305 cm^{-1} can be caused by the presence of Cu–Au ordering [48-49] and a signal at 337 cm^{-1} assigned to the contribution from the modes of the $CuIn_5S_8$ spinel-type structure. This type of structure is more likely to be present in an In-rich composition [47]. The high frequency signal at 495 cm^{-1} has been attributed before to the S-S vibration mode [50]. This confirms the replacement of Se atoms by S atoms in the lattice of the chalcopyrite phase.

CONCLUSION

$CuInSe_2$ (CISe) has been obtained using one-step electrodeposition, either fixing or cycling the potential in time.Different substrates such as tin oxide coated glass (TCO), and TCO covered with TiO_2 were explored. The effect of a single layer of compact TiO_2 or a duplex layer composed of dense and nanoporous TiO_2 was investigated. The incorporation of TiO_2 films to the substrate appears to influence the potential interval chosen to optimize the electro deposition and also controls the morphology of the films. In contrast, the nature of the substrate (conductor or semiconductor) does not seem to affect the formation mechanism of $CuInSe_2$. When the electrodeposition is carried out at constant potential, CISe seems to be the predominant phase. However, some binary compounds of the type Cu_XSe $(0 \leq x \leq 2)$ can be found, as well as some ordered vacancy compounds. The predominant morphology resembles a "cauliflower" similarly to results in other studies where various substrates were employed. A subsequent heat treatment at temperatures greater than 300 ° C is required to improve the crystallinity of the material. After annealing, the density of defects and decreases, but secondary phases such as Cu_XSe remain in the film and its crystallinity also increases after annealing.

The atmosphere during the heat treatment strongly influences the composition and properties of the resulting deposit:

- an inert atmosphere prevents oxidation of the material, while the high temperatures during annealing may favor the evaporation of Se from the film and to a lesser extent, that of In.
- the presence of sulfur in the heat treatment (sulfurizing) affects the film morphology producing a more homogeneous films, as compared to samples treated in argon.

It has been shown that a chemical attack (etching) in KCN solution is an appropriate method to remove both binary phases (Cu_XS and Cu_XSe). The results show that the time of attack appears to influence to a greater or lesser extent, depending on the deposition method used to obtain the chalcopyrite (potentiostatic or pulsed). The film morphology and the substrate coverage seem to improve by pulsing the potential. EDS and micro-Raman maps demonstrate that the composition is spatially homogeneous when the potential is pulsed.

Also, a lower proportion of secondary phases was present when films were prepared with potential pulses. The amount of residual elemental Se is greatly decreased during the thermal treatment while Cu_xSe can only be efficiently removed by etching in KCN solution. Photoelectrochemical tests confirm p-type conduction. These, together with absorption spectra enable the calculation of relevant semiconductor parameters such as the band gap energy and the flat band potential.

The optical characterization suggests that electrodeposition is suitable to prepare photovoltaic devices based on TiO_2 and CISe/CISeS.

ACKNOWLEDGMENTS

The authors acknowledge the financial support received from institutions in Argentina, in particular Consejo Nacional de Investigaciones Científicas y Técnicas (CONICET), Agencia Nacional de Promoción Científica y Tecnológica and Universidad Nacional de Mar del Plata (UNMdP). We are also thankful to Dr. Albert Goossens, the Technological University of Delft (TUDelft) and Senter/Novem from The Netherlands

REFERENCES

[1] Luque A, Hegedus S, *Handbook of Photovoltaic Science and Engineering*, John Wiley and Sons Ltd, West Sussex (2003).
[2] Lincot D, Guillemoles JF, Taunier S, Guimard D, Sicx-Kurdi J, Chaumont A, Roussel O, Ramdani O, Hubert C, Fauvarque JP, Bodereau N, Parissi L, Panheleux P, Fanouillere P, Naghavi N, Grand PP, Benfarah M, Mogensen P, Kerrec O, *Sol. Energy.* 77, 725 (2004).
[3] Kemell M, Ritala M, Leskelá M, Crit. Rev. *Solid State Mater. Sci.* 30,1 (2005)
[4] Pandey RK, Sahu SN, Chanda S, Handbook of Semiconductor Electrodeposition, Marcel Dekker, New York (1996).
[5] Araujo J, Ortíz R, López-Rivera A, Ortega JM, Montilla M, Alarcón D, *Journal of Solid State Electrochemistry.* 11, 407 (2007).
[6] Kemell M, Ritala M, Leskelä M, *J. Mater. Chem.* 11, 668 (2001).
[7] De Silva KTL, Priyantha WAA, Jayanetti JKDS, Chithrani BD, Siripala W, Blake K, Dharmadasa IM, *Thin Solid Films.* 382,158 (2001).
[8] Volobujeva O, Kois J, Traksmaa R, Muska K, Bereznev S, Grossberg M, Mellikov E, *Thin Solid Films.* 516, 7105 (2008).
[9] Hibberd CJ, Chassaing E, Liu W, Mitzi DB, Lincot D, Tiwari AN, Prog. in Photovolt.: *Res. and Appl.* 18, 434 (2010).
[10] Dhere NG, *Sol. Energy Mater. Sol. Cells*, 95, 277 (2011).
[11] Valdés M, Frontini MA, Vázquez M, Goossens A, *Appl. Surf. Sci..* 254, 303 (2007).
[12] Valdés M, Vázquez M, Goossens A, *Electrochim. Acta.* 54, 524 (2008).
[13] Weast RC, CRC Handbook of Chemistry and Physics, CRC Press, Inc, Boca Raton (1987).
[14] Valdés M, Goossens A, Vázquez M,. *Mat. Chem. and Phys.* 125 860 (2011).
[15] International Centre for Difraction Data (ICCD): Powder Difraction File Database. Newtown Square, *EEUU*, (1998).
[16] Kemell M, Ritala M, Saloniemi H, Leskelá M, Sajavaara T, Rauhala E. J, *Electrochem. Soc.* 147,1080 (2000).
[17] Ramdani O, Canava B, Chassaing E, Grand P-P, Lamirand M, Roussel O, A.Etcheberry, Guillemoles J-F, Lincot D, Kerrec O, ECS Trans.. 2, 369 (2007).
[18] Roussel O, Ramdani O, Chassaing E, Grand PP, Lamirand M, Etcheberry A, Kerrec O, Guillemoles JF, Lincot D, *J. Electrochem. Soc.* 155, D141 (2008).
[19] N. B. Chaure, J. Young, A. P. Samantilleke, I. M. Dharmadasa, *Solar Energy Materials and Solar Cells*, 81, 125 (2004).

[20] S. Jost, F. Hergert, R. Hock, J. Schulze, A. Kirbs, T. VoÃŸ, M. Purwins, M. Schmid, *Solar Energy Materials and Solar Cells*, 91, 636 (2007).
[21] J. Yang, Z. Jin, C. Li, W. Wang, Y. Chai, *Electrochemistry Communications*, 11, 711 (2009).
[22] K. Tang, U. Künecke, F. Oehlschläger, A. Hölzing, R. Schurr, R. Hock, P. W. Wellmann, *Solar Energy Materials and Solar Cells*, 94, 1875 (2010).
[23] Calixto, K. D. Dobson, B. E. McCandless, R. W. Birkmire, *Journal of the Electrochemical Society*, 153 G521 (2006).
[24] S. H. Kang, Y. K. Kim, D. S. Choi, Y. E. Sung, *Electrochimica Acta*, 51, 4433 (2006).
[25] T. J. Whang, M. T. Hsieh, Y. C. Kao, S. J. Lee, *Applied Surface Science*, 255, 4600 (2009).
[26] M. H. Valdés, M. Vázquez, *Electrochimica Acta*, 56, 6866 (2011)
[27] J. Kois, S. Bereznev, E. Mellikov, A. Öpik, *Thin Solid Films*, 511-512, 420 (2006).
[28] J. Weszka, P. Daniel, A. Burian, A. M. Burian, A. T. Nguyen, *Journal of Non-Crystalline Solids*, 265, 98 (2000).
[29] D. Y. Lyu, T. Y. Lin, T. W. Chang, S. M. Lan, T. N. Yang, C. C. Chiang, C. L. Chen, H. P. Chiang, *Journal of Alloys and Compounds*, 499, 104 (2010).
[30] P. J. Dale, A. P. Samantilleke, G. Zoppi, I. Forbes, L. M. Peter, *Journal of Physics D: Applied Physics*, 41, (2008).
[31] Ramdani O, Chassaing E, Canava B, Grand PP, Roussel O, Lamirand M, Rzepka E, Etcheberry A, Guillemoles JF, Lincot D, Kerrec O, *J. Electrochem. Soc.* 154, D383 (2007).
[32] Chassaing E, Ramdani O, Grand PP, Guillemoles JF, Lincot D, *Phys. Status Solidi* C, 5, 3445 (2008).
[33] Ramdani O, Guillemoles JF, Lincot D, Grand PP, Chassaing E, Kerrec O, Rzepka E, *Thin Solid Films*, 515, 5909 (2007)
[34] Rincón C, Ramírez FJ. *J. Appl. Phys*. 72, 4321 (1992).
[35] Tanino H, Maeda T, Fujikake H, Nakanishi H, Endo S, Irie T, Phys. Rev. B: Condens. *Matter Mater. Phys*. 45,13323 (1992).
[36] Palacios-Padrós A, Caballero-Briones F, Sanz F, *Electrochem. Comm*. 12, 1025 (2010).
[37] Noufi R, Axton R, Herrington C, Deb SK. *Applied Physics Letters*. 45, 668 (1984).
[38] Zhang SB, Wei S-H, Zunger A, *Phys. Rev. Lett*. 78, 4059 (1997).
[39] Edamura T, Muto J, *J. Mater. Sci.*: *Mater. Electron*. 5, 275 (1994).
[40] Djellal L, Bouguelia A, Kadi Hanifi M, Trari M, Sol. *Energy Mater. Sol. Cells*. 92, 594 (2008).
[41] Sato N, *Electrochemistry at Metal and Semiconductor Electrodes*, Elsevier, Amsterdam (1998).
[42] M. Nanu, J. Schoonman, A. Goossens, *Nano Letters*, 5, 1716 (2005).
[43] V. Izquierdo-Roca, J. Alvarez-Garcia, A. Pérez-Rodríguez, L. Calvo-Barrio, A. Romano-Rodríguez, J. R. Morante, O. Ramdani, V. Bermudez, P. P. Grand, L. Parissi, O. Kerrec, "Electrodeposited $CuIn(S,Se)_2$ films for low cost high efficiency solar cell applications: Microstructural analysis", Spanish Conference on Electron Devices, 31 January 2007, (Madrid) pp 146-149.
[44] J. H. Park, I. S. Yang, H. Y. Cho, *Applied Physics A Solids and Surfaces*, 58, 125 (1994).

[45] N. Stratieva, E. Tzvetkova, M. Ganchev, K. Kochev, I. Tomov, *Solar Energy Materials and Solar Cells*, 45, 87 (1997).

[46] V. Izquierdo-Roca, X. Fontané, L. Calvo-Barrio, A. Pérez-Rodríguez, J. R. Morante, J. Álvarez-García, F. Duault, L. Parissi, V. Bermúdez, *Thin Solid Films*, 517, 2264 (2009).

[47] E. Saucedo, V. Izquierdo-Roca, C. M. Ruiz, L. Parissi, C. Broussillou, P. P. Grand, J. S. Jaime-Ferrer, A. Pérez-Rodríguez, J. R. Morante, V. Bermúdez, *Thin Solid Films*, 517, 2268 (2009).

[48] C. Rincón, S. M. Wasim, G. Marín, J. M. Delgado, J. R. Huntzinger, A. Zwick, J. Galibert, *Applied Physics Letters*, 73, 441 (1998).

[49] J. Álvarez-García, A. Pérez-Rodríguez, B. Barcones, A. Romano-Rodríguez, J. R. Morante, A. Janotti, S. H. Wei, R. Scheer, *Applied Physics Letters*, 80, 562 (2002).

[50] V. Izquierdo-Roca, A. Pérez-Rodríguez, A. Romano-Rodríguez, J. R. Morante, J. Alvarez-García, L. Calvo-Barrio, V. Bermudez, P. P. Grand, O. Ramdani, L. Parissi, O. Kerrec, *Journal of Applied Physics*, 101, (2007).

In: Electrodeposition: Properties, Processes and Applications ISBN: 978-1-61470-826-1
Editor: Udit Surya Mohanty

Chapter 13

THE ELECTROCHEMISTRY OF TIN IN AN ADDITIVES-FREE ACID METHANESULFONATE ELECTROLYTE: VOLTAMMETRY, NUCLEATION, GROWTH AND MORPHOLOGY

Chee Tong John Low* and Frank. C. Walsh
Electrochemical Engineering Laboratory, Faculty of Engineering and the Environment, University of Southampton, Highfield, Southampton, United Kingdom

ABSTRACT

This chapter discusses the electrochemical deposition of tin on copper in an additives-free acid methanesulfonate electrolyte, in a controlled flow condition using a rotating disc electrode. The electrochemistry of Sn^{2+}/Sn are characterised by electrochemical measurements: cyclic voltammograms (at a static electrolyte), linear sweep voltammograms (at a controlled rotating flow condition) and electrodeposition at a constant electrode potential. Effects of fluid flow and concentration of stannous ion (Sn^{2+}) on the electrodeposition of tin are investigated. Surface microstructures of tin deposits are imaged using a scanning electron microscope. Randles-Sevčik, Levich and Koutécky-Levich equations are used to estimate the diffusion coefficient of Sn^{2+}. It was found that the electrodeposition of tin on copper occurred on a very fast kinetic in a diffusion controlled environment. The charge transfer, mixed control and mass transport controlled region were dependent on the electrolysis process parameters. The electrodeposition of tin on copper followed the expressions based on the Scharifker-Hills equations for a three dimensional instantaneous nucleation mechanisms; the growth of tin deposits followed a diffusion controlled process. Tin coatings showed a matte surface appearance and adherent to the copper substrate. A wide range of surface microstructures of tin deposits, starting from compact to dendritic deposits and porous morphology were readily electrodeposited and dependent on the applied electrode potential (or current ~ 5 to 50 mA cm^{-2}). It was found that the electrode potential window for the limiting mass transport controlled region became narrower and the hydrogen evolution on tin occurred at a more

*E-mail: C.T.J.Low@soton.ac.uk.

electro-positive potential at a higher concentration of stannous ion and an increasing rotating flow condition.

INTRODUCTION

Electrodeposited tin coatings on metallic surfaces are used for many functional applications, eg. to provide a low contact resistance surface, to protect against corrosion (ie. tin is protective on iron, steel, nickel, copper and their alloys), to facilitate soldering in electronics components, to provide anti-galling surface, non-seizure properties and novel electrode materials for lithium battery. A number of electrolytes are used to electrodeposit tin, eg. sulfuric acid, fluorosilicate, tetrafluoroboric acid, phenolsulfonic acid, etc, but the current trend is the adoption of acidic tin methanesulfonate electrolyte. This is evident from our recent research effort in our laboratory at Southampton and a series of research publications in the electrodeposition of tin, copper and their alloys from acidic methanesulfonate electrolytes [1-6].This electrolyte is being by used the industry since it provides a high metal salt solubility and solution stability. The rate of spontaneous oxidation of stannous ion (Sn^{2+}) to stannic ion (Sn^{4+}) is slower. The electrolyte conductivity of methanesulfonic acid is comparable to other commodity acids. It is also a less aggressive environment for materials processing, ease in materials handling and wastewater treatment [7, 8]. It is bio-degradable, an environmentally friendly solution and waste methanesulfonic acid can be readily recovered. The increasingly adoption of methanesulfonate acid is support by the price drop to a level, which is more economically feasible for large volume applications. The market for methanesulfonate acid continues to grow. Methanesulfonic acid has been demonstrated to electrodeposit various functional materials including energy conversion metal oxides [9], optically reflective metal oxides [10] and conducting polymers [11]. It has also been investigated for the anodizing of valve metals to form metal oxides nanotubes [12], electrowinning of metals [7], electropolishing and surface colouring of light weight metals [13, 14]. More recently, this electrolyte has been deployed for a number of modern electrochemical technologies including clean energy storage in flow batteries e.g. Zn-Ce system 15, 16], Pb-PbO_2 system [17-20] and electrochemical degradation of colour dyes for wastewater treatment [21, 22]. Almost all acidic tin methanesulfonate electrolyte contains two essential components, namely (a) an antioxidant to maintain electrolyte stability, ie. to prevent spontaneous oxidation of Sn^{2+} to Sn^{4+} and (b) a mixture of brighteners, grain refiners, stress relievers, surfactants, etc. to produce good quality coating and tailored surface appearance, e.g. compact and adherent to substrate, pore-free coating, matte or bright surface finish. Most of the reported investigations have been carried out in the presence of these additives. There is very little information on the electrodeposition of tin from additives-free acidic methanesulfonate electrolytes. The use of additives-free electrolytes is highly desirable for simple electrolyte maintenance and the avoidance of chemicals impurity in the deposits. Pure acidic methanesulfonate electrolyte is hoped to provide a high quality tin deposits which are important for tinplate, electronics applications and battery materials. This chapter investigates the electrodeposition of tin on copper from an additive-free acidic methanesulfonate electrolyte, at room temperature and under a controlled flow condition. This chapter reports the electrochemistry of tin and its electrochemical studies were investigated by cyclic voltammograms, linear sweep voltammograms, etc. The nucleation, growth mechanisms and surface microstructure of tin deposits are also studied in details. The influence of fluid flow and concentration of stannous ion on the electrodeposition of tin are investigated. Surface roughness of tin deposits is also included. This chapter also reports the effect of mass transport phenomenon and hydrogen evolution on tin.

The experimental conditions used in this study are described below:

Tin sulfate, $SnSO_4$ and methanesulfonate acid, CH_3SO_3H (analytical reagent grade, Sigma Aldrich UK) are used as-received. The electrolyte compositions contained 0.006 to 0.078 mol dm^{-3} $SnSO_4$ and 1.93 mol dm^{-3} CH_3SO_3H. The electrolyte volume was approximately 250 cm^3 and temperature of the electrolyte was maintained at 298 K. All the experiments were performed at a copper disc electrode (active surface area: 0.28 cm^2) in a three-electrode and divided two compartments glass cell, see Figure 13.1. The electrode potential was reported with respect to a reference electrode Ag|AgCl (3.5 mol dm^{-3} KCl). A platinum mesh (active dimension: 1 cm length × 1 cm width, 0.025 cm aperture) was used as the counter electrode. Al the electrochemical measurements were made with an EcoChemie Autolab (PGSTAT20) computer controlled potentiostat, Electrochemical Software (GPES) Version 4.5. Before each experiment, the copper disc electrode was wet polished using 600, 1200, and 4000 grade SiC paper then on LeCloth® polishing cloth using 3 μm Al_2O_3 and 0.05 μm SiO_2 particles. Cyclic voltammograms were recorded in a range of operating process parameters: (a) potential sweep rates: 4, 8, 16, 32, 64 and 128 mV s^{-1}, (b) electrode rotation speeds: 200, 400, 600, 800, 1200, 1600, 2400, 3200 and 4800 rpm and (c) stannous ion concentrations: 6, 10, 15, 23, 31, 38, 44, 55, 60, 70 and 78 × 10^{-3} mol dm^{-3}. Chrono-amperometric experiments were carried out at a constant electrode potential: – 0.440, – 0.444, – 0.446, – 0.448, – 0.452, – 0.454, – 0.456, – 0.458, – 0.460, – 0.462, – 0.470 V *vs.* Ag/AgCl for 10 seconds electrodeposition under a static electrode condition. For the surface microstructure analysis, tin was electrodeposited at a constant potential: – 0.65, – 0.75, – 0.85 and – 0.95 V vs. Ag|AgCl for 5 minutes electrodeposition under a rotating electrode at about 400 rpm.

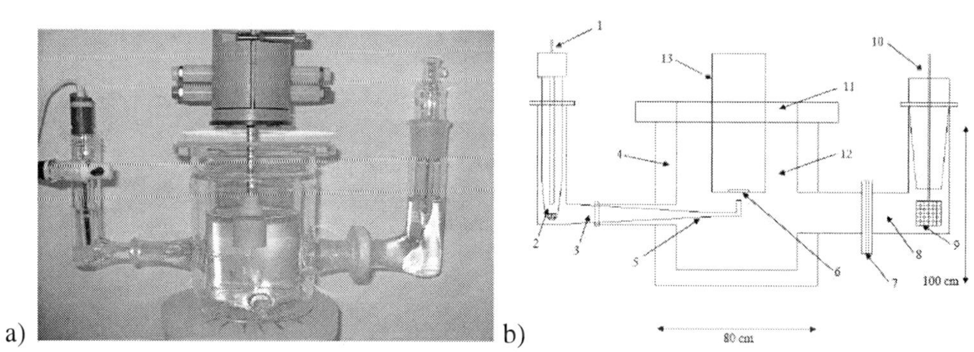

Figure 13.1. Three-electrode electrochemical cell for the investigation of tin electrodeposition. The setup consists of a Cu rotating disc electrode, a Pt/Ti counter electrode and a Ag|AgCl reference electrode. The electrolyte volume ~ 250 cm^3. (a) Actual setup. (b) Pictorial presentation. (1) Electrical connection on the reference electrode, (2) Ag|AgCl reference electrode, (3) Compartment for reference electrode, (4) Reservoir for water circulation, (5) Lugging capillary, (6) Disc working electrode, (7) Nafion® NF115/H^+ cation membrane, (8) Compartment for counter electrode, (9) Platinum mesh, (10) Electrical connection on the counter electrode, (11) PTFE top cover, (12) Compartment for electrolytic baths, (13) PTFE top support with electrical connection to the rotating motor and potentiostat/ galvanostat.

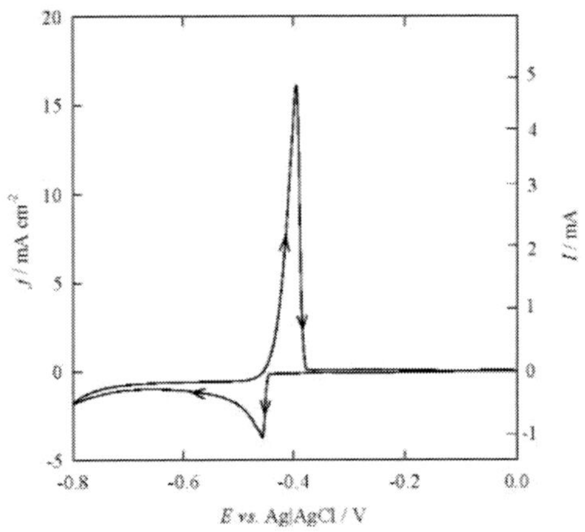

Figure 13. 2. Cyclic voltammogram of Sn^{2+}/Sn in an additives-free tin methanesulfonate acid electrolyte. The electrolyte contained 0.014 mol dm^{-3} Sn^{2+} in 12.5 % vol. CH_3SO_3H. A copper disc electrode was used (surface area *ca.* 0.28 cm^2). The electrode potential was swept from 0 to – 0.8 V vs. Ag|AgCl using 16 mV s^{-1} at 295 K electrolyte temperature and under a static electrolyte condition.

Surface profiles were obtained using a 5 μm diamond tip on Talysurf® instrument (Taylor Hobson Precision). A scanning electron microscope (XL30ESEM, Philips) was used to image the surface microstructure and the operating conditions were set at 15 kV accelerating voltage, 10 mm working distance and 2 minute collection time.

13.1. Cyclic Voltammogram of Sn^{2+}/Sn

Figure 13.2. shows the cyclic voltammogram recorded for an additives-free acidic tin methanesulfonate electrolyte on a copper disc electrode at 298 K and under a static solution. The electrolyte composition contained 0.006 mol dm^{-3} Sn^{2+} and 1.93 mol dm^{-3} CH_3SO_3H. As the electrode potential swept to a more electro-negative direction (using a potential sweep rate ~ 16 mV s^{-1}), a sharp increase in the cathodic current was observed, which is indicative of a very fast kinetic for the electrodeposition of tin on copper. A cathodic peak current (~ 4 mA cm^{-2}) was observed at approximately – 0.44 V *vs.* Ag|AgCl. This corresponds to a single step, two-electron electrochemical reduction reaction of stannous ion, Sn^{2+} to tin metal, Sn as follows:

$$Sn^{2+} + 2e^- \rightarrow Sn \quad (1)$$

After reaching the peak current, the cathodic current was levelled off to a lower value as a result of a diffusion controlled phenomenon on the surface of the copper disc electrode. However, at a more electro-negative potential, the cathodic current rose again which can be associated with hydrogen evolution on tin deposits as follows:

$$2H^+ + 2e^- \rightarrow H_2 \quad (2)$$

At the very electro-negative potentials, the electrodeposition of tin was accompanied by the hydrogen evolution reaction. The linearity rise in the cathodic current is therefore a result of both these reactions. On the reverse potential swept from – 0.8 V to 0 V *vs.* Ag|AgCl, a single anodic oxidation peak current (~ 16 mA cm^{-2}) was observed at approximately – 0.4 V *vs.* Ag|AgCl. This corresponds to a single step, two-electron electro-stripping of tin metal to stannous ion. The stripping efficiency was < 100 %. Both the forward and reverse potential curves were not overlapped; hence no significant overpotential was required for the electrodeposition of tin on copper.

13.2. ELECTRODEPOSITION OF TIN UNDER A CONTROLLED FLOW CONDITION

Figures 13.3.(a) and (b) show the linear sweep voltammograms for the electrodeposition of tin on copper under a controlled flow condition. As the electrode potential swept from 0 V to – 0.8 V *vs.* Ag|AgCl, all the curves showed an abrupt increased in the cathodic current as a result of a very fast kinetic electrodeposition of tin on copper. Figure 13.3.(a) shows that the onset electrode potential for the rise in the cathodic current, started at approximately – 0.44 V *vs.* Ag|AgCl for all electrode rotation speeds (from 200 rpm to 4800 rpm). Figure 13.3. (b) shows that the onset electrode potential gradually shifted to a more electro-positive direction at an increasing concentration of stannous ion (from 6 to, 78 × 10^{-3} mol dm^{-3} Sn^{2+}), which agreed with Nernstian expression. All the curves show that both the charge transfer and mixed controlled region are limited to only a narrow range of electrode potential window. For the mass transport controlled region, it was clearly defined by a limiting plateau.

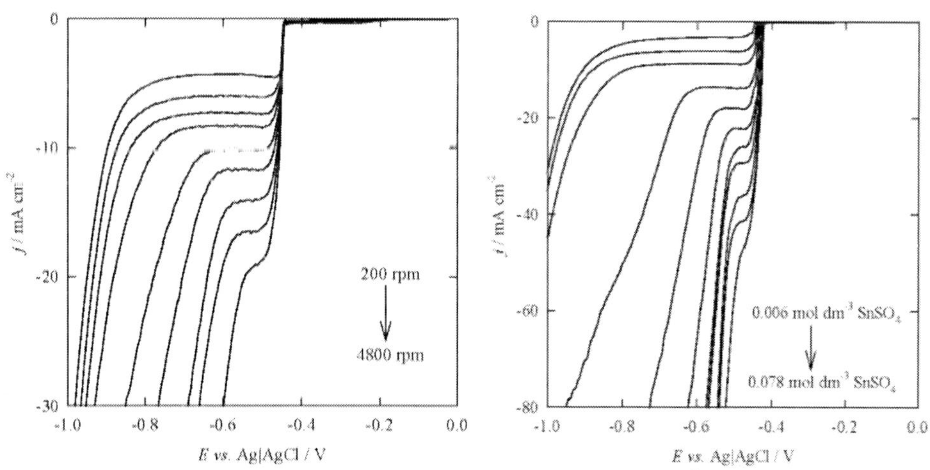

a b

Figure 13.3.Linear sweep voltammograms for the electrodeposition of tin on copper. (a) controlled rotating flow condition: 200, 400, 600, 800, 1200, 1600, 2400, 3200 and 4800 rpm. (b) concentration of stannous ion: 6, 10, 15, 23, 31, 38, 44, 55, 60, 70 and 78 × 10^{-3} mol dm^{-3}. Unless stated, the electrolyte composition was 0.014 mol dm^{-3} $SnSO_4$ and 12.5 % vol. CH_3SO_3H. A copper disc electrode was used (surface area *ca.* 0.28 cm^2). The electrode potential was swept from 0 to – 1.0 V vs. Ag|AgCl using 16 mV s^{-1} at 295 K electrolyte temperature and under 400 rpm rotating electrode flow condition.

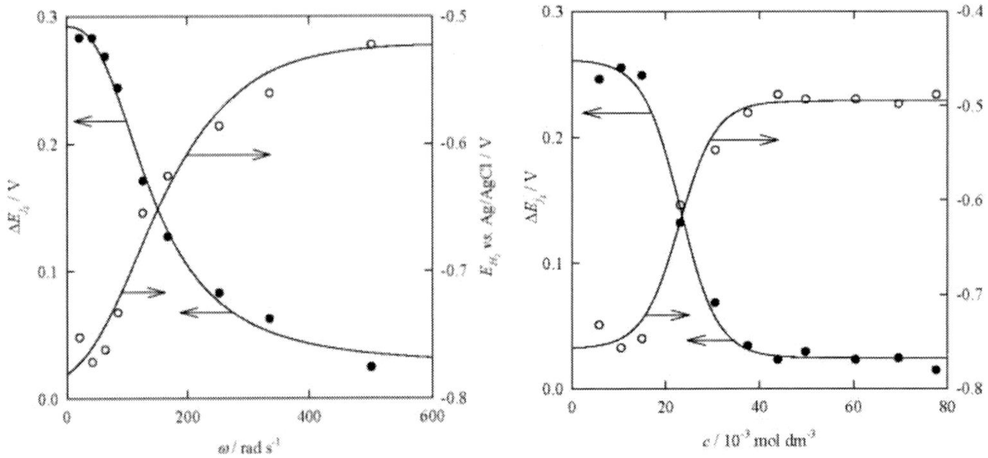

Figure 13.4. Effects of electrolysis process parameters on the potential window of the mass transport controlled region, ● and the electrode potential for hydrogen evolution, O. (a) controlled rotating flow condition. (b) concentration of stannous ion. Data points were taken from Figuire 3.

Figures 13.4.(a) and (b) show the relationships of the potential window for mass transport controlled and the electrode potential for hydrogen evolution as a function of electrode rotation speed and concentration of stannous ion.

The plots showed that the potential window for mass transport controlled became narrower at an increasing electrode rotation speed and a higher concentration of stannous ion. It was found that the hydrogen evolution occurred at a more electro-positive potential and at an increasing flow condition and a high concentration of stannous ion. The limiting current density at the mass transport controlled region was also found to increase at an increasing electrode rotation speed and a higher concentration of stannous ion. All these phenomena demonstrated the importance of controlled flow conditions for electrodeposition at a high current density.

A high value of limiting current density is therefore practical for high-speed electrodeposition and avoidance of roughened or powdery tin deposits. The electrochemistry of Sn^{2+}/Sn shows that the limiting current density can be readily raised by deploying a higher concentration of stannous ions. For example, an increase in the concentration of stannous ion from 6 to 60 × 10^{-3} mol dm^{-3} gave an approximately ten-fold increase in the current density of mixed control region. However, in the presence of a high concentration of stannous ion in the electrolyte, the spontaneous oxidation of stannous ions to stannic ions became more apparent; solid particles of sludge was visible at the bottom of the electrolyte tank after a storage time ~ 1 to 2 hours. The sludge deposits were a mixture of tin oxides or hydroxides.

13.2.1. Diffusion Coefficient of Sn^{2+} in an Additives-Free Acidic Methanesulfonate Electrolyte

Three different approaches were used to estimate the diffusion coefficient of stannous ion in an additives-free acidic methanesulfonate electrolyte. These approaches were applied using different experimental data recorded on the voltammograms on different regions. The equations are: Randles-Sevčik, Levich and Koutécky-Levich as follows [23-26]:

Randles-Sevčik (3)

Levich (4)

Koutécky-Levich $$\frac{1}{j}=\frac{1}{j_{ct}}+\frac{1}{0.62\,z\,F\,D^{2/3}\,\nu^{-1/6}\,c\,\omega^{1/2}}$$ (5)

where j_p is the peak current density, z is the number of electrons involved in the electrode process, D is the diffusion coefficient of Sn^{2+}, ν is the linear potential sweep rate, c is the concentration of stannous ion. is the limiting current density, is the kinematic viscosity of the electrolyte, is the electrode rotation speed and j_{ct} is the pure charge transfer data. Randles-Sevčik equation uses data from the cyclic voltammogram recorded under a static electrolyte. Levich equation assumes a complete mass transport controlled for the electrodeposition of tin under a controlled flow condition (via rotating the copper disc electrode at a centrally fixed axis). Koutécky-Levich equation uses data from the mixed control region, in the presence of a controlled flow condition, and does not require an estimation of the limiting current density for data analysis.

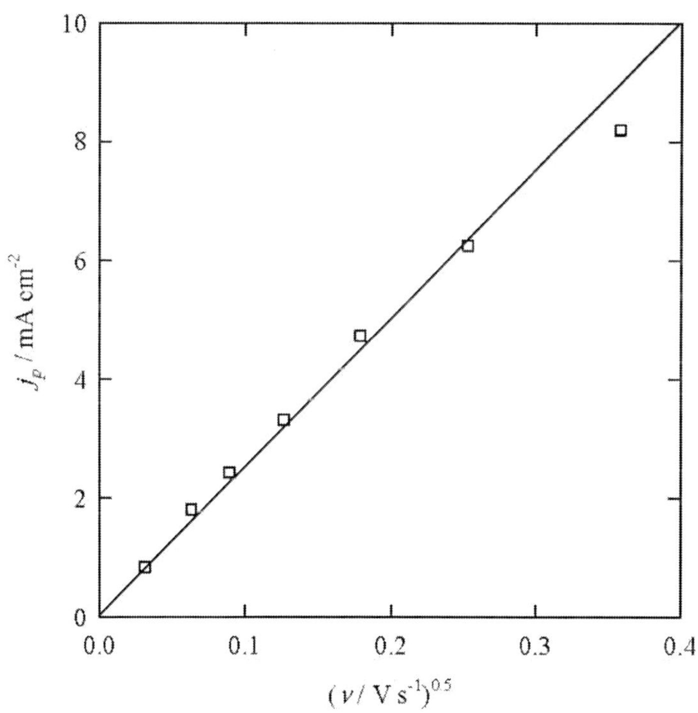

Figure 13.5. The effect of cathodic peak current density *vs.* square root of the electrode potential sweep rate. Data points were taken from the cyclic voltammograms of Sn^{2+}/Sn in an additives-free tin methanesulfonate acid electrolyte. The electrolyte contained 0.014 mol dm^{-3} Sn^{2+} in 12.5 % vol. CH_3SO_3H. The electrode potential was swept from 0 to – 0.8 V vs. Ag|AgCl using 16 mV s^{-1} at 295 K electrolyte temperature and under a static electrolyte condition.

Table 13.1. Literature values for the diffusion coefficients of stannous ion under various electrolyte conditions [27-31]

Type of tin compound	Type of acidic electrolyte	Diffusion coefficient of Sn^{2+}, D / 10^{-6} cm^2 s^{-1}	Ref.
0.2 mol dm^{-3} $SnSO_4$	0.3 mol dm^{-3} H_2SO_4	3.6	[5]
0.01 mol dm^{-3} SnO	1 mol dm^{-3} HBF_4	6.0	[15]
0.05 mol dm^{-3} $Sn(CH_3SO_3)_2$	2 mol dm^{-3} CH_3SO_3H	9.8	[16]
0.14 mol dm^{-3} $SnSO_4$	2 mol dm^{-3} H_2SO_4	7.5	[17]
0.01 mol dm^{-3} $SnSO_4$	1.88 mol dm^{-3} H_2SO_4	11.7	[18]

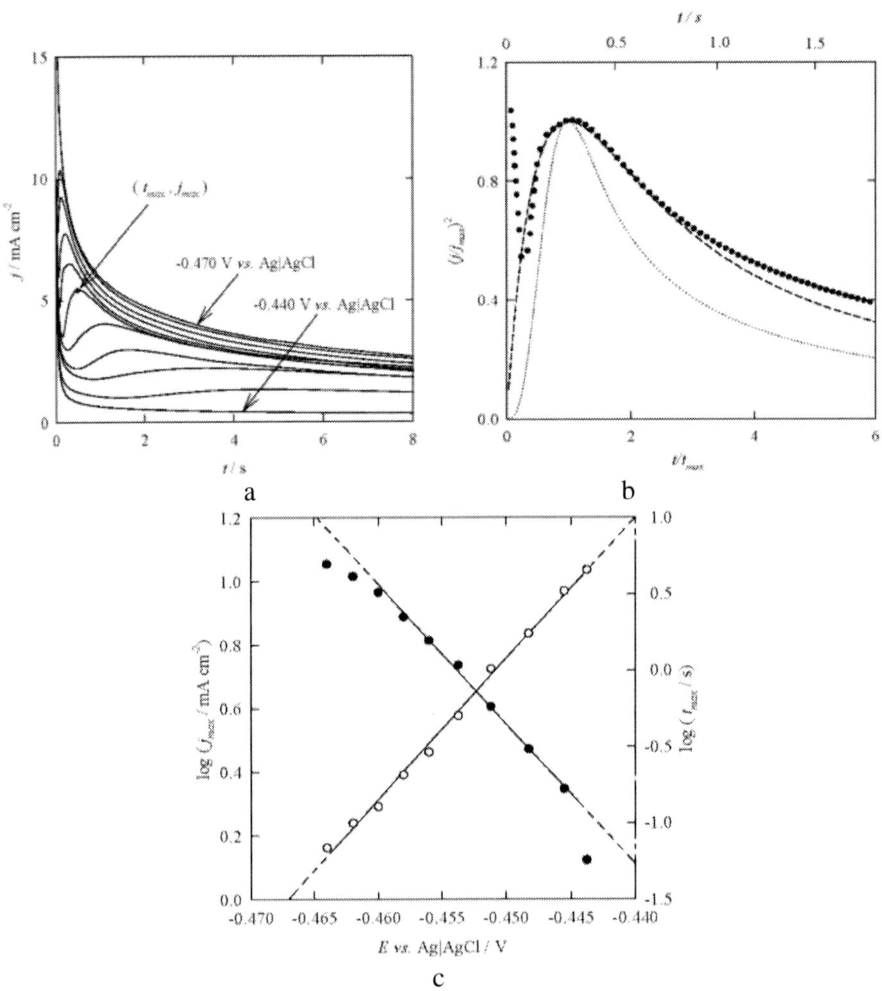

Figure 13.6. Relationship plots showing the nucleation and growth mechanisms of tin deposits. (a) Current transient curves showing the effect of potentiostatic deposition of tin on copper. (b) Simulation of the current transient curve at –0.456 V *vs.* Ag|AgCl. ———— instantanetous nucleation model. ……………… **progressive nucleation model. ••••••••••• experimental data points. (c) Logarithmic plots of** ● j_{max} and O t_{max} *vs.* electrodeposition electrode potential. The electrolyte compositions were 0.014 mol dm^{-3} Sn^{2+} and 12.5 % vol. CH_3SO_3H. The electrolyte temperature was maintained at about 295 K and under a static electrolyte condition.

It was found that the electrochemistry of Sn^{2+}/Sn in an additives-free methanesulfonate acid electrolyte followed all the above expressions. For example, Figure 13.5. shows the Randles-Sevčik plot of j_p vs. $\nu^{0.5}$ followed a linear relationship; the diffusion coefficient of Sn^{2+} was estimated from the slope of the plot to be approximately 5.5 × 10^{-6} cm^2 s^{-1}. The Levich plot of j_L vs. $w^{0.5}$ was a linear relationship and the diffusion coefficient of Sn^{2+} was approximately 5.2 × 10^{-6} cm^2 s^{-1} (using data from Figure 13.3.(a): the influence of electrode rotation speed) or 5.7 × 10^{-6} cm^2 s^{-1} (using data from cyclic voltammograms recorded in Figure 13.3. (b): the influence of concentration of stannous ion). This observation is comparable with the Koutécky-Levich plot of j_L^{-1} vs. $w^{-0..5}$ where the diffusion coefficient of Sn^{2+} was approximately 5.4 × 10^{-6} cm^2 s^{-1}.All these linearity expressions suggested that the electrodeposition of tin on copper is diffusion controlled; the diffusion coefficient of Sn^{2+} was averaged to be approximately 5.5 × 10^{-6} cm^2 s^{-1}. For comparison, literature values for the diffusion coefficients of stannous ions in various electrolytes are shown in Table 13.1. [27-31].

13.2.2. Nucleation and Growth of Tin on Copper

Figure 13.6.(a) shows the current transient curve for the electrodeposition of tin on copper in the absence of additives. The shape of the recorded cathodic current as a function of time was dependent on the applied electrode potential. All curves started with a momentary decrease in the current as a result of charging of the electrical double layer. At – 0.44 V *vs.* Ag|AgCl, the current continued to decrease to a plateau value at steady-state. At – 0.454 V *vs.* Ag|AgCl, following the momentary decrease in the current, it rose sharply to reach a maximum value and then decreased to a plateau value at steady-state, which is a typical characteristic curve for the growth of tin nuclei on copper. At a more electro-positive potential, no peak current value can be seen and the current continued to decrease to a steady-state which is indicative of a diffusion limited phenomenon. Cottrell's expression was used to analyze these curves as follows [26]:

$$j = \frac{z F D^{0.5} c}{\pi^{0.5} t^{0.5}} \tag{6}$$

where t is the duration for electrodeposition. It was found that a linear relationship exists in the logarithmic plot of the cathodic current *vs.* time profile. The slopes of the plots had values in the range between 0.45 to 0.47; supporting the validity of the Cottrell's expression for the electrodeposition tin on copper. This suggested that the electrodeposition reaction proceeded through a rapid cluster three dimensionless nucleation mechanism then followed by a diffusion limited growth mechanism.

Nucleation mechanism of tin nuclei on copper was analysed using the Scharifker-Hills expressions [32]. Figure 13.6.(b) shows the dimensionless current transient curves. It was found that the experimental current transient curve (data collected at – 0.456 V *vs.* Ag|AgCl) closely matched the theoretical current transient curve for an instantaneous nucleation expression as follows [32]:

$$\left(\frac{j}{j_{max}}\right)=\frac{1.9542}{t/t_{max}}\left[1-exp\left(-1.2546\left(\frac{t}{t_{max}}\right)\right)\right]^2 \quad (7)$$

Figure 13.6.(c) illustrates the logarithmic plots of j_{max} and t_{max} as a function of electrode potential. The plots showed a linear relationship in the potential window between – 0.446V V and – 0.46 V *vs.* Ag|AgCl. This is consistent with an instantaneous nucleation expression as follows [32]:

$$j_{max}^2\, t_{max} = 0.1629\, z^2\, F^2\, c_{Sn^{2+}}^2\, D \quad (8)$$

Equation (8) did not apply for electrode potentials more electro-positive than – 0.446 V *vs.* Ag|AgCl or more electro-negative than – 0.46 V *vs.* Ag|AgCl, which suggested that the electrodeposition of tin did not precede via an instantaneous nucleation in the earliest deposition stages. At a more electro-negative potential, the nucleation of tin on copper was dominated by diffusion controlled phenomenon and kinetic mechanisms, hence, the peak current value for the Cottrell expression and Scharifker-Hills expressions could not be clearly characterized.

To further illustrate the validity of a 3-D instantaneous nucleation and growth mechanisms of tin on copper, a mathematical differentiation of the logarithmic of equation (8) with respect to the applied electrode potential follows:

$$2\frac{\partial \log j_{max}}{\partial E}=\frac{\partial \log t_{max}}{\partial E} \quad (9)$$

Indeed, the slope for the two curves in Figure 6.(c): $\frac{\partial \log j_{max}}{\partial E}$ = – 44 decade mV^{-1} and $\frac{\partial \log t_{max}}{\partial E}$ = – 92 decade mV^{-1}. This agrees with the expression in equation (8) and confirms that three-dimensional hemispherical tin nuclei was electrodeposited and grew randomly on the copper surface in a diffusion controlled phenomenon.

13.3. Surface Microstructures of Tin Deposits

Figures 13.7.(a) to 13.7(d) show the surface microstructures of tin deposits on copper; tin was electrodeposited under constant electrode potentials. Since the region for charge transfer and mixed control were limited, all tin deposits were electrodeposited from the mass transport region and beyond. At the initial part of the mass transport region (– 0.45 to – 0.65 V *vs.* Ag|AgCl), non-dendritic tin deposits was electrodeposited under a complete mass transport controlled process. The tin coating was compact, adherent to the copper substrate and showed a matte surface appearance. At a more electro-negative electrode potential (– 0.75 V to – 0.95 V *vs.* Ag|AgCl), needle-like tin dendrites was electrodeposited. The tin coating appeared to have porous morphology with visible pin-holes distributed on the surface coating. It was found that these needle-like tin dendrites, ranging from 10 to 20 μm length and 0.5 to 3 μm diameter, were observed when tin electrodeposition was accompanied by the hydrogen evolution reaction.

Figure 13. 7. Scanning electron micrographs of tin deposits. The electrolyte composition was 0.014 mol dm^{-3} Sn^{2+} and 12.5 % vol. CH_3SO_3H. The electrolyte temperature was maintained at about 295 K. The rotating flow condition was set at 400 rpm. Tin was electrodeposited for 5 minutes at (a) – 0.65 V, (b) – 0.75 V, (c) – 0.85 V and (d) – 0.95 V *vs.* Ag|AgCl.

At an even more electro-negative potential, the needle-like dendrites formed tree-like branches with a solid stem as the support; with length ranging from 100 to 200 μm. The growth of different microstructures of tin deposits is a result of the magnitude of electrode potential (or current) deployed for the electrodeposition. At a more electro-negative electrode potential (or higher cathodic current), tin was electrodeposited at a higher growth rate, which eventually led to the formation of tin dendrites. The electrodeposition of tin at a higher current was always accompanied by co-evolution of hydrogen gaseous which led to the formation of non-compact but porous microstructures.

Figure 13.8. shows the surface roughness of tin deposits as a function of applied electrode potential. The roughness factor for a freshly prepared copper substrate was approximately 0.092 μm. It can be clearly seen that the roughness factor of tin deposits did not vary significantly from the copper substrate when tin was electrodeposited under a complete mass transport controlled process.

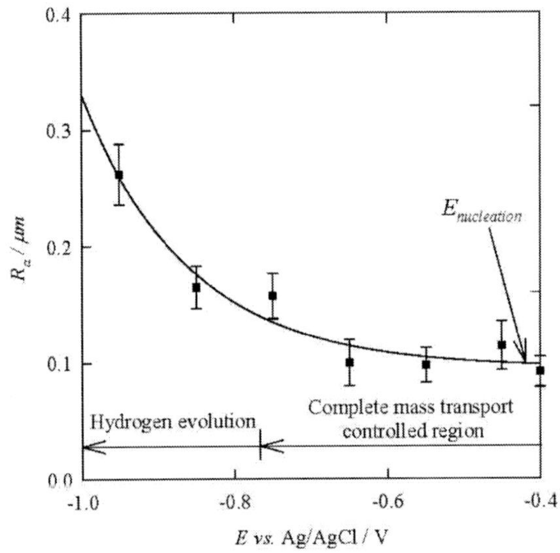

Figure 13.8. Surface roughness of tin deposits. Tin was electrodeposited at a constant potential for 5 minutes under 400 rpm. The electrolyte compositions were 0.014 mol dm^{-3} Sn^{2+} and 12.5 % vol. CH_3SO_3H. The electrolyte temperature was maintained at about 295 K.

As soon as the hydrogen evolution reaction sets in to accompany the tin electrodeposition reaction, the roughness factor of tin deposits increased steeply due to the presence of tin dendrites. It should be noted that some loose tin dendrites were scraped off by the diamond tip of the Talysurf® during measurement. In this case, the actual roughness factor would be significantly higher.

CONCLUSION

This chapter has shown that tin deposits can be readily electrodeposited on copper in an additives-free acidic methanesulfonate electrolyte. A wide range of electrolyte compositions, eg. 0.006 to 0.078 mol dm^{-3} Sn^{2+} in 1.93 mol dm^{-3} CH_3SO_3H was successfully demonstrated for the electrodeposition of tin at room temperature (295 K) and under controlled flow conditions (via a rotating disc electrode ~ 200 to 4800 rpm). The electrochemistry of Sn^{2+}/Sn was clearly shown to be highly dependent on the concentration of stannous ion and the rotating flow conditions introduced to the electrolyte. Levich, Koutécky-Levich and Randles-Sevcik equations were used to characterise the electrochemistry of tin. All the findings suggested that no significant overpotential was required for the electrodeposition of tin on copper; the diffusion coefficient of Sn^{2+} in additives-free acidic methanesulfonate electrolytes was approximately 5.4×10^{-6} cm^2 s^{-1}.Electrodeposition of tin on copper occurred on a very fast kinetic over a narrow range of charge transfer and mixed controlled region. A well-defined limiting mass transport controlled region and co-evolution of hydrogen were shown to be dependent on the concentration of stannous ion and the rotating flow conditions. The experimental results suggested that a high-speed electrodeposition of tin on copper can be achieved by deploying a high concentration of stannous ion in the electrolyte composition and introducing a high speed fluid flow hydrodynamics to the electrolyte tank. Both these process parameters must be carefully chosen to minimize the undesirable hydrogen evolution reaction on tin, since this will affect the microstructures of tin deposits.

A wide range of surface morphologies of tin deposits were readily electrodeposited. Compact and adherent tin coatings were deposited under a complete mass transport control process; while porous tin coatings consisted of needle-like dendrites and tree-like branches were deposited in the presence of hydrogen evolution. Tin coatings appeared matte in surface appearance. The surface roughness of tin coatings and the growth of different microstructures was related to the magnitude of the applied electrode potential (or current density) deployed for the electrodeposition. Scharifker-Hills expressions were used to identify the nucleation and growth mechanisms of tin deposits. It was found that three-dimensional hemispherical tin nuclei was electrodeposited and grew randomly on the copper surface in a diffusion controlled phenomenon.

ACKNOWLEDGMENTS

Dr. John Low would like to acknowledge the support provided by Professor Frank Walsh during his PhD training at The University of Southampton.

REFERENCES

[1] C.T.J. Low, F.C. Walsh, *Surface and Coatings Technology*, 615, 91 (2008).
[2] C.T.J. Low, F.C. Walsh, *Surface and Coatings Technology*, 202, 1339 (2008).
[3] C.T.J. Low, F.C. Walsh, *Surface and Coatings Technology*, 202, 3050 (2008).
[4] C.T.J. Low, C. Kerr, B.D. Barker, J.R. Smith, S.A. Campbell, F.C. Walsh, *Transactions of the Institute of Metal Finishing*, 86(3), 148 (2008).
[5] C.T.J. Low, F.C. Walsh, *Transactions of the Institute of Metal Finishing*, 86(6), 315 (2008).
[6] C.T.J. Low, F.C. Walsh, *Electrochimica Acta*, 53, 5280 (2008).
[7] M. D. Gernon, M. Wu, T. Buszta, P. Janney, *Green Chemistry*, 1, 127 (1999).
[8] R. Balaji, M. Pushpavanam, *Transactions of the Institute of Metal Finishing*, 81(5), 154 (2003).
[9] D. Egan, C.T.J. Low, F.C. Walsh, *Journal of Power Sources, accepted* 6-01-2011, doi:10.1016/j.jpowsour.2011.01.008.
[10] C.T.J. Low, D. Pletcher, F.C. Walsh, *Electrochemistry Communications*, 11, 1301 (2009).
[11] C.T.J.Low, 'Anodic electrodeposition of polypyrroleol from methanesulfonic acid', unpublished research work at Southmapton, 15/10/2010.
[12] C.T.J. Low, M. de la Toba Corral, F.C. Walsh, Transactions of the Institute of Metal Finishing, 89(1), 44 (2011).
[13] Daiwa Kasei Kenkyujo, US4128460, 1977.
[14] BASF, WO2002/04717, 2000.
[15] P.K. Leung, C. Ponce-de-León, C.T.J. Low, F.C. Walsh, Journal of Power Sources, accepted 25-01-2011, doi:10.1016/j.jpowsour.2011.01.095.

[16] P.K. Leung, C. Ponce-de-León, C.T.J. Low, F.C. Walsh, *Electrochimca Acta*, 56(5), 2145 (2011).

[17] J. Collins, G. Kear, X. Li, C.T.J. Low, D. Pletcher, R. Tangirala, D. Stratton-Campbell, F. C. Walsh, C. Zhang, *Journal of Power Sources*, 195(6), 1731 (2010).

[18] D. Pletcher, H. Tao, G. Kear, C.T.J. Low, R.G.A. Wills, F.C. Walsh, *Journal of Power Sources*, 180, 621 (2008).

[19] D. Pletcher, H. Tao, G. Kear, C.T.J. Low, R.G.A. Wills. F.C. Walsh, *Journal of Power Sources*, 180, 630 (2008).

[20] R. G. A. Wills, J. Collins, D. Stratton-Campbell, C. T. J. Low, D. Pletcher, F. C. Walsh, *Journal of Applied Electrochemistry*, 40(5), 955 (2008).

[21] Sirés, C.T.J. Low, C. Ponce-de-León, F.C. Walsh, *Electrochemistry Communications*, 12(1), 70 (2010).

[22] Sirés, C.T.J. Low, C. Ponce-de-León, F.C. Walsh, *Electrochimica Acta*, 55(6), 2163 (2010).

[23] D. Pletcher, 'A First Course in Electrode Processes', *The Electrochemical Consultancy*, Romsey (1991).

[24] F.C. Walsh, 'A First Course in Electrochemical Engineering', T*he Electrochemical Consultancy*, Romsey (1993).

[25] C.M.A. Brett, A.M.O. Brett, '*Electrochemistry: Principles, Methods, and Applications*', Oxford University Press Inc., New York (1993).

[26] J. Bard, L. R. Faulkner, '*Electrochemical Methods: Fundamentals and Applications*', 2nd Edition, John Wiley and Sons, New York (2001).

[27] S. Wen, J. Szpunar, *Electrochimica Acta*, 50(12), 2393 (2005).

[28] Petersson, E. Ahlberg, *Journal of Electroanalytical Chemistry*, 485(2), 166 (2000).

[29] C.L. Rinne, J.J. Hren, P.S. Fedkiw, *Journal of the Electrochemical Society*, 149(3), C150 (2002).

[30] F.J. Barry, V.J. Cunnane, *Journal of Electroanalytical Chemistry*, 537(1-2), 151 (2002).

[31] G.S. Tzeng, S.H. Lin, Y.Y. Wang, C.C. Wan, *Journal of Applied Electrochemistry*, 26(4), 419 (1996).

[32] B. Scharifker, G. Hills, *Electrochimica Acta*, 28(7), 879 (1983).

In: Electrodeposition: Properties, Processes and Applications ISBN: 978-1-61470-826-1
Editor: Udit Surya Mohanty

Chapter 14

ELECTRODEPOSITION, CHARACTERIZATION AND APPLICATIONS OF NEW LAYERED MANGANESE OXIDES

Masaharu Nakayama*
Department of Applied Chemistry, Yamaguchi University

ABSTRACT

Manganese oxide materials, especially manganese dioxide (MnO_2), are very attractive because of their distinctive structures and physicochemical properties as well as environmental compatibility and cost effectiveness. Among various MnO_2 polymorphs, birnessite has attracted particular attention for its unique adsorptive, catalytic, ion-exchange, and electrochemical properties.

Birnessite (δ-MnO_2) is characterized by a two-dimensional layered structure that consists of edge shared MnO_6 octahedra with cations and water molecules occupying the interlayer space.

In our previous studies we have presented an electrochemical route to construct birnessite-type layered manganese oxides intercalated with alkaline metals or alkylammonium ions in a thin film form. The process involves anodic oxidation of aqueous manganese (II) ions in the presence of guest cations. The deposition mechanism consists of anodic formation of MnO_2 and simultaneous assembly of electrolyte cations to balance the negative charges on the deposited MnO_2 layers. This method is quite versatile because the inorganic host (MnO_2) can adjust itself to accommodate guest molecules during electrodeposition. This has enabled us to design various types of new MnO_2-based layered structures.

The guest molecules influence the stacking structure of MnO_2 layers, resulting in unique catalytic, ion-exchange, and electrochemical properties. This chapter reviews the electrochemical formation, properties, and applications of a variety of layered manganese oxides, mainly derived from our own studies during the last eight years.

* E-mail address: nkymm@yamaguchi-u.ac.jp.

INTRODUCTION

Manganese oxides, typically in the form of MnO_2, with various oxidation states and crystalline structures such as α-, β-, γ-, and δ-types have been intensively studied on their synthesis, structures, and physicochemical properties. Microporous manganese oxides including layered and tunnel structures have been reviewed by Suib *et al.* with respect to their synthetic strategies, properties, and applications [1-3]. Like other electroactive transition metal oxides, MnO_2 are able to store electrical charge by simultaneous injection of electrons and charge-compensating cations and are, therefore, potentially useful for charge storage applications such as electrodes in secondary lithium batteries [4], electrochromic devices [5-13], and recently supercapacitors [14,15]. Charge storage properties of transition metal oxides are governed largely by electrical conductivity of the solid and ionic transport within the pores [16, 17]. In this regard, layered MnO_2 having bicontinuous networks of solid are attractive candidates for application as active electrode materials. Birnessite, a mixed valent manganese oxide, has a two-dimensional layered structure consisting of edge-shared MnO_6 octahedra. The MnO_2 layer is charged negative as a result of the involvement of Mn^{3+} ions displaced with some of the Mn^{4+} ions. Small cations (Na^+, K^+, H^+, etc.) are usually located in the gallery to balance negative charges on the MnO_2 layers, typically giving an interlayer spacing of about 0.7 nm [18-20]. A variety of strategies have so far been developed for the synthesis of birnessite, such as oxidation of Mn^{2+} [18], reduction of Mn^{7+} [19], and redox reactions between these two species [20]. Moreover, birnessite-type MnO_2 with large interlayer space were prepared with alkylammonium ions, where the organic cations were inserted between MnO_2 layers by an ion-exchange mechanism. For example, the intercalation of $(C_4H_9)_4N^+$ and $(C_{12}H_{25})NMe_3^+$ ions led to the formation of layered oxides with interlayer distances of 1.28 and 2.41 nm, respectively [21]. The interlayer spacing of birnessite varies depending not only on the type of guest cations, but also on the synthetic and drying conditions because these factors influence the distribution of cations and water molecules [22-24]. Thin film is a desirable form of metal oxides for use in electronic, optical, magnetic, and electrochemical applications or studies. Among various techniques for the fabrication of thin metal oxide films, electrodeposition is one of the most promising ways, particularly for obtaining thin and uniform films on a substrate of complex shape with a high degree of reproducibility [25]. Thin conducting films adhered well to a substrate need no binders or conductive additives in their electrochemical use. More importantly, one can precisely control the film thickness as desired by simply tuning the electrical charge delivered for electrodeposition. Research on electrodeposition of MnO_2 has a long history [26], and it is generally accepted that the reaction in acidic solution obeys the following equation (1), yielding γ-type crystal structure in acidic solution.

$$Mn^{2+} + 2H_2O \rightarrow MnO_2 + 4H^+ + 2e^- \qquad (1)$$

In 2004, we presented a one-step route to build birnessite-type layered MnO_2 intercalated with alkaline metal or alkylammonium ions in a thin film form [27]. The process includes anodic oxidation of aqueous Mn (II) ions by applying a constant potential in the presence of other cations that can play a role of intercalators. This methodology is extremely simple, low cost, environmentally friendly from the viewpoint that no heating is required, and remarkably versatile because the inorganic host can adjust itself to accommodate guest molecules during electrodeposition. In fact, the developed electrochemical technique could be extended to the use of various types of cationic molecules and polymers, as shown below.

14.1. ELECTRODEPOSITION OF THIN FILMS OF LAYERED MnO_2 INTERCALATED WITH GUEST MOLECULES

14.1.1. General Procedures

Electrodeposition of Mn oxides was carried out using a conventional three-electrode system in an undivided glass cell. A platinum sheet and a Ag/AgCl electrode (in saturated KCl) were used as the counter and reference electrodes, respectively. A potentiostat was used to control the electrode potential. Various metals or transparent conducting glass plates could be utilized as the working electrode to fabricate the films. Prior to electrodeposition, the electrode surface was ultrasonically cleaned in acidic solution and then rinsed thoroughly with water. Deposition baths contains $MnSO_4$ and chloride salts of desired guest cations with controlled concentrations. The solution was deoxygenated by the bubbling of purified nitrogen gas prior to use. A constant potential was applied to the working electrode for a certain period of time while monitoring the charge delivered. After electrodeposition, the resulting film coated on a substrate was rinsed thoroughly with water, dried under vacuum in a dessicator and was then analysed by spectroscopic measurements. Characterization of the surface morphology, microstructure, bonding structure, and crystal structure of the deposited coatings were carried out using scanning electron microscope (SEM), X-ray diffraction (XRD), X-ray photoelectron spectroscopy (XPS), transmission electron microscope (TEM), and Fourier transform infrared spectroscopy (FTIR). Electrochemical tests were conducted in the same configuration as that for the film preparation.

14.1.2. Intercalation of Tetraalkylammonium Ions

Figure 14.1 shows XRD patterns of the films deposited on a Pt substrate from 2 mM $MnSO_4$ and 50 mM of ammonium or various chlorides salts of tetraalkylammonium (R_4N^+) ions with different carbon chain lengths, where R = CH_3 (Me), C_2H_5 (Et), C_3H_7 (Pr), C_4H_9 (Bu) [28]. In the presence of NH_4Cl (Figure 14.1a), two peaks appear at 2θ = 12.3 and 24.7°, along with the feature due to the Pt substrate. According to Bragg's equation, these 2θ values correspond to d spacings of 0.72 and 0.36 nm, which can be indexed to 001 and 002 reflections of a layered structure, respectively. This pattern with d_{001} (=interlayer spacing) of 0.72 nm is consistent with that observed for bulk birnessite with hydrated small cations (Na^+, K^+, H^+, etc.) [18]. Since ammonium ionic radius (0.143 nm) is as small as potassium (0.133 nm) [29], the interlayer is maintained by water molecules.

All films except for the film deposited with Me_4N^+ present the patterns arising from a single phase consisting of the 00l reflections. In contrast, the pattern of the film with Me_4N^+ consists of two phases; i. e., one consisting of two peaks at 0.98 and 0.49 nm, and the other at 0.75 and 0.37 nm. With the exception of the 0.98 nm phase of the Me_4N^+ film, the set of diffraction lines shifts to lower angles, correspondingly larger *d* spacings, with an increase in the size of the used cations. The patterns of the films with Et_4N^+ and Pr_4N^+ were not able to be assigned to any known structures. On the other hand, the pattern observed for the Bu_4N^+ film agrees with that of the Bu_4N^+-intercalated birnessite prepared by Omomo *et al.* via an ion-exchange route which involves the equilibration of synthesized proton birnessite in aqueous Bu_4NOH solution [30]. They pointed out that the obtained interlayer spacing (1.25 nm) corresponds to the accommodation of one molecular layer of unhydrated Bu_4N^+ ion with its C_2 axis perpendicular to the plane of MnO_2 sheets.

The obtained interlayer spacing (d_{001}) is plotted in Figure 14.2 against the size of the corresponding tetraalkylammonium ion from the literature data [31]. Plots for the films prepared with Bu_4N^+, Pr_4N^+, and Et_4N^+, and the 0.75 nm plot of the Me_4N^+ sample are fitted to a straight line. This allows us to consider that the unknown patterns are assignable in the same manner as the Bu_4N^+-incorporated product. The other plot for the Me_4N^+ film (0.98 nm), deviated from the straight line, can be ascribed to hydration of Me_4N^+ ions probably because of the less hydrophobicity of Me_4N^+ compared to the other alkylammonium ions. Figure 14.3 displays XRD patterns of the films grown anodically in the presence of Bu_4N^+ by applying different constant potentials.

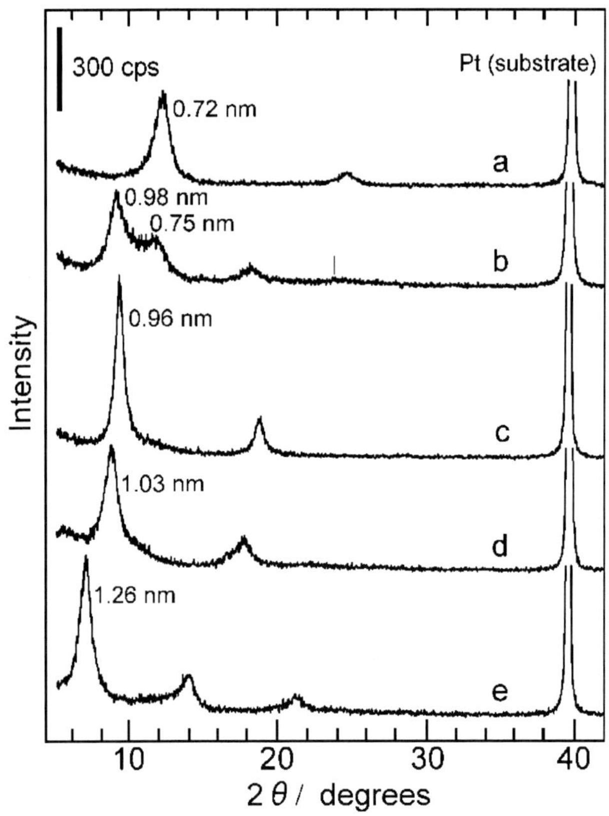

Figure 14.1. XRD patterns of MnO_2 films deposited on a Pt electrode from 2 mM $MnSO_4$ solutions containing (a) NH_4Cl, (b) Me_4NCl, (c) Et_4NCl, (d) Pr_4NCl, and (e) Bu_4NCl at a concentration of 50 mM. The films were prepared by applying a constant potential of +1.0 V for 30 min. (Reprinted from [28], © 2005, with permission from American Chemical Society).

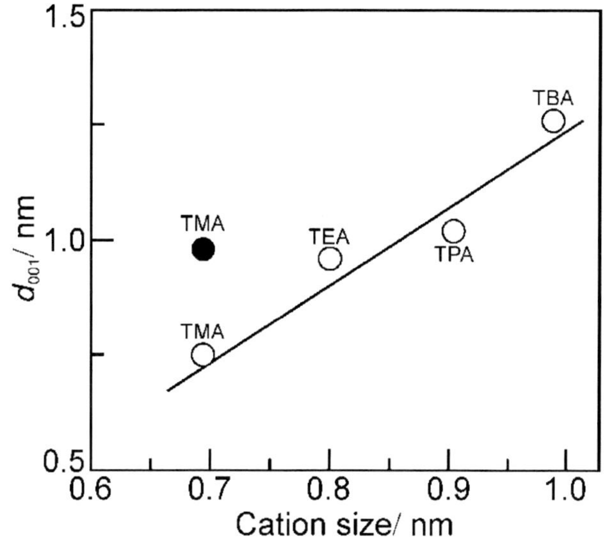

Figure 14.2. Plots of the d_{001} spacings obtained in Figure 14.1 versus the dimensions of alkylammonium ions from the literature data [31]. (Reprinted from [28], © 2005, with permission from American Chemical Society).

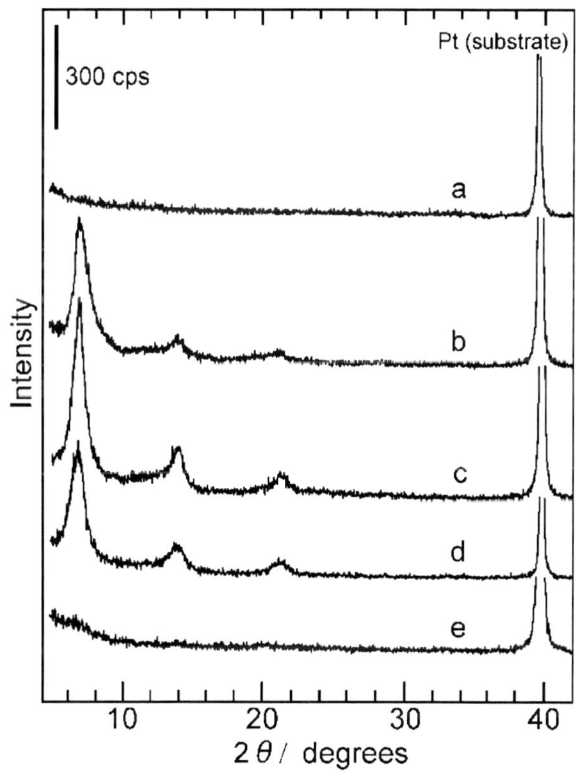

Figure 14.3. XRD patterns of MnO_2 films on a Pt electrode prepared from a 2 mM $MnSO_4$ and 50 mM Bu_4NCl solution by applying constant potentials of (a) +0.9, (b) +0.95, (c) +1.0, (d) +1.05, and (e) +1.1 V for 30 min. (Reprinted from [28], © 2005, with permission from American Chemical Society).

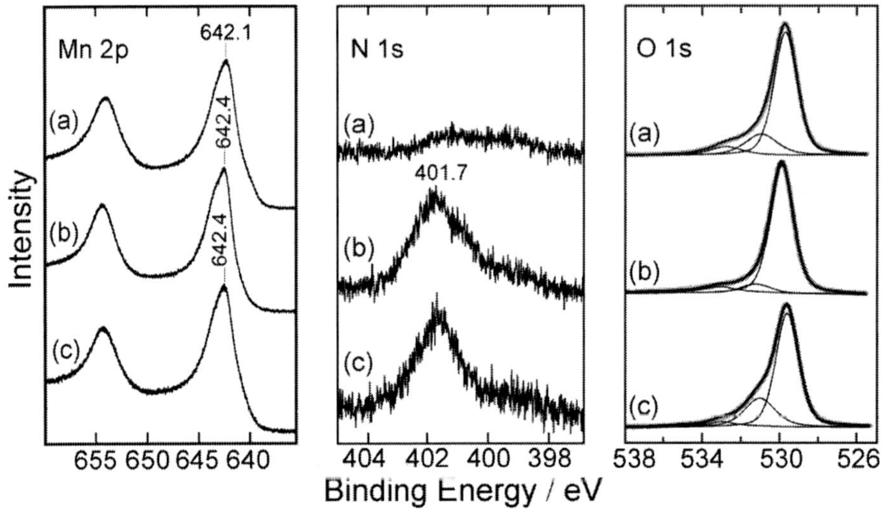

Figure 14.4. XPS spectra in the Mn 2p, N 1s, and O 1s regions of MnO_2 films on a Pt electrode prepared from a 2 mM $MnSO_4$ and 50 mM Bu_4NCl solution by applying constant potentials of (a) +0.9, (b) +1.0, and (c) +1.1 V for 30 min. (Reprinted from [28], © 2005, with permission from American Chemical Society).

Diffraction lines due to the layered structure are visible in the potential region from +0.95 to +1.05 V. For the films made at +0.9, +1.0, and +1.1 V, XPS analysis was conducted in order to explore the crystalline growth process. Figure 14.4 shows XPS spectra for the Mn 2p, N 1s, and O 1s core electronic transitions, where all spectra were normalized in intensity with respect to the Mn 2p peaks for easy comparison. The Mn 2p region is composed of a spin-orbit doublet attributed to $2p_{1/2}$ and $2p_{3/2}$ states around 654 and 642 eV. Binding energy (BE) values of the Mn $2p_{3/2}$ peaks are sometimes used to evaluate the valence state of Mn [32-34]. According to the literature, the BE value of 642 eV is generally attributed to Mn^{4+}, and those of 641 and 640 eV to Mn^{3+} and Mn^{2+}, respectively. All the Mn $2p_{3/2}$ peaks in Figure 14.4 are located at about 642 eV. Hence, the films are mainly composed of Mn^{4+}, though the limited resolution of the XPS instrument prevents the precise determination on the oxidation state of Mn. A slight peak shift to the lower BE side at +0.9 V may reflect an increased fraction of Mn^{3+}.

The N 1s core-level spectra exhibit a peak at 401.7 eV, which can be associated with the positively charged nitrogen belonging to tetraalkylammonium [35]. The absence of the nitrogen peak at lower potential (+0.9 V) indicates that Bu_4N^+ cations have not intercalated into the film, resulting in an amorphous product. On the other hand, the films prepared at +1.0 and +1.1 V show the incorporation of Bu_4N^+. Wide-range XPS spectra of the films yielded no peaks in the Cl 2p (around 200 eV) region (not shown). This indicates that the incorporated Bu_4N^+ exists in a cation form, not as a chloride salt. Based on the peak area ratio and the sensitivity factor, the Bu_4N^+ to Mn molar ratio was calculated to be 0.35 for the film deposited at +1.0 V.

In the O 1s energy region, three different constituents are observed at 529.6-529.9 eV, 531.0-531.3 eV, and 532.8-533.2 eV, which are assignable to oxide (Mn-O-Mn), hydroxide (Mn-OH), and structural water, respectively [36]. The area contributions of O^{2-} and OH^- for the layered product (curve b) are much larger and smaller, respectively, than those for the amorphous counterparts (curves a and c). Surface of metal oxides is easily adsorbed by water in air. The dissociative chemisorption of water causes a higher surface concentration of hydroxide [37,38]. Thus, the high degree of O^{2-} content in the crystalline product can reflect that most of the Mn oxide surface is associated with hydrophobic Bu_4N^+ cations. This suggests that the layered structure is formed over the whole film.

Figure.14.5A shows diffraction patterns of the films deposited with Bu_4N^+ at +1.0 V for different periods of time. In Figure 14.5B, the intensity of the 001 peak and the charge passed during electrodeposition are plotted against the electrolysis time. Obviously, the peak intensities increase with increasing the delivered charge without any change in the pattern, which evidences that the multilayering process proceeds electrochemically and the bulk of the electrodeposited film consists of MnO_2 layers.

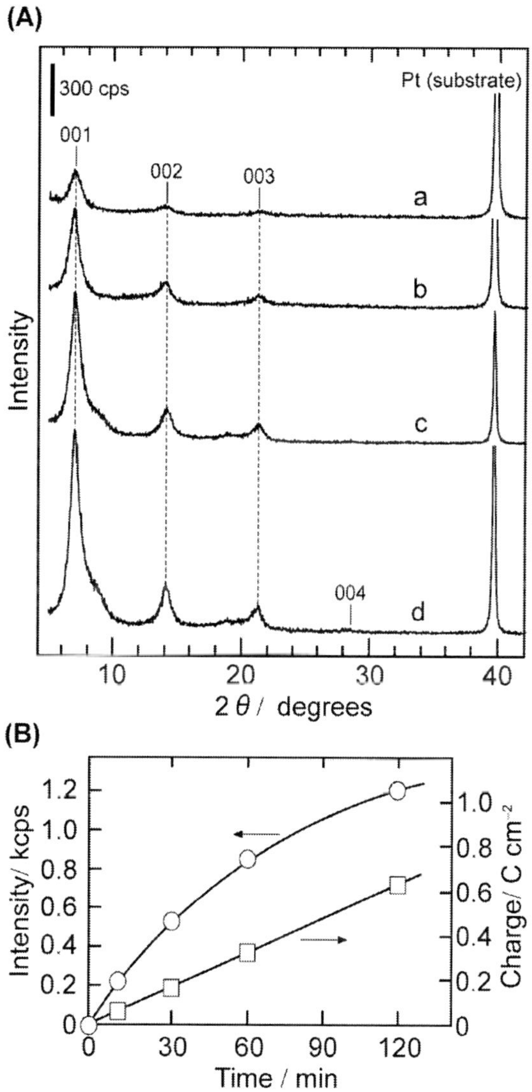

Figure 14.5. (A) XRD patterns of MnO_2 films on a Pt electrode prepared from a 2 mM $MnSO_4$ and 50 mM Bu_4NCl solution by applying a constant potential of +1.0 V for (a) 10, (b) 30, (c) 60, and (d) 120 min. (B) Plots of the intensity of the (001) peak and the charge density passed during deposition versus electrolysis time. (Reprinted from [28], © 2005, with permission from American Chemical Society).

All the above results support the electrochemical formation of birnessite-type MnO_2 in a thin film form, whose thickness can be fine-tuned by changing the delivered electrical charge. Birnessite is denoted as a formula of $A_xMnO_2 \cdot nH_2O$, where A^+ corresponds to a cation. On the basis of an assumption that the electrochemical formation of birnessite takes place similarly to that of the conventional electrolytic MnO_2 (eq. 1), the process can be expressed as below.

$$Mn^{2+} + 2H_2O + xA^+ \rightarrow A_xMn^{3+}{}_xMn^{4+}{}_{1-x}O_2 + 4H^+ + (2-x)e^- \quad (2)$$

Taking into account the Bu_4N^+/Mn ratio (0.35) from the XPS data, the composition of the layered product can be estimated to be $Bu_4N^+{}_{0.35}Mn^{3+}{}_{0.35}Mn^{4+}{}_{0.65}O_2$. This leads to an average Mn valence of 3.7, in good agreement with that typically observed for powdery birnessite [19]. In this reaction, the role of guest cation is to balance the negative charges on the MnO_2 layers grown anodically.

14.1.3. Intercalation of Cationic Polymers

Hybrid materials with interactions between inorganic hosts and polymer guests at molecular level are called "nanocomposites", which sometimes offer improved mechanical, catalytic, electronic, and optical properties that are distinct from those of the pure components and those of the related macro- and microcomposites. Layered nanocomposites are typically composed of organic polymers sandwiched between the layers of an inorganic host having a nanoscale repeating periodicity. Great effort has been devoted to develop synthetic routes for such materials, including (1) direct intercalation from polymer solutions or polymer melting, (2) *in situ* intercalative polymerization, (3) template synthesis, and (4) delamination and entrapping-restacking [39,40]. Since layered MnO_2 possesses a high charge density in its interlayer, it has been very difficult to build a composite intercalated with cationic polymers by a conventional ion-exchange route. In 2002, Liu *et al.* provided a breakthrough to overcome this problem [41]. They prepared a layered nanocomposite with poly(diallyldimethylammonium) (PDDA) cations incorporated between MnO_2 layers by an exfoliation/reassembly technique, where the slurry of exfoliated MnO_2 layers was mixed to a PDDA solution. Alternate adsorption of oppositely charged macromolecules is a well-established approach to fabricate multilayered nanocomposites [42]. Wang *et al.* extended this technique to fabricate an ordered multilayered film consisting of MnO_2 nanosheets and PDDA ions [43]. The electrochemical route developed with tetraallylammonium (Section 14.1.2) was found to be applicable to strong (PDDA) and weak (poly(allylamine hydrochloride), PAH) polycations to produce layered nanocomposites. Their molecular structures are given in Figure 14.6. This section deals with the electrodeposition of layered nanocomposites with cationic polymers. Figure 14.7 depicts XRD patterns of the deposited films on a Pt electrode from a 2.0 mM $MnSO_4$ solution with 5.6 mM PDDA (M_w = 400,000-500,000) by applying a constant potential of +1.0 V for indicated periods of time [44]. A series of diffraction peaks appears at 2θ = 9.1, 18.2, and 27.3°, being evenly spaced by 9.1°. These peaks can be indexed to the $00l$ reflections (l = 1, 2, and 3) for the multilayered structure, as has already been described above. According to Bragg's equation, the obtained 2θ values correspond to d-spacings of 0.97, 0.49, and 0.33 nm, respectively. The d_{001} value

(0.97 nm) corresponds to the interlayer spacing. The diffraction peaks increase in intensity with an increase in electrolysis time without any change in the pattern, indicating successful construction of multilayers. As shown in Figure 14.8, diffraction peaks characteristic of the layered structure are visible from +0.9 to +1.0 V, similar to the result in Figure 14.3. The observed interlayer spacing (0.97 nm) is close to that (0.92 nm) reported by Wang *et al.* for a multilayer ultrathin film consisting of MnO_2 nanosheets and PDDA ions [43]. They attributed the obtained repeating periodicity to the accommodation of a monolayer of PDDA in the parallel arrangement between MnO_2 layers, on the basis of the crystallographic thickness of a single MnO_2 nanosheet (0.45 nm) [45] and that of a PDDA monolayer (0.5 nm) [43]. Figure 14.9 shows XRD patterns of the films obtained from $MnSO_4$ solutions containing PDDA (a) and PAH (b) at indicated concentrations. Here, the electric charge passed during electrodeposition was fixed to 330 mC/cm^2. With an increase in the starting PDDA concentration, the intensity of diffraction peaks increases to reach a maximum at 5.6 mM, while their positions remain unchanged. The intensity decreases at higher concentrations can be ascribed to an increase in hydrophobic interaction between PDDA molecules, which relatively weakens the electrostatic force between polycations and the negative charges on the deposited MnO_2. In Figure 14.9b, the film deposited with PAH also provides a similar set of evenly spaced peaks. Although no assignment could be made according to previous literatures, the peaks can be indexed to the 001, 002, and 003 reflections, similarly to the product with PDDA. The strongest intensity appears at 200 mM, in which each peak is broader than that of the PDDA composite, suggesting a lower degree of the ordering of PAH molecules. The interlayer spacing of the product was 1.19 nm. Figure 14.10A displays XPS survey spectra of the MnO_2 films made with PDDA (a) and PAH (b) that exhibited the highest crystallinity in Figure 14.9, a and b, respectively. In both spectra, in addition to the Mn 2p, O 1s, and Mn 3s signals arising from MnO_2, the N 1s peak appears around 400 eV. This peak is attributed to nitrogen atom belonging to the incorporated polycations. XPS spectra of the N 1s core-level are shown in Figure 14.10B. The film deposited with PDDA exhibits one component at 401.4 eV, while the spectrum of the PAH/MnO_2 film consists of two isolated chemical constitutes peked at 400.9 and 398.9 eV. According to the literature, the higher BE component (around 401 eV) can be attributed to positively charged nitrogen, and the lower one to neutral nitrogen [46]. From the ratio of the peak area of $-NH_3^+$ to that of total N 1s, the fraction of ionized nitrogen groups ($[-NH_3^+]/[N]$) was measured to be 0.68. On the basis of the peak area ratios of N 1s and Mn $2p_{1/2}$ and their sensitivity factors, the N/Mn atomic ratios in the PDDA- and PAH-containing films were calculated to be 0.43 and 0.98, respectively. The higher composition of PAH than PDDA can be recognized by a difference in the degree of ionization of the polymers; i.e., PDDA is fully dissociated, while some amine groups in PAH are un-ionized. Assembly of polycations is driven by an electrostatic attraction between the cationic nitrogen groups of polycations and the negative charges on the MnO_2 layers grown anodically. As a result, many PAHs are used to build a layered structure, resulting in a larger interlayer distance than the PDDA composite. For the Cl 2p region (Figure 14.10C), the PDDA-containing film provides no feature, confirming the intercalation of PDDA as its cationic form, and not as a chloride salt. In contrast, the PAH/MnO_2 film presents a doublet that can be curve-fitted with two peaks assignable to the Cl $2p_{1/2}$ and $2p_{3/2}$ states. This finding enables us to realize that the $-NH_3^+$ groups in the interlayer PAH are associated with Cl^- to produce $-NH_3^+Cl^-$ ionic pairs as well as with the negatively charged MnO_2 layers. The Cl/Mn atomic ratio was calculated to be 0.29 based on the area of the Cl peak. This means that the fraction interacting with negative charges on MnO_2 is 0.38 (0.98×0.68–0.29). This value is somewhat smaller than the PDDA/Mn molar ratio (0.43) and is comparable to those observed for birnessites in a powder form. The structural arrangement of PAH should be affected by various factors such as charge balance, environment pH, hydrophobic interaction, etc. The higher PAH concentration in the deposition bath would cause an increase in the fraction that does not contribute to compensate the negative charges on MnO_2 layers, which results in an expansion of the interlayer space (see Figure 14.9b).

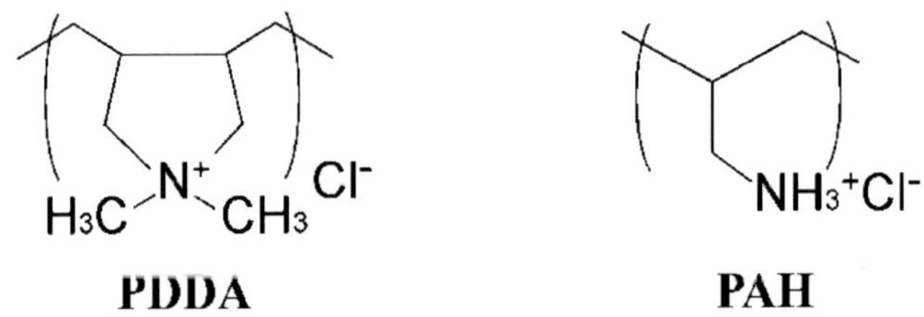

Figure 14.6. Chemical structures of the monomer units of poly(diallyldimethylammonium chloride) (PDDA) and Poly(allylamine hydrochloride) (PAH). (Reprinted from [44], © 2006, with permission from American Chemical Society).

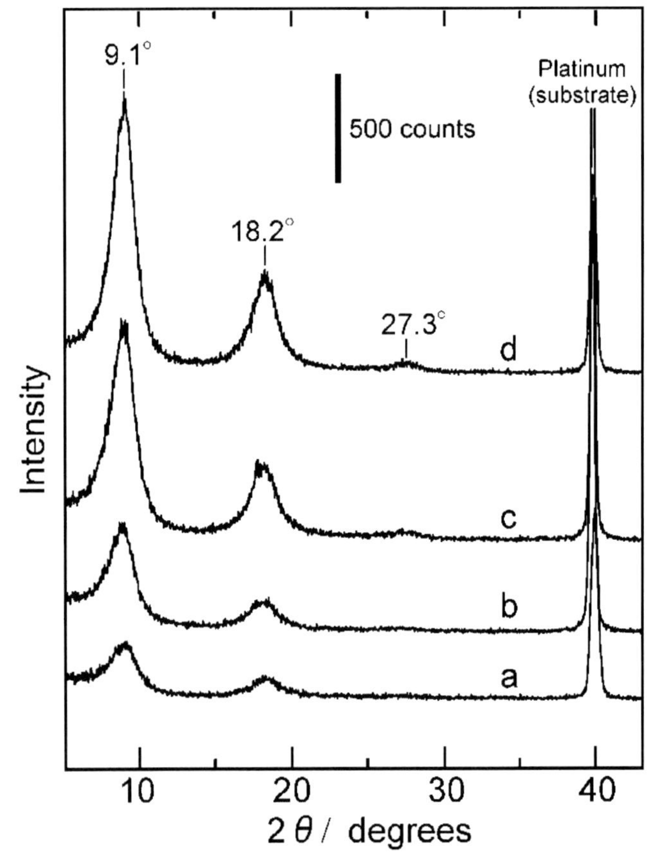

Figure 14.7. XRD patterns of MnO_2 films on a Pt electrode prepared from a 2 mM $MnSO_4$ and 5.6 mM PDDACl solution by applying a constant potential of +1.0 V for (a) 30, (b) 60, (c) 120, and (d) 240 min. (Reprinted from [44], © 2006, with permission from American Chemical Society).

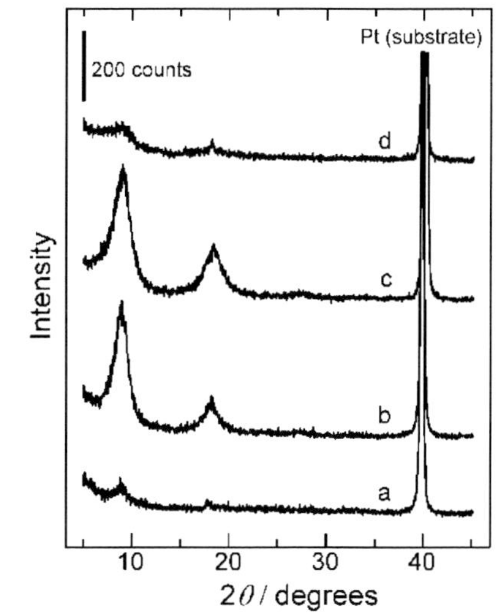

Figure 14.8. XRD patterns of MnO_2 films on a Pt electrode prepared from a 2 mM $MnSO_4$ and 5.6 mM PDDACl solution by applying constant potentials of (a) +0.8, (b) +0.9, (c) +1.0, and (d) +1.1 V for 30 min. (Reprinted from [44], © 2006, with permission from American Chemical Society).

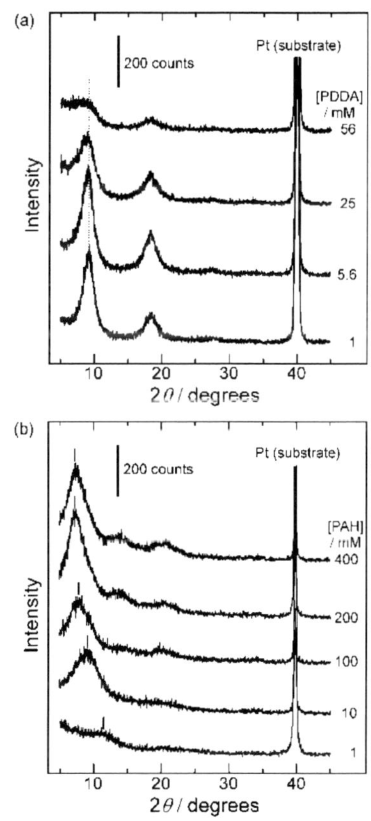

Figure 14.9. XRD patterns of MnO_2 films on a Pt electrode prepared by applying a constant potential of +1.0 V in 2 mM $MnSO_4$ solutions mixed with (a) PDDA and (b) PAH at indicated concentrations. The electric charge passed during electrolysis was always 330 mC cm^{-2}. (Reprinted from [44], © 2006, with permission from American Chemical Society).

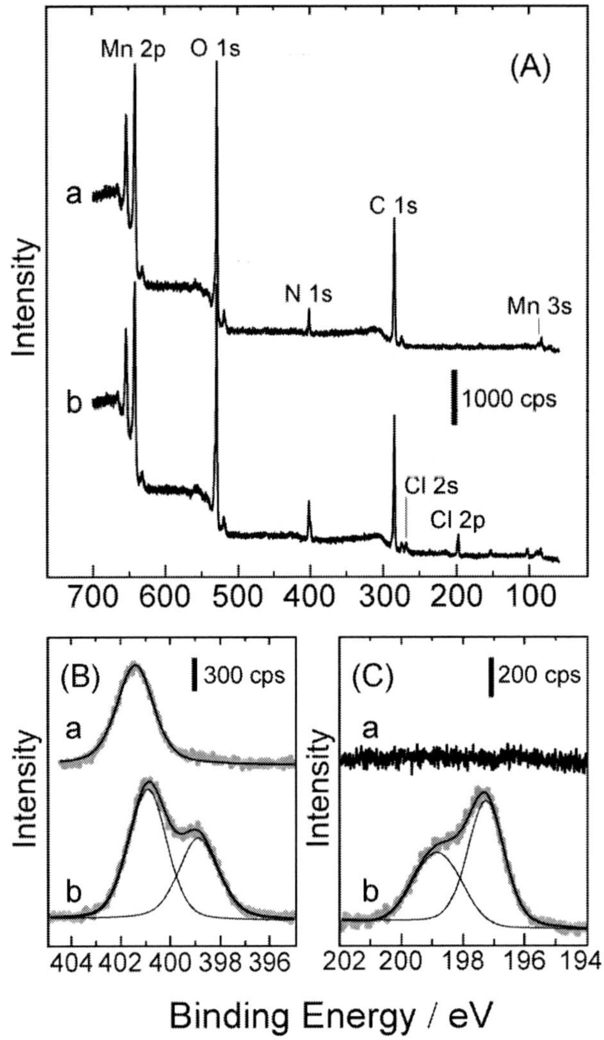

Figure 14.10. (A) XPS survey spectra of MnO_2 films prepared at +1.0 V in 2 mM $MnSO_4$ solutions with (a) 5.6 mM PDDA and (b) 200 mM PAH, along with the core-level spectra in (B) N 1s and (C) Cl 2p regions of the same samples. (Reprinted from [44], © 2006, with permission from American Chemical Society).

14.1.4. Intercalation of Surfactants

A variety of organic/inorganic nanocomposites including ordered surfactants accommodated in silicious frameworks are presented in a review written by Kruk *et al.* [47]. To the best of our knowledge, however, no synthetic strategies for the similar structure associated with conductive MnO_2 layers has be published until we first reported it in 2011 [48]. This section presents our attempt to introduce cationic surfactants to create layered MnO_2 with extra large interlayer which is available for the reaction of an organic compound.

Figure 14.11 displays XRD patterns of the films deposited on an ITO substrate from $MnSO_4$ solutions containing C16 (a), C12 (b), and Brij58 (c). The applied potential was +1.0 V. The molecular structures of the surfactants are depicted in Figure 14.12. The film deposited with Brij58, a nonionic surfactant, gives rise to no peak at all except the ITO peaks (*). Clearly, nonionic surfactants do not take part in the growth mechanism of MnO_2 crystals. In the presence of C16, a set of evenly-spaced seven peaks is observed, which is diagnostic of a highly-ordered layered structure. These peaks are indexed to the 001 plane and its 2nd- (002) to 7th- (007) order diffractions from multilayers of birnessite-type MnO_2 [28]. *d*-spacing of the 001 peak (d_{001}) is equivalent to the interlayer spacing and is estimated to be 3.1 nm according to Bragg's equation. C12 surfactant also provides similar diffractions up to the 3rd order with a smaller d_{001} spacing, 2.4 nm, where each peak becomes broad compared to the corresponding peak in the film deposited with C16. As mentioned above., the electrodeposition with tetraalkylammonium cations ($(C_nH_{2n+1})_4N^+$: n = 0~4) yielded birnessite films with diffraction peaks only up to the 3rd order. Comparison of the product with C16 to those with other cations suggests that the growth mechanism involves a hydrophobic attraction between surfactants tails, as well as the electrostatic interaction between their cationic heads and the negative charges on the MnO_2 layers. IR analysis was performed for the similarly prepared film with C16 on a Pt substrate, and the obtained spectrum is shown in Figure 14.13a, along with that of pristine C16 in a KBr pellet. Absorption peaks at 2926, 2852, 1469, 964, 914, and 721 cm^{-1} in the electrodeposited film can also be detected in the spectrum of the parent reagent. A peak appearing at 634 cm^{-1} can be attributed to the stretching vibrations of Mn-O bond in MnO_2 [49]. Combined with the XRD pattern shown in Figure 14.11a, C16 surfactants are intercalated between the MnO_2 layers. Hereafter, the layered MnO_2 film prepared with C16 surfactants is designated as C16/MnO_2. The growing process of the film can be recognized by the inorganic polymerization of MnO_2 and the simultaneous assembly of organic surfactant cations onto the negative charges of the MnO_2 layers. Considering the crystallographic thickness of a single MnO_2 nanosheet (0.5 nm) [45], the interlayer spacings obtained with C16 (3.1 nm) and C12 (2.4 nm) mean gallery heights of 2.6 and 1.9 nm, respectively. Since the molecular sizes of C16 and C12 are 2.17 and 1.67 nm, respectively [50], it is reasonable to assume that the surfactants are accommodated in a form of monolayer. XPS wide-scan spectrum of the C16/MnO_2 film provided Mn 2p, O 1s, and Mn 3s signals arising from MnO_2 (not shown). In the O 1s core-level spectrum, the contribution of O^{2-} (528.7 eV) was much larger than that of OH^- (530.2 eV) or H_2O (532.1 eV) [51], confirming the formation of oxide network. In the N1s energy region, a peak at 401.6 eV attributable to positively charged nitrogen [44,52] was observed, belonging to the cationic portion of C16 surfactant. Besides, the incorporation of Cl^- anions into the interlayer was recognized by doublet peaks assignable to the Cl $2p_{1/2}$ (197.8 eV) and Cl $2p_{3/2}$ (196.3 eV) states [44, 53]. This finding enables us to realize that cationic heads of the incorporated surfactants are not only in coupling with negative charges on the MnO_2 layers, but also in coupling with Cl^- anions. The atomic ratios of Cl/Mn and N/Mn were evaluated to be 0.42 and 0.74, respectively, using the corrected areas of N 1s, Cl $2p_{3/2}$, and Mn $2p_{1/2}$ peaks. The fraction of the surfactants that interact with the negative charges on MnO_2 layers can be calculated to be 0.32 by subtracting the Cl/Mn (0.42) ratio from the N/Mn (0.74) ratio. The obtained value, equivalent to x in eq 2, is typical of chemically-prepared K-birnessite [19-21].

Top-surface SEM (a) and cross-sectional TEM (b and c) images of the C16/MnO_2 film are shown in Figure 14.14. The relatively smooth surface in Figure 14. 14a is characteristic of the birnessite films grown anodically [28]. Stripes at equal intervals can be seen over the whole image, thus strongly evidencing the electrochemical multilayer assembly. The dark stripes in Figure 14.14c correspond to MnO_2 layers, while the white stripes to the layers of the assembled surfactants. The 001 planes are indicated in the picture, where the interlayer spacing is roughly estimated to be 3 nm, in good agreement with the value obtained from the XRD data. On the basis of XRD data for the films prepared at various surfactant concentrations (not shown), the best products with respect to crystallinity were formed at 150 and 50 mM of C16 and C12, respectively. At these concentrations, both surfactants are present as micelles in bulk solutions [54], which means that their assembly forms in bulk solution do not affect the film formation process. This coincides with the finding reported by Huo *et al.*, where surfactants pack parallel to provide high possible charge density to balance inorganic ions adsorbed on the electrode surface [55].

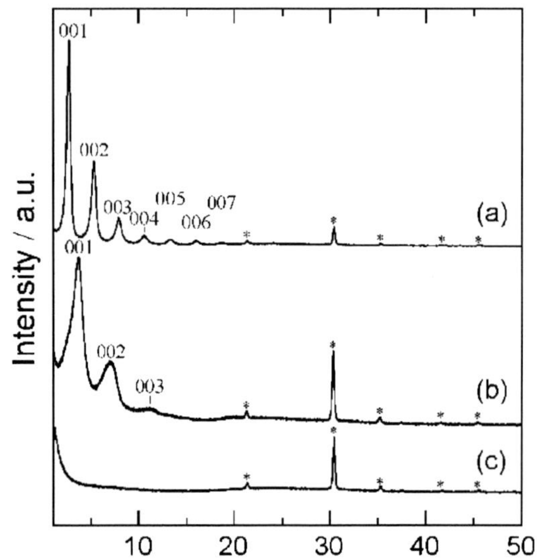

Figure 14.11. XRD patterns of the films obtained from 2 mM $MnSO_4$ solutions containing 150 mM of (a) C16, (b) C12, and (c) Brij58 by applying a constant potential of +1.0 V. The electric charge passed during electrolysis was always 340 mC/cm^2.

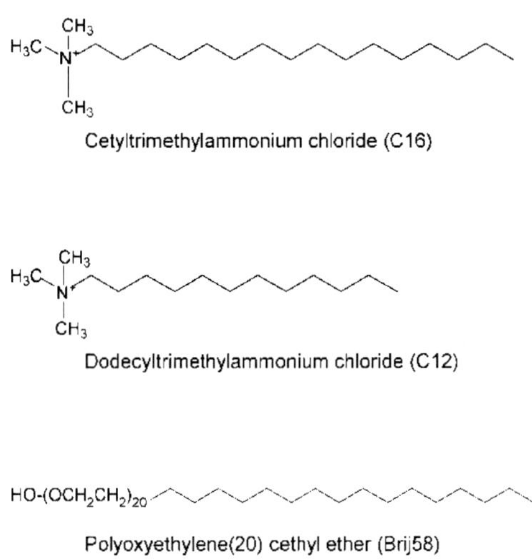

Figure 14.12. Structures of the surfactants used. (Reprinted from [48], © 2010, with permission from American Chemical Society).

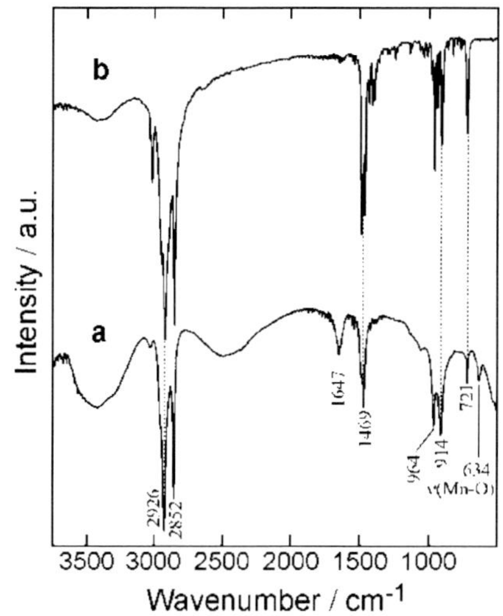

Figure 14.13. FTIR spectra of (a) the electrodeposited film from a 2 mM $MnSO_4$ solution with 150 mM C16 and (b) the C16 reagent in a KBr pellet. (Reprinted from [48], © 2010, with permission from American Chemical Society).

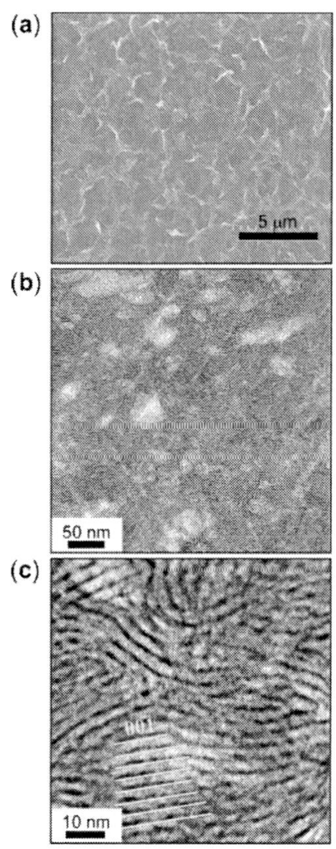

Figure 14.14. (a) Top-surface SEM and (b, c) cross-sectional TEM images of electrodeposited C16/MnO_2 film. The film was prepared by the same method as that of film a in Figure 14.11. (Reprinted from [48], © 2010, with permission from American Chemical Society).

14.2. ION EXCHANGE AND ELECTROCHEMICAL PROPERTIES

14.2.1. Tetraalkylammonim Ion/MnO_2

A Bu_4N^+-intercalated birnessite film, designated as Bu_4N^+/MnO_2, grown anodically on a Pt electrode was immersed in dilute KCl solution, and the dried sample was subjected to XRD measurements. The patterns obtained at different immersion times are shown in Figure 14.15. A set of three peaks due to the Bu_4N^+-intercalated phase decreases in intensity with increasing time, disappearing at 10 min and being replaced by two peaks at 2θ=12.3 and 24.7°. The newly obtained pattern with d_{001} of 0.72 nm is attributed to birnessite with hydrated potassium ions, indicating that the layered structure is compressed as a result of replacement of bulky Bu_4N^+ ions in the interlayer by small K^+ ions in solution.

Electrochemical tests were made for the Bu_4N^+/MnO_2-coated electrode. Figure 14.16 compares voltammetric responses during the first 20 cycles between in aqueous electrolytes of KCl and Bu_4NCl. Both CVs were taken as soon as the electrode was immersed. In KCl solution (Figure 14.16a), the current response gradually increases during initial cycles. As expected from Figure 14.15, this increase suggests the incorporation of K^+ ions from the electrolyte solution by ion-exchange and the involvement of the interlayer K^+ ions in the electrochemical processes of the film. At latter cycles, the voltammetric response is rather capacitive, with a very broad reversible peak around +0.45 V, and exhibits no defined reduction or oxidation peaks. This type of response involves continuous change of the oxidation state of Mn ions following the anodic and cathodic scans [56]. In the potential region examined, the conversion between Mn^{4+} and Mn^{3+} can take place. As seen from Figure 14.16b, the CV in Bu_4NCl solution is totally different. An anodic current appears from +0.5 V and decreases in intensity with increasing the number of cycles. Note that no current appears on the cathodic scan.

The films obtained after 20 CV cycles in KCl and Bu_4NCl solutions were subjected to XRD and XPS measurements (Figure 14.17). Both films provide the same diffraction pattern of birnessite with an interlayer spacing of 0.72 nm, indicating a removal of the intercalated Bu_4N^+ ions in both solutions while maintaining the framework structure. As shown in Figure 14.17B(a), the N 1s peak disappears completely with occurrence of K 2s and 2p signals, confirming that Bu_4N^+ cations are quantitatively replaced with K^+.

A very weak N 1s signal in Figure 14.17B (b) indicates the loss of the intercalated Bu_4N^+ ions. Since no ion-exchange substantially takes place in the Bu_4NCl solution, the decrease in the Bu_4N^+ content suggests that the interlayer Bu_4N^+ ions are excluded electrochemically to balance the charge of the deposit upon oxidation. As seen from Figure 14.17A(b), the electrochemical deintercalation of Bu_4N^+ ions is accompanied by the shrinkage of the interlayer and the incorporation of H_2O. The compressed interlayer can no longer accommodate bulky Bu_4N^+ ions from the electrolyte upon the cathodic scan. This is the reason for no cathodic current in Figure 14.16b. Such an obvious difference in electrochemical behavior between two electrolytes strongly indicates that most of the electroactive surface in the film is present not on the external surface but inside the interlayer space.

To better understand the CV response in KCl, further XPS experiments were made for the films polarized at oxidation (+0.8 V) and reduction (0 V) potentials for 20 min following the 20 potential cycles. The resulting spectra are displayed in Figure 14. 18. The oxidized film gives the Mn $2p_{3/2}$ peak which is almost the same as that of the as-deposited film (Figure 14.4).

However, the reduced film exhibits a small but distinct shoulder around 241 eV, probably as a result of an increase in the Mn^{3+} fraction. On the other hand, doublet peaks at 295.3 and 292.5 eV can be assigned to K $2p_{1/2}$ and $2p_{3/2}$ energy states, respectively, being larger for the reduced film (K/Mn=0.38) than for the oxidized one (K/Mn=0.23). In the O 1s energy region (Figure 14. 18c), the contribution of OH^- species becomes larger in the reduced film.

Probably, a certain portion of oxides are associated with protons from solution to produce hydroxides. In general, the redox reactions of conventional MnO_2 in neutral electrolytes have been associated with homogeneous insertion/deinsertion of protons [57]. However, the above electrochemical and spectroscopic results demonstrate that not only protons but also alkaline metals are responsible for the electrochemical reactions of the anodically grown birnessite-type MnO_2.

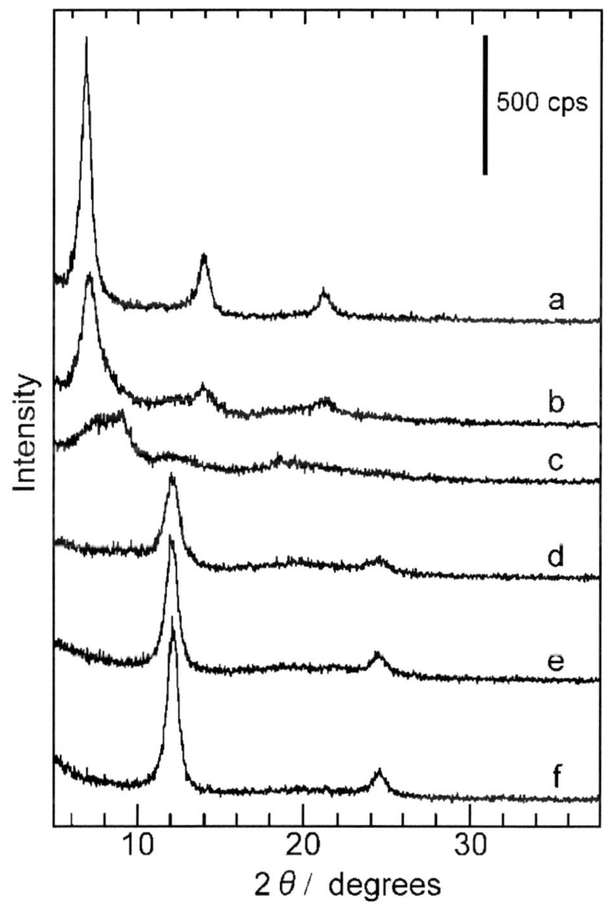

Figure 14.15. XRD patterns of a Bu_4N^+-intercalated MnO_2 film on a Pt electrode after being immersed in a 0.001 M KCl solution for (a) 0, (b) 0.25, (c) 1, (d) 10, (e) 60, and (f) 720 min. The film was prepared by the same way as that in Figure 14.5. (Reprinted from [48], © 2010, with permission from American Chemical Society).

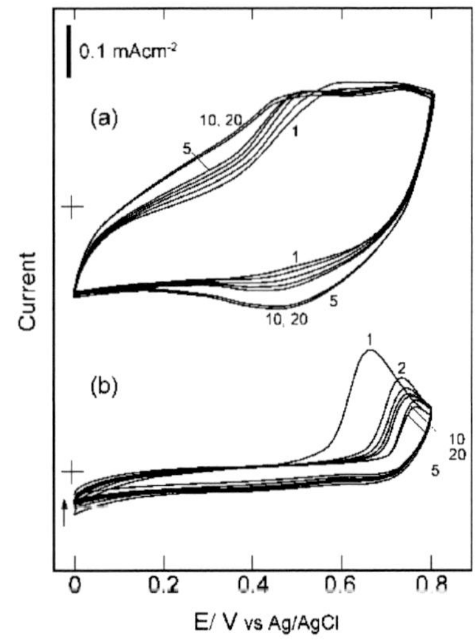

Figure 14.16. CVs of a Bu_4N^+-intercalated MnO_2 film-coated Pt electrode in 0.1 M (a) KCl and (b) Bu_4NCl solutions. Scan rate, 20 mV/s. (Reprinted from [28], © 2005, with permission from American Chemical Society).

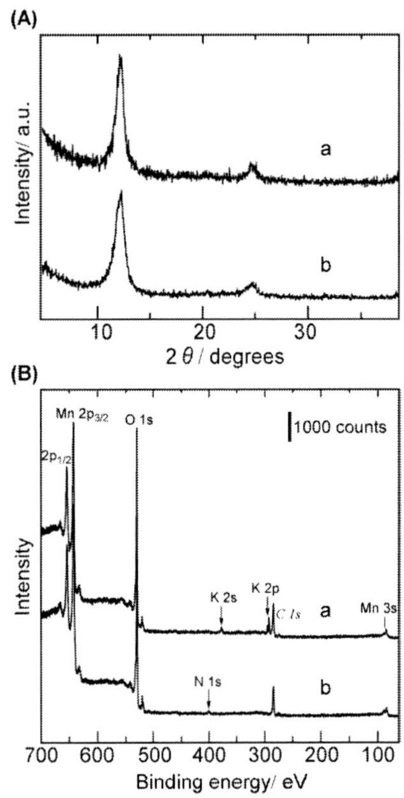

Figure 14.17. (A) XRD and (B) XPS data of a Bu_4N^+-intercalated MnO_2 film on a Pt electrode taken after 20 potential cycles between 0 and +0.8 V in 0.1 M (a) KCl and (b) TBACl solutions. The potential cycling was stopped at +0.8 V. (Reprinted from [28], © 2005, with permission from American Chemical Society).

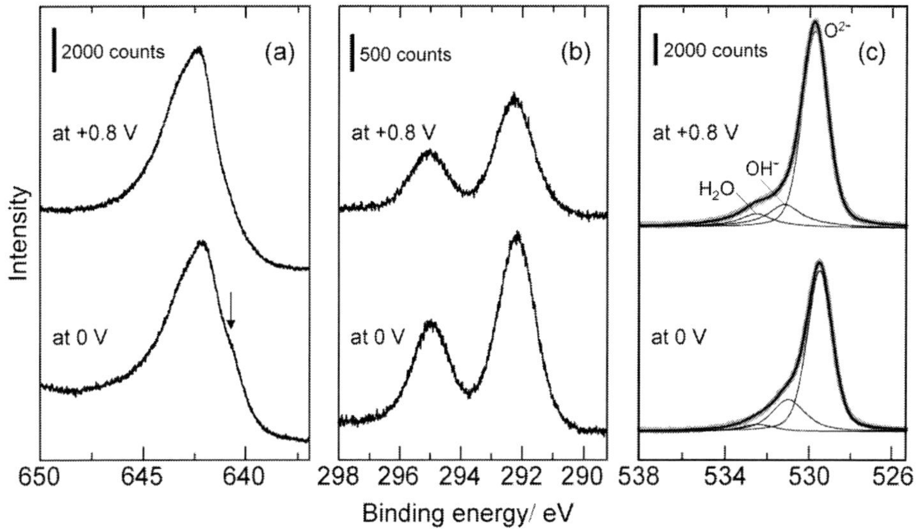

Figure 14.18. XPS spectra in (a) Mn 2p, (b) K 2p, and (c) O 1s regions of the MnO_2 film-coated electrode taken after poralized at +0.8 and 0 V for 20 min following the 20 potential cycles in a 0.1 M KCl solution. (Reprinted from [28], © 2005, with permission from American Chemical Society).

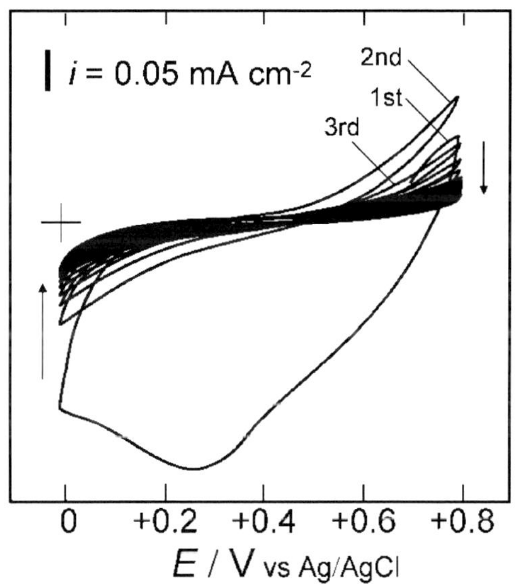

Figure 14.19. Cyclic voltammograms of the layered PDDA/MnO_2 film with a layered structure in a 0.1 M KCl solution at a scan rate of 20 mV/s. Twenty consecutive cycles are shown. The film was prepared on a Pt electrode prepared from a 2 mM $MnSO_4$ and 5.6 mM PDDA solution at a constant potential of +1.0 V with an electric charge of 330 mC/cm^2. (Reprinted from [44], © 2006, with permission from American Chemical Society).

14.2.2. Cationic Polymer/MnO_2

XRD and XPS data of the layered PDDA/MnO_2 film after being immersed in a 0.1 M KCl solution for 12 h were the same as those of the as-deposited film. Obviously, the PDDA polycations once incorporated are not ion-exchanged with small cations (K^+) in solution. This observation is apparently different from the case of tetraalkylammonium ions. Figure 14.19 shows initial 20 consecutive voltammograms recorded on the layered PDDA/MnO_2 film in 0.1 M KCl solution at a scan rate of 20 mV/s. The first cathodic scan gives a large reductive current, whereas no corresponding oxidation current appears upon the reversed scan. The response decreases with consecutive scans. This behavior was quite different from that observed for the similarly-prepared amorphous counterpart which yielded a stable response associated with the oxidation/reduction of Mn^{3+}/Mn^{4+} couple in the oxide layers. The amorphous surface is accessible to charge compensating cations in solution.

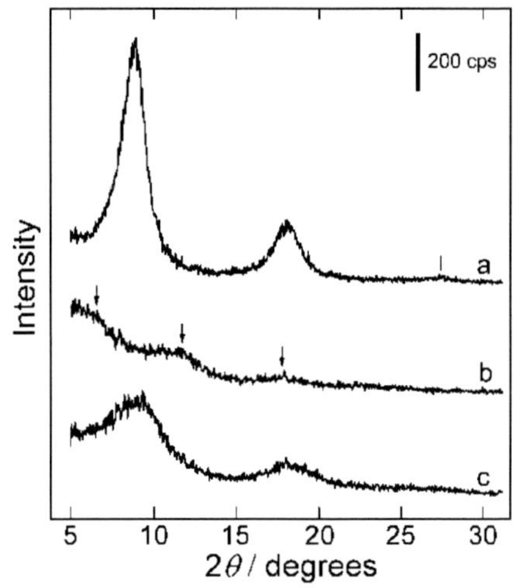

Figure 14.20. XRD patterns of the layered PDDA/MnO_2 film taken after scanning the electrode potential in a 0.1 M KCl solution and stopping at (a) +0.8 and (b) 0 V on the 1st cycle and at (c) +0.8 V on the 2nd cycle. The film was prepared in the same way as that in Figure 14.19. (Reprinted from [44], © 2006, with permission from American Chemical Society).

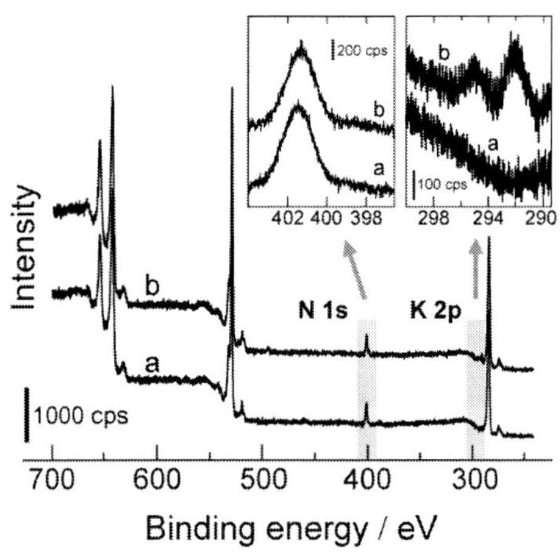

Figure 14.21. XPS spectra of the layered PDDA/MnO_2 film taken after scanning the electrode potential in a 0.1 M KCl solution and stopping at (a) +0.8 and (b) 0 V on the 1st cycle. The film was prepared in the same way as that in Figure 14.19. (Reprinted from [44], © 2006, with permission from American Chemical Society).

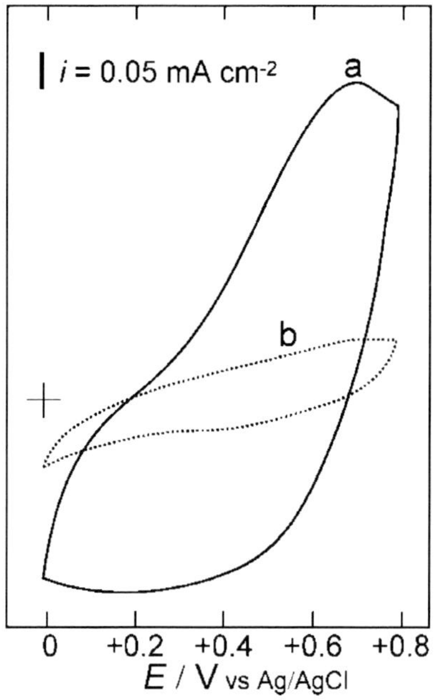

Figure 14.22. Voltammeric behaviors of (a) layered and (b) amorphous films of PAH/MnO_2 taken in a 0.1 M KCl solution at a scan rate of 20 mV/s. The films were prepared on a Pt electrode at 1.0 V from 2 mM $MnSO_4$ solutions containing (a) 200 and (b) 1 mM PAH, respectively. (Reprinted from [44], © 2006, with permission from American Chemical Society).

To interpret the voltammetric behavior of the layered PDDA/MnO_2 film, XRD analysis was carried out for the films that had been polarized in 0.1 M KCl solution. The patterns in Figure 14.20. were taken after scanning the electrode potential and stopping at +0.8 (a) and 0 (b) V on the 1st cycle and at +0.8 V (c) on the 2nd cycle. Pattern a is substantially the same as that of the as-deposited sample, but it dramatically decreases in intensity at 0 V. A closer look reveals that the pattern at 0 V consists of evenly-spaced three peaks diagnostic of a layered structure with d_{001} = 1.38 nm. This indicates an expansion of the interlayer upon reduction while lowering the crystallinity. When the electrode is reoxidized at +0.8 V, the pattern is recovered only in part. Figure 14.21 displays XPS spectra of the sample films prepared in Figure 14.19a and 14.19b. The N 1s spectra of both samples are the same as that of the as-deposited film, indicating that the intercalated PDDA polycations are stable enough to remain in the interlayer. In the K 2p region, a doublet of peaks at 295.3 and 292.0 eV assignable to $2p_{1/2}$ and $2p_{3/2}$ states, which is missing in the oxidized film, appears in the reduced film. Taking into account the fact that PDDA cations remain intercalated during reduction, injection of electrons must be balanced by incorporation of external electrolyte cations (K^+). The incorporation of K^+ into the tightly packed interlayer would bring about expansion and collapse of the structure. The separated MnO_2 sheets cannot communicate electrochemically with the underlying electrode, which is the most likely origin of the attenuated CV response in Figure 14.19.

On the other hand, electrochemistry of the layered PAH/MnO_2 film was totally different from that of the PDDA/MnO_2 film. Curve a in Figure 14.22 shows a cyclic voltammogram of the layered PAH/MnO_2 film in the same electrolyte solution taken when it reached a steady state, along with that (curve b) of the amorphous counterpart. The CV of the layered product is much larger in size than that of the amorphous one.

XRD measurements were made for the films polarized for 20 min at oxidation (+0.8 V) and reduction (0 V) potentials after the steady state had been achieved (Figure 14.23). Unlike the PDDA case, the layered structure is maintained during the potential cycling and further polarization. These CV and XRD results strongly suggest that the electrochemical process including the redox of Mn^{3+}/Mn^{4+} couple takes place at the interior surface of the multilayered film.

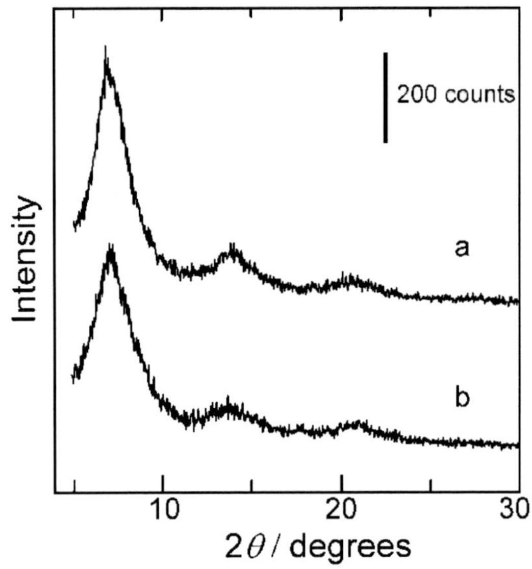

Figure 14.23. XRD patterns of the layered PAH/MnO_2 film taken after being polarized at (a) +0.8 and (b) 0 V for 20 min following 40 CV cycles. (Reprinted from [44], © 2006, with permission from American Chemical Society).

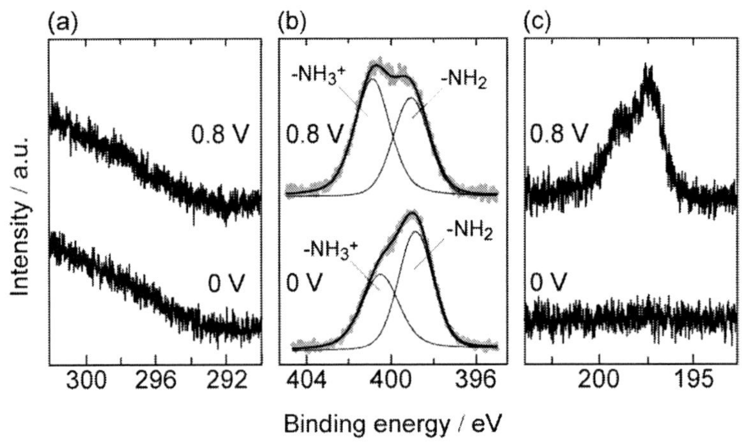

Figure 14.24. XPS core-level spectra in (a) K 2p, (b) N 1s, and (c) Cl 2p regions of the layered PAH/MnO_2 film taken after being polarized at +0.8 and 0 V for 20 min following 40 CV cycles. (Reprinted from [44], © 2006, with permission from American Chemical Society).

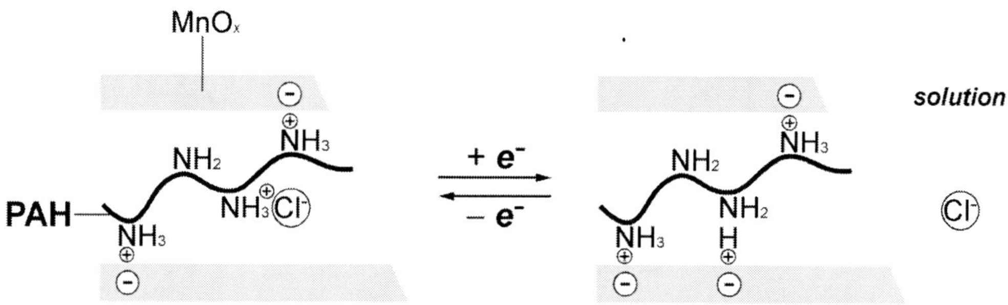

Figuure 14.25. Schematic illustration of the redox process that takes place in the layered PAH/MnO_2 film. (Reprinted from [44], © 2006, with permission from American Chemical Society).

To gain more information on the redox process of the PAH/MnO_2 film, XPS analysis was further conducted for the polarized films, and the spectra are presented in Figure 14.24. There is no feature in the K 2p region, removing the possibility of the participation of K^+ ions in the redox process. Note that in the N 1s core-level spectra the higher BE fraction (i.e., -NH_3^+) decreases at the reduced state. This phenomenon cannot be ascribed to a decrease in the proton content of the film because the motion of protons should be opposite to the direction of the charge compensation. As shown in Figure 14.24c, the Cl 2p doublet appearing at the oxidation potential completely disappears in the reduced film. Since the net charge of MnO_2 layers is negative, the charge balancing mechanism cannot be explained only by the transfer of Cl^-. That is, the observed behavior would not happen individually. All the above results can be explained schematically in Figure 14.25. As already mentioned, the oxidized film contains neutral (-NH_2) and positively charged (-NH_3^+) amine groups in the interlayer space, and some of the -NH_3^+ groups are associated with Cl^- anions, and not only with the negatively charges on MnO_2 layers. Injection of electrons in the oxide layer upon reduction can be balanced by the protons released from the -$NH_3^+Cl^-$ sites, which gives rise to both an increase and decrease in the -NH_2 and -NH_3^+ fractions, respectively. This situation is probably favored over the H^+ incorporation from external electrolyte. At this time, the Cl^- anions become unnecessary and are inevitably excluded from the interlayer to solution.

14.2.3. Surfactant/MnO_2

Figure 14.26 shows XRD patterns of the C16/MnO_2 film taken before (as-deposited) (a) and after being immersed in aqueous Na_2SO_4 solution (b) and ethanol (c). In Na_2SO_4 solution, the diffraction pattern itself is essentially the same as that of the as-deposited film. Clearly, the C16 cations intercalated during electrodeposition are not replaced with small Na^+ ions in solution but remain immobilized in the interlayer, probably due to the lateral interaction between hydrophobic tails of the surfactants. This observation is quite different from the layered MnO_2 film with tetrabutylammonium (Bu_4N^+) ions (Figure 14.15) where the intercalated Bu_4N^+ cations are rapidly replaced with smaller cations in solution, leading to a shrinkage of the interlayer spacing. The interlayer spacing of the MnO_2 film after immersion in Na_2SO_4 solution was 2.8 nm. On the other hand, the immersion in ethanol led to a complete disappearance of the series of peaks, indicating that the crystalline structure of the film was lost.

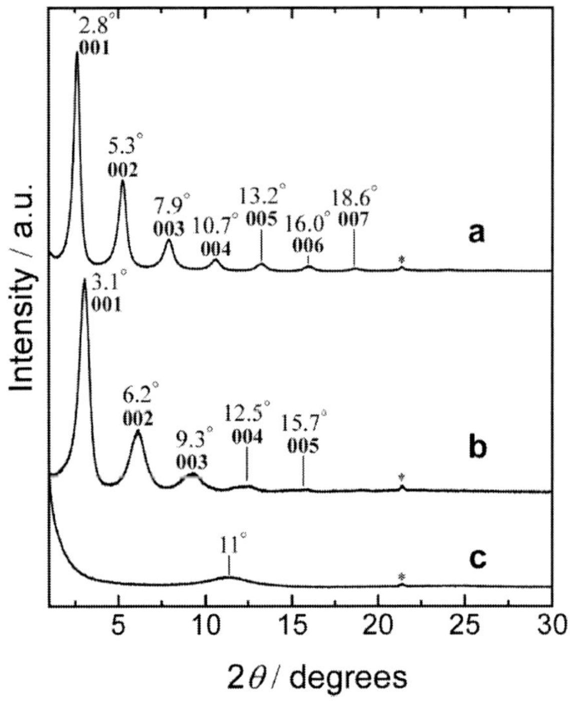

Figure 14.26. XRD patterns of the electrodeposited films (a) before and (b, c) after immersion for 24h in (b) 0.5 M Na_2SO_4 and (c) ethanol. The film was prepared by the same method as that of film a in Figure 14.11. (Reprinted from [48], © 2010, with permission from American Chemical Society).

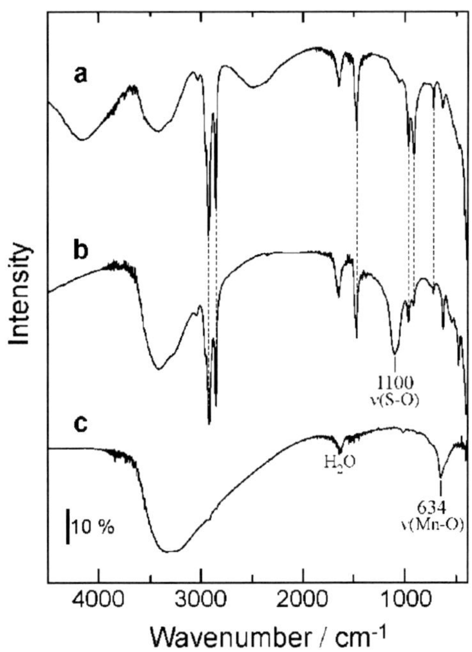

Figure 14.27. FTIR spectra of the electrodeposited films (a) before and (b, c) after immersion for 24 h in (b) 0.5 M Na_2SO_4 and (c) ethanol. (Reprinted from [48], © 2010, with permission from American Chemical Society).

The same samples were subjected to reflection absorption FTIR analysis, and the obtained spectra are displayed in Figure 14. 27. As expected from the XRD data, the absorptions due to C16 molecules remain unchanged after immersion in Na_2SO_4 solution (curve b), as indicated by dotted lines. The only difference is the occurrence of a peak at 1100 cm^{-1}, which can be attributed to the ν(S-O) vibrations in SO_4^{2-} [58]. This clearly indicates the replacement of Cl^- ions by SO_4^{2-}, probably as a result of strong and specific affinity of SO_4^{2-} to the C16 surfactant [59]. As expected from the XRD data, the C16 absorptions completely disappear in ethanol (curve c), leaving the broadened peak at 634 cm^{-1} due to the Mn-O stretching vibrations. Clearly, the anodically-incorporated C16 surfactants are extracted into the organic phase.

Figure 14.28A shows CVs of the layered C16/MnO_2 film when it was cycled repeatedly in aqueous Na_2SO_4 electrolyte. After 30 cycles, the film reached a steady-state where the oxidation/reduction of Mn^{3+}/Mn^{4+} ions in the oxide can take place reversibly [60,61]. The steady-state voltammogram exhibits a roughly rectangular shape without any distinct redox peaks, which corresponds to a typical pseudocapacitive behavior where the current flow is independent of the electrode potential.

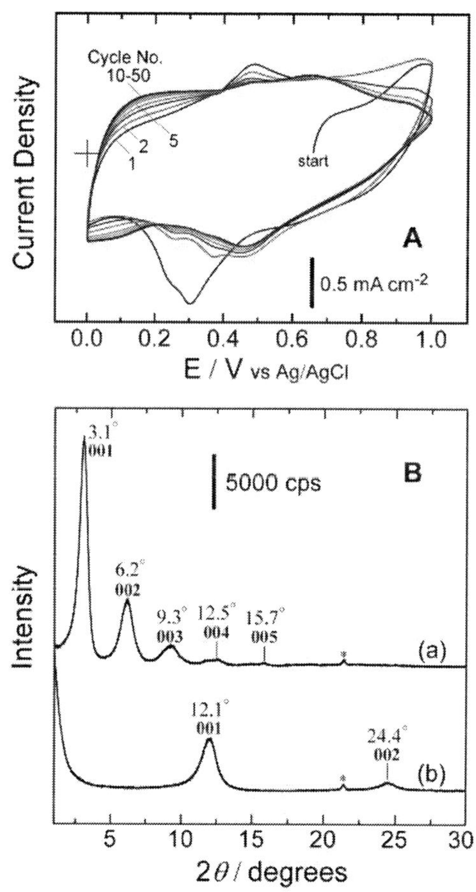

Figure 14.28. (A) Voltammograms of the C16/MnO_2 film electrode in a 0.5 M Na_2SO_4 solution as a potential scan rate of 20 mV/s. (B) XRD patterns of the C16/MnO_2 film taken after being immersed in a 0.5 M Na_2SO_4 solution (a) and then cycled in the same solution (b). (Reprinted from [48], © 2010, with permission from American Chemical Society).

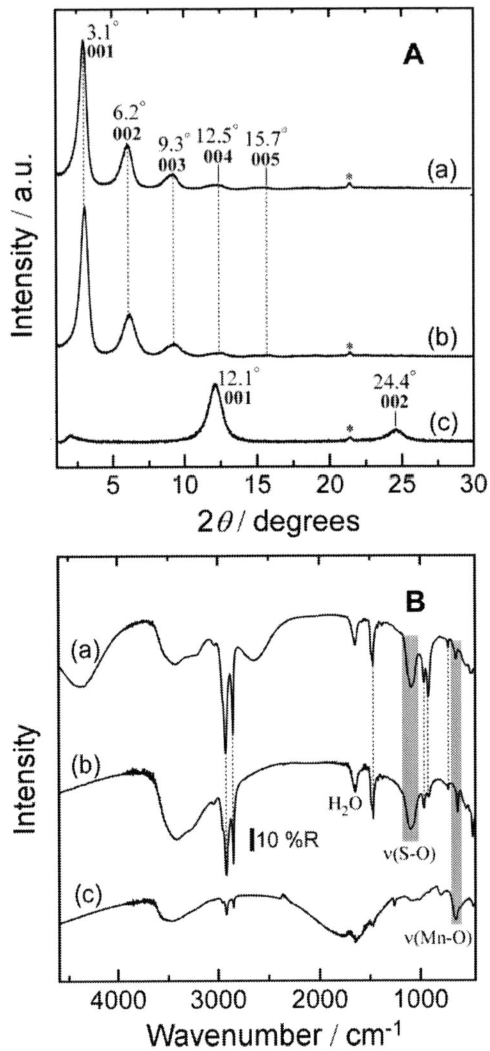

Figure 14.29. (A) XRD patterns and (B) FTIR spectra of the C16/MnO_2 film taken after being held for 30 min at (a) +1.0, (b) OCP, and (c) 0 V in a 0.5 M Na_2SO_4 solution. (Reprinted from [48], © 2010, with permission from American Chemical Society).

XRD pattern of the C16/MnO_2 film taken after 50 CV cycles is shown in Figure 14.28B(b), along with that before cycling; i.e., after being immersed in Na_2SO_4 solution. After the potential cycling, two new peaks appear at 12.1° and 24.4°, corresponding to *d*-spacings of 0.73 (d_{001}) and 0.36 (d_{002}) nm, respectively, while the characteristic peaks of the C16/MnO_2 structure disappear completely. Clearly, all the intercalated C16 cations are replaced with smaller Na^+ ions (0.19 nm) [62]. In this case, the interlayer spacing (d_{001}) is determined by the size of water molecule (0.28 nm), resulting in a shrinkage of the interlayer from 2.8 to 0.73 nm. Since the surfactants are not replaced with Na^+ ions in solution through a simple ion-exchanged mechanism, as shown in Figure 14.26 and 14.27, they must be expelled electrochemically to balance the charge on the MnO_2 film during the redox reactions. The fact that all surfactants were replaced with Na^+ ions strongly suggests that the MnO_2 layers are electrochemically active and their internal surface is available to be intercalated with charge-compensating ions (Na^+) during redox processes. To further elucidate the electrochemical properties, the film structure was investigated as a function of the applied potential. Figure 14.29 displays XRD patterns (A) and FTIR spectra (B) of the C16/MnO_2 films taken after being held at open circuit potential (OCP, b) and anodic (a) and cathodic (c) potentials for 30 min in Na_2SO_4 electrolyte. Unexpectedly, in Figure 14.29A, the diffraction pattern of the film polarized at +1.0 V is the same as that taken at OCP. This can also be confirmed from the infrared absorptions arising from C16 and SO_4^{2-} (curves a and b in Figure 14.29B). On the other hand, the XRD pattern at 0 V is essentially the same as that obtained after CV cycling (Figure 14.28), indicating the occupation of the interlayer with Na^+ ions. The FTIR spectrum also reveals the loss of both C16 and SO_4^{2-}.

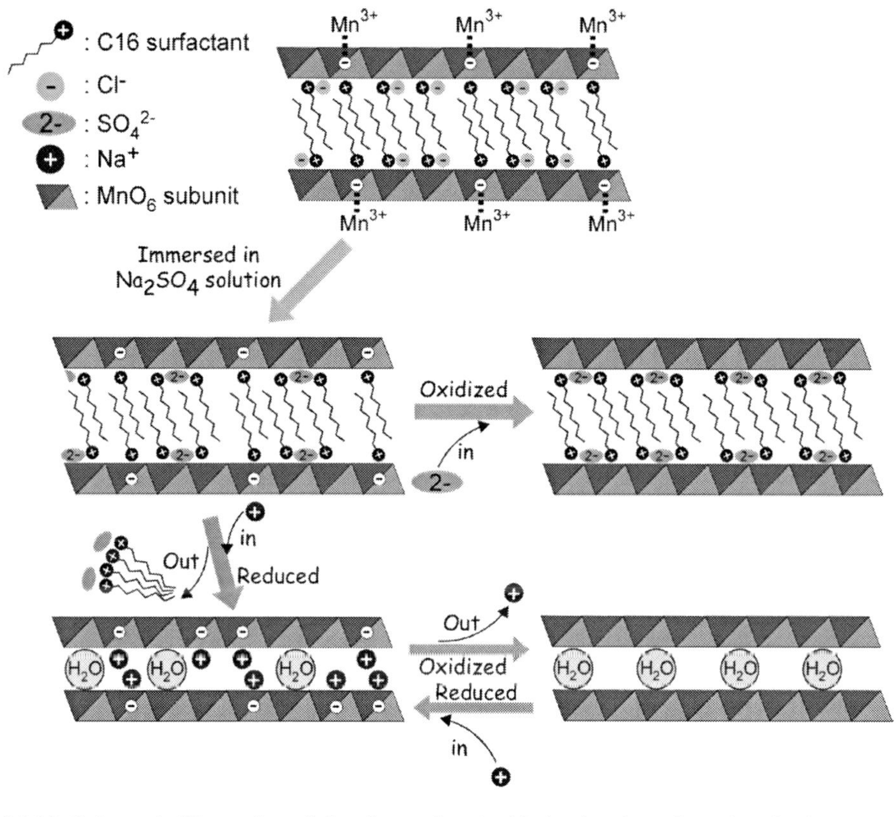

Figure 14.30. Schematic illustration of the electrochemical behavior that takes place in the C16/MnO_2 film. (Reprinted from [48], © 2010, with permission from American Chemical Society).

All the above results can be explained schematically in Figure 14.30. The Cl^- anions incorporated during electrodeposition are ion-exchanged with SO_4^{2-} when the C16/MnO_2 film is immersed in Na_2SO_4 solution. Upon oxidation, Mn^{3+} ions in the oxide sheets are converted to Mn^{4+} ions, leading to a decrease in the negative charge of MnO_2 layers, which can be compensated by the insertion of SO_4^{2-} anions, and not by the deintercalation of individual C16 cations. This phenomenon can be recognized by a unique property of surfactant molecules; i.e., they act as aggregates owing to their hydrophobic interaction and do not act as individual ions above the critical aggregation (micelle) concentration (1.35 mM in the absence of additional salt [63]). In this situation, the newly-inserted SO_4^{2-} anions are coupled with the cationic heads of the surfactants that became unnecessary to balance the negative charges on the MnO_2 layers. Upon reduction, there are two possibilities to balance the increased negative charge on the MnO_2 layers due to the conversion of some Mn^{4+} ions to Mn^{3+} ions: (i) the SO_4^{2-} ions intercalated in the interlayer go out to leave C16 cations, or (ii) Na^+ ions in solution are inserted. The latter process coincides with the obtained results, probably because of the specific and strong affinity of SO_4^{2-} to C16. The insertion of Na^+ ions leads to simultaneous deintercalation of all the surfactants together with SO_4^{2-}. Once Na^+ ions are inserted, they can act as the charge-compensating ions during the redox processes of MnO_2 [61], which is responsible for the steady state voltammetric response after 30 cycles in Figure 14.28A. As will be described in Section 14.5.1., the fact that C16 cations remain immobilized between the MnO_2 layers at anodic potentials prompted us to examine the use of the resulting film electrode for the oxidation of hydroquinone (HQ) within the large interlayer space.

14.3. APPLICATIONS

14.3.1. MnO_2-Based Supercapacitor Electrodes

Supercapacitors have recently attracted world-wide research interest because of their higher power density and longer cycle life than rechargeable batteries and higher energy density than traditional electrostatic capacitors. Supercapacitors can be classified into two types according to their charge storage mechanism: electrical double-layer capacitors (EDLCs) and pseudocapacitors (or redox supercapacitors). Pseudocapacitors are able to store energy through fast and reversible redox (faradaic) reactions of electroactive materials having multiple valence states, such as transition metal oxides or conductive polymers. Capacitive behavior of amorphous hydrated manganese dioxide (a-$MnO_2 \cdot nH_2O$) mixed with activated carbon as a conductive agent was first recognized in 1999 by Lee and Goodenough [64]. Since then, a wide variety of MnO_2-based materials have been reported with their pseudocapacitive properties. Among all the transition metal oxides investigated, MnO_2 is one of the most promising electrode materials for pseudocapacitor applications because of its specific capacitance as well as environmental compatibility and cost effectiveness. Transition of Mn^{III}/Mn^{IV} involving a single electron transfer is responsible for the pseudocapacitance of MnO_2. As indicated in the following equation, both cations (C^+) and electrons take part in the charge/discharge processes.

$$MnO_2 + C^+ + e^- \Leftrightarrow MnOOC \qquad (3)$$

If one Mn atom in MnO_2 is assumed to store one electron, the specific capacitance of MnO_2 should be 1370 F/g [61]. In fact, however, a specific capacitance of only one-fifth or one-sixth of the theoretical value is delivered. Such low practical specific capacitance of MnO_2 can be ascribed to its intrinsic poor electrical conductivity and dense morphology. To avoid these problems, mainly three strategies have been developed: (1) fabrication of micro- and nanostructured MnO_2 [65-69], (2) doping of other transition metals [70-75], and (3) combination with nanocarbons such as carbon nanotubes (CNTs) [76-82]. Supercapacitors based on MnO_2 have been recently reviewed by Wei *et al.* [15]. Relationship between the crystallographic structures and their pseudocapacitive properties has been investigated by Brousse *et al.*[83] and Devaraj *et al.*[84]. Both research groups pointed out that birnessite has higher specific capacitance than 1D and 3D crystals. This can be ascribed to bicontinuous network of solid and pores in birnessite, which enables both electrons and cations to move fast and reversibly.

14.3.2. Pseudocapacitive Performance of Layered MnO_2 Thin Films

This section will review our two approaches to draw out the potential pseudocapacitive ability of birnessite films [85,86].

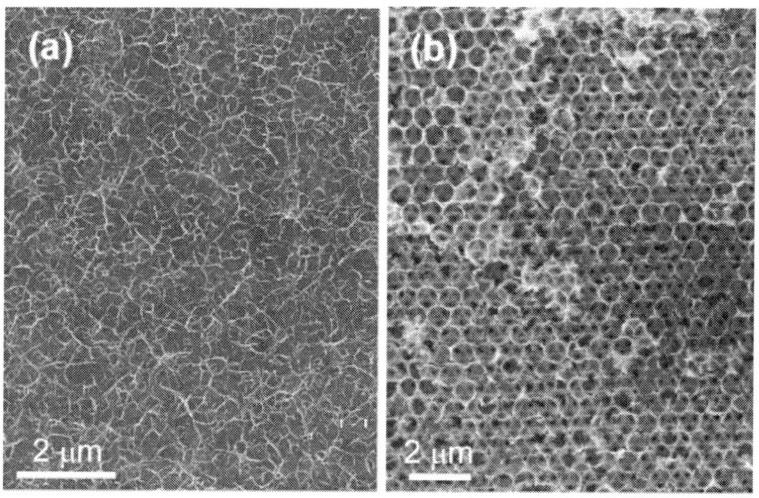

Figure 14.31. SEM images of the MnO_2 films deposited on ITO electrodes without (a) and with (b) a template assembled from 740 nm PS particles. The films were prepared by applying a constant potential of +1.0 V in a 2 mM $MnSO_4$ and 50 mM KCl solution. The electric charge passed during electrolysis was always 80 mC/cm^2. (Reprinted from [85], © 2007, with permission from Elsevier).

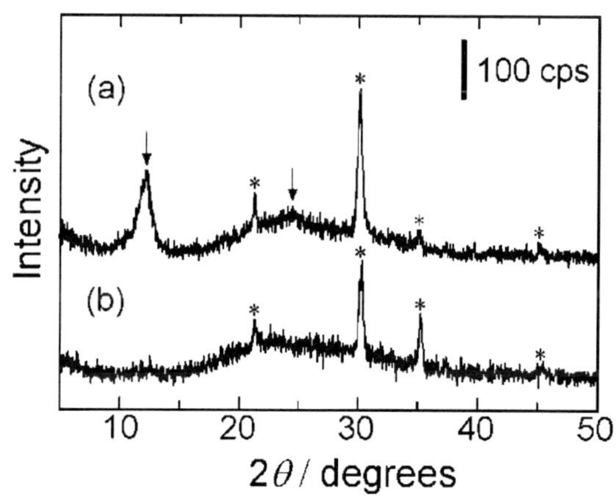

Figure 14.32. XRD patterns of the MnO_2 films deposited on ITO electrodes without (a) and with (b) a template assembled from 740 nm PS particles. The film preparation was made the same way as that in Figure 14.31. (Reprinted from [85], © 2007, with permission from Elsevier).

14.3.2.1. Macroporous Birnessite Film

Birnessite-type MnO_2 was electrodeposited through a colloidal crystal template formed from polystyrene (PS) particles with a diameter of 740 nm in order to fabricate ordered macropores in the MnO_2 film [85]. The template was prepared on an ITO glass substrate by immersing the substrate vertically into a PS suspension and then keeping at 60 °C for evaporation. After electrodeposition, PS particles were removed by dissolving in toluene to leave the structured MnO_2 film. Figure 14. 31 displays SEM images of the MnO_2 films deposited on ITO substrates without (a) and with (b) the PS template by applying a constant potential of +1.0 V to pass the same anodic charge per unit geometric area (80 mC/cm^2). The time necessary for this charge was different in the former (700 s) and latter (1800 s), reflecting a difference in the electroactive surface area. A sheet-like morphology for the former electrode is common for the layered MnO_2 films electrodeposited. On the other hand, the latter presents a three-dimensionaly ordered macroporous structure consisting of regular spherical voids arranged in a close-packed structure. The size of the spherical voids is the same as that of the template particles. A close inspection of the image reveals that the thickness of the MnO_2 film is less than 1+1/2 sphere diameters in thickness, which corresponds to about 1 μm assuming the closed packing of 740-nm PS particles.XRD patterns of the films in Figure 14.31 are depicted in Figure 14.32. In curve a, two peaks appear at 2θ = 12.3 and 24.7°, together with the peaks marked by an asterisk (*) and a broad hump between 15 and 40° arising from the ITO substrate. These peaks can easily be indexed to the 00*l* (*l*=1 and 2) reflections of birnessite, as described above. The d_{001} value of 0.72 nm corresponds to the interlayer spacing and is attributable to birnessite with hydrated K^+ ions.

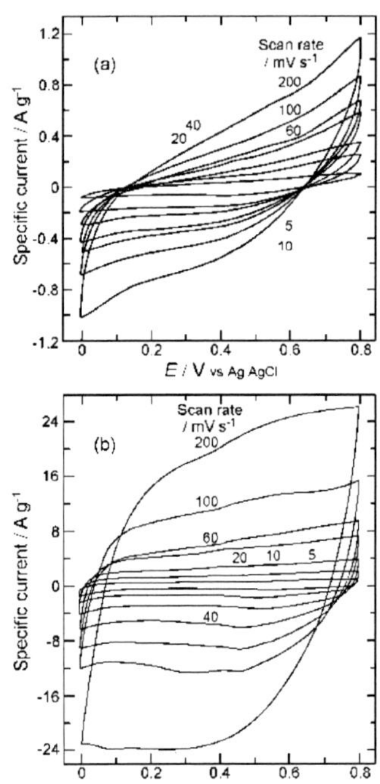

Figure 14.33. CVs of the MnO_2 films deposited on ITO electrodes without (a) and with (b) the PS template when the electrode potential was cycled in a 0.5 M KCl at various scan rates. The film preparation was made the same way as that in Figure 14.31. (Reprinted from [85], © 2007, with permission from Elsevier).

Table 14.1. Specific capacitances of the MnO_2 films deposited on ITO substrates with and without the PS template

	Specific capacitance (F/g)	
Scan rate (mV/s)	with PS	without PS (bare ITO)
2	163.4	21.1
5	152.0	13.9
20	130.4	6.4
60	111.2	3.5
100	98.4	3.4
200	80.7	2.2

As shown in curve b, however, the corresponding peaks cannot be seen for the film deposited through the template. One possible reason is that the crystallites formed inside the interstitial spaces between PS particles are so small, as expected from the SEM photograph, that they could not be detected by XRD.Electrochemical properties of the ITO-supported MnO_2 films were elucidated by cyclic voltammetry in 0.5 M KCl solution at different potential scan rates ranging from 2 to 200 mV/s. Figure 14.33 displays CV curves recorded at the steady-state where the oxidation/reduction of Mn^{3+}/Mn^{4+} in the oxide sheets can take place, concomitantly with the exclusion/inclusion of cations. The film on a bare ITO electrode exhibits smaller voltammograms, while their shape is seriously distorted from the rectangular shape even at very slow scan rates, probably as a result of a large electrical resistance of the film. In contrast, the macroporous film exhibits much larger current than that of the film on bare ITO.

Moreover, the rectangular shape is maintained even at higher scan rates, indicating an ideal capacitive behavior. This can be attributed to the improved ionic diffusion and electron conduction owing to the regularly ordered macropores. Specific capacitance (in F/g) can be calculated by integrating the CV curve to obtain the voltammetric charge (in C), and subsequently dividing it by the mass of the electroactive material (in g) and the potential window (in V). The mass of the electroactive (as MnO_2) was calculated on the basis of the charge consumed for the electrodeposition of MnO_2. Table 14.1 summarizes specific capacitances of birnessite films deposited on ITO electrodes with and without the PS template, which were calculated from the CV data of Figure 14.33a and 14.33b. The values observed at the slow scan rates for the macroporous film are similar to those (150 to 200 F/g) reported previously for a-$MnO_2 \cdot nH_2O$ with conductive carbon [64], despite the less conductive property of ITO substrate.

Figure 14.34 displays the steady-state CV responses of the macroporous film recorded in KCl or Bu_4NCl solution at 20 and 100 mV/s. At 20 mV/s, the voltammetry charge is much larger in KCl solution than that in Bu_4NCl solution, and the difference is enhanced at the higher scan rate. This clearly indicates that electrolyte cations play a role as the charge-compensating ion and that a part of electroactive sites are associated with Bu_4N^+ cations with more difficulty. This finding can be explained schematically, as shown in Figure 14.35. Both K^+ and Bu_4N^+ ions are accessible to the surface of macropores replicated from the PS particles with a diameter of 740 nm.

The interlayer space of the layered MnO_2 is capable of accommodating K^+ ions electrochemically upon reduction, and vice versa upon oxidation [28]. If MnO_2 forms the layered structure, bulkier Bu_4N^+ ions (1.00 nm in diameter) [87] are unable to reach the internal surface with the interlayer distance of 0.72 nm; however, smaller K^+ ions (0.26 nm) [87] make it possible. Therefore, the difference between the CV responses in KCl and Bu_4NCl solutions demonstrates that multilayers are grown anodically in the void spaces between the PS particles, and their internal surface is available for inclusion/exclusion of K^+ ions upon reduction/oxidation of Mn^{4+}/Mn^{3+} in the oxide sheets.

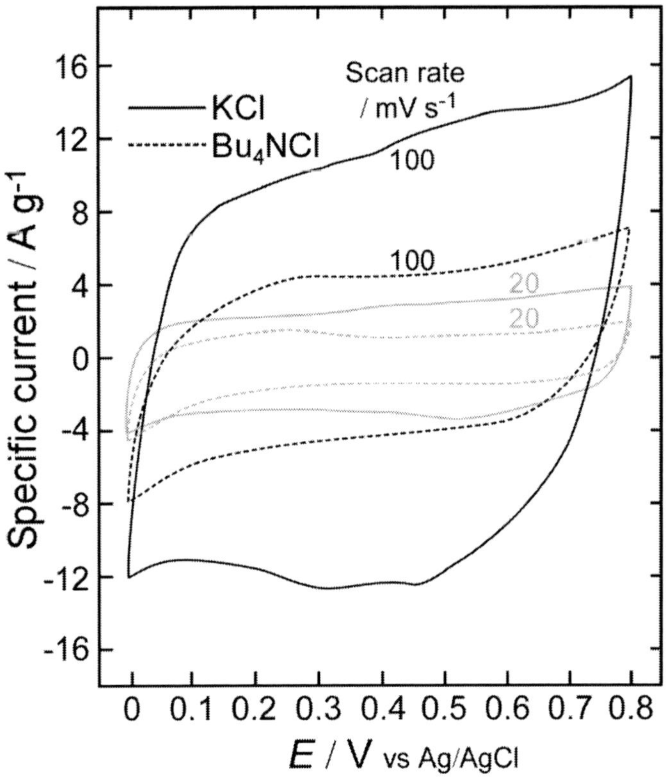

Figure 14.34. CVs of the MnO_2 film deposited on the template-coated ITO electrode when the electrode potential was cycled in 0.5 M KCl and Bu_4NCl solutions at scan rates of 20 and 100 mV/s. The film preparation was made the same way as that in Figure 14.31. (Reprinted from [85], © 2007, with permission from Elsevier).

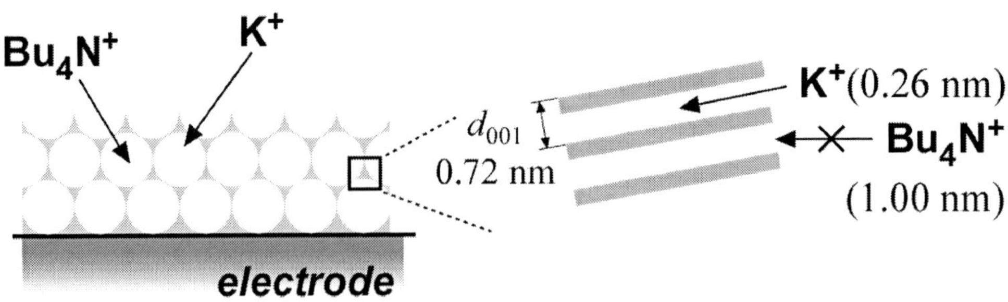

Figure 14.35. Schematic illustration of the electrochemical behavior that takes place in the layered MnO_2 film. (Reprinted from [85], © 2007, with permission from Elsevier).

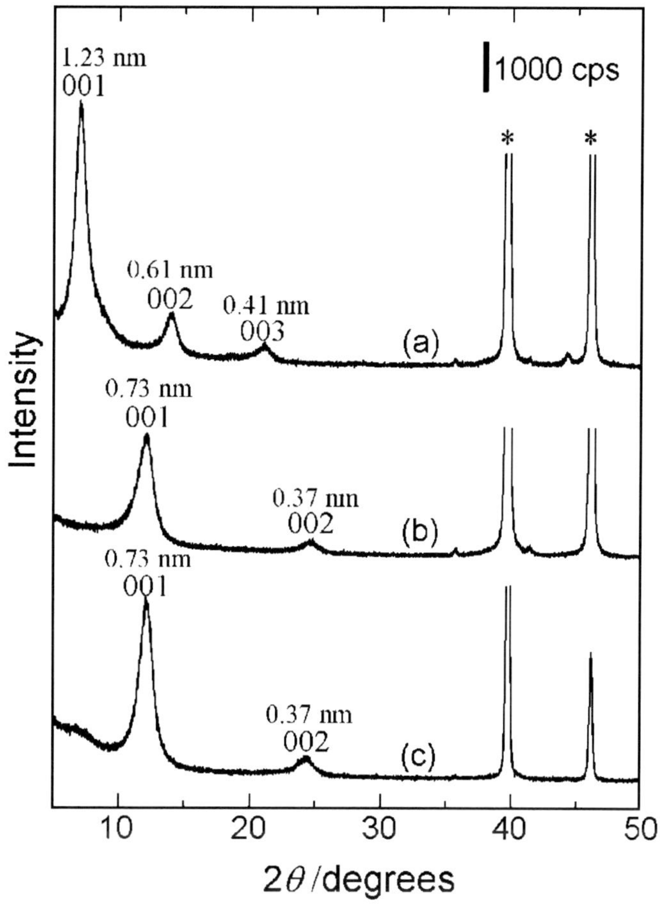

Figure 14.36. XRD patterns of MnO_2 films on a Pt electrode taken (a) before and (b, c) after immersing the electrode in 0.5 M (b) $CoSO_4$ and (c) Na_2SO_4 solutions for 24 h. (*) represents the contributions from the underlying Pt substrate. (Reprinted from [86], © 2009, with permission from The Electrochemical Society).

14.3.2.2. Birnessite Film with Vertically Aligned Multilayers

Figure 14.36 depicts XRD patterns of the films taken before (a) and after immersion in Co^{2+} (b)- and Na^+ (c)-containing solutions [86]. As described in Section 14.1.2, peaks at 2θ = 7.0, 14.0, and 21.2° in curve a can be indexed to the 001 diffraction, and its second (002) and third (003) order diffractions from multilayers of birnessite-type MnO_2 film. The interlayer spacing (=d_{001}) was estimated to be 1.23 nm according to Bragg's equation. In curve b, the 001 and 002 peaks shift to higher angles, correspondingly smaller *d*-spacings. Clearly, bulky Bu_4N^+ ions were replaced with small Co^{2+} ions (0.14 nm) [62], resulting in shrinkage of the interlayer spacing. The intercalated Co ions are hydrated, and the interlayer spacing is determined by the size of water molecule (0.28 nm) and the crystallographic thickness of a single MnO_2 nanosheet (0.45 nm) [45]. In curve c, the same pattern was obtained for the film ion-exchanged with Na^+ (0.19 nm) which is also smaller than water molecule. Figure 14. 37 displays electrochemical cycling processes recorded for the Co/MnO_2 (a) and Na/MnO_2 (b) films at a scan rate of 20 mV/s. As shown in Figure 14.37b, the Na/MnO_2 film provides very small current. This observation is quite different from the fact that chemically-synthesized Na-birnessite mixed with carbon and binder exhibits a pseudocapacitance as high as 200 F/g

[88-90]. In contrast, the Co/MnO_2 film exhibits larger current which gradually increases with the number of cycles. Although the voltammograms acquired at initial cycles are distorted, they gradually become a rectangular mirror image, characteristic of an ideal pseudocapacitor. Transition of Mn^{4+}/Mn^{3+} involving a single electron transfer is responsible for the pseudocapacitance of MnO_2 in the potential region examined.

Note that the capacitive current is superimposed by anodic and cathodic peaks at 0.55 and 0.52 V. With increasing the cycle number until 200 cycles, the redox peaks become more prominent, which can also be associated with the redox of MnO_2 itself [91]. An anodic current appears from +0.6 V and decreases with the cycle number. This current was not observed for the Na^+/MnO_2 film. Wei *et al.* reported that Co^{2+} ions cannot be oxidized at less positive potentials than +1.2 V vs SCE at 25 °C and become oxidizable at +0.6 V at 70 °C as a result of dissociation of $CoSO_4$ [92]. For the same reason, the Co^{2+} ions isolated in the interlayer may be oxidized to Co^{3+} at potentials more positive than +0.6 V [93].

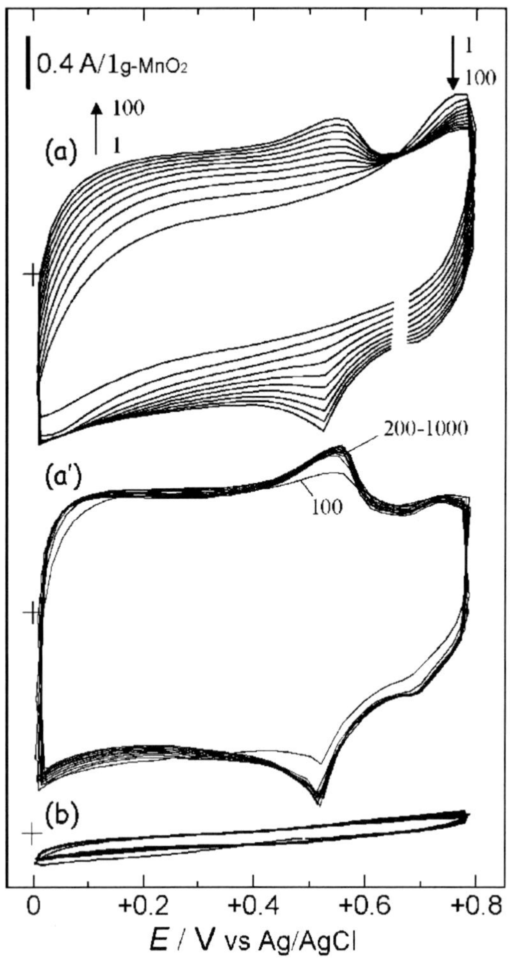

Figure 14.37. CVs of (a) Co/MnO_2 and (b) Na/MnO_2 films deposited on a Pt electrode. These CVs were taken in an aqueous electrolyte of 0.5 M Na_2SO_4 at a scan rate of 20 mV/s. (Reprinted from [86], © 2009, with permission from The Electrochemical Society).

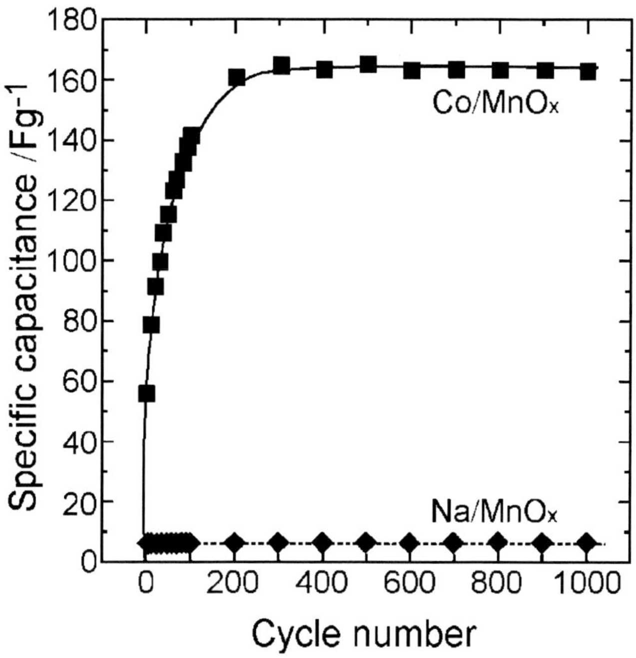

Figure 14.38. Specific capacitance obtained in Figure 14.36 versus cycle number. (Reprinted from [86], © 2009, with permission from The Electrochemical Society).

Table 14.2. Dependence of specific capacitances on scan rate for the Co/MnO_2 film electrochemically activated in a 0.5 M Na_2SO_4 solution

Scan rate (mV/s)	2	5	10	20	50	100	200
Capacitance (F/g)	223.0	206.2	186.5	165.5	147.1	132.9	114.7

On the basis of the above CVs, specific capacitance was calculated and is plotted against the number of cycles in Figure 14.38. The specific capacitance gradually increases with the cycle number to reach the highest value of 165 F/g at 300th cycle and then becomes constant until at least 1000 charge/discharge cycles. The capacitance value of the Na/MnO_2 film was 8.4 F/g. The dependence of capacitance on scan rate was examined for the film after 500 charge/discharge cycles (Table 14.2). The specific capacitance was measured to be 223 F/g at 2 mV/s, in good agreement with those reported for birnessites where conductive and binding reagents were added [88-90]. Similar behavior can be found in previous studies [94,95], where specific capacitances of MnO_2 films were enhanced by electrochemical cycling. Both phenomena were commonly accompanied by drastic morphology changes.

Figure 14.39 shows SEM images of the Co/MnO_2 film before and after 500 cycles. Sheet-like morphology of the uncycled film is typical of the anodically grown layered MnO_2 film (Figure 14.31). In this film, MnO_2 multilayers are lying parallel to the surface of substrate. Amazingly, after 500 cycles plate-like deposits of nearly-uniform size appear, standing vertically to the substrate surface. In an EPMA spectrum taken after CV cycling (not shown), a peak due to Na^+ was detected which was missing in the uncycled film. On the other hand, peaks due to Co ions decreased in intensity although they remained present even after 1000 cycles. Because it was confirmed that the intercalated Co^{2+} ions were not ion-exchanged with Na^+ ions in solution, the Co^{2+} ions are repulsed electrochemically upon oxidation of the

MnO_2 layers. The electroneutrality of the film can be maintained insertion/removal of electrolyte Na^+ ions. Although the state of Co ions remains uninvestigated, some of the Co ions intercalated may move to defects in MnO_2 layers [96], causing a change in morphology.

Figure 14.40 shows XRD patterns for the uncycled and cycled Co/MnO_2 films. Both patterns exhibit the characteristic 001 and 002 peaks at the same positions, confirming no essential change in the crystal structure. It is worth noting that the peaks are sharpened after the CV cycling; i. e., the FWHM (full width at half maximum) was 0.015 and 0.009° for the uncycled and cycled films, respectively. This corresponds to an increase of the crystallite size along the 001 direction from 9.2 to 15.5 nm which can be deduced by Scherrer's equation (k=0.89). The crystallite size estimated for the cycled film is in very good agreement with a thickness of the plate-like deposit in the SEM image of Figure 14.39b. Therefore, it is natural to consider that each plate-like deposit consists of multilayers of MnO_2 which are no longer parallel to the electrode substrate but aligned vertically. This structure allows transfer of electrons directing normal to the electrode surface and easy access of charge-compensating cations to the redox active sites in the internal surface of MnO_2, giving rise to a good pseudocapacitive performance.

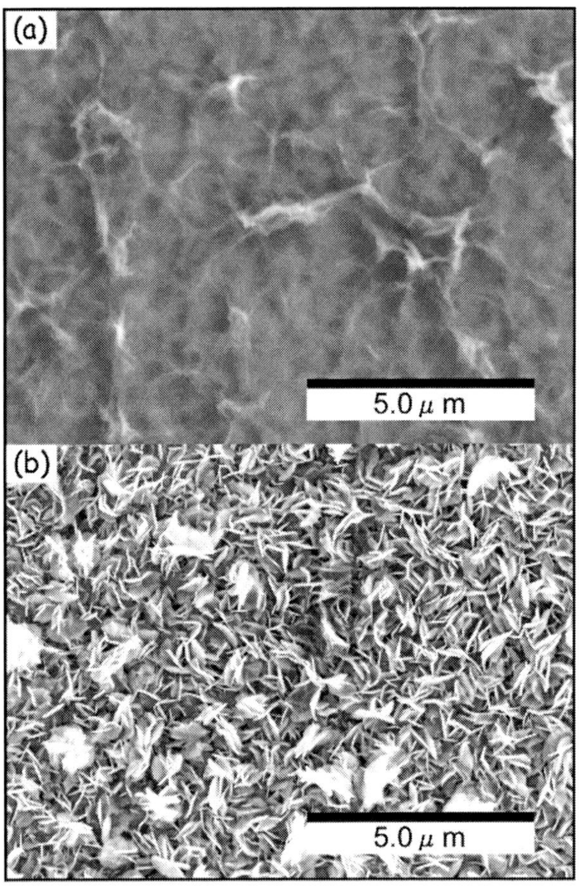

Figure 14.39. SEM images of the Co/MnO_2 film (a) before and (b) after 500 CV cycles.

14.4. Electrochromic Materials

14.4.1. MnO_2-Based Electrochromic Materials

Electrochromic (EC) materials can reversibly change their optical characteristics upon charge injection and extraction induced by the application of an external electric field. Because of their low operation voltage and low power consumption, EC materials have found a variety of applications including large area displays, antiglare mirrors, and energy-saving smart windows. Among various inorganic oxides having EC characteristics, cathodically coloring tungsten oxide (WO_3) is recognized as one of the highest-efficiency material for smart windows [97-99]. In order to assemble an electrochromic device based on WO_3, complementary coloring materials are needed as a counter electrode to enhance the color change. MnO_2 which is colored at anodic potentials are one of the promising candidates, and their electrochromic properties have been examined for various crystal phases and morphologies in both aqueous [5-8] and nonaqueous [9-13] electrolytes. The performance of EC materials depends on their crystalline structure and morphology. As with other transition metal oxides, the electrochromism of MnO_2-based materials can be achieved by simultaneous insertion/extraction of electrons and electrolyte cations into/from the inorganic matrix. The Mn^{4+} state of Mn oxide has an optical absorption due to the *d–d* transition in the visible region, resulting in a brown color. Upon reduction of Mn^{4+} to Mn^{3+}, the color changes from brown to pale yellow as a result of elimination of the absorbing species.

In recent years, nanotechnologies have been introduced to the fabrication of metal oxides, allowing facile mass transfer of cations and solvent molecules to the electroactive sites. Long *et al.* prepared a birnessite-type layered MnO_2 film, $Na_\delta MnO_2 \cdot xH_2O$, on ITO substrate by sol-gel reaction and found a high coloration efficiency of 36 cm^2/C in propylene carbonate solution with $LiClO_4$ [12]. Almost the same efficiency (37 cm^2/C) was also reported by Sakai *et al.* for a multilayered film composed of MnO_2 nanosheets and polycations that was formed by layer-by-layer assembly [13]. In such layered structures, all of the material can be regarded as surfaces. This aspect is quite different from bulk materials, in which not all of the metal ions can be involved in the redox reactions, making it difficult to control the color change with the thickness of electrochromic materials.

14.4.2. Electrochromism of Anodically-Deposited Layered MnO_2

As shown above, the layered PAH/MnO_2 film provides a good voltammetric response as a result of the redox reaction of Mn^{3+}/Mn^{4+} couple without any changes in the structure. XPS studies revealed that the oxidation/reduction of the MnO_2 layers is neutralized by insertion/extraction of Cl^- anions into/from -NH_3^+ sites belonging to the intercalated PAH. This mechanism can be represented schematically in Figure 14.25 [44]. In brief, injection of electrons into the MnO_2 layers upon reduction is balanced by the protons that are released from -$NH_3^+Cl^-$ sites. This situation is probably preferred to the incorporation of cations from the electrolyte solution. At this time, the Cl^- anions become unnecessary and are inevitably excluded from the interlayer to the solution. In this section, the layered PAH/MnO_2 films were deposited on a transparent ITO glass electrode using the anodic procedure and submitted to *in situ* ultraviolet-visible spectroscopy to evaluate their electrochromic performance [100] .

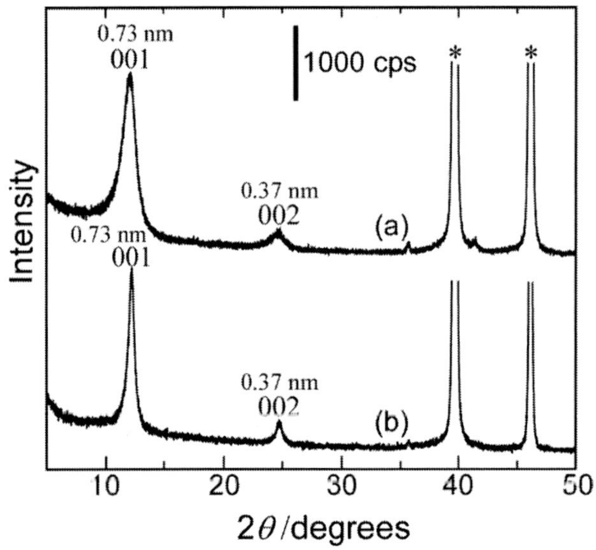

Figure 14.40. XRD patterns of the Co/MnO_2 film (a) before and (b) after 500 CV cycles. (*) represents the contributions from the underlying Pt substrate. (Reprinted from [86], © 2009, with permission from The Electrochemical Society).

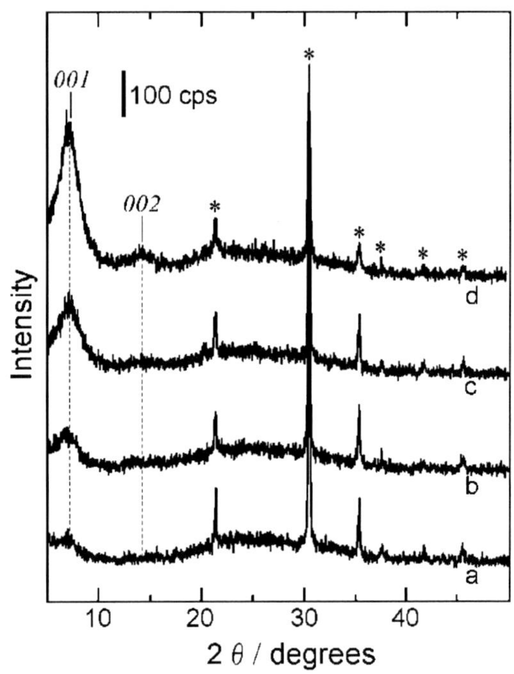

Figure 14.41. XRD patterns of films deposited on an ITO substrate from an aqueous mixture of 2 mM $MnSO_4$ and 200 mM PAH by applying a constant potential of +1.0 for (a) 5, (b) 10, (c) 30, and (d) 60 min. (*) represents the contributions from the underlying ITO substrate.

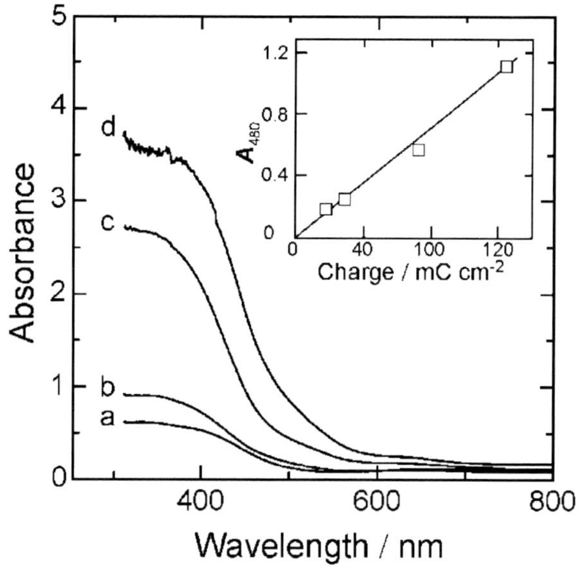

Figure 14.42. UV-vis spectra of films deposited on an ITO substrate from an aqueous mixture of 2 mM $MnSO_4$ and 200 mM PAH by applying a constant potential of +1.0 for (a) 5, (b) 10, (c) 30, and (d) 60 min. Inset shows plots of the absorbance at the wavelength of 480 nm as a function of the charge density passed during electrodeposition. (Reprinted from [100], © 2009, with permission from The Electrochemical Society).

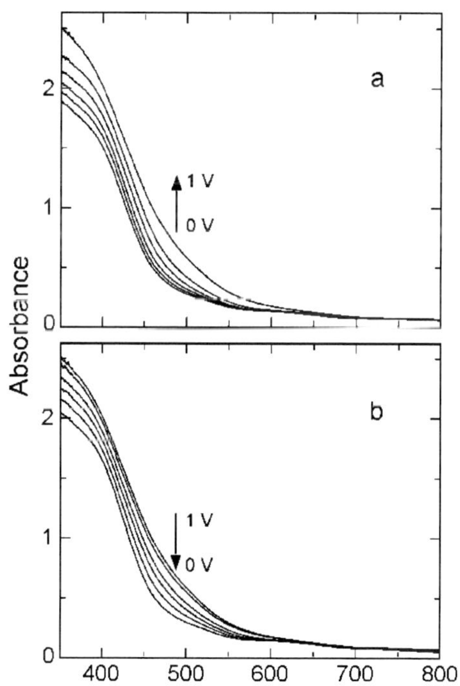

Figure 14.43. *In situ* absorption spectra of a PAH/MnO_2 film-coated ITO electrode taken from (a) 0 to +1.0 V and (b) +1.0 to 0 V at 0.2 V intervals. (Reprinted from [86], © 2009, with permission from The Electrochemical Society).

Figure 14.41 shows XRD patterns of the films on ITO substrate prepared at a constant potential of +1.0 V for different periods of time. Peaks marked with an asterisk (*) are due to the ITO substrate. In addition, two diffraction peaks appear at 7.4 and 14.6°, corresponding to *d*-spacings of 1.19 and 0.61 nm, respectively. These peaks are identical to those observed previously on Pt substrate (Section 14.2.2) and can be indexed on the 00*l* reflections ($l = 1, 2$) of a multilayered structure consisting of MnO_2 layers and the PAH cations sandwiched between them, where the layers and polycations are stacked regularly with a repeating periodicity of 1.19 nm [44]. Figure 14.42 displays *ex situ* UV-vis absorption spectra of the films in Figure 14.41. Broad absorptions lie in the wavelength region from 300 to 700 nm with no distinct peak. This absorption can be ascribed to *d–d* transitions of the Mn^{4+} state ($t_{2g}^3e_g^0$) which presents brown color [13]. The inset shows a relationship between the absorbance at the wavelength of 480 nm (A_{480}) and the charge delivered for electrodeposition. The observed linear relationship reveals the accumulation of MnO_2 with the same electronic state.

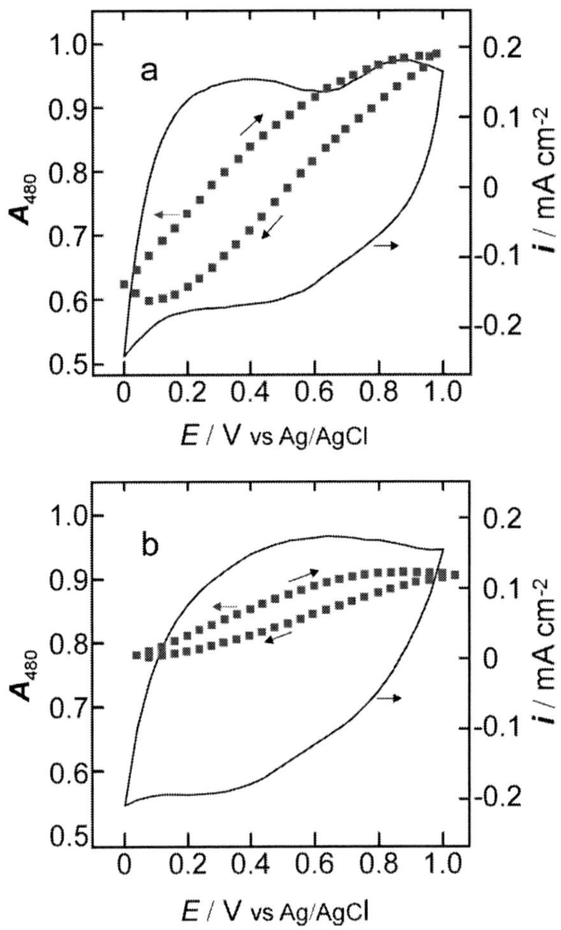

Figure 14.44. CVs and concurrently taken absorbance changes of PAH/MnO_2 films with (a) layered and (b) amorphous structures. Scan rate of 20 mV/s. (Reprinted from [86], © 2009, with permission from The Electrochemical Society).

Figure 14.43 depicts *in situ* absorption spectra of the layered PAH/MnO_2 film recorded when the electrode potential was stepped gradually to anodic (a) and cathodic (b) directions at 0.2 V intervals between 0 and +1.0 V. All the spectra were recorded after a two-minute equilibration period. The absorption due to Mn^{4+} ions in the film increases with an increase in the anodic potential, making the film brown. Upon reduction, the content of Mn^{4+} decreases, going back to the bleached state (pale yellow). Figure 14.44. shows the steady-state CVs of PAH/MnO_2 films with layered (a) and amorphous (b) structures and their accompanying absorbance changes at the wavelength of 480 nm. The amorphous counterpart was made from a 2 mM $MnSO_4$ and 1 mM PAH solution with the same applied charge. Both CVs present almost rectangular profiles characteristic of pseudocapacitor, while the absorbance changes reversibly as a result of increase/decrease of Mn^{4+} ions in the film. Clearly, the coloration/bleaching is attributed to charging/discharging of the pseudocapacitor of MnO_2. Electrochromic phenomena induced by the same mechanism have been reported on thin films of Ni oxides [101,102].

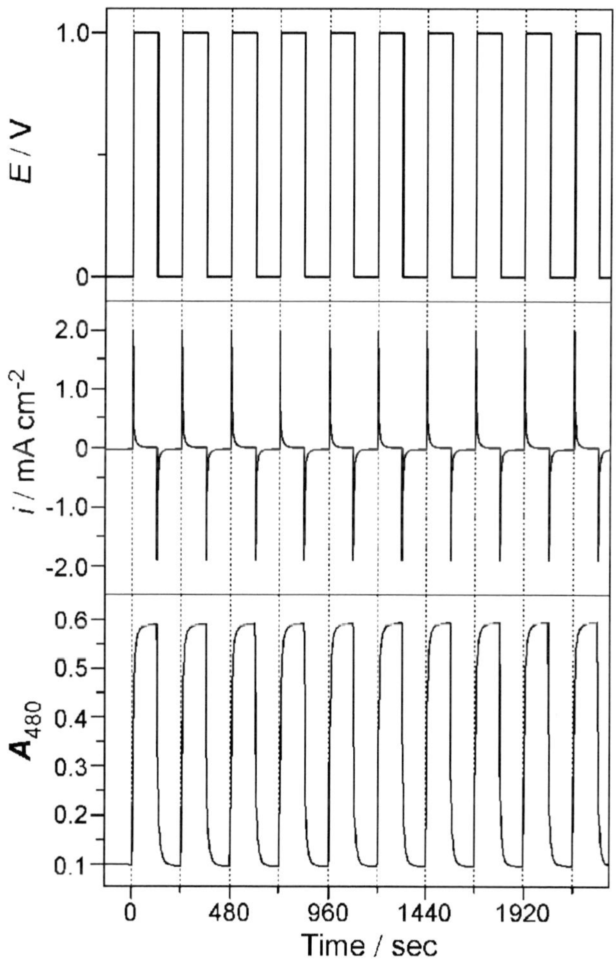

Fugure 14.45. Applied potential, current density, and absorbance at 480 nm of a PAH/MnO_2 film during repetitive potential steps between 0 and +1.0 V. (Reprinted from [86], © 2009, with permission from The Electrochemical Society).

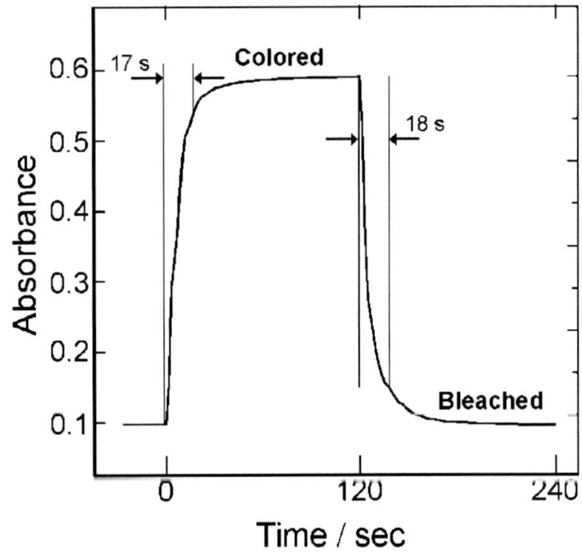

Figure 14.46. One coloration/bleaching cycle extracted from Figure 14.45. (Reprinted from [86], © 2009, with permission from The Electrochemical Society).

From another viewpoint, the color change occurring concurrently with the capacitive behavior can be regarded as an optical evidence that the pseudocapacitance of MnO_2 arises from the faradaic redox reactions of Mn^{4+}/Mn^{3+} couple, which has so far been demonstrated by *ex situ* XPS data [103,61] or *in situ* X-ray absorption spectroscopy [104,105]. No significant difference is seen in their charges; i.e., 6.2 and 6.0 mC/cm^2 for the layered (a) and amorphous (b) films respectively. This suggests that the similar numbers of electroactive sites are available for the redox of Mn^{3+}/Mn^{4+} in the oxide. However, there is a marked difference between their absorption changes. The difference in absorbance between the colored and bleached states is much larger in the layered MnO_2 than in the amorphous one. This implies that MnO_2 layers can absorb visible light more effectively than particles in amorphous phase where the scattering of light will be more serious. Repetitive potential steps between 0 and 1.0 V were applied to the layered PAH/ MnO_2 film at 2 min intervals. The applied potential, the current, as well as the absorbance recorded at 480 nm are shown in Figure 14.45. An anodic current appears immediately after the potential step to +1.0 V and decays rapidly, resulting in an increase in absorbance. The reverse step similarly leads to a rapid reduction of the colored film to the bleached state. The difference in A_{480} between the breached state and the colored state, which corresponds to "optical contrast", was estimated to be 0.49. Such current and absorbance responses did not change noticeably even after 100 cycles. Taken from Figure 14.45, one coloration/bleaching cycle is depicted in Figure 14.46. The time required for 90 % modulation can be estimated to be 17 and 18 s for coloration and bleaching, respectively. The observed ΔA_{480} value (0.49) is much larger than that (<0.3) obtained for the birnessite $Na_{\delta}MnO_2{\cdot}xH_2O$ ambigel film in $LiClO_4$/propylene carbonate during the potential steps between -1.0 and +1.4 V (vs. Ag/AgCl), where the absorbance change was completed after 60 and 240 s for the coloration and bleaching processes, respectively [12]. Besides, a spinel-type $LiMn_2O_4$ film provided a contrast value of 0.4 in an aqueous mixture of LiCl and borate buffer (pH 7.5) for the potential steps between +0.4 and +1.0 V (vs. Ag/AgCl), in which 700 s and 1000 s are required to reach the steady states upon coloration and bleaching processes [6]. Generally, the bleaching (cathodic) process of MnO_2 is considerably slower than the coloration (anodic) process because the insertion of cations upon reduction requires longer time than the extraction upon oxidation. However, such a feature cannot be found in the anodically deposited PAH/MnO_2 film, as indicated above, which may also reflect the unique charge-compensation process associated with anions. The coloration efficiency is defined by $\eta = \Delta A/\Delta Q$, where ΔQ is the corresponding charge consumed for the absorbance change ΔA. Our product was measured to be 63.2 cm^2/C. To the best of our knowledge, this is the highest value among MnO_2-based materials irrespective of the solvent being aqueous or nonaqueous; i.e., it is superior to that of the multilayered MnO_2 film prepared chemically (36 cm^2/C at 500 nm in nonaqueous electrolyte) and that of the γ-Mn_2O_3-based film grown anodically (41.7 cm^2/C at 400 nm in aqueous electrolyte). Good adhesion between the Mn oxide film and the substrate owing to electrodeposition as well as the nanolayered structure can be responsible for the improved EC performance.

14.5. Catalysts

14.5.1. Catalytic Properties of Layered MnO_2

Catalytic activity of birnessite, originating from its higher valant sites (Mn^{4+} ions), has been found in the oxidation of alcohols [106], iodine [107], phenolic compounds [108-110], etc. In such cases, leaching of the reduced Mn species must be avoided in order to sustain the catalytic ability. Phenolic compounds such as hydroquinone, catechol, and resorcinol are discharged to the environment from a variety of industries because they are used in the production of rubber, dyes, pesticides, colors, plastics, and cosmetics. Thus, electrochemical oxidation of phenolic compounds is an important process in the analytical detection of phenols and in the oxidative decomposition of phenol impurities in aqueous systems [111-113].As mentioned in the above sections, surfactants intercalated between MnO_2 layers remain immobilized at anodic potentials in sodium electrolyte. This fact has motivated us to examine catalytic ability of the electrodeposited C16/MnO_2 film toward anodic oxidation of phenol, as will be described below. *In situ* absorption spectra of hydroquinone (HQ) were acquired in 0.5 M Na_2SO_4 electrolyte, while the C16/MnO_2 film-coated ITO electrode was polarized at a constant potential of +1.0 V. Time courses of the UV-vis absorption spectra are shown in Figure 14. 47, taken with (a) and without (b) polarization. HQ provides an absorption peak at 288 nm due to π-π^* transition [114]. Regardless of the polarization state, the absorption at 255 nm attributable to *p*-benzoquinone (*p*-BQ) increases with time, indicating two-electron oxidation of HQ with the high valent Mn sites (Mn^{4+}) to *p*-BQ. In the inset, the residual HQ and the conversion of HQ to BQ are plotted against reaction time. The initial rate of HQ oxidation is apparently enhanced when polarized at +1.0 V. Figure 14.48 depicts *i-t* plots of the C16/MnO_2/ITO electrode when polarized at +1.0 V in 0.5 M Na_2SO_4 solution with (a) and without (b) HQ. As a control experiment, the oxidation of HQ was similarly conducted using a bare ITO electrode (curve c). The film-modified electrode exhibits an enhancement of oxidation current 684 times larger than the unmodified ITO electrode. The current is missing in the absence of HQ, thus indicating that HQ is oxidized catalytically at the C16/MnO_2 film electrode.

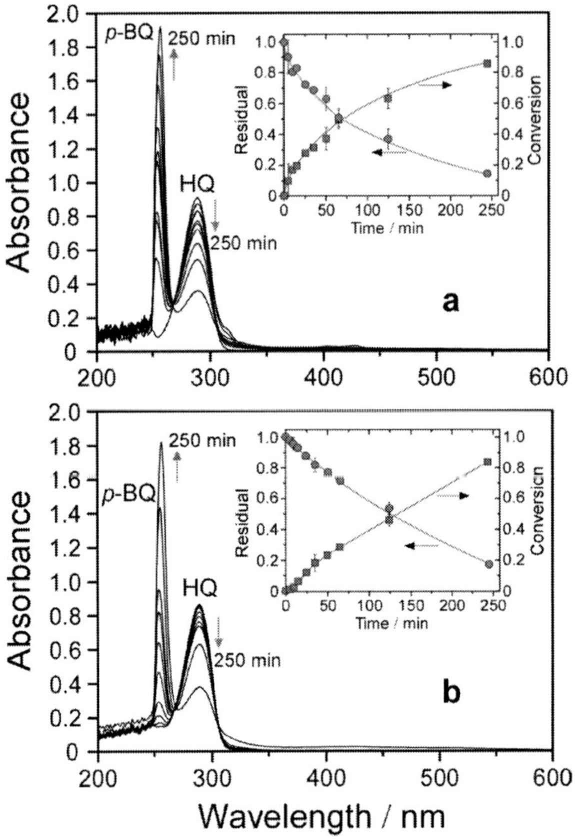

Figure 14.47. Time courses of UV-vis absorption spectra of HQ at 0.3 mM in 0.5 M Na_2SO_4 solution while the C16/MnO_2 film was kept at (a) +1.0 V and (b) OCP. (Reprinted from [48], © 2010, with permission from American Chemical Society).

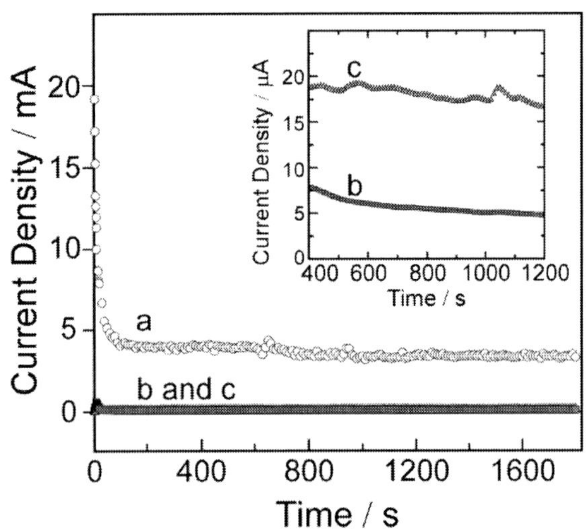

Figure 14.48. *i-t* plots obtained when (a, b) C16/MnO_2 film-modified and (c) unmodified ITO electrodes were polarized at +1.0 V in 0.5 M Na_2SO_4 solution (a, c) with and (b) without 0.3 mM HQ. (Reprinted from [48], © 2010, with permission from American Chemical Society).

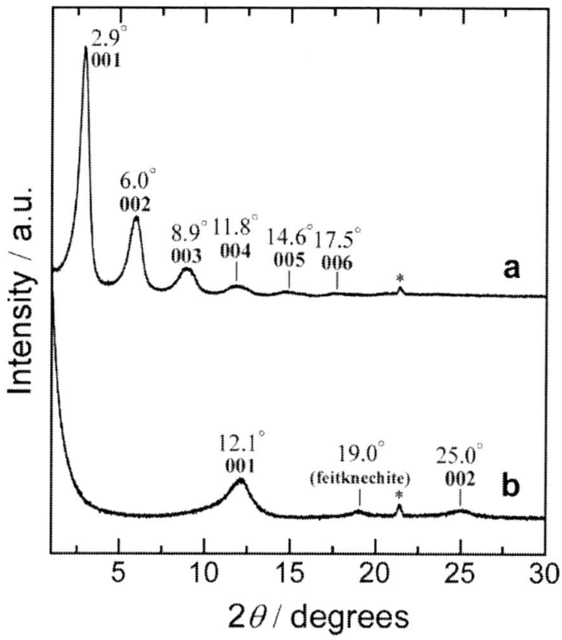

Figure 14.49. XRD patterns of the C16/MnO_2 films on an ITO electrode after being polarized for 250 min at (a) +1.0 V and (b) OCP in a Na_2SO_4 solution containing 0.3 mM HQ. (Reprinted from [48], © 2010, with permission from American Chemical Society).

XRD patterns shown in Figure 14.49 were taken after the HQ oxidation at +1.0 V (a) and OCP (b). Amazingly, although the reaction at OCP led to a collapse of the C16-intercalated structure, the anodic polarization retains the structure. Since the surfactants cannot be ion-exchanged with Na^+ ions, the removal of C16 cations at OCP strongly suggests that the HQ oxidation takes place within the interlayer space. The HQ oxidation proceeds accompanied with the reduction of Mn^{4+} to Mn^{3+}, leading an increase of the negative charge to be compensated by electrolyte Na^+ ions. The surfactants are expelled as soon as they become unnecessary, as in the case when the C16/MnO_2 film is reduced (curves c in Figure 14.29). A peak at 19.0° can be attributed to β-Mn^{III}OOH (feitknechite) [115], indicating the presence of the reduced species. When the electrode is polarized positively, however, the generated Mn^{3+} sites are reoxidized to Mn^{4+}, keeping the net negative charge in the MnO_2 layers so that the surfactants are not necessary to be deintercalated. That is, electrons are transferred from HQ to the substrate electrode through the MnO_2 layers, which provides catalytic current. In other words, the MnO_2 layers can act as mediators for the HQ oxidation taking place in the large interlayer space composed of the assembled C16 surfactants.

14.6. Biosensors

14.6.1. MnO_2-Based Biosensors

MnO_2 exhibits outstanding catalytic abilities in various oxidation and reduction reactions, and the activity for the oxidation of H_2O_2 has found several applications as sensors. In such cases, H_2O_2 is oxidized catalytically on an electrode [116-120], leading to current flow depending on the concentration of H_2O_2. H_2O_2 is one of the products in a variety of biochemical reactions catalyzed by oxidases. Enzyme-based biosensors are usually fabricated by immobilizing enzymes to the surface of an electrode. If an enzyme such as glucose oxidase (GOD) can be incorporated into a matrix that can act as an electron mediator for the oxidation of H_2O_2, the resulting electrode functions as a biosensor such as glucose sensor [121,122]. In such applications, manganese oxides are often employed by mixing with conductive carbon paste [116,118,121]. Otherwise, thin films of manganese oxides are prepared by casting [117,119], electrostatic layer-by-layer (LBL) assembly [120], and electrodeposition [122]. Furthermore, nanostructures of MnO_2 are able to improve their electrochemical properties as a result of reduced diffusion paths and increased accessible active sites. As a matter of fact, one-dimensional tunnel [116,121] and two-dimensional layered [119,120] MnO_2 have been employed with incorporated enzymes, providing excellent sensing performances. Next, we electrodeposited the PAH/MnO_2 film on an ITO substrate (see Section 14.2.2) and the resulting electrode was applied to the electrochemical quantification of H_2O_2 [123].

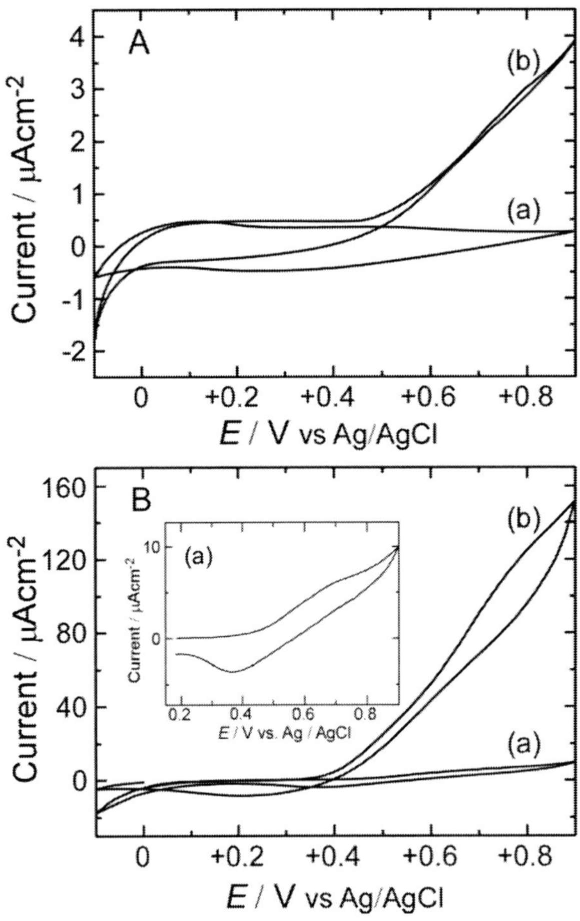

Figure 14.50. Cyclic voltammograms of ITO electrodes modified (A) without and (B) with the PAH/MnO2 film. The electrolytes were 0.1 M phosphate buffer solutions of pH 7.4 (a) without and (b) with 0.9 mM H_2O_2. The PAH/MnO_x film was prepared by applying a constant potential of +1.0 V for 30 min in an aqueous bath containing 2 mM $MnSO_4$ and 200 mM PAH. (Reprinted from [123], © 2010, with permission from Wiley-VCH Verlag GmbH & Co. KGaA).

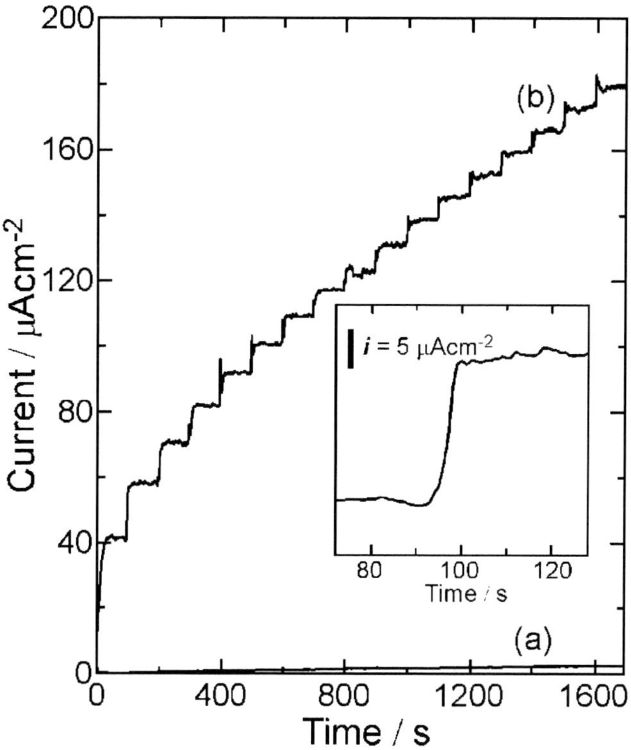

Figure 14.51. Current-time responses of ITO electrodes modified without (a) and with (b) the PAH/MnO_2 film to successive additions of 10 μmol H_2O_2, which were taken 0.1 M phosphate buffer (pH 7.4) at an applied potential of +0.75 V.

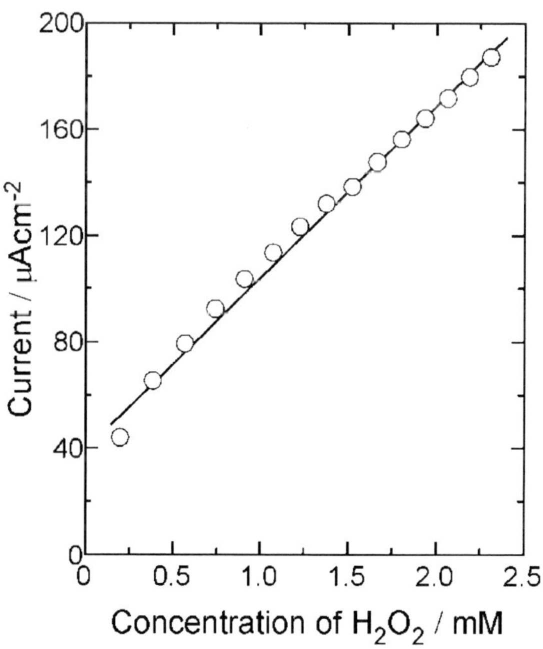

Figure 14.52. Calibration curve of the PAH/MnO_2 film-modified electrode for H_2O_2. The supporting electrolyte was 0.1 M phosphate buffer (pH 7.4).

Figure 14.50 shows CV responses of the ITO electrode without (A) and with (B) PAH/MnO_2 film, which were recorded in phosphate buffer solutions without (a) and with (b) H_2O_2. As shown in Figure 14.50A, oxidation of H_2O_2 starts at +0.45 V on the unmodified electrode. For the PAH/MnO_2 electrode (Figure 14.50B), there is a pair of broad redox waves peaked around +0.7 V and at +0.35 V in the absence of H_2O_2. They can be attributed to the oxidation of the low oxidation state (3+) of Mn in the oxide layers and the reduction of Mn^{4+}, respectively [28]. With the addition of H_2O_2 to the solution, the oxidation current starts to appear at +0.35 V (curve b), which is about 40 times larger in magnitude than that of the unmodified electrode. This clearly indicates a catalytic ability of the PAH/MnO_2 for oxidation of H_2O_2.Figure 14.51. illustrates the current-time responses of ITO electrodes modified without (a) and with (b) PAH/MnO_2 film upon successive additions of 10 μmol H_2O_2 at an applied potential of +0.75 V. The unmodified electrode provides only insignificant response for the concentration changes applied. In contrast, the modified electrode responds rapidly to the change in H_2O_2 concentration. As shown in the inset, it takes less than 10 s for the PAH/MnO_2 electrode to reach 95 % of the maximum current. Figure 14.52 shows the calibration curve of the PAH/MnO_2 film-modified ITO electrode to the H_2O_2 concentration, based on the data in Figure 14.51. The observed anodic currents are in proportion to the concentration of H_2O_2 from 1.96×10^{-1} to 2.3 mM. The fitted linear equation is $y = 64.10x + 41.35$ ($R^2 = 0.9933$), where y and x stand for the peak current (μA) and the concentration (mM) of H_2O_2, respectively. The limit of detection was estimated to be 1.7×10^{-1} mM at a signal/noise ratio of 3. In order to study the selectivity of the PAH/MnO_2 film electrode, current-time responses were tested for the additions of 1 mmol glucose, 0.01 mmol uric acid, 0.1 mmol $NaNO_2$, and 0.1 mmol ascorbic acid (Figure 14.53). In general, $NaNO_2$ is used as an anti-corrosion reagent in antiseptic solution for cleaning medical instruments. Uric acid exists in blood. One form of ascorbic acid is commonly known as vitamin C. No obvious current was observed in the presence of former three species, while ascorbic acid led to an anodic current, implying the possibility of interference from ascorbic acid.

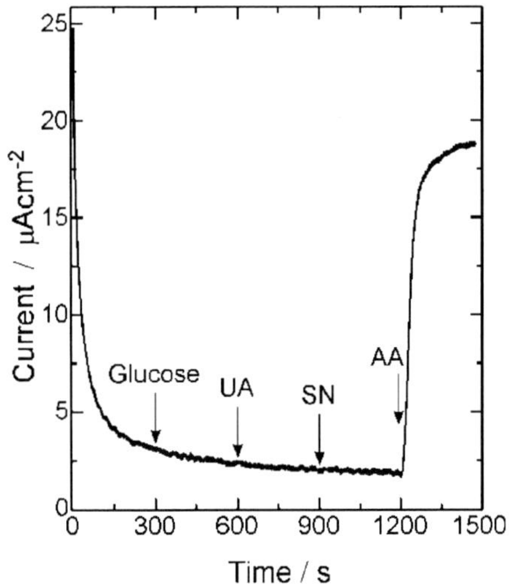

Figure 14.53. Amperometric response of the PAH/MnO_2-modified ITO electrode measured upon additions of 1 mmol glucose, 0.01 mmol uric acid (UA), 0.1 mmol $NaNO_2$ (SN), and 0.1 mmol ascorbic acid (AA). (Reprinted from [123], © 2010, with permission from Wiley-VCH Verlag GmbH & Co. KGaA).

CONCLUSION

This chapter has highlighted the electrodeposition of new types of layered manganese oxides intercalated with cationic guest molecules or polymers. The synthetic strategy is quite versatile because the inorganic host (MnO_2) can adjust itself to accommodate guest molecules during electrodeposition. The products presented unique properties depending on the intercalated molecules, which were applicable to pseudocapacitor electrodes, electrochromic materials, catalysts, and biosensors.

REFERENCES

[1] S. L. Suib, *Current Opinion in Solid State and Materials Science* 3, 63 (1998).
[2] S. L. Brock, N. Duan, Z. R. Tian, O. Giraldo, H. Zhou, and S. L. Suib, *Chem. Mater.* 10, 2619 (1998).
[3] S. L. Suib, *Accounts of Chemical Research* 41, 479 (2008).
[4] Y. Chabre and J. Pannetier, *Prog. Solid State Chem.* 23, 1 (1995).
[5] L. D. Burke and O. J. Murphy, *J. Electroanal. Chem.* 109, 373 (1980).
[6] H. Kanoh, T. Hirotsu, and K. Ooi, *J. Electrochem. Soc.* 143, 905 (1996).
[7] S. I. Cordoba de Terresi and A. Gorenstien, *Electrochim. Acta* 37, 2015 (1992).
[8] M. Chigane, M. Ishikawa, and M. Izaki, *J. Electrochem. Soc.* 148, D96 (2001).
[9] C.-H. Lee, B. Cahan, and E. Yeager, *J. Electrochem. Soc.* 120, 1689 (1973).
[10] T. Seike and J. Nagai, *Solar Energy Mater.* 22, 107 (1991).
[11] T. Murayama and Y. Osaki, *J. Electrochem. Soc.* 142, 3137 (1995).
[12] J. W. Long, L. R. Qadir, R. M. Stroud, and D. R. Rolison, *J. Phys. Chem. B* 105, 8712 (2001).
[13] N. Sakai, Y. Ebina, K. Takada, and T. Sasaki, *J. Electrochem. Soc.* 152, E384 (2005).
[14] D. Bélanger, T. Brousse, and J. W. Long, *ECS Interface* 17, 49 (2008).
[15] W. Wei, X. Cui, W. Chen, and D. G. Ivey, *Chem. Soc. Rev.* 40, 1697 (2011).
[16] B. B. Owens, S. Passerini, and W. H. Smyrl, *Electrochim. Acta* 30, 215 (1999).
[17] J. W. Long, R. M. Stroud, and D. R. Rolison, *J. Non-Crystalline Solids* 285, 288 (2001).
[18] D. C. Golden, C. C. Chen, and J. B. Dixon, *Clays Clay Miner.* 35, 271 (1987).
[19] S. Ching, D. J. Petrovay, M. L. Jorgensen, and S. L.Suib, *Inorg. Chem.* 36, 883 (1997).
[20] Y. F. Shen, S. L. Suib, and C. L. O'Young, *J. Am. Chem. Soc.* 116, 11020 (1994).
[21] J. Luo and S. L. Suib, *Chem. Commun.* 1031 (1997).
[22] S. L. Brock, M. Sanabria, S. L. Suib, V. Urban, P. Thiyagarajan, and D. I. Potter, *J. Phys. Chem. B* 103, 7416 (1999).
[23] Z. Liu, K. Ooi, H. Kanoh, W. Tang, and T. Tomida, *Langmuir* 16, 4154 (2000).
[24] Q. Gao, O. Giraldo, W. Tong, and S. L. Suib, *Chem. Mater.* 13, 778 (2001).
[25] G. H. Threse and P. V. Kamath, *Chem. Mater.* 12, 1195 (2000).
[26] E. Preisler, *J. Appl. Electrochem.* 6, 301 (1970).
[27] M. Nakayama, S. Konishi, A. Tanaka, and K. Ogura, *Chem. Lett.* 33, 670 (2004).
[28] M. Nakayama, S.Konishi, H. Tagashira, and K. Ogura, *Langmuir* 21, 354 (2005).
[29] E. Graf, J. P. Kintzinger, J. M. Lehn, and J. LeMoigne, *J. Am. Chem. Soc.* 104, 1672

(1982).
[30] Y. Omomo, T.Sasaki, L. Wang, and M. Watanabe, *J. Am. Chem. Soc.* 125, 3568 (2003).
[31] P. S. Nikam and A. B. Nikumbh, *J. Chem. Eng. Data* 47, 400 (2002).
[32] T. Hasemi and A. W. Brinkman, *J. Mater. Res.* 7, 1278 (1992).
[33] B. J. Aronson, C. F. Blanford, and A. Stein, *J. Phys. Chem. B* 104, 449 (2000).
[34] M. Nakayama, A. Tanaka, S. Konishi, and K. Ogura, *J. Mater. Res.* 19, 1509 (2004).
[35] Y. Lvov, B. Munge, O. Giraldo, I. Ichinose, S. L. Suib, and J. F. Rusling, *Langmuir* 16, 8850 (2000).
[36] M. Chigane and M. Ishikawa, *J. Electrochem. Soc.* 147, 2246 (2000).
[37] N. S. McIntyre and D. G. Zetaruk, *Anal. Chem.* 49, 1521 (1977).
[38] M. Nakayama, T. Ii, H. Komatsu, and K. Ogura, *Chem. Commun.* 1098 (2004).
[39] K. A. Carrado, L. Xu, S. Seifert, R. Csencsits, and C. A. A. Bloomquist, Polymer Clay Nanocomposites Derived from Polymer-Silicate Gels. in *Polymer-Clay Nanocomposites*, T. J. Pinnavaia, G. W. Beall (Eds), Wiley and Sons, New York, pp 47-63 (2000).
[40] K. A. Carrado, Polymer-Clay Nanocomposites. in *Advanced Polymeric Materials: Structure Property Relationships*, S. G. Advani, G. O. Shonaike (Eds), CRC Press, Boca Raton, FL, pp 349-396 (2003).
[41] Z. Liu, X. Yang, Y. Makita, and K. Ooi, *Chem. Mater.* 14, 4800 (2002).
[42] G. Decher, *Science* 277, 1232 (1997).
[43] L. Wang, Y. Omomo, N. Sakai, K. Fukuda, I. Nakai, Y. Ebina, K. Takada, M. Watanabe, and T. Sasaki, *Chem. Mater.* 15, 2873 (2003).
[44] M. Nakayama and H. Tagashira, *Langmuir* 22, 3864 (2006).
[45] J. E. Post and D. R. Veblem, *Am. Mineral.* 75, 477 (1990).
[46] Y. Lvov, K. Ariga, I. Ichinose, and T. Kunitake, *Langmuir* 12, 3038 (1996).
[47] M. Kruk and M. Jaroniec, *Chem. Mater.* 13, 3169 (2001).
[48] M. Nakayama, M. Shamoto, and A. Kamimura, *Chem. Mater.* 22, 5887 (2010).
[49] G. C. Allen, M. T. Curtis, A. J. Hopper, and P. M. Tucker, *J. Chem. Soc., Dalton Trans.* 14, 1525 (1974).
[50] T. Wallin and P. Linse, *J. Phys. Chem. B* 28, 5506 (1997).
[51] I. G. Casella and M. Gatta, *Anal. Chem.* 72, 2969 (2000).
[52] A. S. Zahr, M. Villiers, and M. V. Pishko, *Langmuir* 21, 403 (2005).
[53] N. Usha, B. Viswanathan, V. R. K. Murthy, and J. Sobhanadri, *Spectrochim. Acta Part A* 53, 1761 (1997).
[54] D. P. Acharya, K. Hattori, T. Sakai, and H. Kunieda, *Langmuir* 19, 9173 (2003).
[55] Q. S. Huo, D. I. Margolese, U. Cielsa, D. G. Demuth, P. Y. Feng, T. E. Gier, P. Sieger, A. Firouzi, B. F. Chmelka, F. Schüth, and G. D. Stucky, *Chem. Mater.* 6, 1176 (1994).
[56] B. E. Conway, *J. Electrochem. Soc.* 138, 1539 (1991).
[57] P. Ruetschi, *J. Electrochem. Soc.* 135, 2657 (1988).
[58] V. P. Mahadevan Pillai, V. U. Nayar, and V. B. Jordanovska, *J. Sold State Chem.* 133, 407 (1997).
[59] E. Feitosa, M. R. S. Brazolin, R. M. Z. G. Naal, M. P. F. M. D. Lama, J. R. Lopes, W. Loh, and M. Vasilescu, *J. Colloid. Interface Sci.* 299, 883 (2006).
[60] M. Toupin, T. Brousse, and D. Bélanger, *Chem. Mater.* 14, 3946 (2002).
[61] M. Toupin, T. Brousse, and D. Bélanger, *Chem. Mater.* 16, 3184 (2004).

[62] R. D. Shannon, *Acta Crystallogr., Sect. A: Cryst. Phys., Diffr. Theor. Gen. Crystallogr.* A32, 751 (1976).
[63] L. Sepulveda and J.Cortes, *J. Phys. Chem.* 1985, *89*, 5322.
[64] H. Y. Lee and J. B. Goodenough, *J. Solid State Chem.* 144, 220 (1999).
[65] B. Babakhani and D. G. Ivey, *J. Power Sources* 195, 2110 (2010).
[66] M. Nakayama, Y. Kanaya, and R. Inoue, *Electrochem. Commun.* 9, 1154 (2007).
[67] K. R. Prasad and N. Miura, *J. Power Sources* 135, 354 (2004).
[68] E. Beaudrouet, A. Le Gal La Salle, and D. Guyomard, *Electrochim. Acta* 54, 1240 (2009).
[69] V. Subramanian, H. Zhu, and B. Wei, *J. Power Sources* 159, 361 (2006).
[70] H. Kim and B. N. Popov, *J. Electrochem. Soc.* 150, D56 (2003).
[71] K. R. Prasad and N. Miura, *Electrochem. Commun.* 6, 1004 (2004).
[72] P.–Y. Chuang and C.–C. Hu, *Mater. Chem. Phys.* 92, 138 (2005).
[73] J.–K. Chang, M.–T. Lee, C.–H. Huang, and W.–T. Tsai, *Mater. Chem. Phys.* 108, 124 (2008).
[74] M. Nakayama, A. Tanaka, Y. Sato, T. Tonosaki, and K. Ogura, *Langmuir* 21, 5907 (2005).
[75] M.–T. Lee, J.–K. Chang, and W.–T. Tsai, *J. Electrochem. Soc.* 154, A875 (2007).
[76] C. Y. Lee, H. M. Tsai, H. J. Chuang, S. Y. Li, P. Lin, and T. Y. Tseng, *J. Electrochem. Soc.* 152, A716 (2005).
[77] Y. T. Wu and C. C. Hu, *J. Electrochem. Soc.* 151, A2060 (2004).
[78] H. Zhang, G. P. Cao, Z. Y. Wang, Y. S. Yang, Z. J. Shi, and Z. N. Gu, *Nano Letters* 8, 2664 (2008).
[79] Z. Fan, J. Chen, B. Zhang, F. Sun, B. Liu, and Y. Kuang, *Mater. Res. Bull.* 43, 2085 (2008).
[80] Z. Fan, J. Chen, B. Zhang, B. Liu, X. Zhou, and Y. Kuang, *Diam. Relat. Mater.* 17, 1943 (2008).
[81] Y. H. Wang, H. Liu, X. L. Sun, and I. Zhitomirsky, *Scr. Mater.* 61, 1079 (2009).
[82] K. Okamura, R. Inoue, T. Sebille, K. Tomono, and M. Nakayama, *J. Electrochem. Soc.* 158, A711 (2011).
[83] T. Brousse, M. Toupin, R. Dugas, and L. Athouël, *J. Electrochem. Soc.* 153, A2171 (2006).
[84] S. Devaraj and N. Munichandraiah, *J. Phys. Chem. C* 112, 4406 (2008).
[85] M. Nakayama, T. Kanaya, and R. Inoue, *Electrochem. Commun.* 9, 1154 (2007).
[86] R. Inoue and M. Nakayama, *Electrochem. Solid-State Lett.* 12, A203 (2009).
[87] C. M. Criss and M. J. Mastroianni, *J. Phys. Chem.* 75, 2532 (1971).
[88] V. Subramanian, H. Zhu, R. Vajtai, P. M. Ajayan, and B. Wei, *J. Phys. Chem. B* 109, 20207 (2005).
[89] L. Athouël, F. Moser, R. Dugas, O. Crosnier, D. Bélanger, and T. Brousse, *J. Phys. Chem. C* 112, 7270 (2008).
[90] S. Komaba, A. Ogata, and T. Tsuchikawa, *Electrochem. Commun.* 10, 1435 (2008).
[91] W. Wei, X. Cui, W. Chen, and D. G. Ivey, *J. Power Sources* 186, 543 (2009).
[92] W. Wei, W. Chen, and D. G. Ivey, *Chem. Mater.* 19, 2816 (2007).
[93] M. Nakayama, K. Suzuki, K. Okamura, R. Inoue, L. Athouël, O. Crosnier, and T. Brousse, *J. Electrochem. Soc.* 157, A1067 (2010).
[94] Y. Dai, K. Wang, and J. Xie, *Appl. Phys. Lett.* 90, 104102 (2007).

[95] W. Wei, X. Cui, W. Chen, and D. G. Ivey, *Eroctrochim. Acta* 54, 2271 (2009).
[96] W. Wei, W. Chen, and D. G. Ivey, *Chem. Mater.* 20, 1941 (2008).
[97] M. Deepa, A. G. Joshi, A. K. Srivastava, S. M. Shivaprasad, and S. A. Agnihotry, *J. Electrochem. Soc.* 153, C365 (2006).
[98] C. C. Liao, F. R. Chen, and J. J. Kai, *Sol. Energy Mater. Sol. Cells* 91, 1282 (2007).
[99] H. J. Ahn, H. S. Shim, Y. E. Sung, T. Y. Seong, and W. B. Kim, *Electrochem. Solid-State Lett.* 10, E27 (2007).
[100] M. Nakayama, Y. Kashiwa, and K. Suzuki, *J. Electrochem. Soc.* 156, D125 (2009).
[101] S. H. Lee, C. E. Tracy, Y. Yan, J. R. Pitts, and S. K. Deb, *Electrochem. Solid-State Lett.* 8, A188 (2005).
[102] Y. Abe, S. H. Lee, E. O. Zayim, C. E. Tracy, J. R. Pitts, and S. K. Deb, *Electrochem. Solid-State Lett.* 9, G17 (2006).
[103] M. Nakayama, A. Tanaka, Y. Sato, T. Tonosaki, and K. Ogura, *Langmuir* 21, 5907 (2005).
[104] K. –W. Nam, M. G. Kim, and K. –B. Kim, *J. Phys. Chem. C* 111, 749 (2007).
[105] J. –K. Chang, M. –T. Lee, and W. –T. Tsai, *J. Power Sources* 166, 590 (2007).
[106] Y.–C. Son, V. D. Makwana, A. R. Howell, and S. L. Suib, *Angew. Chem. Int. Ed.* 40, 4280 (2001).
[107] S. Allard, U. Gunten, E. Sahli, R. Nicolau, and H. Gallard, *Water Research* 43, 3417 (2009).
[108] A. G. Hardie, J. J. Dynes, L. M. Kozak, and P. M. Huang, *J. Mol. Catal. A* 308, 114 (2009).
[109] S. W. Chan Chien, H. L. Chen, M. C. Wang, and K. Seshaiah, *Chemosphere* 74, 1125 (2009).
[110] M. A. Rao, G. Iamarino, R. Scelza, F. Russo, and L. Gianfreda, *Sci. Total Environ* 407, 438 (2008).
[111] C. Prado, G. G. Murcott, F. Marken, J. S. Foord, and R. G. Compton, *Electroanalysis* 14, 975 (2002).
[112] P. Canizares, F. Martinez, M. Diaz, J. Garcia-Gomez, and M. A. Rodrigo, *J. Electrochem. Soc.* 149, D118 (2002).
[113] Y. Wang, B. Gu, and W. Xu, *J. Hazardous Materials* 162, 1159 (2009).
[114] M. K. Carter, *J. Mol. Struct.* 831, 26 (2007).
[115] J. Luo, Q. Zhang, and S. L. Suib, *Inorg. Chem.* 39, 741 (2000).
[116] Y. Lin, X. Cui, and L. Li, *Electrochem. Commun.* 7, 166 (2005).
[117] S. Yao, S. Yuan, J. Xu, Y. Wang, J. Luo, and S. Hu, *Applied Clay Sci.* 33, 35 (2006).
[118] B. Šljukić and R. G. Compton, *Electroanalysis* 19, 1275 (2007).
[119] X. Yang, X. Chen, X. Zhang, W. Yang, and D. G. Evans, *Sensor and Actuators B* 129, 784 (2008).
[120] X. Chen, X. Zhang, W. Yang, and D. G. Evans, *Mater. Sci. and Eng. C* 29, 284 (2009).
[121] X. Cui, G. Liu, and Y. Lin, *Nanomedicine* 1, 130 (2005).
[122] J.-J. Xu, J.-J. Feng, X. Zhong, and H.-Y. Chen, *Electroanalysis* 20, 507 (2008).
[123] M. Yoshimoto, C. Iida, A. Kariya, N. Takaki, and M. Nakayama, *Electroanalysis* 22, 653 (2010).

INDEX

C

D

F

G

H

I

J

K

L

M

N

O

P

Q

R

S

T

U

V

W

X

Y

Z